LAB MANUAL

Automotive Service: Inspection, Maintenance, Repair

Sixth Edition

Chuck Rockwood

Tim Gilles

Cengage

Australia • Brazil • Canada • Mexico • Singapore • United Kingdom • United States

***Automotive Service: Inspection, Maintenance, Repair Lab Manual*, Sixth Edition**
Chuck Rockwood and Tim Gilles

SVP, Higher Education & Skills Product: Erin Joyner

Product Director: Matthew Seeley

Senior Product Manager: Katie McGuire

Product Assistant: Kimberly Klotz

Director, Learning Design: Rebecca von Gillern

Senior Manager, Learning Design: Leigh Hefferon

Learning Designer: Mary Clyne

Marketing Director: Sean Chamberland

Marketing Manager: Andrew Ouimet

Director, Content Creation: Juliet Steiner

Content Creation Manager: Alexis Ferraro

Senior Content Manager: Cheri Plasse

Digital Delivery Lead: Amanda Ryan

Art Director: Jack Pendleton

Designer: Erin Griffin

Cover image(s): Ciprian Stremtan/ Shutterstock.com

Library of Congress Control Number: 2019909579

Book Only ISBN: 978-1-337-79404-6

Cengage
200 Pier 4 Boulevard
Boston, MA 02210
USA

Notice to the Reader

Publisher does not warrant or guarantee any of the products described herein or perform any independent analysis in connection with any of the product information contained herein. Publisher does not assume, and expressly disclaims, any obligation to obtain and include information other than that provided to it by the manufacturer. The reader is expressly warned to consider and adopt all safety precautions that might be indicated by the activities described herein and to avoid all potential hazards. By following the instructions contained herein, the reader willingly assumes all risks in connection with such instructions. The publisher makes no representations or warranties of any kind, including but not limited to, the warranties of fitness for particular purpose or merchantability, nor are any such representations implied with respect to the material set forth herein, and the publisher takes no responsibility with respect to such material. The publisher shall not be liable for any special, consequential, or exemplary damages resulting, in whole or part, from the readers' use of, or reliance upon, this material.

Printed at CLDPC, USA, 08-24

Contents

Preface

This *Automotive Service, Inspection, Maintenance, Repair Lab Manual, Sixth Edition,* is designed to help students build automotive maintenance and light repair skills. It contains two parts: Part I is made up of Activity Sheets to reinforce the student's automotive vocabulary, identify common automotive tools and equipment, as well as an understanding of the various systems required for vehicle service and light repair. The Activity Sheet exercises include tool and parts identification, matching exercises, and fill-in sheets. They are designed to help reinforce students' understanding of the operation of the automobile and its systems.

Part II includes a wide variety of hands-on ASE Education Foundation Worksheets that emphasize practical, real-life skills needed to service today's automobiles. Worksheets are presented in order of increasing difficulty, and students should complete one task before progressing to the next one. Each project or lab assignment is built upon the next in a logical sequence in much the same manner as lab science instructional programs are constructed. The Worksheets also include references to the ASE Education Foundation's MLR, AST, and MAST task lists for programs that wish to track students' progress against these criteria. A set of supplemental worksheets is available at the Instructors Companion Website. These supplemental sheets are organized by ASE system and ensure that instructors have worksheet coverage of every ASE Education Foundation task at every program level. Worksheets include these features:

Objective: A description of the learning outcome(s) for the lab assignment.

ASE Education Foundation Task Correlation: A cross-reference to the ASE Education Foundation tasks (MLR, AST, and MAST) addressed in the worksheet.

Directions: Specific instructions for completing the lab assignment.

Tools and Equipment Required: A list of the most important components and materials needed to complete the lab assignment.

Procedure: A step-by-step description of the work to be performed in the lab assignment.

Notes: Inserted where helpful hints will facilitate the completion of the lab assignment.

Cautions: Warnings about potentially hazardous situations that could cause personal injury or damage to the vehicle.

Shop Tips: Suggestions and shortcuts with careful instructions for implementing them.

Environmental Notes: Proper disposal methods for hazardous materials and product expiration information.

Illustrations: The lab manual is fully illustrated to facilitate learning and enhance lab discussion.

Part I

Activity Sheets

Section One

The Automobile Industry

Instructor OK ______________ Score ____________

Activity Sheet #1

IDENTIFY FRONT- AND REAR-WHEEL-DRIVE COMPONENTS

Name ______________________________ Class ______________________

Directions: Identify the front- and rear-wheel drive components in the drawings. Place the identifying letter next to the name of each component.

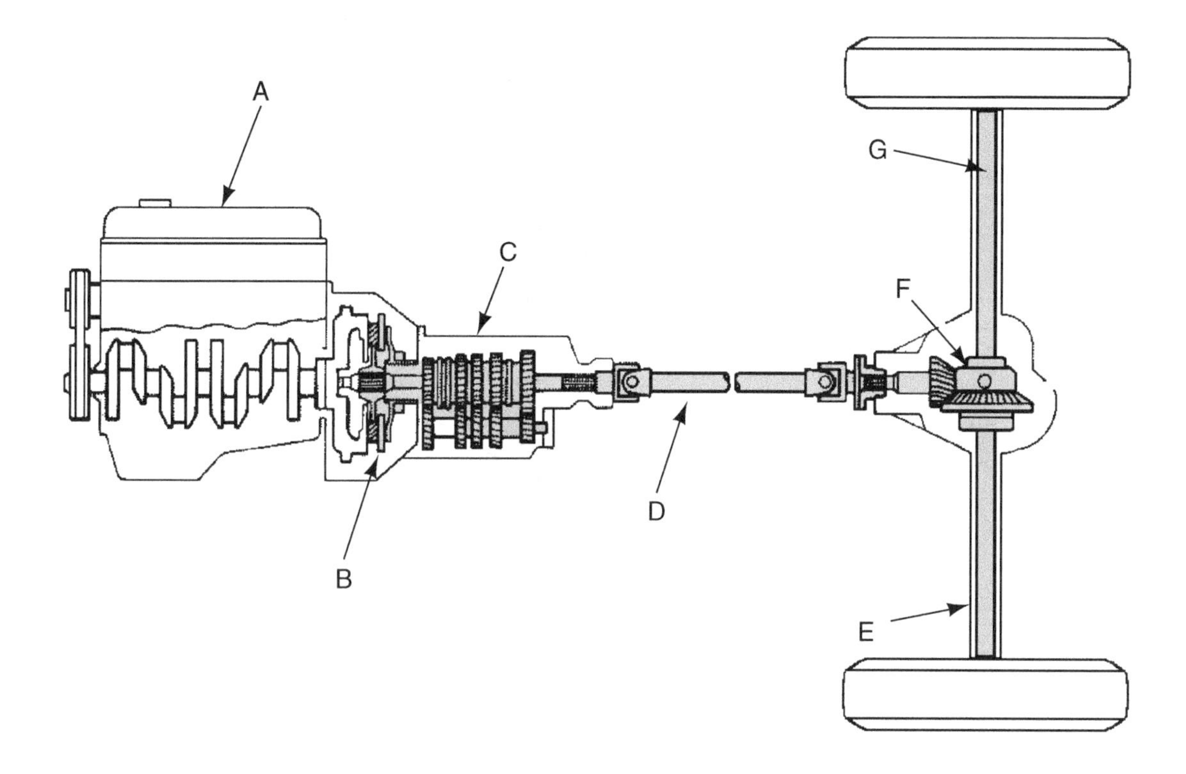

Axle Housing ______	Propeller Shaft ______	Rear Axle Shaft ______
Clutch ______	Transmission ______	Engine ______
Differential ______		

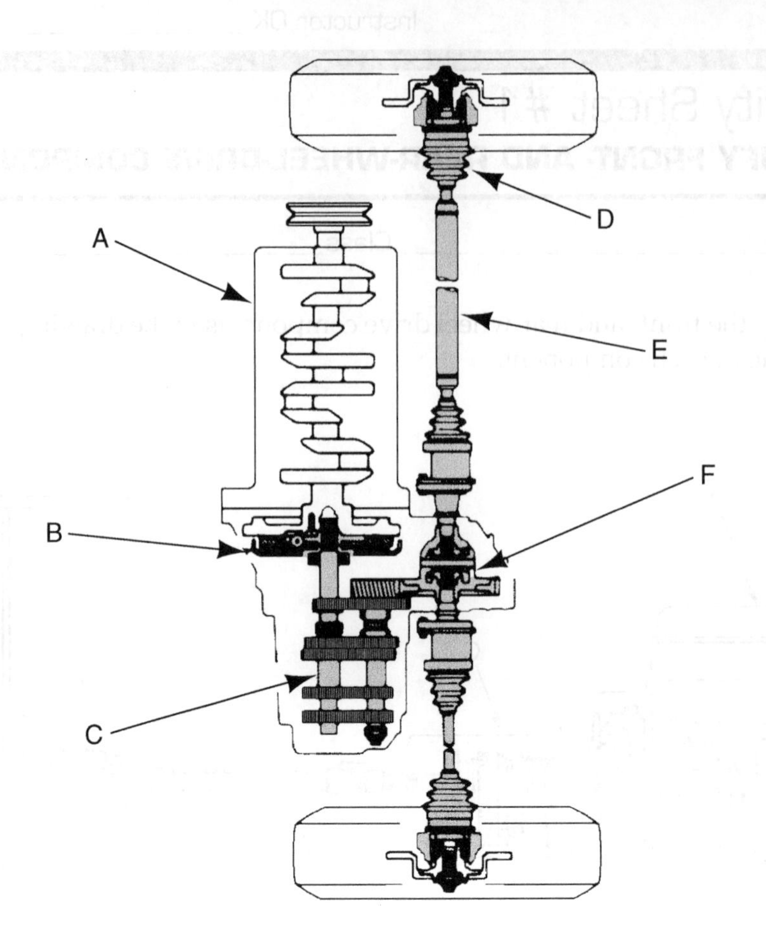

Differential	______	Drive Axle	______	Transaxle	______
CV Joint	______	Engine	______	Clutch	______

STOP

Instructor OK ____________ Score ____________

Activity Sheet #2

THE AUTOMOTIVE SERVICE INDUSTRY

Name ____________________ Class ____________

Directions: Answer the questions in the spaces provided.

1. List four careers in the automotive repair industry.

   a. ____________________

   b. ____________________

   c. ____________________

   d. ____________________

2. List four different types of automotive repair shops.

   a. ____________________

   b. ____________________

   c. ____________________

   d. ____________________

3. List the eight ASE areas of specialization required to become a Master Automotive Technician.

   a. ____________________

   b. ____________________

   c. ____________________

   d. ____________________

   e. ____________________

   f. ____________________

   g. ____________________

   h. ____________________

Courtesy of ASE

4. List two additional ASE automotive certifications.

   a. ____________________

   b. ____________________

5. List any special licenses required to perform certain types of automotive work in your area.

   a. ____________________

   b. ____________________

6. ASE Education Foundation accreditation (formerly NATEF) could be used by a prospective student to evaluate a ____________________.

TURN

7. How would it benefit a student to achieve ASE certification?

____

____

8. What is the work experience requirement to receive ASE certification? _______________

9. How much ASE work experience credit is given for completing a college automotive program?

____

____

10. What type of questions are used on the ASE certification tests? _______________

11. Additional information about ASE can be found on the _______________.

12. What is the address for the ASE website? _______________

13. How many times a year are the ASE certification tests available? _______________

14. ASE certifications expire after a period of _______________.

15. What is the difference between being certified and having a license?

____

____

STOP

Instructor OK ______________ Score ____________

Activity Sheet #3

CREATE A JOB RESUME

Name ______________________________ Class ____________________

Directions: Follow the format below to create a resume. Read this assignment completely before starting to work on the assignment. Use a computer to create your resume. Your resume should not be longer than one page and include the following:

a. Name, Address, and Contact Information

b. Objective

c. Education

d. Experience

e. Activities and Honors

f. Special Skills

g. Certifications and Licenses

1. After completing your resume, save the file for future revision.
2. Print your resume and turn it in to the instructor.

STOP

Instructor OK ______________ Score __________

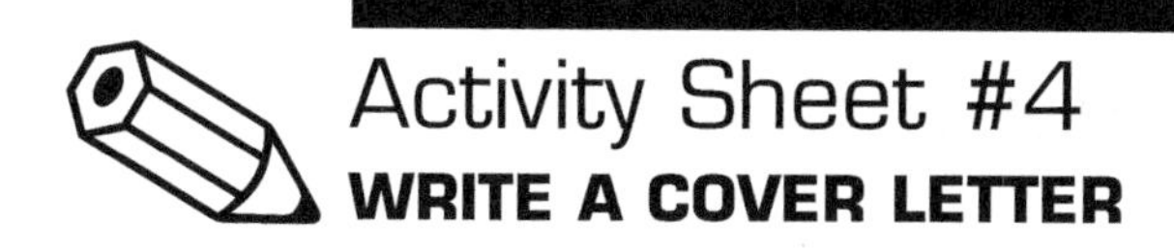

Activity Sheet #4

WRITE A COVER LETTER

Name ____________________________ Class ____________________

A cover letter is an important tool to use during a job search. It is your introduction and explains why you are seeking a particular job. This assignment will help you to write a convincing cover letter.

Directions: Read this assignment completely before starting to work on your cover letter. Use a computer to write your cover letter. Print your letter and turn it in to the instructor. Save the computer file in safe place for future revisions.

When writing a cover letter there are a few simple rules:

- Keep it brief—Three paragraphs are enough to explain why you are seeking the position. Your letter should not exceed one page.
- Be specific—Explain why you are applying for a specific job.
- Explain—Tell about yourself and your qualifications. Include any special awards or talents.
- Hit the highlights—Do not include details. The details will be part of your resume.
- Set the stage—Conclude the letter by telling the employer that you will follow up within the next week to discuss the opportunities.
- Sign the letter—Sign your letter above your typewritten name

1. Use the standard business letter format to format your letter.
   a. Your name, address, and contact in formation
   b. Date
   c. The name, title, address, and contact information of the prospective employer
2. In the first paragraph explain why you are writing. Be specific about the position you are interested in and why working for this particular company is a good place for you.
3. The second paragraph should explain why you should be hired. Summarize your qualifications. Explain why you are interested in the position and why you are the best person for the job. Keep it short, no more than five or six sentences.
4. The third paragraph is your chance to ask for an interview. Suggest a day and let them know that you will call in the next week.
5. End the letter with "Respectfully yours" or "Sincerely" and provide several spaces, and then type your name.
6. After printing your letter, do not forget to sign it.

Sample Cover Letter

Timothy Gilles
1000 Oak Ave
Los Angeles, CA 90030
Phone: 714-555-1212
E-mail: gilles@sbceo.org

May 3rd, 20XX
Mr. Bob Olsen
Service manager
Professional Automotive Service
14233 Wilshire Blvd.
Los Angeles, CA 93030

Dear Mr. Olsen,
I saw your advertisement in the local newspaper would like to be considered for the position of an automotive service technician. I have completed the automotive programs at Los Angeles High School and Los Angeles City College where I earned a 3.8 grade point average in my automotive studies.

As my attached resume shows, I have worked part time in the automotive industry during my studies in high school and college. My current experience includes providing general and preventative maintenance on automobiles. Other work experience has given me customer service skills and the ability to work well in a team setting or as an individual. I am seeking work in a company where my technical and mechanical skills can be utilized to provide quality car repair services.

Your business has an excellent reputation and I would be honored to be selected for an interview.

Thank you for your consideration.

Respectfully yours,

Timothy Gilles

Instructor OK ______________ Score __________

Activity Sheet #5

COMPLETING A JOB APPLICATION

Name ____________________________ Class ____________________

Directions: Complete the job application.

Application for Employment	
Name:	Home Phone:
Email Address:	Cell Phone:
Mailing Address: Number and street	
City	State /Zip Code
How did you learn about our company?	
What position are you applying for?	
Part-time? [ ] Y or [ ] N	Full-time work? [ ] Y or [ ] N
Days and hours you are available to work?	
When are you available to begin work? ___ / ___ / ___	Are you available to work weekends? ☐ Yes ☐ No
Are you available for evening work? ☐ Yes ☐ No	Are you available on Saturdays? ☐ Yes ☐ No
Hourly pay desired $	
Do you have a driver's license? ☐ Yes ☐ No	Do you have transportation to work? ☐ Yes ☐ No
Are you over 18? ☐ Yes ☐ No	(Under 18 is subject to legal regulations)
Do you have evidence of citizenship or legal status to work? ☐ Yes ☐ No	
Are you willing to take a drug test? ☐ Yes ☐ No	Are you physically able to perform all of the required duties of the job? ☐ Yes ☐ No
Have you been convicted of a felony or misdemeanor? If so, describe.	

Education and Work experience	
High School graduate? ☐ Yes ☐ No	Name of School:
Grade level if still in school:	
College or Trade School? ☐ Yes ☐ No Certificate/Diploma or Degree? ☐ Yes ☐ No If yes, which? ______________	School #1: Years ____
	School #1: Years ____
Please list any classes completed that are pertinent to this position?	
Are you a military veteran? ☐ Yes ☐ No If so, describe branch, rank, and years of service:	
References:	

STOP

Part I

Activity Sheets

Section Two

Shop Procedures, Safety, Tools, and Equipment

Instructor OK ______________ Score __________

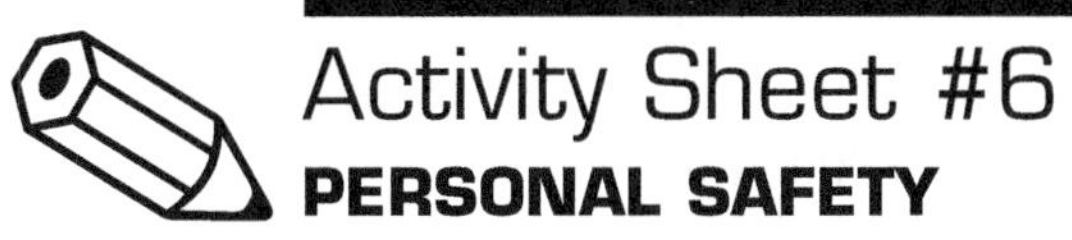

Activity Sheet #6

PERSONAL SAFETY

Name ________________________ Class ________________

ASE Education Foundation Correlation

This activity sheet addresses the following **Required Supplemental Tasks (RSTs):**

Shop and Personal Safety:

Task 1. Identify general shop safety rules and procedures.

Preparing Vehicle for Service:

Task 2. Identify purpose and demonstrate proper use of fender covers, mats.

We Support ASE Education Foundation

Directions: Match the words on the left to the descriptions on the right. Write the letter for the correct word on the line next to the term. If you are uncertain of the meaning of a term, use the glossary in your textbook.

Term		Description
A. Safety glasses	______	Covers the seat when driving a customer's vehicle
B. Safety goggles	______	Not worn in the shop area
C. Face shield	______	Protect the steering wheel
D. Nitrile gloves	______	Proper dress when working in the shop
E. Mechanic gloves	______	Worn to protect hands from hot parts
F. Shoes with open toes	______	Worn when using a press
G. Sleeved shirts and long pants	______	Protect the vehicle's carpet
H. Belt buckle	______	Could save eyes from hazardous liquids
I. Fender cover	______	Identifies SRS/air bag wiring
J. Refrigerant	______	Identifies high-voltage wiring
K. Seat protectors	______	Protect hands from oil and coolant
L. Rings	______	Should be in place when working on a vehicle
M. Yellow wire insulation	______	Should not be worn in the shop
N. Watch	______	Worn when working with chemicals
O. Ignition coil	______	Worn at all times in the shop area
P. Paper floor mats	______	Could result in a burned wrist
Q. Orange wire covers	______	Could cause scratches in paint

TURN

R.	Short pants (shorts)	______	Should be removed from fingers before working in the shop
S.	Steering wheel covers	______	Induces 5,000 to 100,000 volts
T.	Shirt tail	______	When unpressurized, this fluid will boil vigorously
U.	Eye wash fountain	______	Must be tucked in at all times

STOP

Instructor OK ______________ Score __________

Activity Sheet #7

SERVICE INFORMATION

Name ____________________________ Class ____________________

ASE Education Foundation Correlation

This activity sheet addresses the following **MLR** tasks:

I.A.1, II.A.1, III.A.1, IV.A.1, V.A.1, VI.A.1, VII.A.1, VIII.A.1	Research vehicle service information including fluid type, vehicle service history, service precautions, and technical service bulletins. **(P-1); (P-1); (P-1); (P-1); (P-1); (P-1); (P-1); (P-1)**

This worksheet addresses the following **AST** and **MAST** tasks:

I.A.1	Complete work order to include customer information, vehicle identifying information, customer concern, related service history, cause, and correction. **(P-1)**
I.A.2	Research vehicle service information including fluid type, internal engine operation, vehicle service history, service precautions, and technical service bulletins. **(P-1)**
VI.A.1	Research applicable vehicle and service information including vehicle service history, service precautions, and technical service bulletins. **(P-1)**
II.A.2, III.A.2, IV.A.1, V.A.2, VII.A.2	Research vehicle service information including fluid type, vehicle service history, service precautions, and technical service bulletins. **(P-1); (P-1); (P-1); (P-1); (P-1)**
VIII.A.2	Research vehicle service information including vehicle service history, service precautions, and technical service bulletins. **(P-1)**
VIII.B.2	Access and use service information to perform step-by-step (troubleshooting) diagnosis. **(P-1)**

We Support ASE Education Foundation

Directions: Match the words on the left to the descriptions on the right. Write the letter for the correct word on the line next to the term. If you are uncertain of the meaning of a term, use the glossary in your textbook.

A. Owner's manual	_______	Information on an under-hood label
B. First character of VIN	_______	Automotive Engine Rebuilders Association
C. Interchange manual	_______	Technical service bulletin
D. Flat rate	_______	Number to identify a particular vehicle
E. Labor guide	_______	Repair shop that repairs older vehicles
F. Estimates repair cost	_______	Vacuum diagram label
G. TSB	_______	Magazine for the automotive professional
H. VIN	_______	Found in a lubrication service manual
I. IATN	_______	Sells and repairs new vehicles
J. Emission requirements	_______	Manual that is the most comprehensive one for a particular vehicle

TURN

K. Manufacturers	_______ Sells parts for most vehicles
L. Lubrication fittings	_______ VIN year identifier
M. Vacuum hose routing	_______ Automatic Transmission Rebuilders Association
N. Trade journal	_______ Parts and time guide
O. AERA	_______ A service writer
P. 10th character	_______ Used by salvage yards
Q. ATRA	_______ Booklet that comes with a new car
R. Dealership	_______ VIN character for country of origin
S. Independent	_______ Estimated time to perform a repair
T. Parts store	_______ Remove and replace
U. Dealership parts department	_______ Sells parts for only one make of vehicle
V. R&R	_______ International Automotive Technicians Network
W. Made in Brazil	_______ First under-hood emission label
X. 1972	_______ When the first character of the VIN is a 9

STOP

Instructor OK ______________ Score __________

Activity Sheet #8

IDENTIFY MEASURING INSTRUMENTS

Name ____________________________ Class ____________________

ASE Education Foundation Correlation

This activity sheet addresses the following **Required Supplemental Tasks (RSTs):**

Tools and Equipment:

Task 2. Identify standard and metric designation.

Task 5. Demonstrate proper use of precision measuring tools (i.e., micrometer, dial-indicator, and dial-caliper).

This worksheet addresses the following **AST/MAST** tasks:

I.C.6 Inspect and measure camshaft bearings for wear, damage, out-of-round, and alignment; determine needed action. **(P-3)**

I.C.7 Inspect crankshaft for straightness, journal damage, keyway damage, thrust flange and sealing surface condition, and visual surface cracks; check oil passage condition; measure end play and journal wear; check crankshaft position sensor reluctor ring (where applicable); determine needed action. **(P-1)**

We Support ASE | Education Foundation

Directions: Identify the measuring instruments in the drawing. Place the identifying letter next to the name of the tool.

A

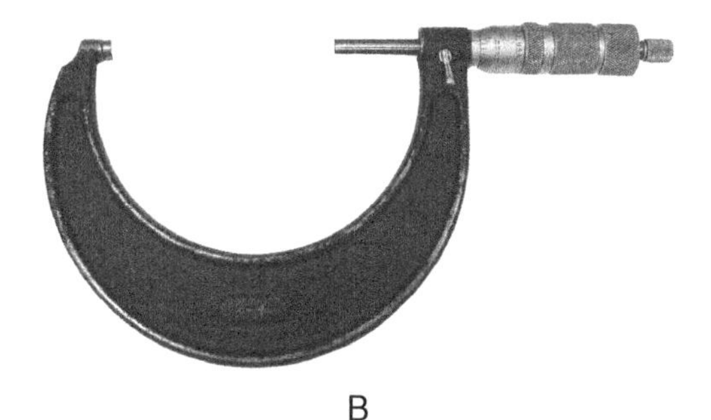

B

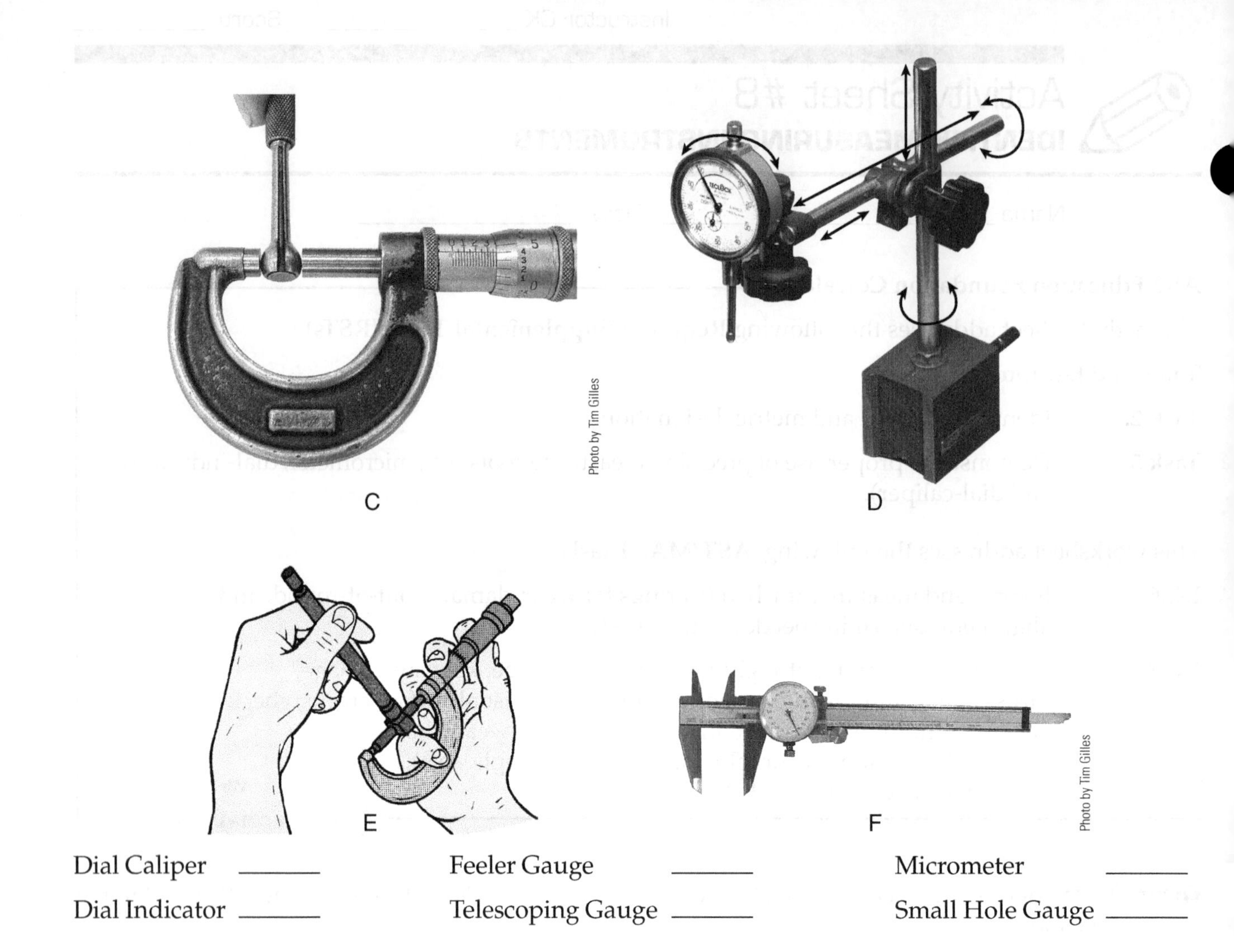

Dial Caliper	______	Feeler Gauge	______	Micrometer	______
Dial Indicator	______	Telescoping Gauge	______	Small Hole Gauge	______

STOP

Instructor OK ______________ Score __________

Activity Sheet #9

MEASURING

Name ______________________________ Class ____________________

Directions: Match the words on the left to the descriptions on the right. Write the letter for the correct word on the line next to the term. If you are uncertain of the meaning of a term, use the glossary in your textbook.

Term		Description
A. British Imperial system	______	Inside diameter
B. Metric system	______	Used to calibrate a micrometer
C. Foot-pounds	______	Cubic inches equal 2.7 liters
D. Newton-meters	______	How 1/1000 inch is expressed with decimals
E. Plastigage	______	5 inches convert to how many millimeters?
F. O.D.	______	Measuring system that uses fractions and decimals, based on inches, feet, and yards
G. I.D.	______	Used to measure small holes
H. LCD	______	A strip of plastic that deforms when crushed; used to measure oil clearance
I. 0.001 inch	______	How torque readings are expressed in the metric system
J. Gauge block	______	Extra gauge that long-range dial indicators have
K. Vernier	______	Outside diameter
L. Ball gauge	______	What one revolution of the micrometer thimble measures
M. Dial indicator	______	Scale on the micrometer used to measure to 0.0001 inch
N. Feeler gauge	______	How torque readings are expressed in the English system
O. Revolution counter	______	One revolution of the micrometer thimble equals what fraction of an inch?
P. 127	______	Liquid crystal display
Q. 60	______	One liter equals ______ cubic centimeters.
R. 162	______	Gauge used to measure valve clearance
S. 1,000	______	Approximate cubic inches in 1 liter
T. Vernier caliper	______	Easier to read than a vernier caliper
U. 1/40 of an inch	______	Instrument that would be used to measure end play
V. 0.025 inch	______	Measuring system based on the meter
W. Dial caliper	______	One tool that can be used to measure I.D., O.D., and depth

STOP

Activity Sheet #10

FASTENER GRADE AND TORQUE

Name______________________________ Class______________________

ASE Education Foundation Correlation

This activity sheet addresses the following **MLR** task:

I.A.6 Perform common fastener and thread repair, to include: remove broken bolt, restore internal and external threads, and repair internal threads with thread insert. **(P-1)**

This worksheet addresses the following **AST/MAST** task:

I.A.7 Perform common fastener and thread repair, to include: remove broken bolt, restore internal and external threads, and repair internal threads with thread insert. **(P-1)**

We Support Education Foundation

Directions: Record your answers in the spaces provided.

1. The size of the following bolt is 3/8" × 16 × 1. List the dimensions in the correct boxes. Locate a bolt of this size and list the size of the wrench that correctly fits the bolt head.

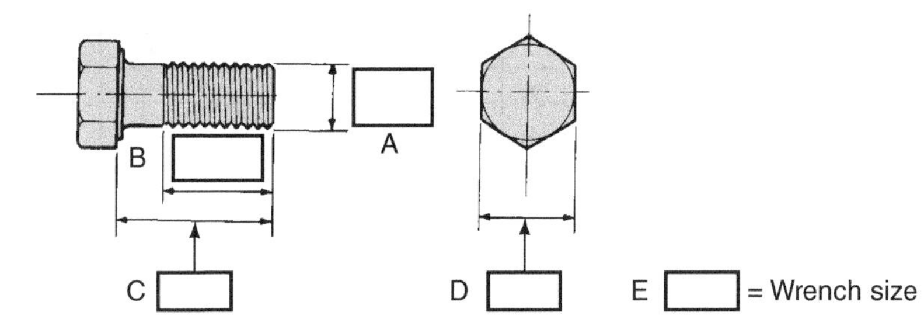

2. Locate fasteners of the following sizes and determine which size wrench fits the fastener head.

   a. 1/2" ______ c. 6 mm ______

   b. 1/4" ______ d. 10 mm ______

3. List the grade under each of these SAE inch standard bolts.

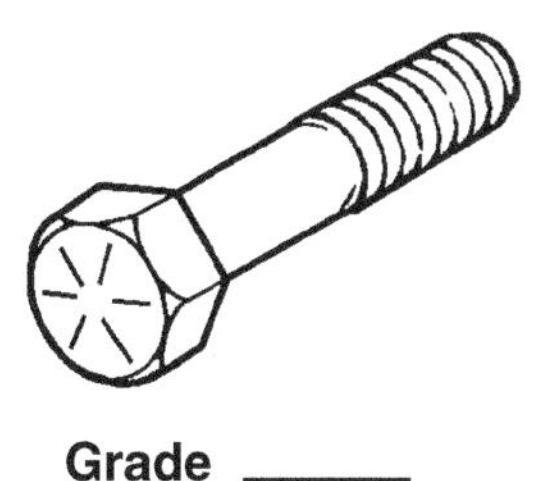

Grade ______

A

Grade ______

B

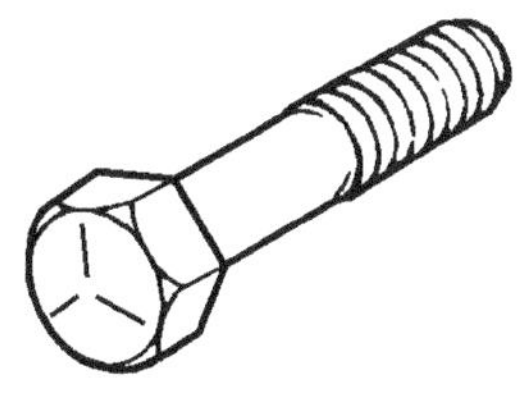

Grade ______

C

TURN

4. Place an X under the metric bolt shown here that is closest in strength to an SAE grade 8 bolt.

5. A bolt is usually torqued to ______% of its elastic limit.
6. A nut can be turned easily onto a bolt for the first few threads and then begins to turn hard. What is the most likely cause? ____________
7. Refer to the bolt torque chart below. What is the recommended torque for a 3/8" diameter grade 5 fastener? ____________

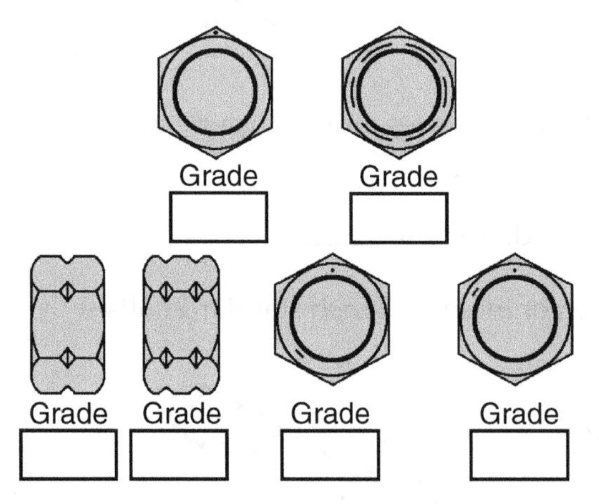

Material	SAE 2 Mild Steel		SAE 5		SAE 8	Socket Head Cap Screws
Minimum Tensile P.S.I. Strength	74,000	60,000	120,000	105,000	150,000	160,000
Proof P.S.I. Load	55,000	33,000	85,000	74,000	120,000	136,000
Steel Grade Symbols						
Bolt Diameter Inches	Torque: pound-foot					
1/4	7	—	10	—	14	16
5/16	14	—	21	—	30	33
3/8	24	—	37	—	52	59
7/16	39	—	60	—	84	95
1/2	59	—	90	—	128	145

8. List the grade under each of these SAE inch standard nuts.

Grade ☐ Grade ☐

Grade ☐ Grade ☐ Grade ☐ Grade ☐

9. Name the five washers pictured below.

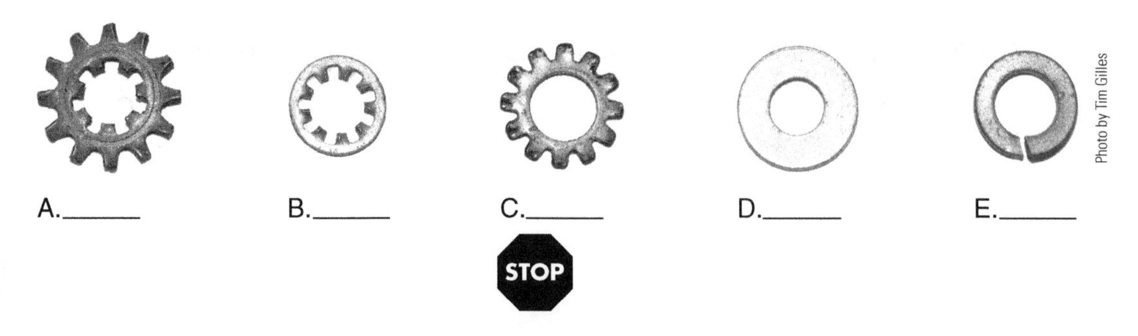

A.______ B.______ C.______ D.______ E.______

STOP

Instructor OK ______________ Score __________

Activity Sheet #11

TAPS AND DRILLS

Name________________________ Class________________

ASE Education Foundation Correlation

This activity sheet addresses the following **MLR** task:

I.A.6 Perform common fastener and thread repair, to include: remove broken bolt, restore internal and external threads, and repair internal threads with thread insert. **(P-1)**

This worksheet addresses the following **AST/MAST** task:

I.A.7 Perform common fastener and thread repair, to include: remove broken bolt, restore internal and external threads, and repair internal threads with thread insert. **(P-1)**

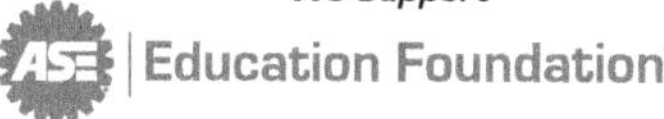

Directions:

1. Identify each tap below by listing the correct name in the space provided.

a. ____________

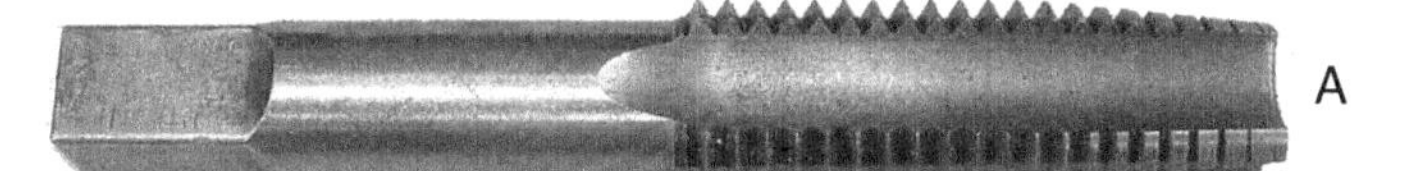

A

b. ____________

B

c. ____________

C

d. ____________

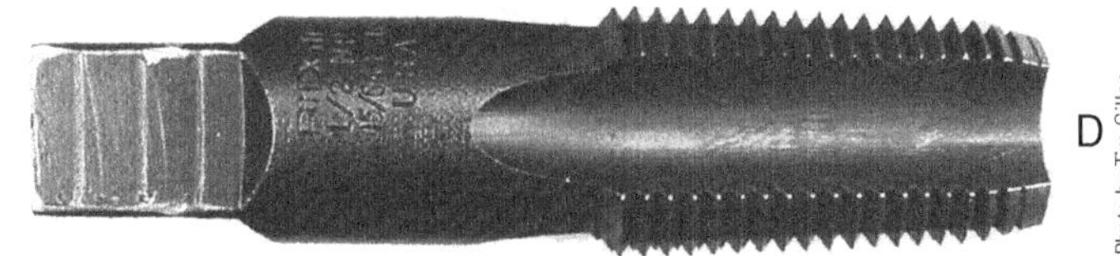

D

Photo by Tim Gilles

2. Refer to the tap drill chart below to answer the next three questions.

   a. What is the correct size tap drill for a 1/4 × 20 screw thread? ____________

   b. What is the closest fractional drill in a 1/64" increment drill index that could safely be used as a tap drill for a 1/4 × 20 thread? ____________

   Note: A decimal equivalent chart is located in the appendix in your textbook.

TURN

c. When referring to a screw thread, what does the number 20 in 1/4 × 20 mean? ____________

Thread diameter	Threads per inch			Decimal equivalent	Tap drill Approx. 75% full thread	Decimal equivalent of tap drill
	NC	NF	NS			
12	...	28	...	.2160	14	.1820
12	...	...	32	.2160	13	.1850
1/4	20	...	...	.2500	7	.2010
1/4	...	28	...	.2500	3	.2130
5/16	18	...	...	.3125	F	.2570
5/16	...	24	...	.3125	I	.2720
3/8	16	...	...	.3750	5/16	.3125
3/8	...	24	...	.3750	Q	.3320
7/16	14	...	...	.4375	U	.3680
7/16	...	20	...	.4375	25/64	.3906
1/2	13	...	...	.5000	27/64	.4219
1/2	...	20	...	.5000	29/64	.4531

3. Name two additional types of drill classifications.
   a. fractional
   b. ____________
   c. ____________

4. What is the approximate drill speed for a 1/2" drill bit when drilling into mild steel? ______
5. Generally speaking, the larger the drill bit, the ____________ (slower/faster) the drill speed.
6. List the trade names for these types of replaceable thread inserts.

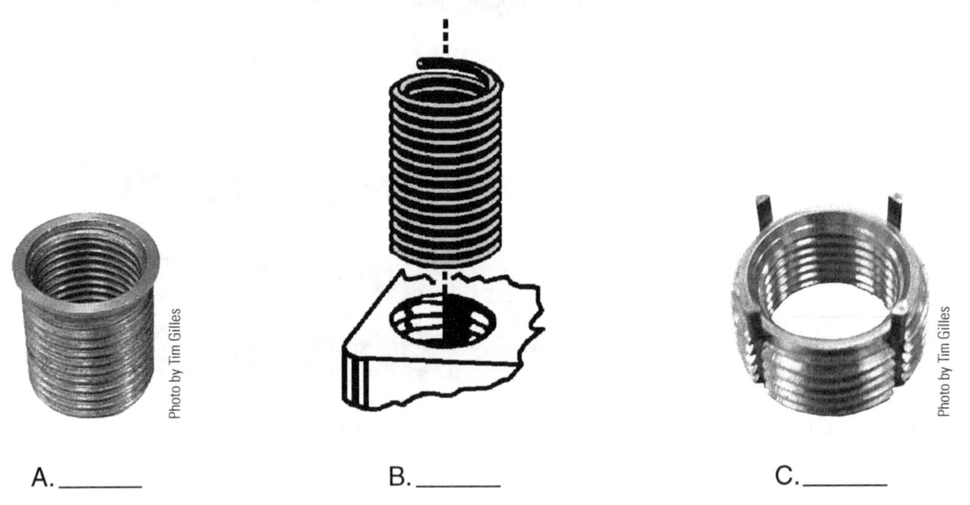

7. Identify the retaining clips pictured below.

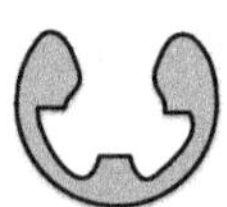

A. ____________ B. ____________ C. ____________ D. ____________

STOP

Instructor OK ________________ Score ____________

Activity Sheet #12
HAND TOOLS

Name_______________________________ Class______________________

ASE Education Foundation Correlation

This activity sheet addresses the following **Required Supplemental Tasks (RSTs):**

Shop and Personal Safety:

Task 2. Utilize safe procedures for handling of tools and equipment.

Tools and Equipment:

Task 3. Demonstrate safe handling and use of appropriate tools.

Task 4. Demonstrate proper cleaning, storage, and maintenance of tools and equipment.

We Support ASE | Education Foundation

Directions: Match the words on the left to the descriptions on the right. Write the letter for the correct word on the line provided. If you are uncertain of the meaning of a term, use the glossary in your textbook.

A. Flare-nut wrench	_______ Hacksaw blade that is the best choice for cutting thick steel
B. Stubby	_______ Type of socket used with air tools
C. Impact driver	_______ Used with a puller to pull a bearing from a shaft
D. Speed handle	_______ Can be used on a 6-point fastener head
E. Dead blow hammer	_______ Measures the turning effort applied to a fastener
F. Slide hammer	_______ Commonly called Allen wrenches
G. Bearing separator	_______ Used to move around under vehicles
H. Breaker bar	_______ What a very short screwdriver is commonly called
I. Coarse tooth	_______ Best socket to use on rusty fasteners
J. 8-point socket	_______ Never used as a prybar
K. 12-point socket	_______ Fits between a socket and a ratchet
L. 6-point socket	_______ Tool used to loosen and tighten fuel line fittings
M. Fine tooth	_______ Wrench used with a ratchet
N. Impact wrench	_______ Always remove mushroom edge before using
O. Impact socket	_______ Hand tool that is used to loosen fasteners that are very tight
P. Vise grips	_______ Puller that uses a heavy weight that is slid against its handle
Q. Creeper	_______ Worn whenever working in the shop

R.	Torque wrench	______	Box that drill bits are stored in
S.	Extension	______	Used to protect the vehicle's finish
T.	Safety goggles	______	Screwdriver that is pounded on with a hammer to loosen a screw
U.	Screwdriver	______	Used on square drive fastener heads
V.	Chisel	______	Air-powered tool used to remove fasteners
W.	Hex wrench	______	Pliers that can be locked to a part
X.	Crowfoot wrench	______	Hacksaw blade that is the best choice for cutting sheet metal
Y.	Drill index	______	A soft-faced hammer that has metal shot in its head
Z.	Fender cover	______	Hand tool used for quickness when assembling parts

STOP

Instructor OK ________________ Score __________

Activity Sheet #13

IDENTIFY SHOP TOOLS

Name ______________________________ Class ____________________

ASE Education Foundation Correlation

This activity sheet addresses the following **Required Supplemental Tasks (RSTs):**

Shop and Personal Safety:

Task 2. Utilize safe procedures for handling of tools and equipment.

Tools and Equipment:

Task 3. Demonstrate safe handling and use of appropriate tools.

Task 4. Demonstrate proper cleaning, storage, and maintenance of tools and equipment.

We Support ASE Education Foundation

Directions: Identify the shop tools and equipment. Place the identifying letter next to the name of the item on page 30.

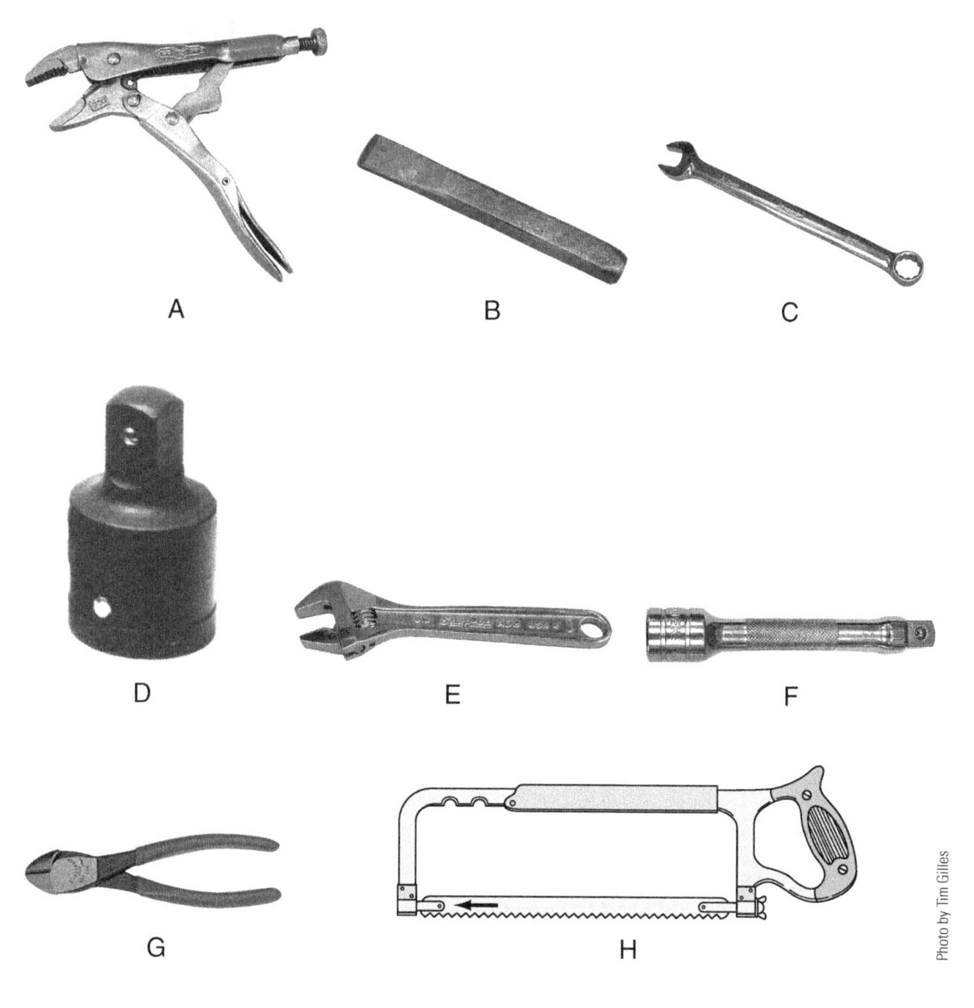

TURN

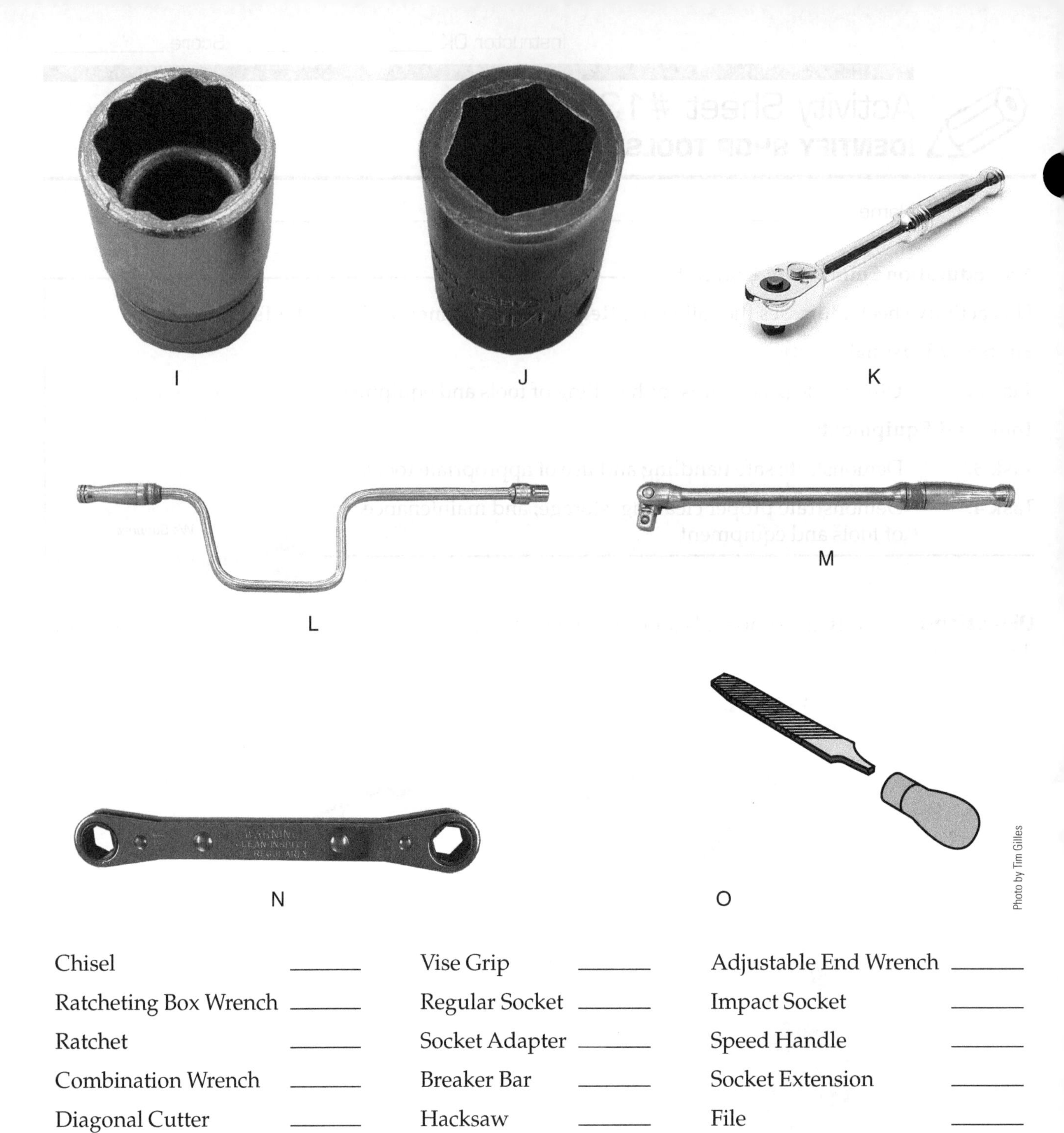

Photo by Tim Gilles

Chisel	______	Vise Grip	______	Adjustable End Wrench	______
Ratcheting Box Wrench	______	Regular Socket	______	Impact Socket	______
Ratchet	______	Socket Adapter	______	Speed Handle	______
Combination Wrench	______	Breaker Bar	______	Socket Extension	______
Diagonal Cutter	______	Hacksaw	______	File	______

STOP

Instructor OK ______________ Score __________

Activity Sheet #14
IDENTIFY COMMON SHOP TOOLS

Name ______________________________ Class ______________________

ASE Education Foundation Correlation

This activity sheet addresses the following **Required Supplemental Tasks (RSTs):**

Shop and Personal Safety:

Task 2. Utilize safe procedures for handling of tools and equipment.

Tools and Equipment:

Task 3. Demonstrate safe handling and use of appropriate tools.

Task 4. Demonstrate proper cleaning, storage, and maintenance of tools and equipment.

We Support ASE Education Foundation

Directions: Identify the common shop tools. Place the identifying letter next to the name of the item on page 32.

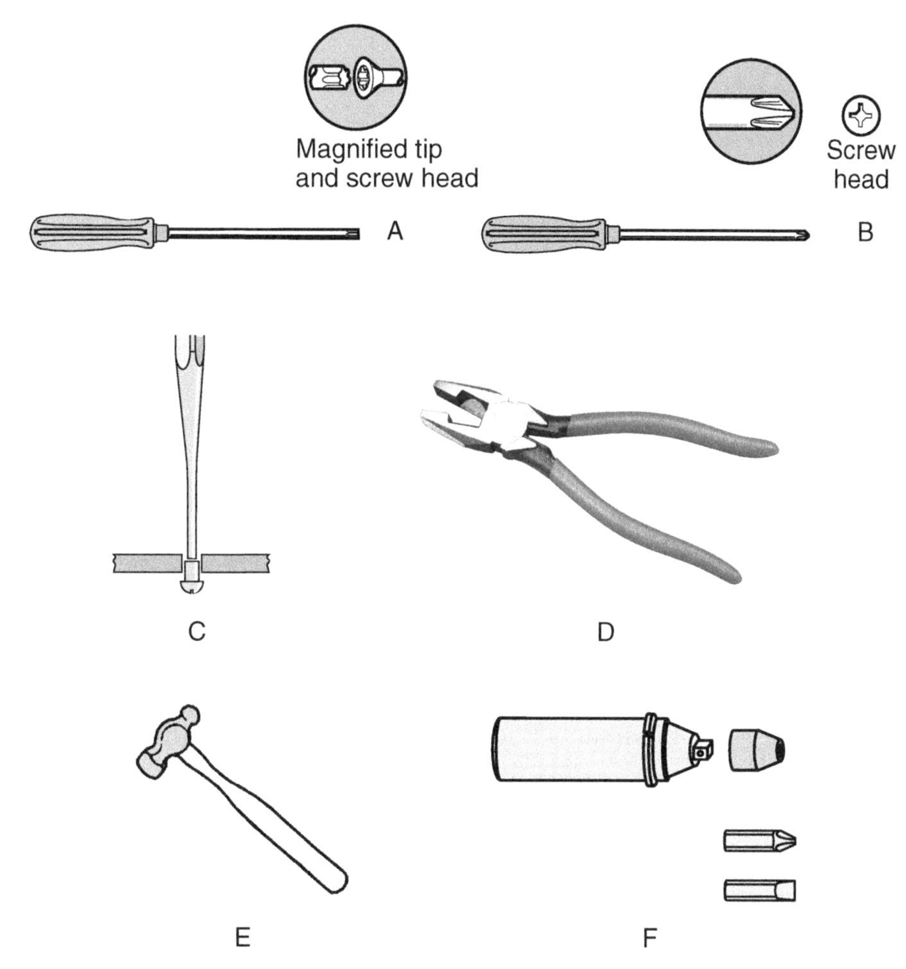

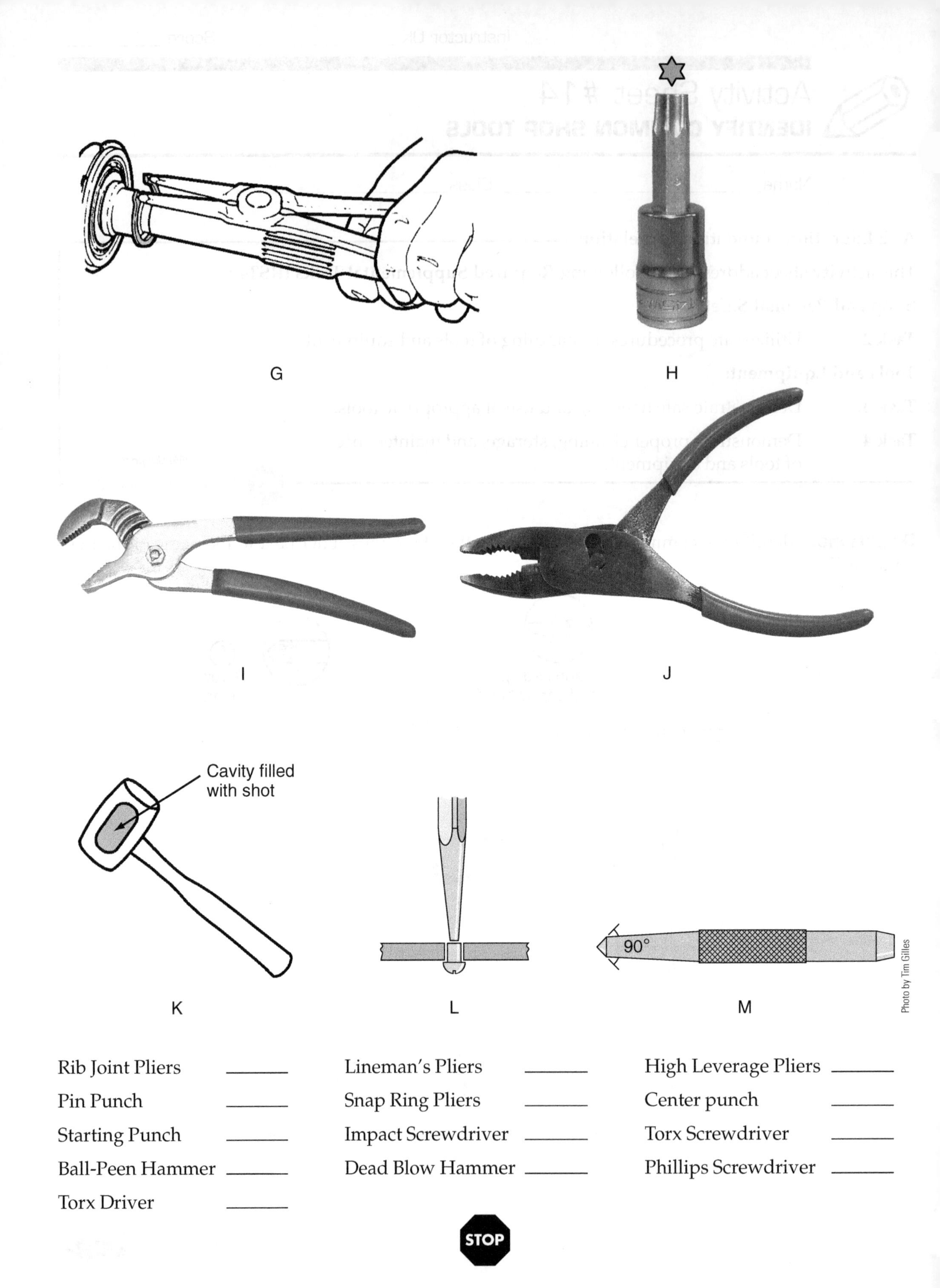

Rib Joint Pliers	______	Lineman's Pliers	______	High Leverage Pliers	______
Pin Punch	______	Snap Ring Pliers	______	Center punch	______
Starting Punch	______	Impact Screwdriver	______	Torx Screwdriver	______
Ball-Peen Hammer	______	Dead Blow Hammer	______	Phillips Screwdriver	______
Torx Driver	______				

STOP

Instructor OK ______________ Score __________

Activity Sheet #15
IDENTIFY SHOP EQUIPMENT

Name ______________________ Class ________________

ASE Education Foundation Correlation

This activity sheet addresses the following **Required Supplemental Tasks (RSTs):**

Shop and Personal Safety:

Task 2. Utilize safe procedures for handling of tools and equipment.

Tools and Equipment:

Task 3. Demonstrate safe handling and use of appropriate tools.

Task 4. Demonstrate proper cleaning, storage, and maintenance of tools and equipment.

We Support ASE Education Foundation

Directions: Identify the shop equipment. Place the identifying letter next to the name of the item on page 34.

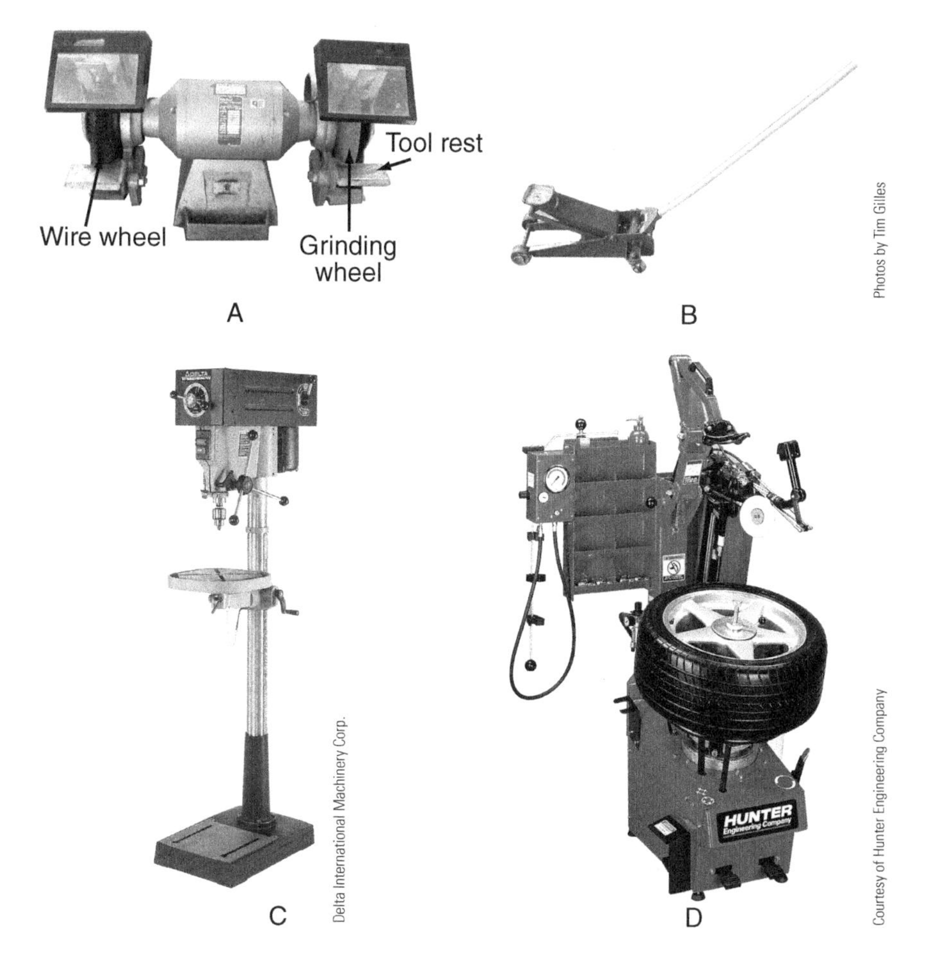

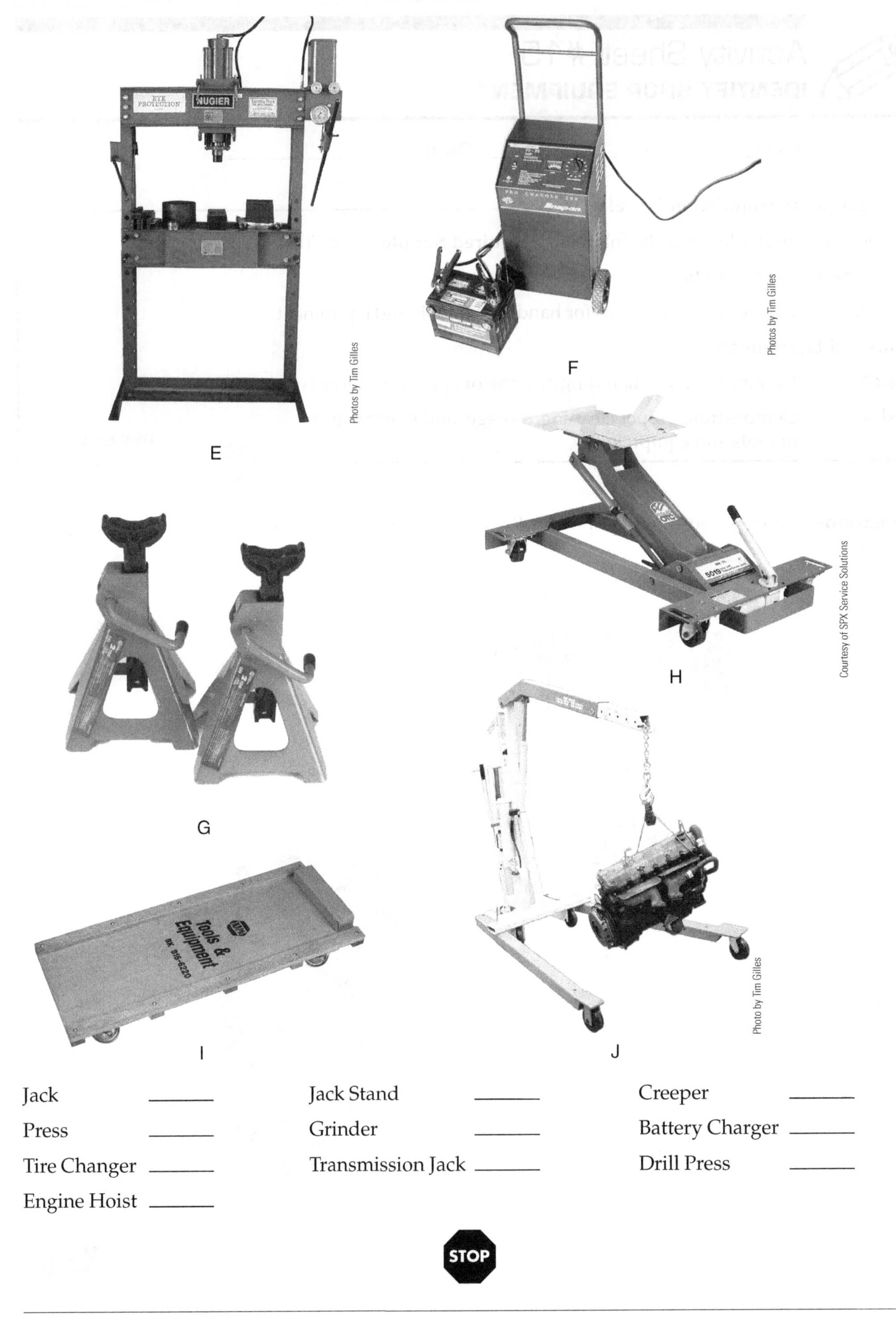

Jack	______	Jack Stand	______	Creeper	______
Press	______	Grinder	______	Battery Charger	______
Tire Changer	______	Transmission Jack	______	Drill Press	______
Engine Hoist	______				

Instructor OK ______________ Score ____________

Activity Sheet #16

IDENTIFY SHOP POWER TOOLS

Name ______________________________ Class ____________________

ASE Education Foundation Correlation

This activity sheet addresses the following **Required Supplemental Tasks (RSTs):**

Shop and Personal Safety:

Task 2. Utilize safe procedures for handling of tools and equipment.

Tools and Equipment:

Task 3. Demonstrate safe handling and use of appropriate tools.

Task 4. Demonstrate proper cleaning, storage, and maintenance of tools and equipment.

We Support ASE Education Foundation

Directions: Identify the shop power tools. Place the identifying letter next to the name of the item at the top of page 36.

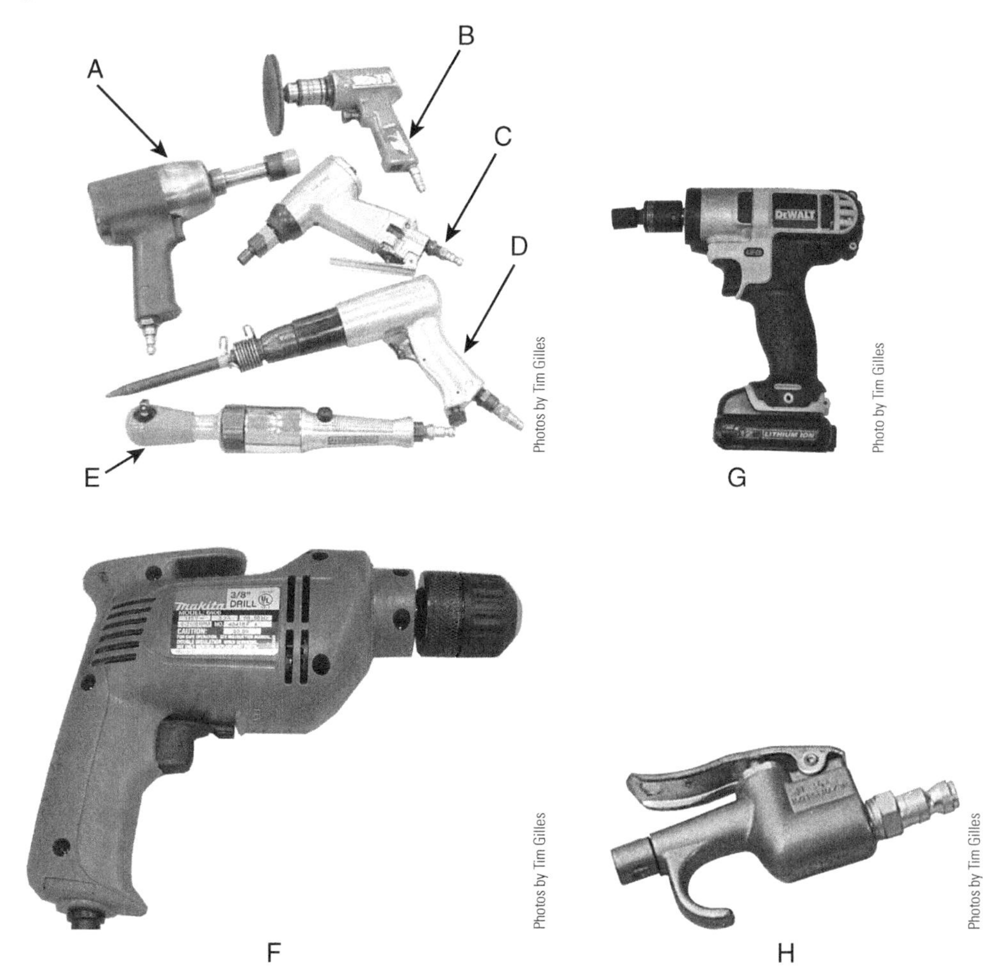

TURN

Photo by Tim Gilles

I

Photo by Tim Gilles

J

Electric Drill Motor	______	1/2" Impact Wrench	______	3/8" Impact Wrench	______
Air Drill	______	Air Blowgun	______	Air Hammer	______
Electric Impact Wrench	______	Air Ratchet	______	Die Grinder	______
Air Grinder	______				

STOP

Instructor OK ________________ Score ____________

Activity Sheet #17

FIRE EXTINGUISHERS

Name ______________________________ Class ____________________

ASE Education Foundation Correlation

This activity sheet addresses the following **Required Supplemental Task (RST):**

Shop and Personal Safety:

Task 7. Identify the location and the types of fire extinguishers and other fire safety equipment; demonstrate knowledge of the procedures for using fire extinguishers and other fire safety equipment.

We Support ASE Education Foundation

	Class of Fire	Typical Fuel Involved	Type of Extinguisher
Class A Fires (green)	**For Ordinary Combustibles** Put out a Class A fire by lowering its temperature or by coating the burning combustibles.	Wood Paper Cloth Rubber Plastics Rubbish Upholstery	Water*[1] Foam* Multipurpose dry chemical[4]
Class B Fires (red)	**For Flammable Liquids** Put out a Class B fire by smothering it. Use an extinguisher that gives a blanketing flame-interrupting effect; cover whole flaming liquid surface.	Gasoline Oil Grease Paint Lighter fluid	Foam* Carbon dioxide[5] Halogenated agent[6] Standard dry chemical[2] Purple K dry chemical[3] Multipurpose dry chemical[4]
Class C Fires (blue)	**For Electrical Equipment** Put out a Class C fire by shutting off power as quickly as possible and by always using a nonconducting extinguishing agent to prevent electric shock.	Motors Appliances Wiring Fuse boxes Switchboards	Carbon dioxide[5] Halogenated agent[6] Standard dry chemical[2] Purple K dry chemical[3] Multipurpose dry chemical[4]
Class D Fires (yellow)	**For Combustible Metals** Put out a Class D fire of metal chips, turnings, or shaving by smothering or coating with a specially designed extinguishing agent.	Aluminum Magnesium Potassium Sodium Titanium Zirconium	Dry powder extinguishers and agents only

*Cartridge-operated water, foam, and soda-acid types of extinguishers are no longer manufactured. These extinguishers should be removed from service when they become due for their next hydrostatic pressure test.

Notes:

(1) Freeze in low temperatures unless treated with antifreeze solution, usually weighs over 20 pounds, and is heavier than any other extinguisher mentioned.

(2) Also called ordinary or regular dry chemical. (solution bicarbonate)

(3) Has the greatest initial fire-stopping power of the extinguishers mentioned for class B fires. Be sure to clean residue immediately after using the extinguisher so sprayed surfaces will not be damaged. (potassium bicarbonate)

(4) The only extinguishers that fight A, B, and C class fires. However, they should not be used on fires in liquified fat or oil of appreciable depth. Be sure to clean residue immediately after using the extinguisher so sprayed surfaces will not be damaged. (ammonium phosphates)

(5) Use with caution in unventilated, confined spaces.

(6) May cause injury to the operator if the extinguishing agent (a gas) or the gases produced when the agent is applied to a fire is inhaled.

TURN

Directions: Use the chart to answer the following questions.

1. What kind of fire extinguisher should be used to put out an oil, a gasoline, or a grease fire?

 ____

2. What kind of fire extinguisher should be used to put out an electrical fire?

 ____

3. What kind of fire extinguisher should be used to put out a wood or paper fire?

 ____

4. What types of fire extinguishers are available in your school's shop?

 ____

5. When was the last time the fire extinguishers in the school's shop were inspected?

 ____

6. How should used shop towels and rags be stored?

 ____

7. Flammable materials should be stored in an approved flammable material storage

 ______________________________.

8. The process of self-ignition is known as ____________ combustion.
9. ____________ should never be used to extinguish a gasoline or an oil fire.
10. Gasoline should never be stored in a ____________ container.

STOP

Part I

Activity Sheets

Section Three

Vehicle Inspection (Lubrication/Safety Check)

Instructor OK ______________ Score __________

Activity Sheet #18

ENGINE OIL

Name ______________________ Class ______________

ASE Education Foundation Correlation

This activity sheet addresses the following **MLR** task:

I.C.5 Perform engine oil and filter change; use proper fluid type per manufacturer specification; reset maintenance reminder as required. **(P-1)**

This activity sheet addresses the following **AST/MAST** task:

I.D.10 Perform engine oil and filter change; use proper fluid type per manufacturer specification. **(P-1)**

We Support ASE | Education Foundation

Directions: Fill in the information in the spaces provided.

1. What does SAE represent? ______________________________
2. What does API represent? ______________________________
3. What does ILSAC represent? ______________________________
4. What does the "W" in the SAE rating mean? ______________________________
5. What symbol identifies that the motor oil meets ILSAC standards?

   ____

6. Which oil is thicker when hot?

   SAE 20W- 40 ______

   SAE 40 ______

7. An engine with variable valve timing generally requires an oil with a lower viscosity. Mark which of the following is the most likely manufacturer recommendation.

   5W/20 __________

   5W/30 __________

   10W/40 __________

TURN

8. Write the following terms in their correct positions in the API donut below.
   - ❑ API service SN
   - ❑ SAE 10W-30
   - ❑ Energy conserving II

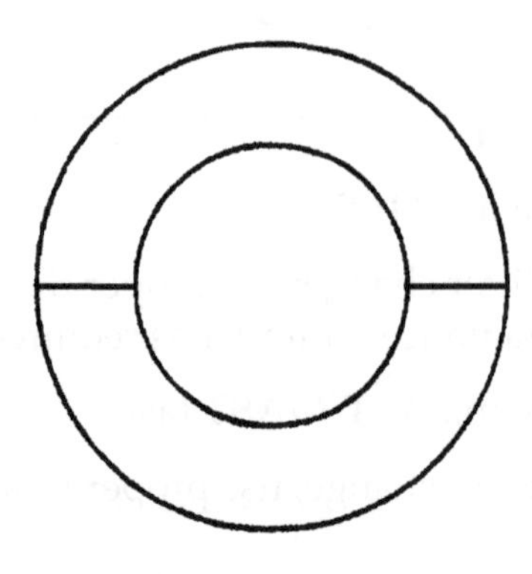

9. Place an X on the line(s) that best identifies the symbol below.

   Multiviscosity oil ______

   Single-viscosity oil ______

   ILSAC GF-4 standards ______

10. Engine oil pressure is a result of ____________ to flow.
11. A lack of maintenance can cause ____________ to accumulate in the engine.
12. The maintenance ____________ light must be reset after an oil change.

STOP

Instructor OK ______________ Score ____________

Activity Sheet #19

IDENTIFY PARTS OF THE OIL FILTER

Name ______________________________ Class ______________________

ASE Education Foundation Correlation

This activity sheet addresses the following **MLR** tasks:

I.C.5 Perform engine oil and filter change; use proper fluid type per manufacturer specification; reset maintenance reminder as required. **(P-1)**

I.C.6 Identify components of the lubrication and cooling systems. **(P-1)**

This activity sheet addresses the following **AST/MAST** task:

I.D.10 Perform engine oil and filter change; use proper fluid type per manufacturer specification. **(P-1)**

We Support
ASE Education Foundation

Directions: Identify the oil filter parts pictured below. Write the identifying letter next to the name of the part.

B C
A
F
E D

Anti-Drainback Valve	______	By-Pass Valve	______	Metal Center	______
Paper Element	______	Gasket	______	Spring	______

1. Draw an arrow(s) to show where the oil enters the filter pictured above.
2. Which way does the oil flow through the filter element?

   outside to inside ______

   inside to outside ______
3. When installing an oil filter, its ______________ must be coated with oil.
4. Always use the correct filter because some filters have a ______________ valve and others do not have this valve.

5. Always use the correct filter because some filters have an anti- ______________ valve, while others do not.
6. Spin-on filters typically are tightened ______________ turn(s) after the filter contacts the block.
7. Which type of oil filter is becoming more common on newer vehicles, spin-on or cartridge? ______________
8. After changing the oil and filter on a vehicle, run the engine and check for leaks. Then shut off the engine and check the oil level on the ______________.

STOP

Instructor OK ______________ Score __________

Activity Sheet #20

IDENTIFY UNDERHOOD COMPONENTS

Name ______________________ Class ______________

ASE Education Foundation Correlation

This activity sheet addresses the following **MLR** task:

I.A.3 Inspect engine assembly for fuel, oil, coolant, and other leaks; determine necessary action. **(P-1)**

This activity sheet addresses the following **AST/MAST** task:

I.A.4 Inspect engine assembly for fuel, oil, coolant, and other leaks; determine needed action. **(P-1)**

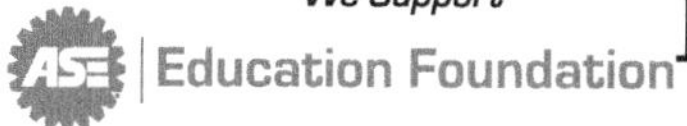

Directions: Identify the underhood components in the photograph below. Write the identifying letter next to the name of each component listed below the photograph.

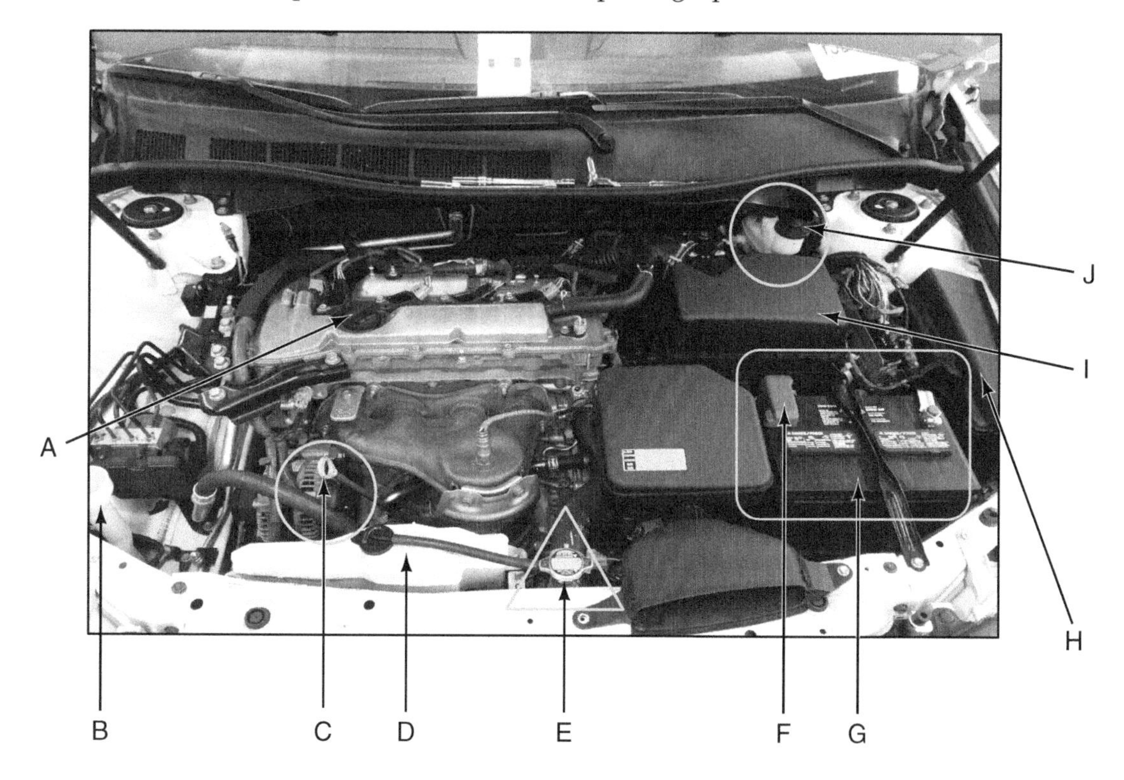

Engine Oil Dipstick	______	Radiator Cap	______	Battery	______
Positive Battery Terminal	______	Coolant Recovery Tank	______	Master Cylinder	______
Fuse Box	______	Air Cleaner	______	Wiper Fluid Reservoir	______
Engine Oil Cap	______				

STOP

Activity Sheet #21

IDENTIFY MAJOR UNDERCAR COMPONENTS

Name ____________________________ Class ____________________

ASE Education Foundation Correlation

This activity sheet addresses the following **MLR** task:

I.A.3 Inspect engine assembly for fuel, oil, coolant, and other leaks; determine necessary action. **(P-1)**

This activity sheet addresses the following **AST/MAST** task:

I.A.4 Inspect engine assembly for fuel, oil, coolant, and other leaks; determine needed action. **(P-1)**

We Support ASE Education Foundation

Directions: Identify the major undercar components of the automobile in the drawing. Place the identifying letter next to the name of each component listed below.

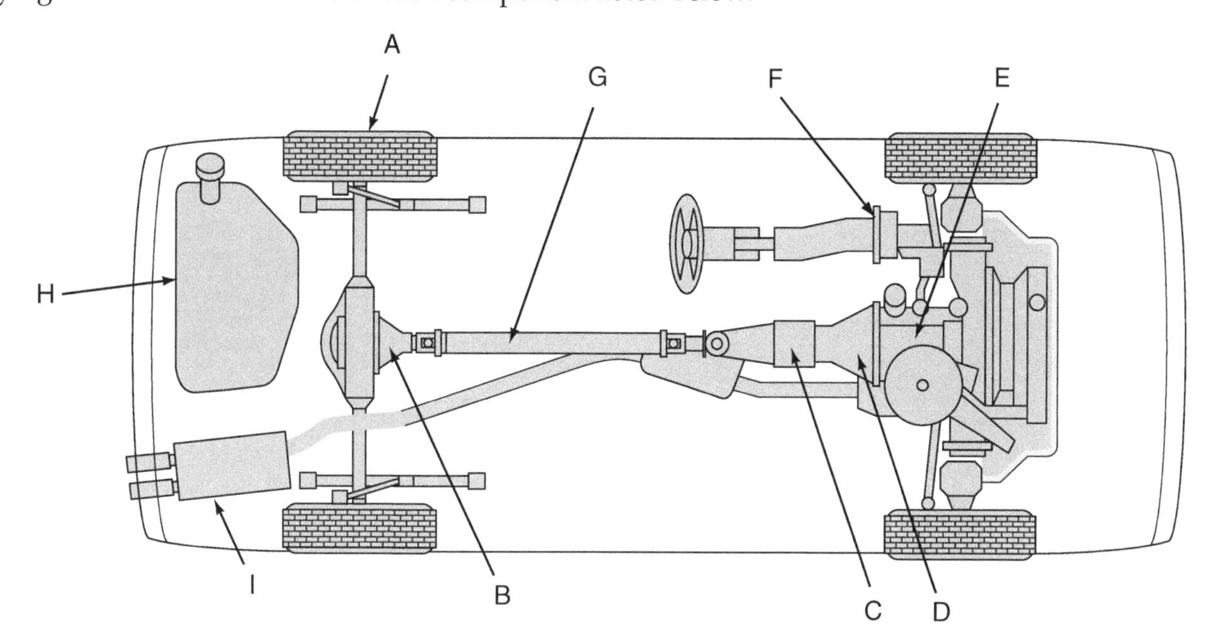

Brakes	______	Steering	______	Transmission	______
Driveshaft	______	Differential	______	Fuel Tank	______
Clutch	______	Engine	______	Muffler	______

Instructor OK ________________ Score ____________

Activity Sheet #22

INSTRUMENT PANEL WARNING LIGHTS

Name ______________________________ Class ______________________

ASE Education Foundation Correlation

This activity sheet addresses the following **MLR** task:

I.A.2 Verify operation of the instrument panel engine warning indicators. **(P-1)**

This activity sheet addresses the following **AST/MAST** task:

I.A.3 Verify operation of the instrument panel engine warning indicators. **(P-1)**

We Support ASE Education Foundation

Directions: Identify the instrument panel warning/indicator lights. Place the correct identifying letter next to its warning/indicator symbol.

A.

B.

C.

D.

E.

F.

G.

______ Seat Belt Reminder

______ Car Door Open

______ Tire Pressure Monitor-Low Tire Warning

______ Low Fuel

______ Engine Warning or Malfunction Indicator

______ Fog Light

______ Electric Power Steering

H.

I.

J.

K.

L.

M.

N.

______ Front Air bag Warning

______ Anti-lock Braking System

______ Parking Brake and Brake System Problem Warning

______ Battery and Charging

______ Low Oil Pressure

______ Engine Overheating

______ High Beam Light

TURN

Directions: Answer the following questions about instrument panel warning and indicator lights.

1. What should you do if a red instrument panel light is illuminated?

   a. Stop and evaluate the problem, or b. Check the problem as soon as possible.

2. What should you do if an amber instrument panel light is illuminated?

   a. Stop and evaluate the problem, or b. Check the problem as soon as possible.

3. A blue instrument panel light is used to indicate when the ______________ lights are on.

4. Green instrument panel lights are used to indicate when the ______________ ______________ lights are on.

STOP

Part I

Activity Sheets

Section Four

Engine Operation and Service

Instructor OK ______________ Score __________

Activity Sheet #23

IDENTIFY FOUR-STROKE CYCLE

Name ____________________ Class ______________

ASE Education Foundation Correlation

This activity sheet addresses the following **AST/MAST** task:

I.A.2 Research vehicle service information including fluid type, internal engine operation, vehicle service history, service precautions, and technical service bulletins. **(P-1)**

We Support ASE Education Foundation

Directions: Identify which stroke is occurring in each sketch. Place the identifying letter next to the name of the stroke.

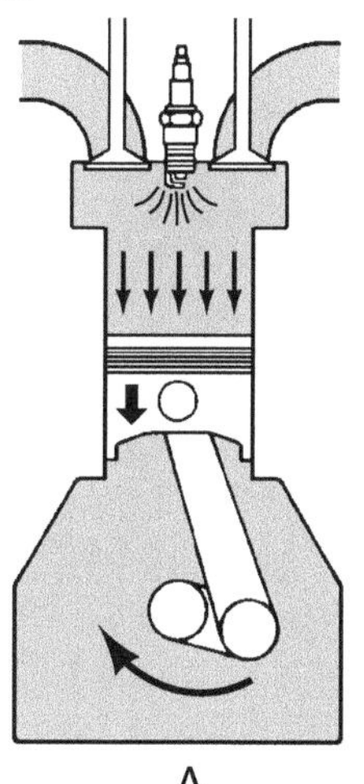

A

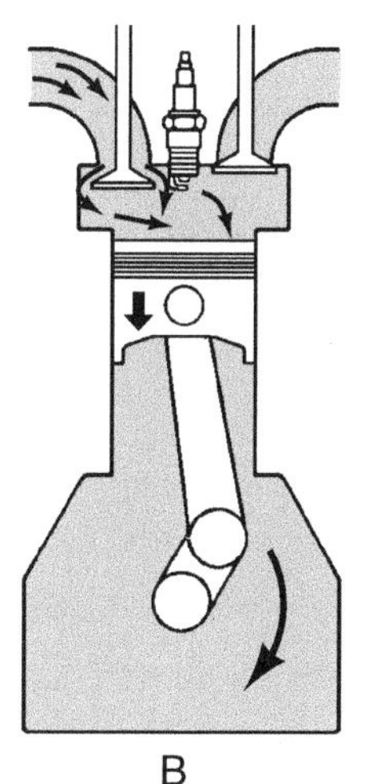

B

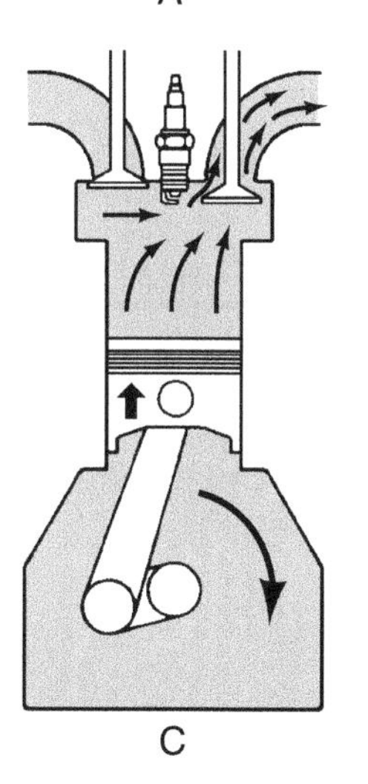

C

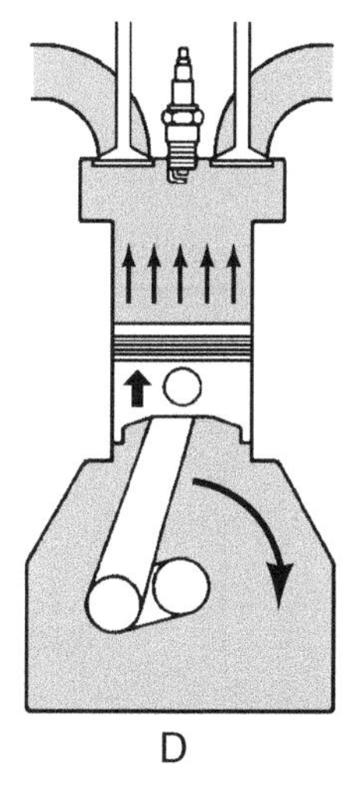

D

Intake ______

Compression ______

Power ______

Exhaust ______

STOP

Instructor OK ______________ Score ____________

Activity Sheet #24

IDENTIFY ENGINE OPERATION

Name ______________________________ Class ______________________

ASE Education Foundation Correlation

This activity sheet addresses the following **AST/MAST** task:

I.A.2 Research vehicle service information including fluid type, internal engine operation, vehicle service history, service precautions, and technical service bulletins. **(P-1)**

We Support ASE Education Foundation

Directions: Identify the missing information in the drawing. Place the identifying letter next to the best answer.

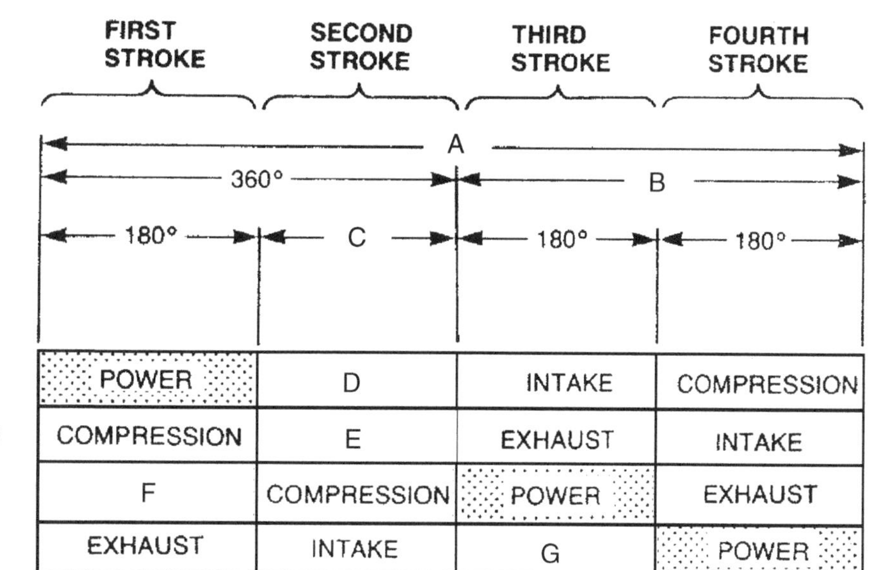

FIRST CYLINDER	POWER	D	INTAKE	COMPRESSION
SECOND CYLINDER	COMPRESSION	E	EXHAUST	INTAKE
THIRD CYLINDER	F	COMPRESSION	POWER	EXHAUST
FOURTH CYLINDER	EXHAUST	INTAKE	G	POWER

Power Stroke ______

Compression Stroke ______

180 degrees ______

Intake Stroke ______

360 degrees ______

720 degrees ______

Exhaust Stroke ______

Related Four-Stroke Cycle Questions

1. A revolution is ______ degrees of crankshaft rotation.
2. The camshaft rotates ______ degrees during one four-stroke cycle.
3. The crankshaft rotates ______ degrees during one four-stroke cycle.
4. How many revolutions does the crankshaft make per second at 3,000 rpm? ______

5. At 3,000 rpm, how many times does each intake valve open in one minute? ______
6. At 3,000 rpm, how many times does each intake valve open in one second? ______
7. At 3,000 rpm, how many times does each exhaust valve open in one minute? ______
8. At 3,000 rpm, how many times does each exhaust valve open in one second? ______
9. At 3,000 rpm, how many four-stroke cycles occur each minute in a cylinder? ______
10. At 3,000 rpm, how many four-stroke cycles occur each second in a cylinder? ______

STOP

Instructor OK ______________ Score __________

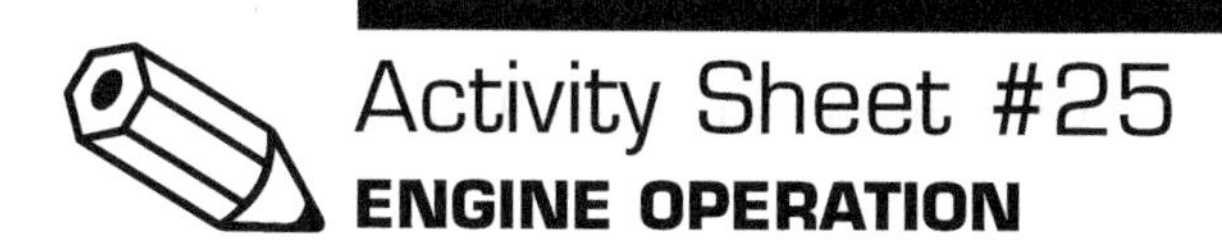

Activity Sheet #25
ENGINE OPERATION

Name ________________________ Class ________________

ASE Education Foundation Correlation

This activity sheet addresses the following **MLR** task:

I.B.2 Identify components of the cylinder head and valve train. **(P-1)**

We Support Education Foundation

Directions: Match the words on the left to the descriptions on the right. Write the letter for the correct word on the line provided. For the terms that you are not certain of, use the glossary in your textbook.

Term		Description
A. Poppet valve	_____	Number of degrees the camshaft turns during one four-stroke cycle
B. Blowby	_____	Number of degrees the crankshaft turns during one four-stroke cycle
C. Upper end	_____	Number of times a valve opens during 720 degrees of crankshaft rotation
D. Valve train	_____	Forces the piston ring against the cylinder wall
E. Pushrod engine	_____	Exhaust valve is open during part of this stroke
F. Overhead cam engine	_____	Job of top two piston rings
G. Long block	_____	Space between bearing and shaft
H. Short block	_____	Intake valve is open during part of this stroke
I. Crankcase	_____	May be larger than the exhaust valve
J. End play	_____	Number of cam lobes for a typical pushrod V8 camshaft
K. Bearing clearance	_____	May be driven by the camshaft
L. Piston slap	_____	Another name for the harmonic balancer
M. Once	_____	Includes the cylinder head(s) and valve train
N. OHC	_____	Leakage of gases past the rings
O. Exhaust	_____	Engine with the cam above the cylinder head
P. 50	_____	Camshaft turns _____ as fast as the crankshaft
Q. Vibration damper	_____	Noise when there is excessive piston-to-cylinder wall clearance
R. Combustion	_____	The parts that open and close the valves
S. Seal compression	_____	Nonmagnetic valve

TURN

T.	720 degrees	______	Overhead cam
U.	360 degrees	______	The style of valve used by four-stroke cycle internal combustion engines
V.	16	______	Back and forth clearance
W.	Half	______	Dual overhead cam
X.	Oil pump	______	Variable valve timing
Y.	Intake valve	______	Number of times each valve opens every second at 6,000 rpm
Z.	Compression stroke	______	Rebuilt engine including heads
AA.	Intake stroke	______	Area surrounding the crankshaft
AB.	VVT	______	Rebuilt engine without heads
AC.	DOHC	______	Engine with the cam in the block
AD.	Companion cylinders	______	V, inline, and opposed
AE.	Cylinder arrangements	______	Pistons that move in pairs

STOP

Activity Sheet #26

IDENTIFY ENGINE PARTS

Name_______________________________ Class_______________________

ASE Education Foundation Correlation

This activity sheet addresses the following **MLR** task:

I.B.2 Identify components of the cylinder head and valve train. **(P-1)**

This activity sheet addresses the following **MAST** task:

I.C.2 Disassemble engine block; clean and prepare components for inspection and reassembly. **(P-1)**

We Support ASE | Education Foundation

Directions: Identify the engine parts in the drawings. Place the identifying letter next to the name of each part listed below the diagram.

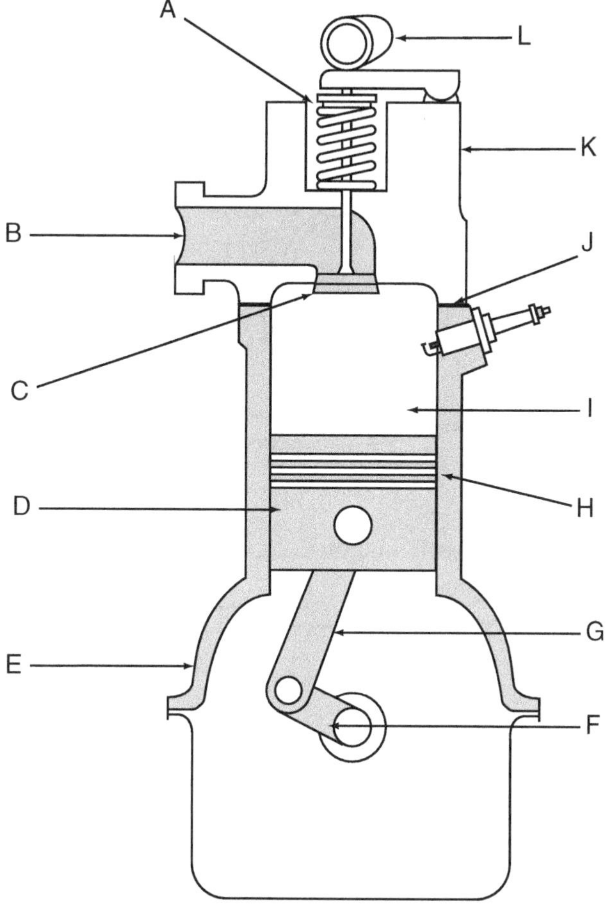

Connecting Rod	______	Cam Lobe	______	Port	______	Crankshaft	______
Block	______	Head Gasket	______	Piston	______	Valve	______
Head	______	Valve Spring	______	Cylinder	______	Piston Rings	______

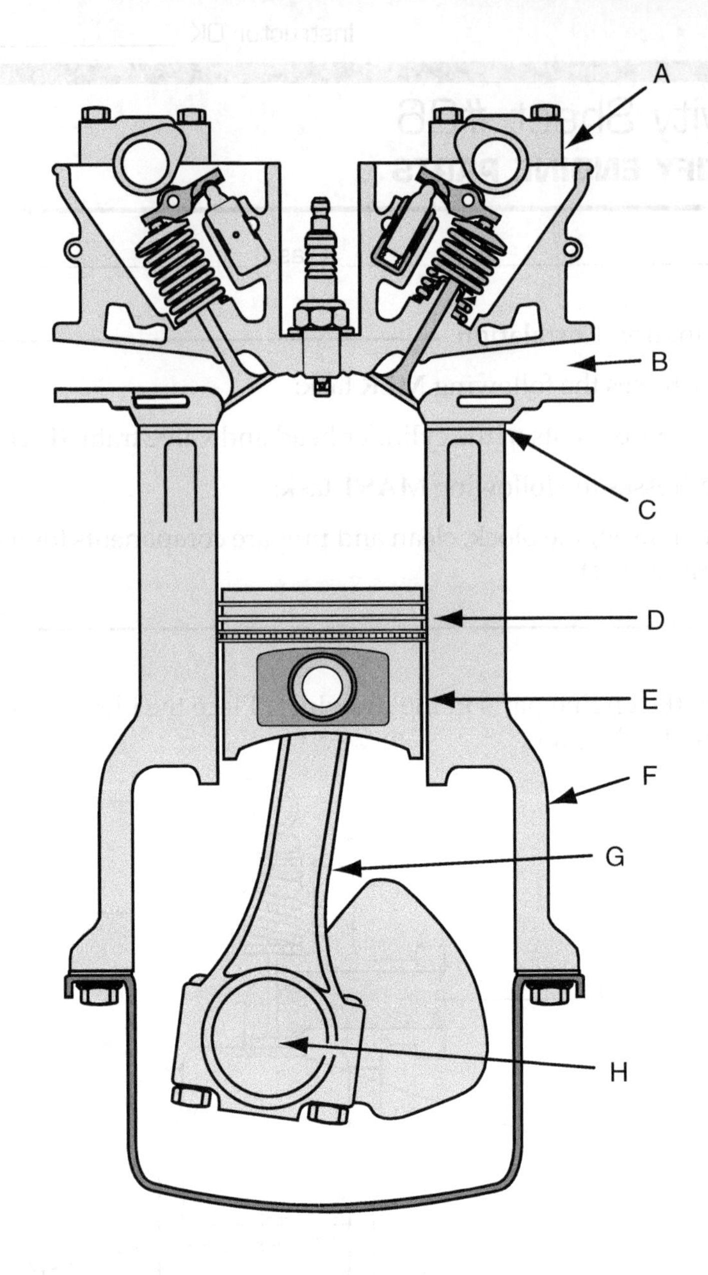

Crankshaft	______	Head Gasket	______	Cylinder Head	______
Block	______	Rings	______	Connecting Rod	______
Piston	______	Valve Port	______		

Instructor OK ________________ Score ____________

Activity Sheet #27

IDENTIFY CYLINDER BLOCK ASSEMBLY

Name ______________________________ Class ____________________

ASE Education Foundation Correlation

This activity sheet addresses the following **MAST** task:

I.C.2 Disassemble engine block; clean and prepare components for inspection and reassembly. **(P-1)**

We Support ASE | Education Foundation

Directions: Identify the cylinder block assembly in the drawing. Place the identifying letter next to the name of each component listed below the diagram.

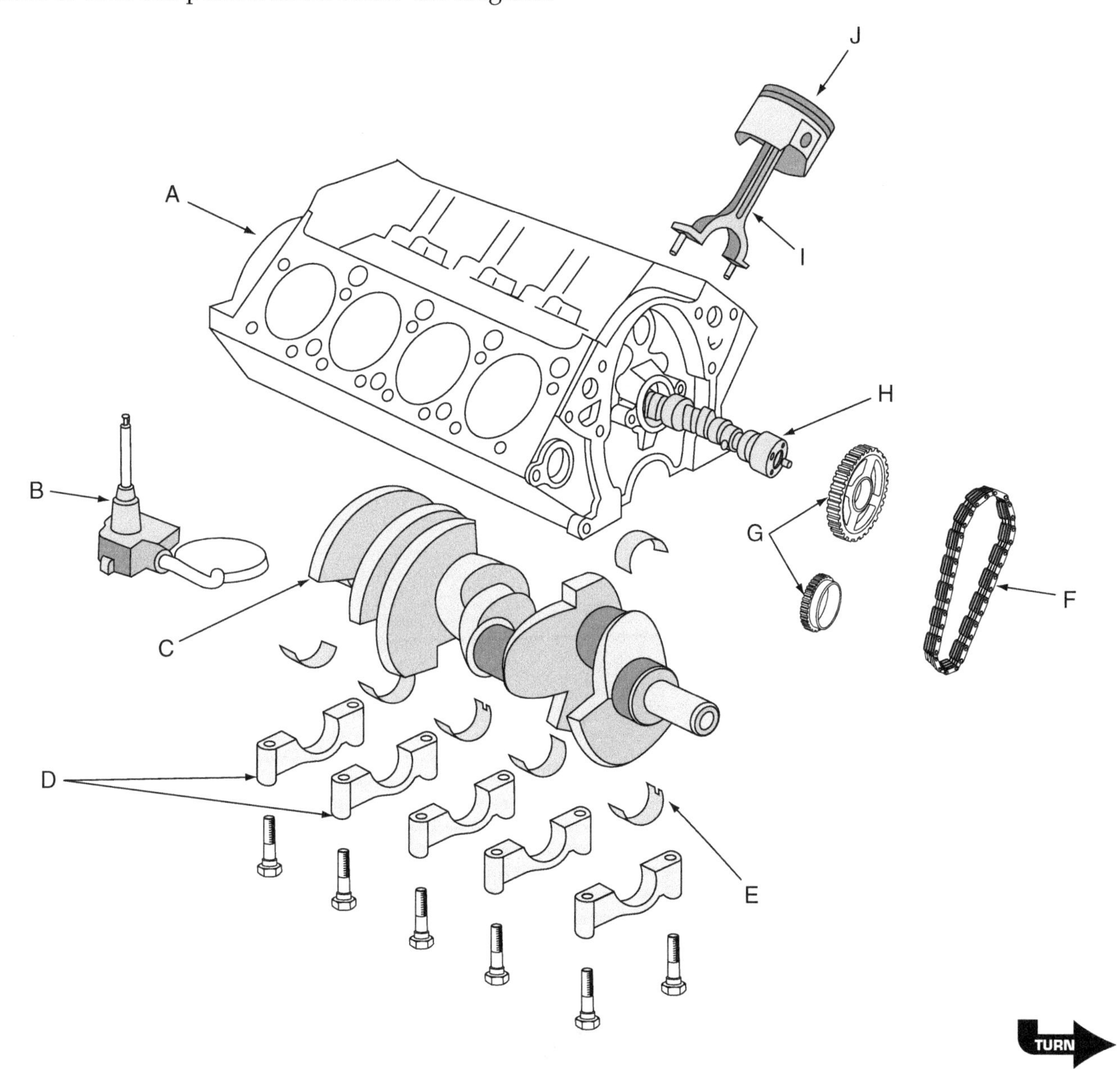

TURN

Crankshaft	______	Piston	______	Camshaft	______
Timing Sprockets	______	Main Bearing	______	Oil Pump	______
Main Caps	______	Connecting Rod	______	Timing Chain	______
Cylinder Block	______				

STOP

Instructor OK ________ Score ________

Activity Sheet #28

IDENTIFY CYLINDER HEAD CLASSIFICATIONS

Name ________________ Class ________________

ASE Education Foundation Correlation

This activity sheet addresses the following **MLR** task:

I.B.2 Identify components of the cylinder head and valve train. **(P-1)**

We Support ASE | Education Foundation

Directions: Identify the cylinder head classification in each drawing. Place the identifying letter next to the correct name at the bottom of the page.

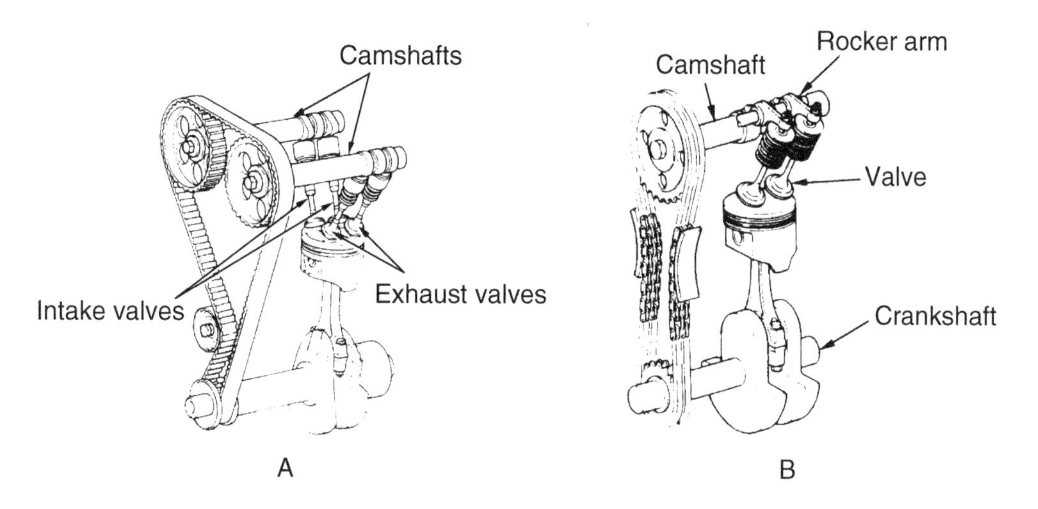

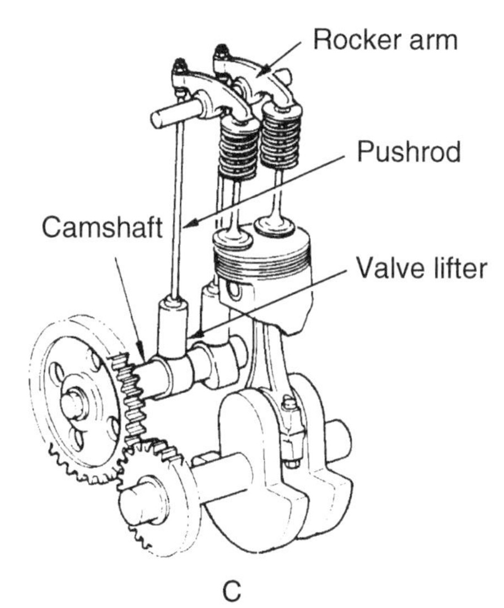

OHC ______ DOHC ______ Cam-in-Block ______

STOP

Instructor OK ______________ Score __________

Activity Sheet #29

IDENTIFY ENGINE BLOCK CONFIGURATIONS

Name ____________________________ Class ____________________

Directions: Identify the engine block configurations in the drawing. Place the identifying letter next to the name of the design listed at the bottom of the page.

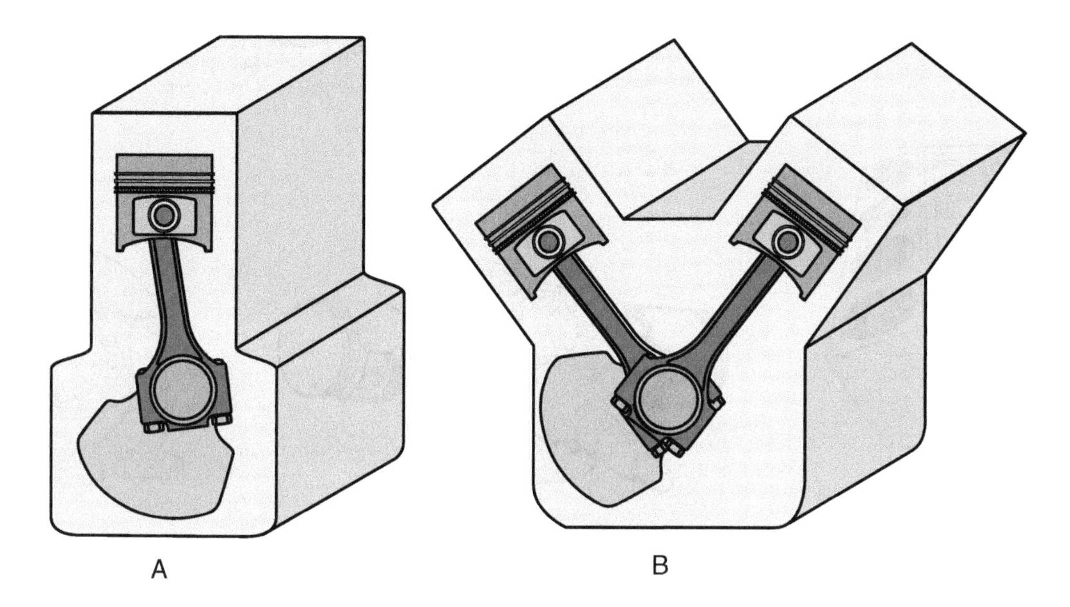

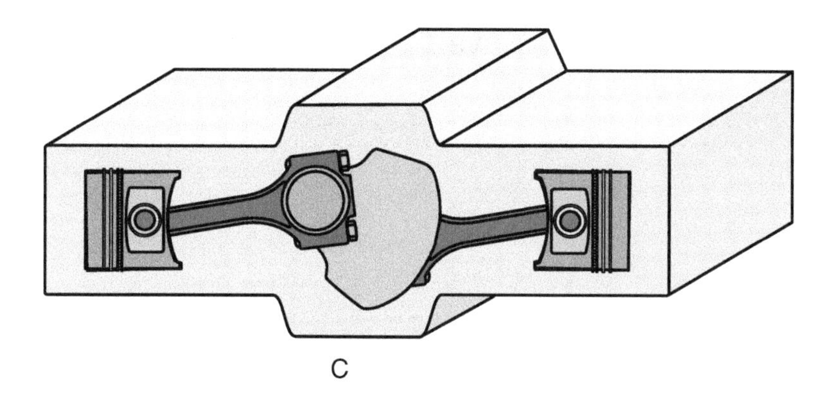

V-type ______ Inline ______ Opposed ______

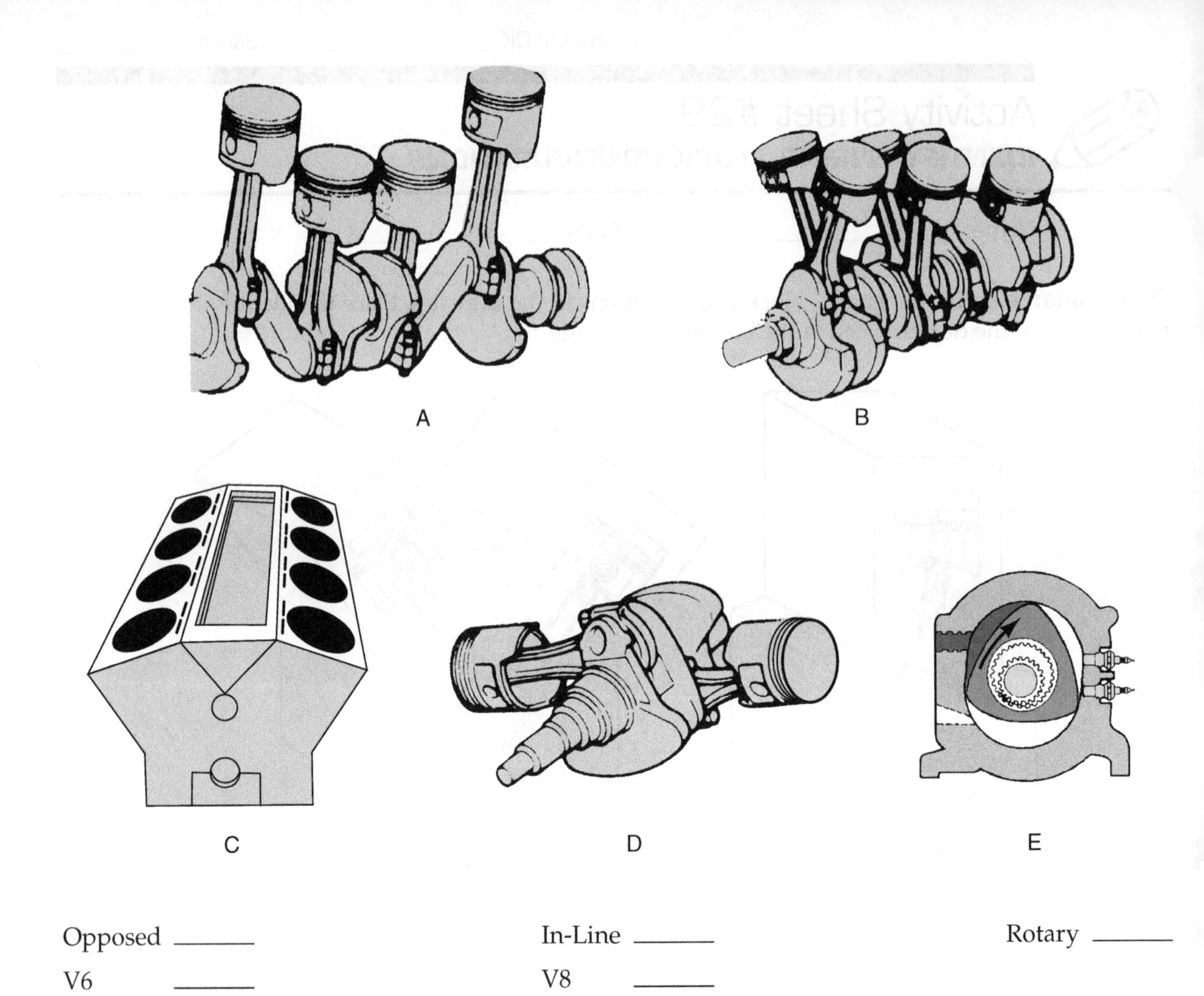

Opposed ______

V6 ______

In-Line ______

V8 ______

Rotary ______

STOP

Instructor OK _______________ Score ___________

Activity Sheet #30

ENGINE CLASSIFICATIONS

Name_______________________________ Class_____________________

Directions: Match the words on the left to the descriptions on the right. Write the letter for the correct word on the line provided. For the terms that you are not certain of, use the glossary in your textbook.

Term		Description
A. Cylinder bank	______	Cam is located in the cylinder block
B. Firing order	______	In-line, V, opposed
C. Companion cylinders	______	Engine has four camshafts
D. Hybrid	______	Turbulent combustion chamber design
E. I-head	______	Volume difference between TDC and BDC
F. Pushrod engine	______	Rich air-fuel mixture starts lean mixture burning
G. OHC	______	Four-stroke spark ignition engine
H. SOHC	______	Four-cylinder engine firing order
I. DOHC	______	Zero emission vehicle
J. Cross-flow head	______	Compression ignition engine
K. Hemi-head	______	A row of cylinders
L. ICE	______	Rows of cylinders in a V-type block
M. Stratified charge	______	Rotary engine
N. Otto-cycle	______	Cam is located in the cylinder head
O. S.I. engine	______	The order in which the spark plugs fire
P. Diesel-cycle	______	Dual overhead cam
Q. FCEV	______	Area between the heads on a V-block
R. Compression ratio	______	Valve placement is in the cylinder head
S. Wankel engine	______	Intake and exhaust ports on opposite sides of the engine
T. Two-stroke engine	______	Ignites its fuel mixture with a spark
U. ZEV	______	Single overhead cam
V. Hybrid vehicle	______	Engine used in chain saws
W. Cylinder arrangements	______	Completes cycle in one revolution
X. Bank	______	Hemispherical combustion chamber design
Y. Valley	______	Rotary engine design
Z. 1–3–4–2	______	Ignites fuel using heat from compression
AA. Wankel	______	The valves are in the cylinder block
AB. Two-stroke	______	Uses more than one type of energy

TURN

AC.	Four-stroke engine	______	Two revolutions to fire all cylinders
AD.	Boxer	______	Pistons come to TDC and BDC together
AE.	V8 DOHC	______	Fuel cell electric vehicle
AF.	L-head	______	Uses electric motor and ICE
AG.	Wedge head	______	Internal combustion engine
AH.	Diesel engine	______	Horizontally opposed engine

STOP

Instructor OK ______________ Score __________

Activity Sheet #31

ENGINE SIZES AND MEASUREMENTS

Name ____________________________ Class ____________________

Objective: After completing this assignment, you should be able to identify and calculate the displacement of an engine. This task will help prepare you to pass the ASE certification examination in engine repair.

Directions: Identify the cylinder displacement terms in the drawing below. Place the identifying letter next to the name of each component.

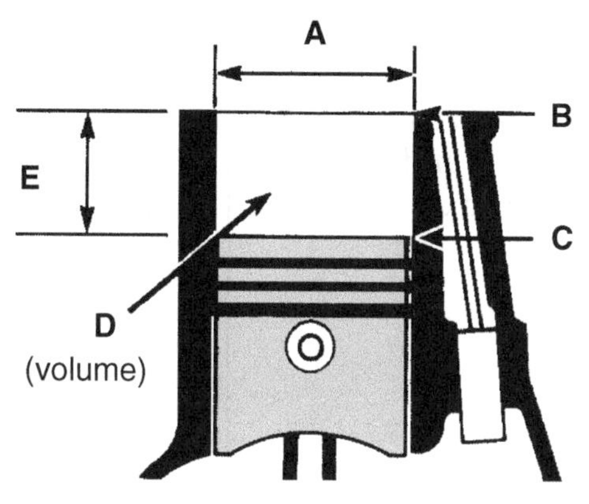

TDC ______ BDC ______ CI, L, or CCs ______

Bore ______ Stroke ______

1. Calculate the displacement of a cylinder that has a 3.5-inch bore and a 3.5-inch stroke. Show your work in the box below.

   Example: Formula for displacement of a cylinder is

   B × B × S × .785

   Bore 4 inches
   × Bore 4 inches
   16 square inches
   × Stroke 4 inches
   64 cubic inches
   × .785
   50.240 cubic inches (displacement for the cylinder)
   50.240 cubic inches

   Show your work

Show your work

2. Calculate the displacement of an eight-cylinder engine that has a 4-inch bore and a 4-inch stroke. Show your work in the box.

   Note: The formula for engine displacement is:

   B × B × S × .785 × number of cylinders

Show your work

3. Calculate the displacement of a six-cylinder engine that has a 4-inch bore and a 4-inch stroke. Show your work in the box.

Show your work

4. Calculate the displacement of a four-cylinder engine that has a 3 ½-inch bore and a 3 9/16-inch stroke. Show your work in the box.

5. What is the displacement in cubic centimeters (cc) of the following engines?

   a. 1.7 liters _______ cc b. 2 liters _______ cc c. 5.7 liters _______ cc

6. What is the displacement in cubic inches (ci) of the following engines?

   a. 1.7 liters _______ ci b. 2 liters _______ ci c. 5.7 liters _______ ci

STOP

Instructor OK ______________ Score ___________

Activity Sheet #32

ENGINE MEASUREMENTS

Name______________________________ Class______________________

Directions: Match the words on the left to the descriptions on the right. Write the letter for the correct word on the line provided. For the terms that you are not certain of, use the glossary in your textbook.

A. Bore	______	Measurement of work in which 1 pound is moved for a distance of 1 foot
B. Stroke	______	Volume displaced by the piston
C. Oversquare	______	Cylinder volume at BDC compared to volume at TDC
D. Cylinder displacement	______	Measurement comparing the volume of airflow actually entering the engine with the maximum that theoretically could enter
E. Engine displacement	______	Ability to do work
F. Compression ratio	______	The tendency of a body to keep its state of rest or motion
G. Compression pressure	______	The diameter of the cylinder
H. Force	______	Usable crankshaft horsepower
I. Work	______	Metric horsepower equivalent
J. Foot-pound	______	The measurement of an engine's ability to perform work
K. Energy	______	Cylinder displacement times number of cylinders
L. Inertia	______	Typical gasoline engine compression pressure
M. Momentum	______	Cylinder bore larger than stroke
N. Power	______	Any action that changes, or tends to change, the position of something
O. Torque	______	The turning force exerted by the crankshaft
P. Btu	______	Pressure in cylinder as piston moves up when the valves are closed
Q. 1 Btu	______	When an object is moved against a resistance or opposing force
R. Horsepower	______	Name of equipment used to test an engine's power
S. Watts	______	Heat required to heat a pound of water by 1°F
T. Brake horsepower	______	Piston travel from TDC to BDC
U. Road horsepower	______	How fast work is done
V. Volumetric efficiency	______	British thermal unit
W. 125–175 psi	______	Body going in a straight line will keep going in the same direction at the same speed if no other forces act on it
X. Dynamometer	______	Horsepower available at the car's drive wheels

STOP

Instructor OK ____________ Score ____________

Activity Sheet #33

IDENTIFY VALVE PARTS

Name____________________ Class____________________

ASE Education Foundation Correlation

This activity sheet addresses the following **MLR** task:

I.B.2 Identify components of the cylinder head and valve train. **(P-1)**

This activity addresses the following **AST/MAST** task:

I.B.4 Adjust valves (mechanical or hydraulic lifters). **(P-1)**

We Support ASE | Education Foundation

Directions: Identify the valve train components in the drawings. Place the identifying letter next to the name of each component listed on the next page.

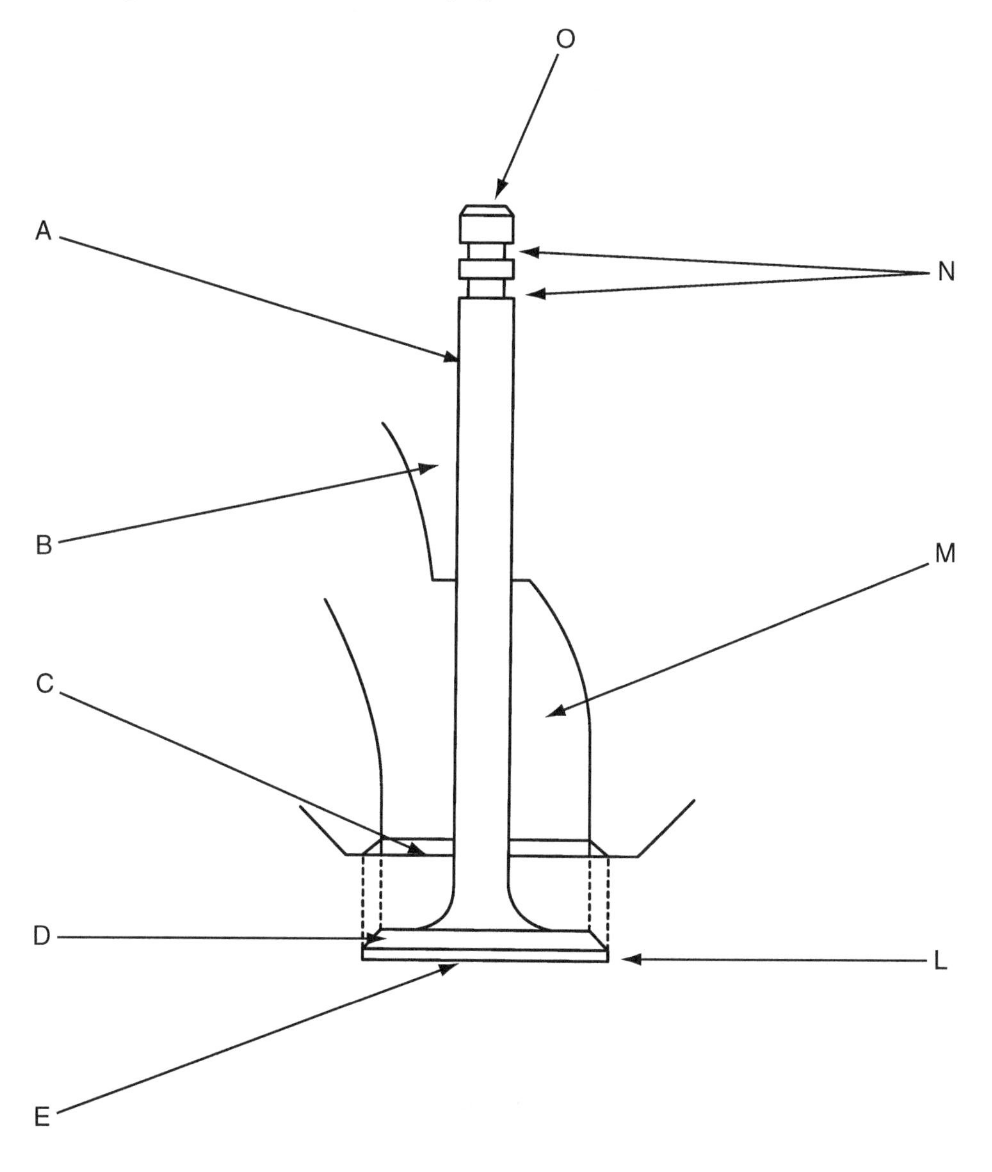

TURN

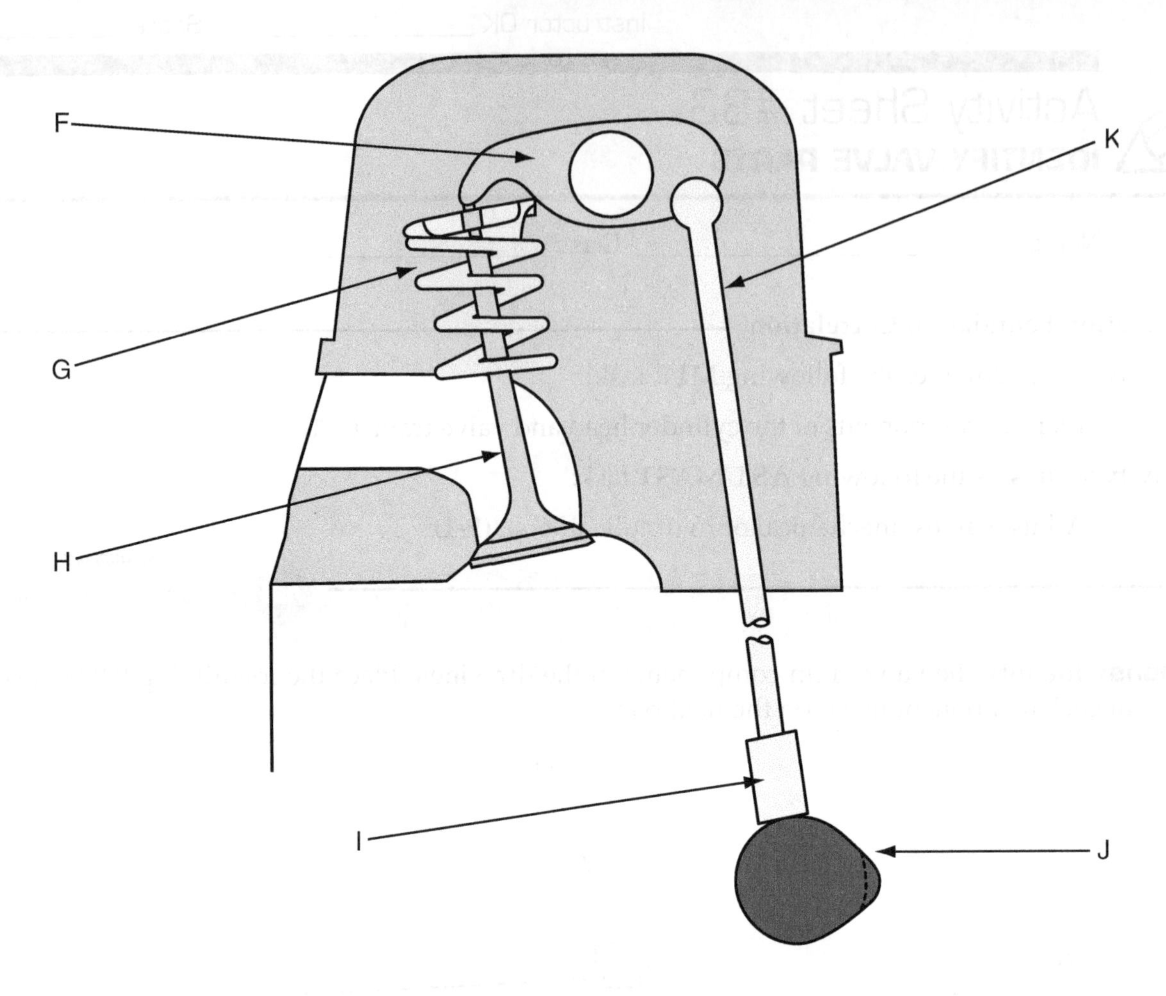

Valve Port	_____	Guide	_____	Lifter	_____
Valve Head	_____	Stem	_____	Spring	_____
Valve Seat	_____	Keeper Groove	_____	Valve	_____
Margin	_____	Stem Tip	_____	Rocker Arm	_____
Face	_____	Pushrod	_____	Cam Lobe	_____

STOP

Instructor OK ______________ Score __________

Activity Sheet #34

IDENTIFY OHC VALVE TRAIN PARTS

Name ______________________________ Class ____________________

ASE Education Foundation Correlation

This activity sheet addresses the following **MLR** task:

I.B.2 Identify components of the cylinder head and valve train. **(P-1)**

This activity addresses the following **AST/MAST** task:

I.B.4 Adjust valves (mechanical or hydraulic lifters). **(P-1)**

We Support ASE Education Foundation

Directions: Identify the OHC valve train parts in the drawings. Place the identifying letter next to the correct name listed below the drawings.

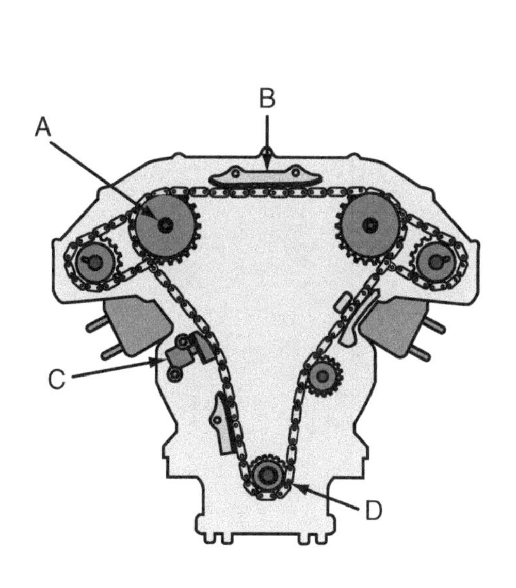

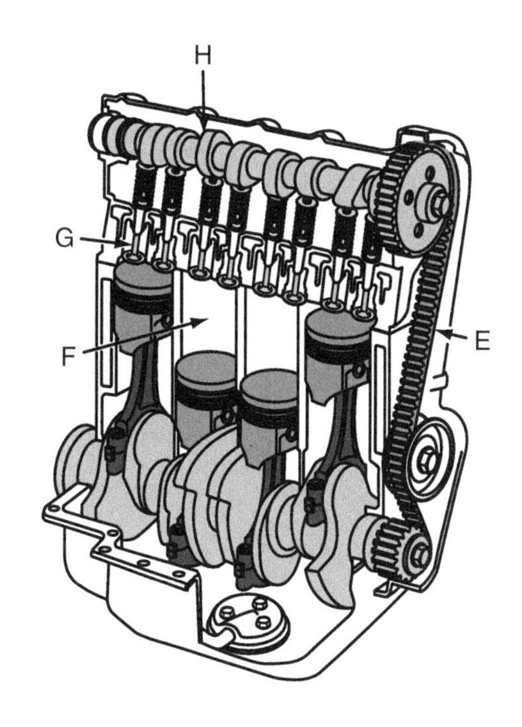

Cam Sprocket ______
Chain Guide ______
Timing Belt ______
Chain Tensioner ______
Crank Sprocket ______
Camshaft ______
Valve ______
Combustion Chamber ______

Instructor OK ______________ Score ____________

Activity Sheet #35

IDENTIFY PUSHROD ENGINE COMPONENTS

Name______________________________ Class______________________

ASE Education Foundation Correlation

This activity sheet addresses the following **MLR** task:

I.B.2 Identify components of the cylinder head and valve train. **(P-1)**

We Support ASE | Education Foundation

Directions: Identify the pushrod engine components in the drawing. Place the identifying letter next to the name of each component.

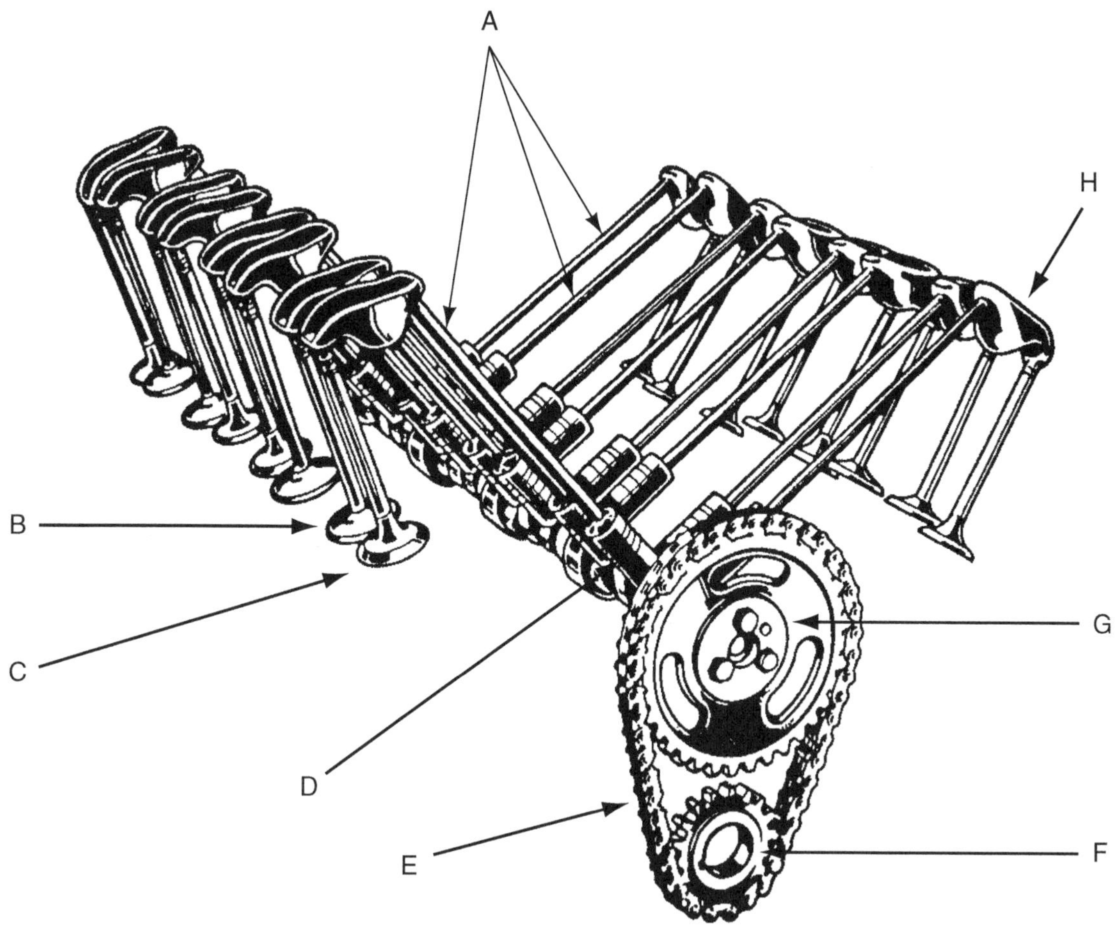

Rocker Arm	______	Pushrods	______	Timing Chain	______
Exhaust Valve	______	Crank Sprocket	______	Lifter	______
Intake Valve	______	Cam Sprocket	______		

STOP

Instructor OK ______________ Score __________

Activity Sheet #36

CYLINDER HEAD

Name ____________________________ Class ____________________

ASE Education Foundation Correlation

This activity sheet addresses the following **MLR** task:

I.B.2 Identify components of the cylinder head and valve train. **(P-1)**

We Support ASE Education Foundation

Directions: Match the words on the left to the descriptions on the right. Write the letter for the correct word on the line provided. For the terms that you are not certain of, use the glossary in your textbook.

A. Cylinder head	______ Engine produced at the factory
B. Integral guide	______ Two types of valve guides
C. Induction hardened	______ Types of valve guide seals
D. Poppet valves	______ Moves up and down with the valve stem
E. Production engine	______ Attached to the top of the valve guide
F. Stock	______ Valve guides that are made as part of the head
G. Iron or aluminum	______ Type of valve guide used in aluminum heads
H. O-ring, positive, and umbrella	______ Style of valve used in cylinder heads
I. Core	______ Original equipment
J. Insert guide	______ Cylinder heads can be made of two materials
K. Umbrella seal	______ How integral valve seats are hardened
L. Positive valve seal	______ Integral seats are integral with the ______
M. Integral or insert	______ Used part that is returned

STOP

Instructor OK ______________ Score __________

Activity Sheet #37

IDENTIFY FOUR-STROKE CYCLE EVENTS

Name_________________________ Class__________________

Directions: Identify the events in the four-stroke cycle on the sketch. Place the identifying letter next to its name.

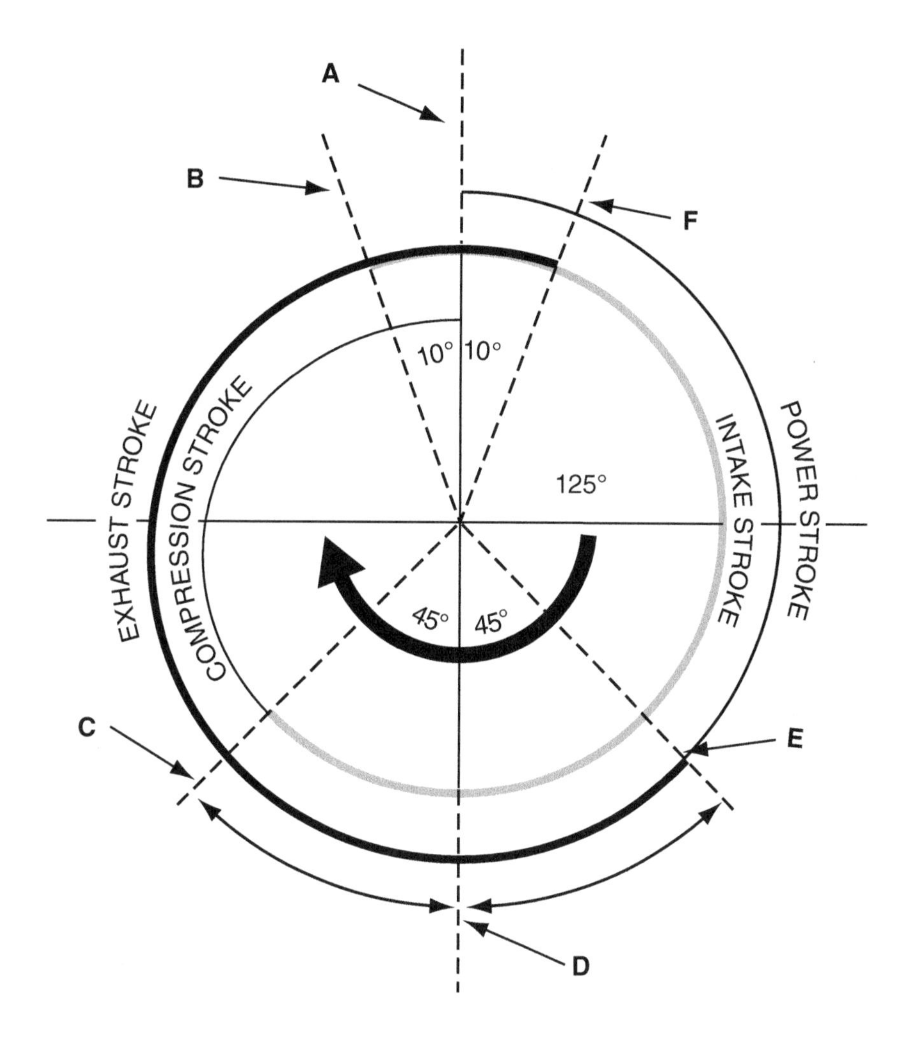

Intake Valve Opens	______	Exhaust Valve Opens	______	BDC	______
Intake Valve Closes	______	TDC	______	Exhaust Valve Closes	______

1. Highlight the intake stroke with a red highlighter.
2. Highlight the exhaust stroke with a blue highlighter.
3. Highlight the compression stroke with a yellow highlighter.
4. Highlight the power stroke with a green highlighter.

Instructor OK ____________ Score ____________

Activity Sheet #38

CAM AND LIFTERS

Name ______________________ Class ______________

Directions: Match the words on the left to the descriptions on the right. Write the letter for the correct word on the line provided. For the terms that you are not certain of, use the glossary in your textbook.

Term		Description
A. Positive stop	_____	Ways that a camshaft can be driven
B. Base circle	_____	When there is no clearance
C. Lift	_____	Automatic adjustment feature used with hydraulic valve lifters
D. Duration	_____	Engine that relies on atmospheric pressure
E. Valve overlap	_____	Parts that the camshaft can drive
F. Zero-lash	_____	Lifter that must be held from turning
G. Variable valve timing	_____	Height the cam lobe raises the lifter
H. Freewheeling engine	_____	What would be left if the cam lobe were removed?
I. Naturally aspirated	_____	The number of degrees of crankshaft travel when the valve is open
J. Roller lifter	_____	An engine that will *not* experience piston-to-valve interference if the timing chain or belt skips or breaks
K. Chain, belt, gears	_____	An engine that will experience piston-to-valve contact if the timing chain or belt skips or breaks
L. Fuel pump, oil pump, and distributor	_____	The time that both the intake exhaust valves are open at the same time
M. Interference engine	_____	Improves performance and reduces harmful emissions
N. Atkinson Cycle Engine	_____	Increases fuel economy in large engines
O. Active Fuel Management	_____	Lowers the operating compression ratio

STOP

Instructor OK ______________ Score ______________

Activity Sheet #39

IDENTIFY LUBRICATION SYSTEM COMPONENTS

Name ______________________________ Class ______________________

ASE Education Foundation Correlation

This activity sheet addresses the following **MLR** task:

I.B.2 Identify components of the lubrication and cooling systems. **(P-1)**

We Support ASE Education Foundation

Directions: Identify the lubrication system components in the drawing. Place the identifying letter next to the name of each component.

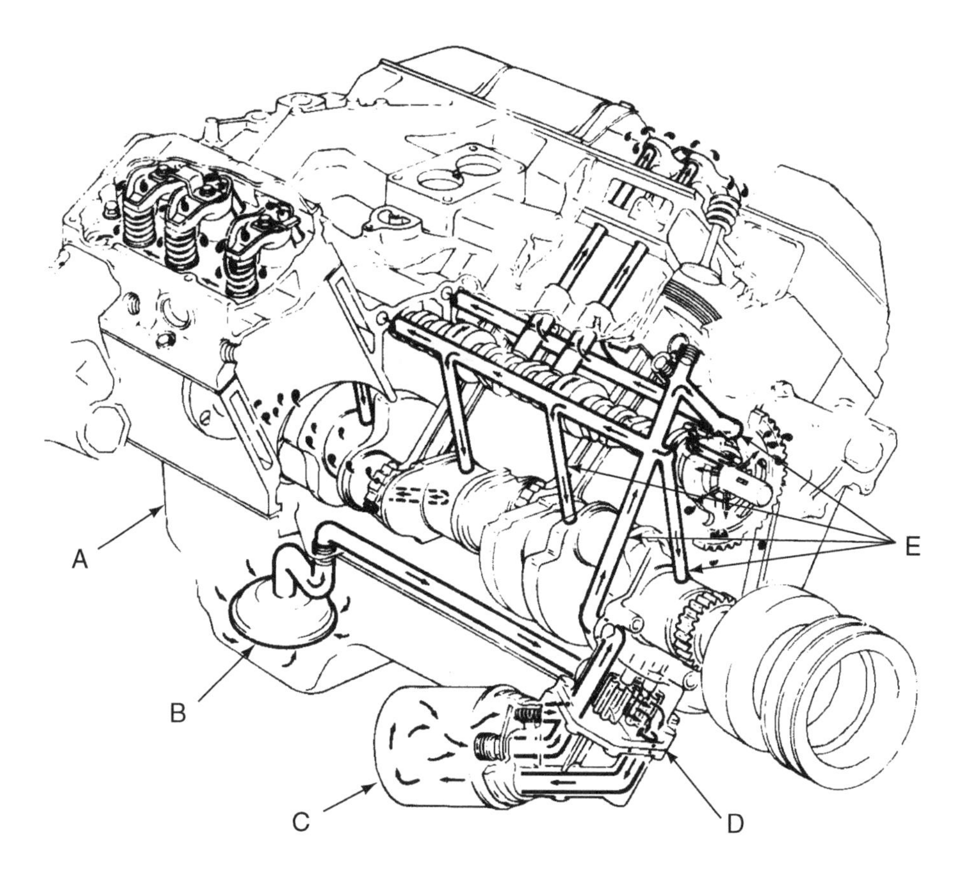

Oil Pump ______

Pick-Up Screen ______

Oil Pan ______

Oil Filter ______

Oil Galleries ______

Color the oil galleries with a highlighter or a colored pen.

Directions:

1. Identify the lubrication system components in the drawing. Place the identifying letter next to the name of the component.
2. Use a highlighter or colored pen to color the oil pressure passages red and oil return galleries blue.

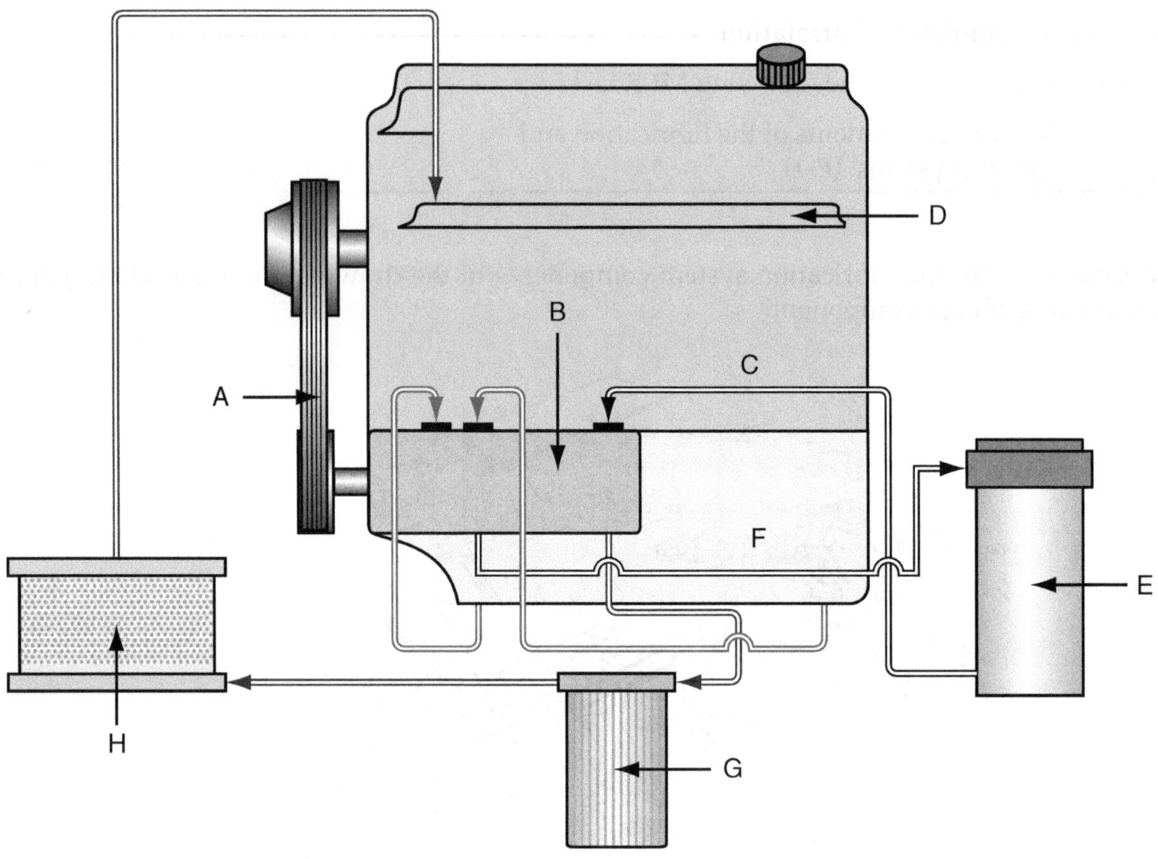

Oil pan	______	Main Oil Gallery	______	Oil Filter	______
Oil Pump	______	Pump Drive Belt	______	Pump Inlet Line	______
Oil Cooler	______	Oil Reservoir	______		

STOP

Instructor OK ______________ Score ____________

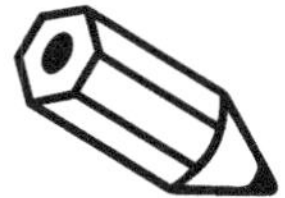

Activity Sheet #40
IDENTIFY OIL PUMP COMPONENTS

Name______________________________ Class____________________

ASE Education Foundation Correlation

This activity sheet addresses the following **MLR** task:

I.B.2 Identify components of the lubrication and cooling systems. **(P-1)**

This activity addresses the following **MAST** task:

I.D.13 Inspect oil pump gears or rotors, housing, pressure relief devices, and pump drive; perform needed action. **(P-2)**

We Support ASE | Education Foundation

Directions: Identify the oil pump components in the drawings. Place the identifying letter next to the name of each component or part listed at the bottom of the page.

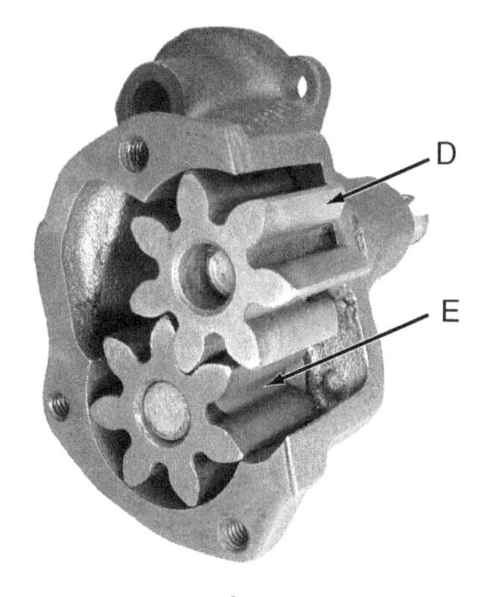

A

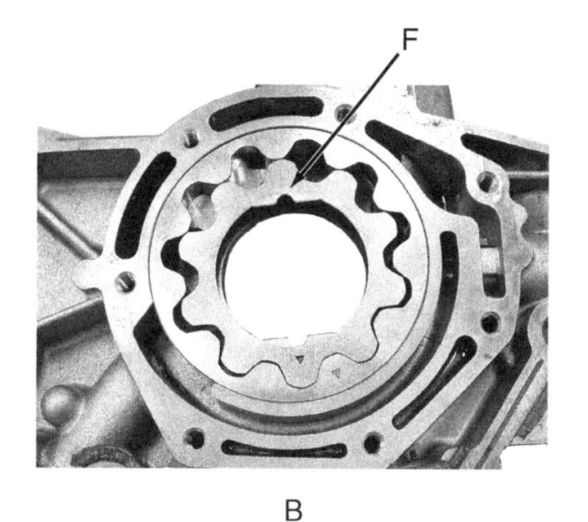

B

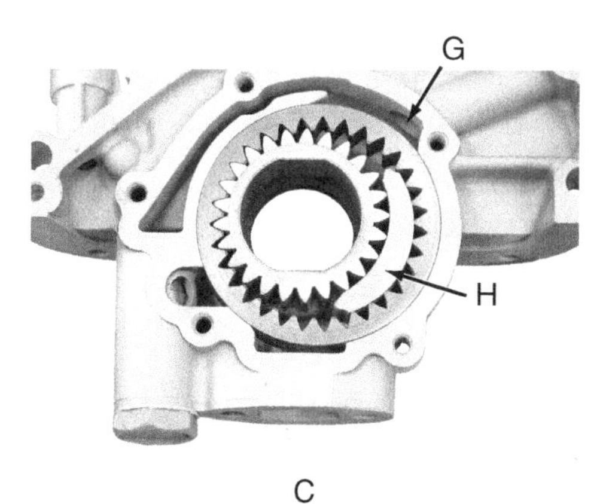

C

Photo by Tim Gilles

Drive Gear	______	Internal Gear Pump	______	Rotor	______
Discharge Port	______	External Gear Pump	______	Rotor Pump	______
Driven Gear	______	Crescent	______		

STOP

Instructor OK ______________ Score __________

Activity Sheet #41

IDENTIFY PISTON COMPONENTS

Name ____________________________ Class ____________________

Directions: Identify the piston components in the drawing. Place the identifying letter next to the name of each component.

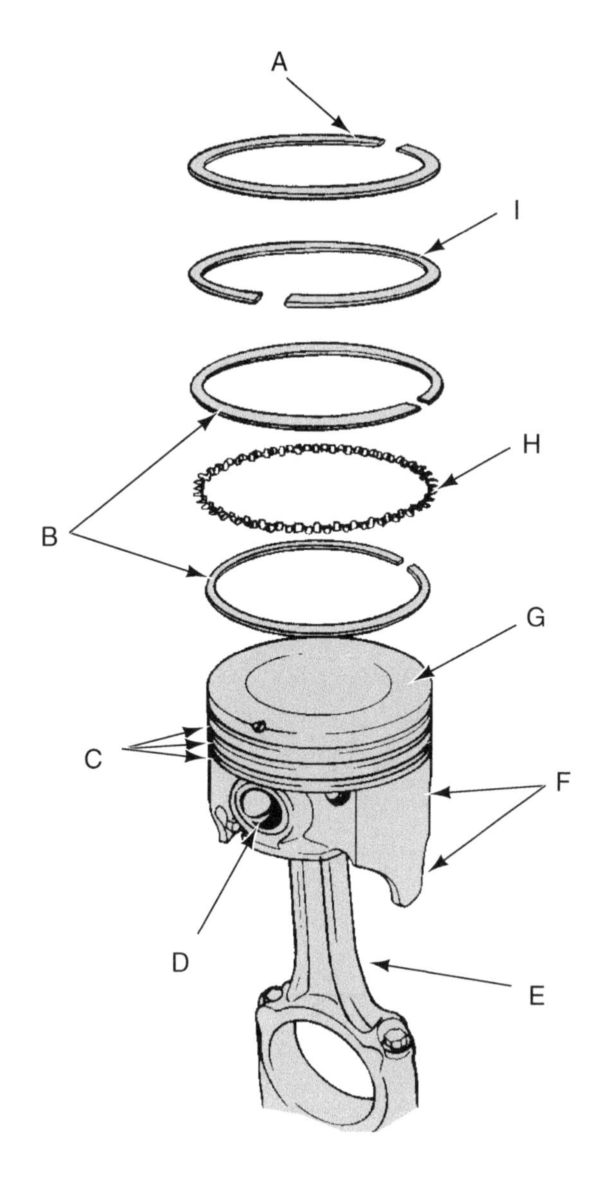

Second Ring	______	Expander Spacer	______	Piston Head	______
Rails	______	Top Ring	______	Skirt	______
Connecting Rod	______	Piston Pin	______	Ring Grooves	______

Instructor OK ______________ Score ___________

Activity Sheet #42
ENGINE BLOCK

Name___________________________ Class____________________

Directions: Match the words on the left to the descriptions on the right. Write the letter for the correct word on the line provided. For the terms that you are not certain of, use the glossary in your textbook.

A. Lower end	_____ A part installed to correct a damaged cylinder
B. Cylinder taper	_____ Prevents the bearing from turning in its bore
C. End thrust	_____ The number of times the piston must start and stop in 1 second at 3,000 rpm
D. Torsional vibration	_____ Two types of ring facings
E. Bearing spread	_____ Front or back force against a shaft
F. Bearing crush	_____ Wear occurring in the top inch of a cylinder wall
G. Cam ground	_____ Result of tapered wear
H. Sleeve	_____ The name that describes all of the parts of a short block
I. One hundred	_____ Oil passages
J. Moly, chrome	_____ Space between bearing and journal
K Cylinder ridge	_____ Piston's cold shape
L. Bearing clearance	_____ Occurs when force on the pistons is imparted to the crankshaft of a V-type engine
M. Open deck	_____ Provides better cooling
N. Galleries	_____ Holds bearing insert in place during engine assembly

STOP

Part I

Activity Sheets

Section Five

Cooling System, Hoses, and Plumbing

Instructor OK _______________ Score __________

Activity Sheet #43

IDENTIFY COOLING SYSTEM COMPONENTS

Name_____________________________ Class____________________

ASE Education Foundation Correlation

This activity sheet addresses the following **MLR** tasks:

I.C.1 Perform cooling system pressure and dye tests to identify leaks; check coolant condition and level; inspect and test radiator, pressure cap, coolant recovery tank, heater core, and galley plugs; determine necessary action. **(P-1)**

I.C.6 Identify components of the lubrication and cooling systems. **(P-1)**

This activity addresses the following **AST/MAST** tasks:

I.D.1 Perform cooling system pressure and dye tests to identify leaks; check coolant condition and level; inspect and test radiator, pressure cap, coolant recovery tank, heater core, and galley plugs; determine needed action. **(P-1)**

I.D.2 Identify causes of engine overheating. **(P-1)**

Directions: Identify the cooling system components in the drawing. Place the identifying letter next to the name of each component.

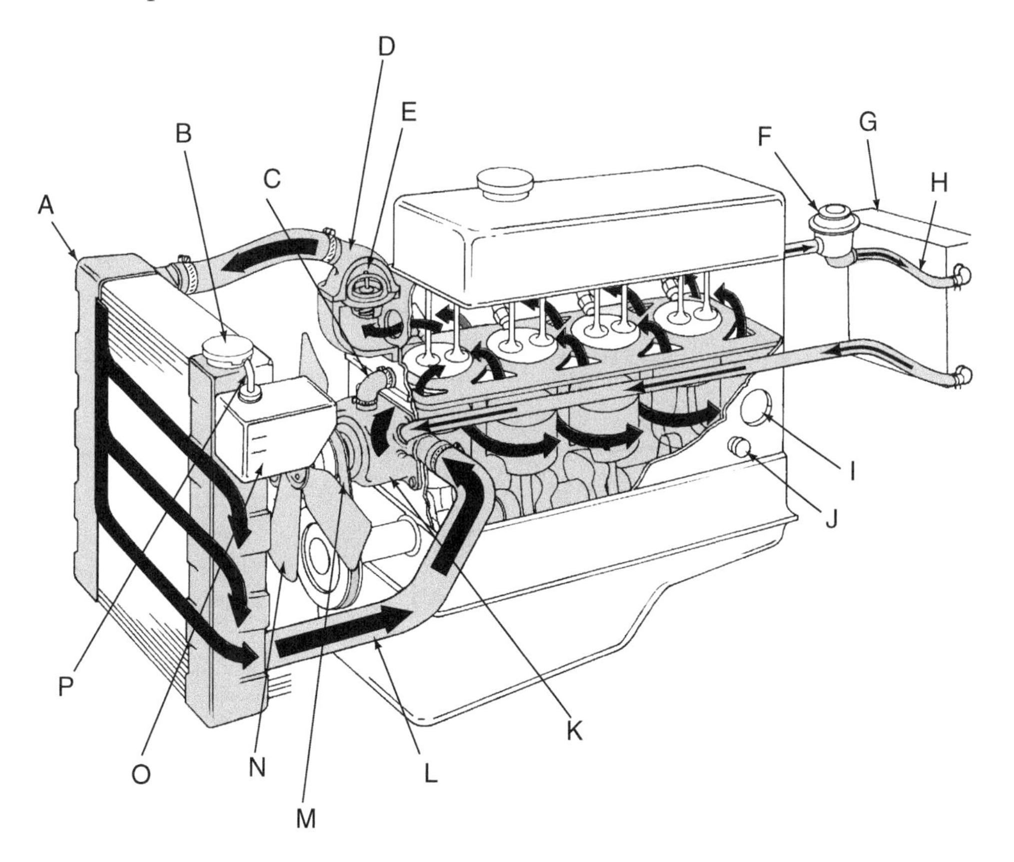

V-Belt	_____	Water Pump	_____	Thermostat	_____
Fan	_____	Heater Supply Hose	_____	Overflow Tube	_____
Drain Plug	_____	Thermostat Housing	_____	Radiator	_____
Coolant Recovery Tank	_____	Heater Core	_____	Pressure Cap	_____
Core Plug	_____	By-Pass Hose	_____	Radiator Hose	_____
Heater Control Valve	_____				

Directions:

1. Identify the cooling system components in the drawing. Place the identifying letter next to the name of each component.
2. Draw arrows in the cooling system to indicate the flow of coolant through the system.

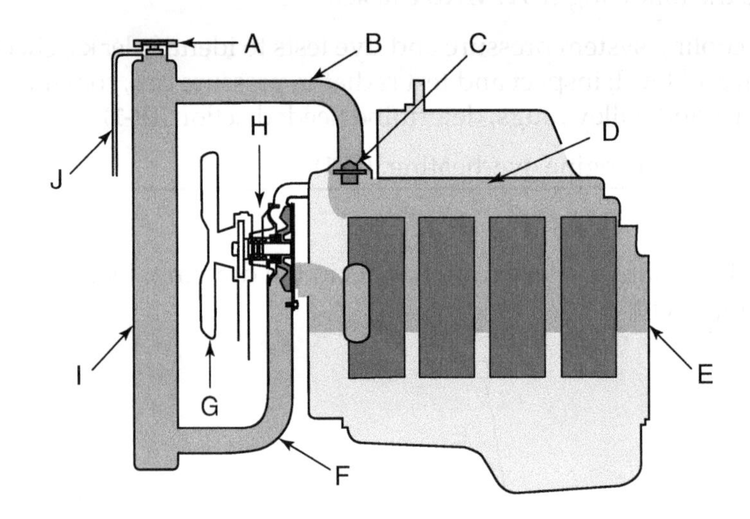

Thermostat	_____	Cylinder Head	_____	Engine Block	_____
Water Pump	_____	Radiator Cap	_____	Cooling Fan	_____
Radiator	_____	Overflow Hose	_____	Outlet Radiator Hose	_____
Inlet Radiator Hose	_____				

STOP

Instructor OK ______________ Score __________

Activity Sheet #44

IDENTIFY RADIATOR CAP COMPONENTS

Name ______________________________ Class ____________________

ASE Education Foundation Correlation

This activity sheet addresses the following **MLR** tasks:

I.C.1 Perform cooling system pressure and dye tests to identify leaks; check coolant condition and level; inspect and test radiator, pressure cap, coolant recovery tank, heater core, and galley plugs; determine necessary action. **(P-1)**

I.C.6 Identify components of the lubrication and cooling systems. **(P-1)**

This activity addresses the following **AST/MAST** tasks:

I.D.1 Perform cooling system pressure and dye tests to identify leaks; check coolant condition and level; inspect and test radiator, pressure cap, coolant recovery tank, heater core, and galley plugs; determine needed action. **(P-1)**

I.D.2 Identify causes of engine overheating. **(P-1)**

We Support ASE Education Foundation

Directions: Identify the radiator cap components in the drawing. Place the identifying letter next to the name of each component.

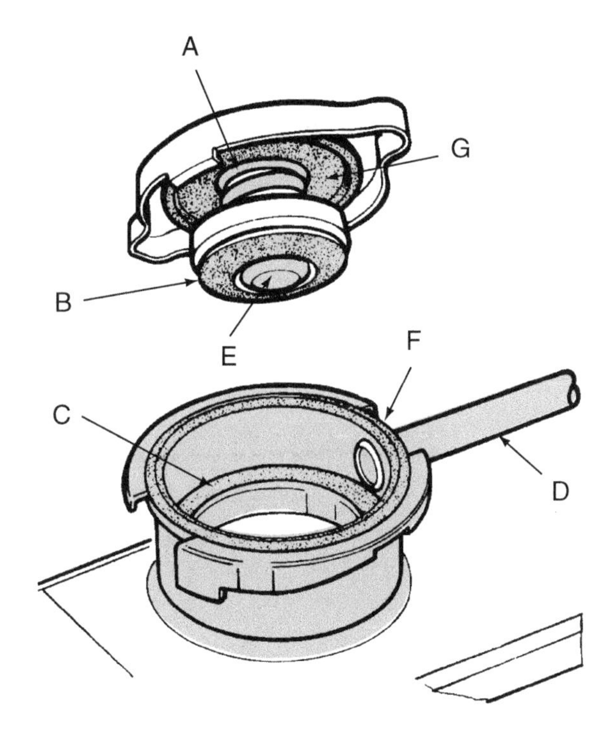

Overflow Hose	_____	Vacuum Valve	_____	Pressure Relief Spring	_____
Upper Sealing Surface	_____	Pressure Seal	_____		
Lower Sealing Surface	_____	Upper Sealing Gasket	_____		

TURN

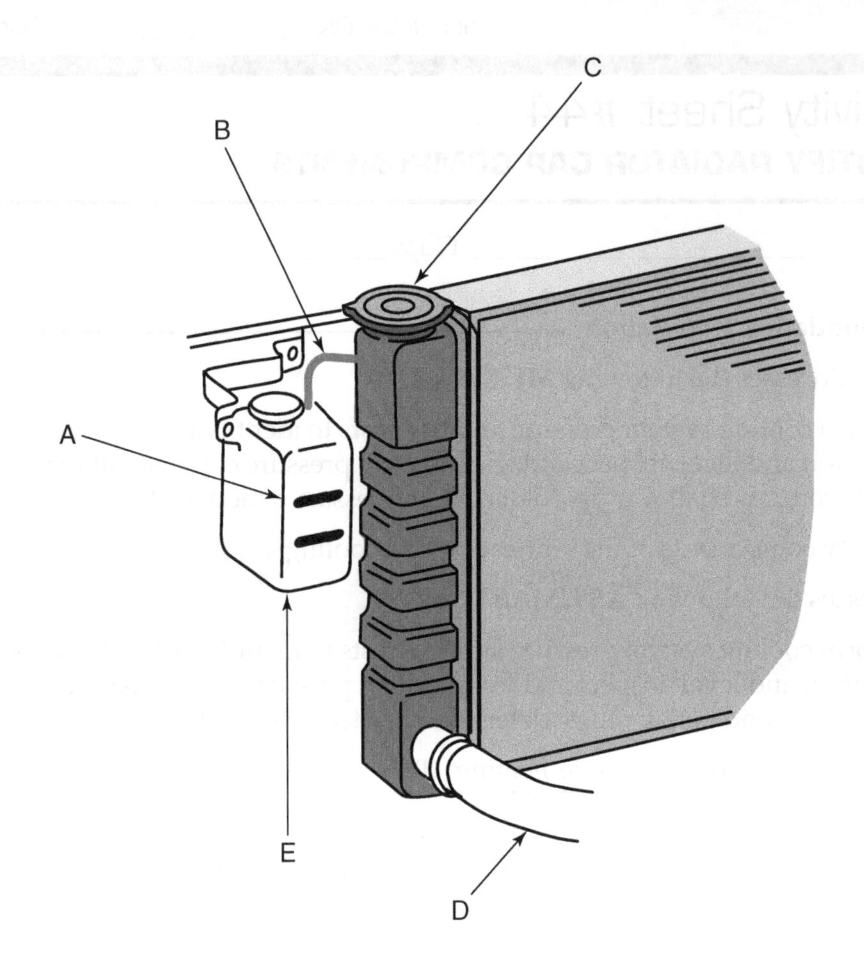

Overflow Tube ______

Overflow Tank ______

Pressure Cap ______

Radiator Hose ______

Full Hot Mark ______

STOP

Instructor OK ____________ Score ____________

Activity Sheet #45

ENGINE COOLING

Name ____________________ Class ____________

ASE Education Foundation Correlation

This activity sheet addresses the following **MLR** task:

I.C.1 Perform cooling system pressure and dye tests to identify leaks; check coolant condition and level; inspect and test radiator, pressure cap, coolant recovery tank, heater core, and galley plugs; determine necessary action. **(P-1)**

This activity addresses the following **AST/MAST** tasks:

I.D.1 Perform cooling system pressure and dye tests to identify leaks; check coolant condition and level; inspect and test radiator, pressure cap, coolant recovery tank, heater core, and galley plugs; determine needed action. **(P-1)**

I.D.2 Identify causes of engine overheating. **(P-1)**

We Support ASE Education Foundation

Directions: Match the words on the left to the descriptions on the right. Write the letter for the correct word on the line provided. For the terms that you are not certain of, use the glossary in your textbook.

Term		Description
A. Down-flow radiator	_____	Temperature- or torque-sensitive clutch attached to a belt-driven cooling fan
B. Cross-flow radiator	_____	Amount boiling point of coolant increases under 1 psi of pressure
C. Heat exchanger	_____	Thermostatic coil consisting of two types of metal wound together
D. Oil cooler	_____	Result of two dissimilar metals in a liquid
E. Thermostat bypass	_____	Controls engine temperature
F. Sending unit	_____	Normal operating temperature of an engine
G. Fan clutch	_____	Automotive coolant
H. Bimetal coil spring	_____	Found in the radiator to cool the transmission
I. Heater core	_____	Pulls air through the radiator when the engine is warm
J. Ethylene glycol	_____	Radiator design where coolant flows from top to bottom
K. Cylinder blocks	_____	Small radiator for passenger heat
L. Electrolysis	_____	Another name for a heat exchanger
M. Silicate	_____	Expands to open the thermostat
N. 180–212°F	_____	Sensing device for gauges

O. 3°F	_____	Radiator design where coolant flows from side to side
P. Vacuum valve	_____	Allows coolant to circulate when the thermostat is closed
Q. Wax	_____	Made of iron or aluminum
R. Fan	_____	Coolant additive that protects aluminum
S. Thermostat	_____	Small valve in the center of a radiator pressure cap

Instructor OK ______________ Score __________

Activity Sheet #46

FLUIDS, FUELS, AND LUBRICANTS

Name ____________________ Class ______________

Directions: Match the words on the left to the descriptions on the right. Write the letter for the correct word on the line provided. For the terms that you are not certain of, use the glossary in your textbook.

A. Gasoline	_____ DOT 5 fluid
B. Ethanol	_____ Causes damage to the ozone
C. Methanol	_____ White lubricant
D. Power steering fluid	_____ Fuel for compression ignition engine
E. HOAT coolant	_____ Blue fluid
F. Coolant	_____ Compressed natural gas
G. Automatic transmission fluid	_____ National Lubricating Grease Institute
H. Motor oil	_____ Red lubricating fluid
I. Windshield washer fluid	_____ Predicted to be the fuel of the future
J. Brake fluid	_____ Refrigerant used in newer vehicles
K. Synthetic brake fluid	_____ Ethylene-glycol-based liquid
L. Diesel fuel	_____ Extra-long-life fluid
M. CNG	_____ ATF occasionally used as substitute
N. Hydrogen	_____ Used as engine lubricant
O. Lithium grease	_____ Not compatible with petroleum-based fluids
P. R-134A	_____ Fuel used in many race cars
Q. R-12	_____ Corn-based gasoline additive
R. API	_____ Oils with the starburst symbol meet this standard
S. SAE	_____ European motor oil standard
T. NLGI	_____ Licenses and certifies oil
U. ACEA	_____ Identifies oil viscosity
V. ILSAC GF-4	_____ Most common automotive fuel

STOP

Instructor OK ______________ Score __________

Activity Sheet #47

IDENTIFY BELT-DRIVEN ACCESSORIES

Name ____________________________ Class ____________________

ASE Education Foundation Correlation

This activity sheet addresses the following **MLR** task:

I.C.2 Inspect, replace, and/or adjust drive belts, tensioners, and pulleys; check pulley and belt alignment. **(P-1)**

This activity addresses the following **AST/MAST** task:

I.D.3 Inspect, replace, and/or adjust drive belts, tensioners, and pulleys; check pulley and belt alignment. **(P-1)**

Directions: Identify the belt-driven accessories in the drawing. Place the identifying letter next to the name of each component listed below.

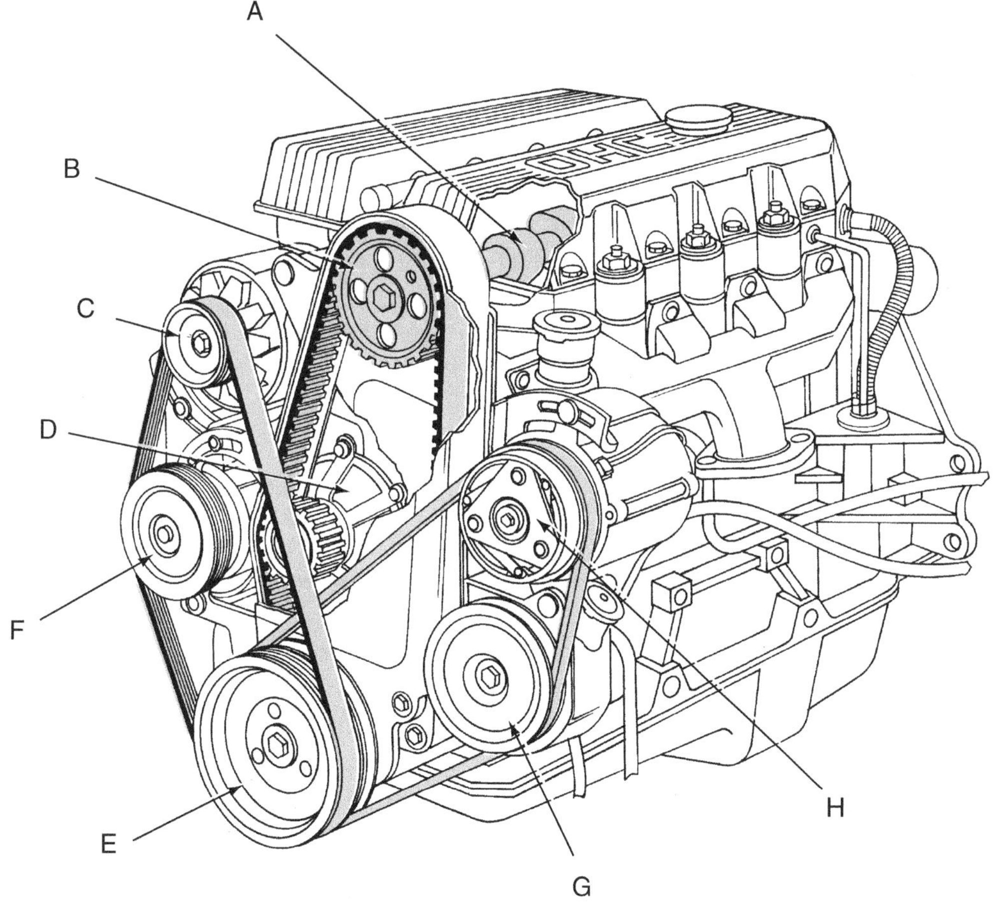

Power Steering Pump	_____	Cam Sprocket	_____	Air-Conditioning Compressor	_____
Air Pump	_____	Water Pump	_____	Alternator	_____
Overhead Cam	_____	Crankshaft Pulley	_____		

Directions: Identify the belt-driven accessories in the drawing. Place the identifying letter next to the name of each component listed below.

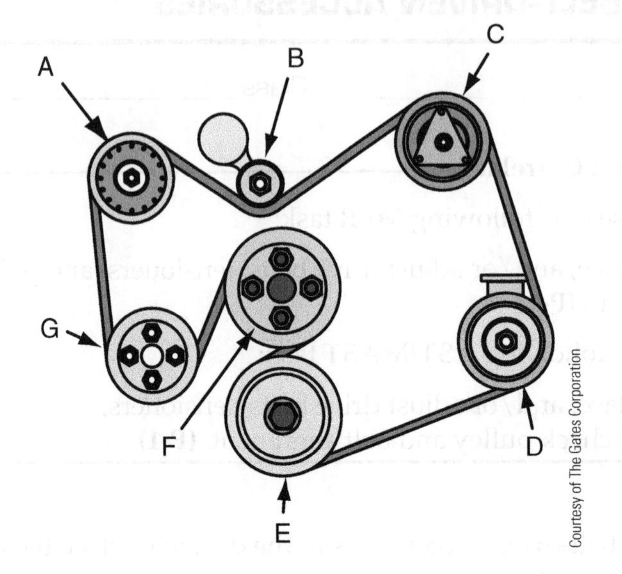

Alternator	_____	Water Pump	_____	A/C Compressor	_____
Air Pump	_____	Power Steering Pump	_____	Tensioner	_____
Crankshaft	_____				

Instructor OK ________________ Score ____________

Activity Sheet #48

BELTS

Name______________________________ Class______________________

ASE Education Foundation Correlation

This activity sheet addresses the following **MLR** task:

I.C.2 Inspect, replace, and/or adjust drive belts, tensioners, and pulleys; check pulley and belt alignment. **(P-1)**

This activity addresses the following **AST/MAST** task:

I.D.3 Inspect, replace, and/or adjust drive belts, tensioners, and pulleys; check pulley and belt alignment. **(P-1)**

We Support ASE Education Foundation

Directions: Match the words on the left to the descriptions on the right. Write the letter for the correct word on the line provided. For the terms that you are not certain of, use the glossary in your textbook.

A. Tensile cords	_____ Belt that has multiple ribs on one side and is flat on the other side
B. Neoprene	_____ Used to apply leverage with a tool when tightening a drive belt
C. High cordline belt	_____ Screw for adjusting belt tension
D. V-ribbed belt	_____ Provide strength to the belts
E. Serpentine belt	_____ Used for checking V-ribbed belt and timing belt tension
F. Jackscrew	_____ Higher quality V-belt with the tensile cord above center
G. V-belt and V-ribbed	_____ Oil-resistant artificial rubber
H. Square bracket hole	_____ Types of accessory drive belts
I. Click-type gauge	_____ Belt that follows a snake-like path

STOP

Instructor OK ______________ Score __________

Activity Sheet #49

IDENTIFY FUEL INJECTION COMPONENTS

Name ____________________________ Class ____________________

Directions: Identify the fuel injection components in the drawing. Place the identifying letter next to the name of each component.

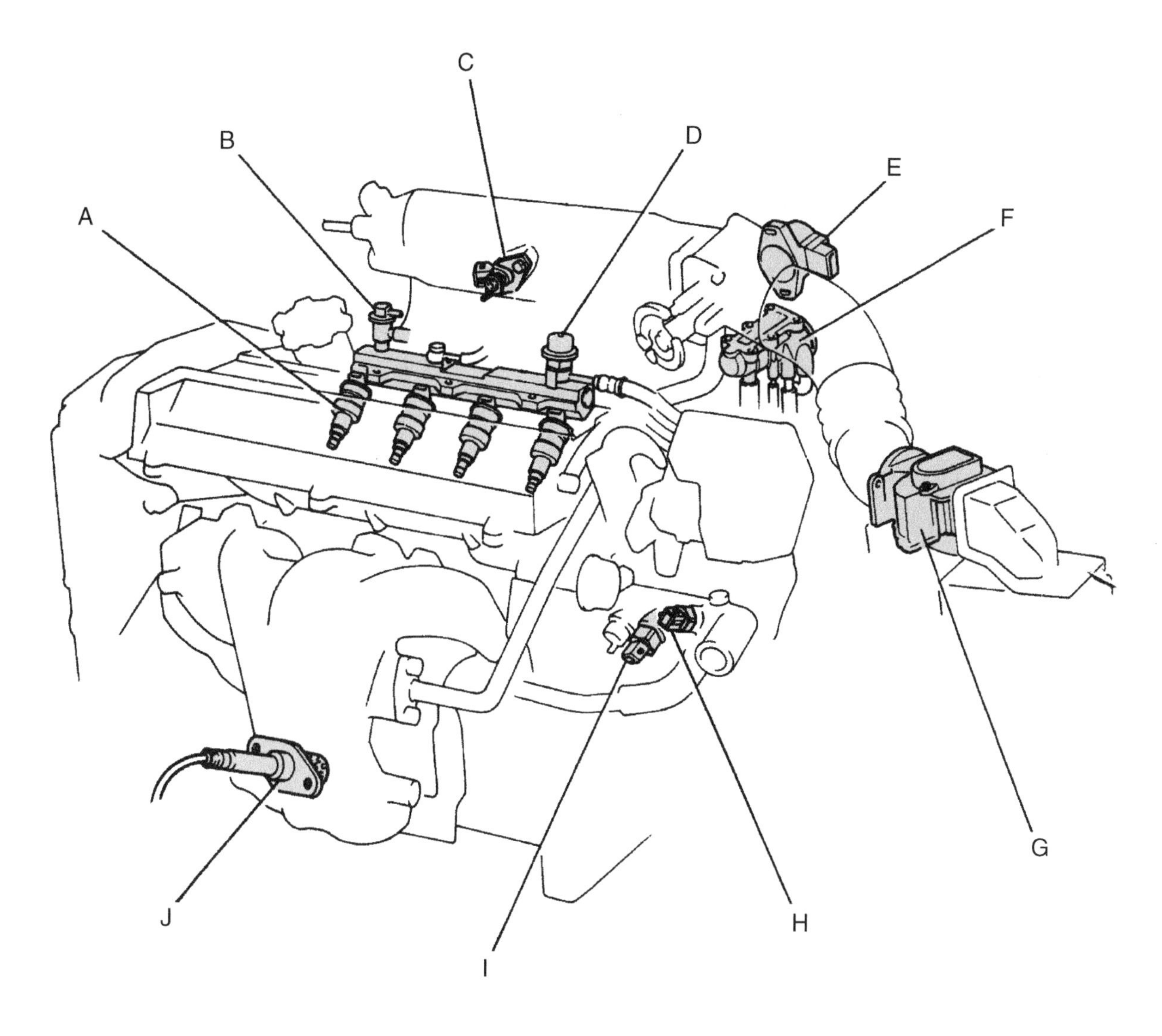

Thermo-Time Switch	____	Engine Temperature Sensor	____	Fuel Pulsation Damper	____
Airflow Meter	____	Fuel Pressure Regulator	____	Idle Speed Control	____
Fuel Injector	____	Oxygen Sensor	____		
Throttle Position Sensor	____	Cold Start Injector	____		

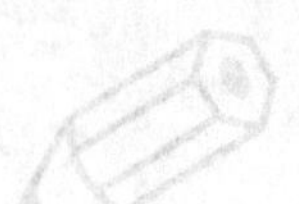

Instructor OK ______________ Score __________

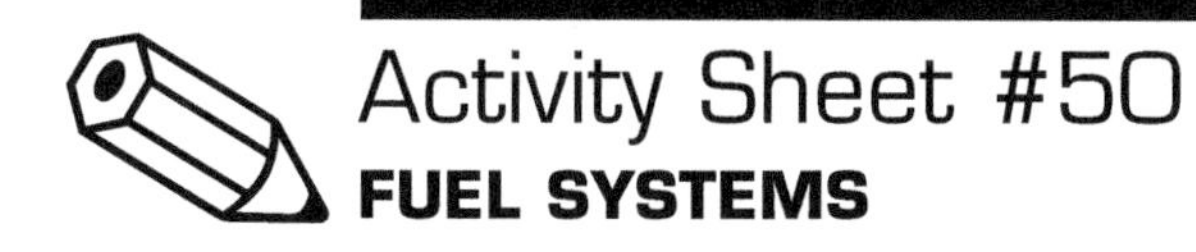

Activity Sheet #50

FUEL SYSTEMS

Name ______________________________ Class ______________________

Directions: Match the words on the left to the descriptions on the right. Write the letter for the correct word on the line provided. For the terms that you are not certain of, use the glossary in your textbook.

Term		Description
A. Filter sock	_____	Name of the process by which the fuel is suspended in the air
B. Manifold vacuum	_____	When the computer uses a manifold absolute pressure (MAP) sensor and engine rpm (tach) signal to calculate the amount of air entering the engine
C. Atomization	_____	Electronic fuel injection
D. Vaporization	_____	Abbreviation for throttle-body injection
E. Venturi	_____	Uses an intake manifold similar to a carbureted system
F. Feedback systems	_____	When the computer is controlling the fuel system
G. Pulse width	_____	Length of time that an injector remains open
H. EFI	_____	When the oxygen sensor does not send signals to the computer
I. TBI	_____	The filter inside the gas tank
J. Throttle-body injection	_____	Devices that relay information to the computer
K. Port injection	_____	Smaller area in the carburetor that restricts airflow
L. Speed density systems	_____	The oxygen sensor signal for rich air-fuel mixture
M. Air density systems	_____	When atomized fuel turns into a gas
N. Sensors	_____	Name of the butterfly valve that controls the amount of air entering the engine
O. Actuators	_____	Computer fuel systems that monitor the oxygen content in the exhaust
P. Closed loop	_____	Fuel injection system that uses individual fuel injectors at each intake port
Q. Open loop	_____	When an airflow sensor measures the volume of air entering the engine
R. Throttle plate	_____	Devices that carry out an assigned change from the computer
S. Higher than 0.5 volt	_____	Pressure inside the intake manifold when the engine is running

STOP

Instructor OK ______________ Score __________

Activity Sheet #51

IDENTIFY EXHAUST SYSTEM COMPONENTS

Name ______________________________ Class ____________________

Directions: Identify the exhaust system components in the drawing. Place the identifying letter next to the name of each component listed below.

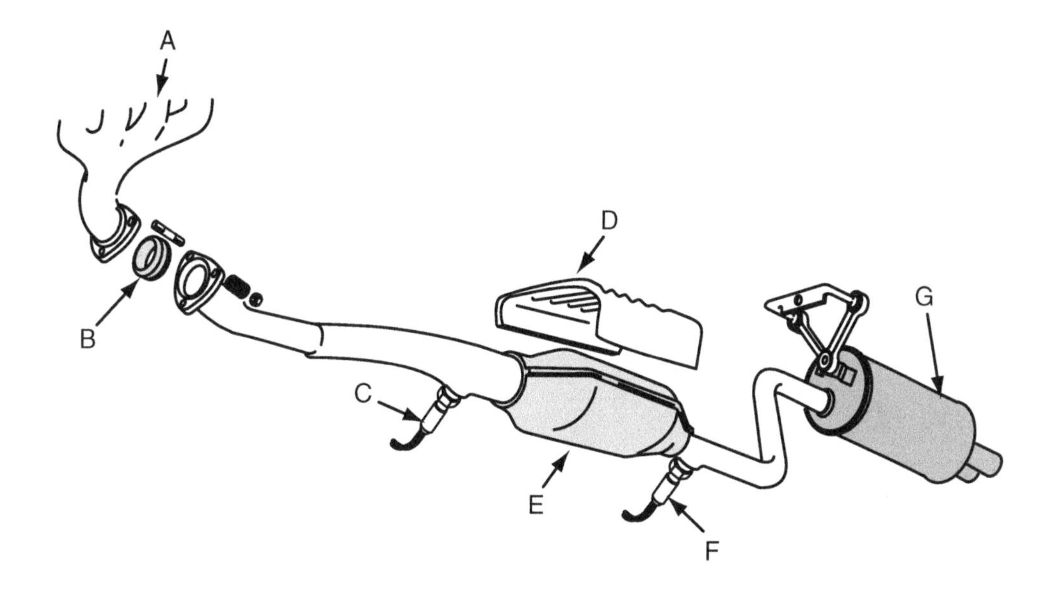

Gasket	_____	Catalytic Converter	_____
Exhaust Manifold	_____	Downstream O_2 Sensor	_____
Muffler	_____	Upstream O_2 Sensor	_____
Heat Shield	_____		

STOP

Instructor OK ______________ Score __________

Activity Sheet #52

INTAKE AND EXHAUST SYSTEMS

Name ________________________ Class ________________

Directions: Match the words on the left to the descriptions on the right. Write the letter for the correct word on the line provided. For the terms that you are not certain of, use the glossary in your textbook.

Term		Description
A. Heat riser	_____	When each barrel of the carburetor supplies half of an engine's cylinders
B. Runners	_____	Butterfly valve that fits between the exhaust manifold and the exhaust pipe
C. Cross-flow head	_____	A second muffler in line with the main muffler
D. Siamese runners	_____	Intake fan driven by the exhaust flow
E. Dual-plane manifold	_____	Exhaust manifolds made of tube steel
F. Headers	_____	When each barrel of the carburetor serves all of an engine's cylinders
G. Sound	_____	When intake and exhaust manifolds are on opposite sides of an in-line engine
H. Resonator	_____	When one runner feeds two neighboring cylinders
I. Catalytic converter	_____	Passages in the intake manifold
J. Turbocharger	_____	Cleans up engine emissions before they leave the tailpipe
K. Backpressure	_____	Caused by restriction in the exhaust system
L. Supercharger	_____	Vibration in the air
M. Single-plane manifold	_____	Belt-driven intake air pump

STOP

Instructor OK ______________ Score ______________

Activity Sheet #53

IDENTIFY HOSES AND TUBING

Name______________________________ Class______________________

Directions: Write the letter that describes each of the hoses and parts in the spaces provided.

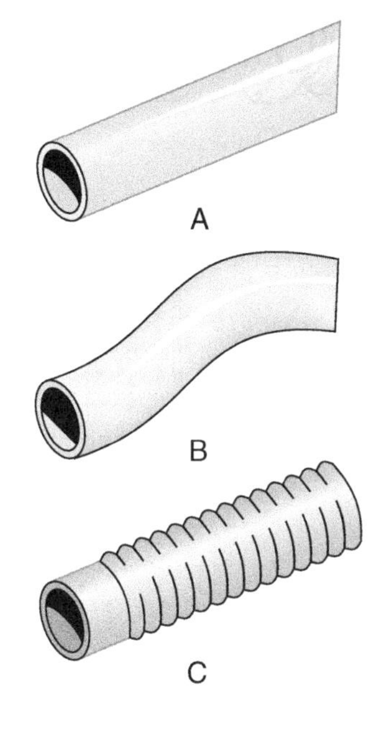

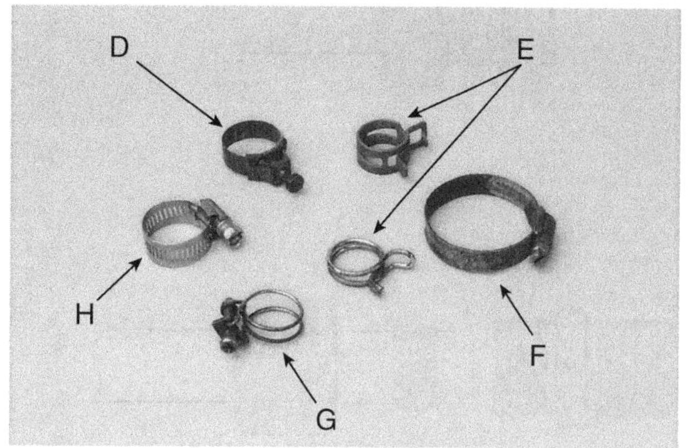

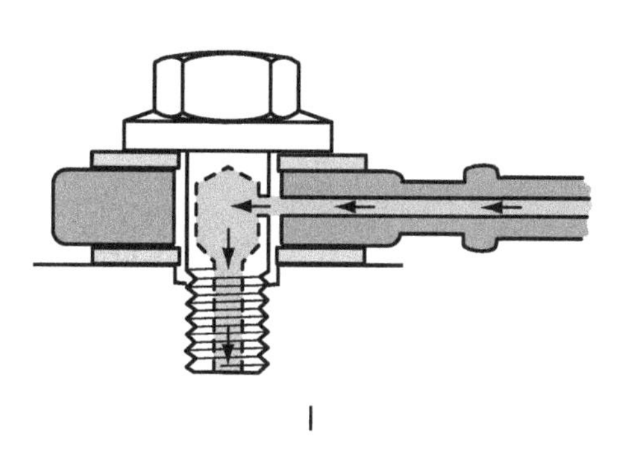

Courtesy of Ford Motor Company

Straight Hose _____

Rolled Edge Clamp _____

Universal Hose _____

Wire Clamp _____

Screw Tower Clamp _____

Inlet Heater Hose _____

Spring Clamp _____

Outlet Heater Hose _____

Worm Gear Clamp _____

Molded Hose _____

Banjo Fitting _____

TURN

Directions: Write the letters that describes each tubing or fitting in the spaces provided.

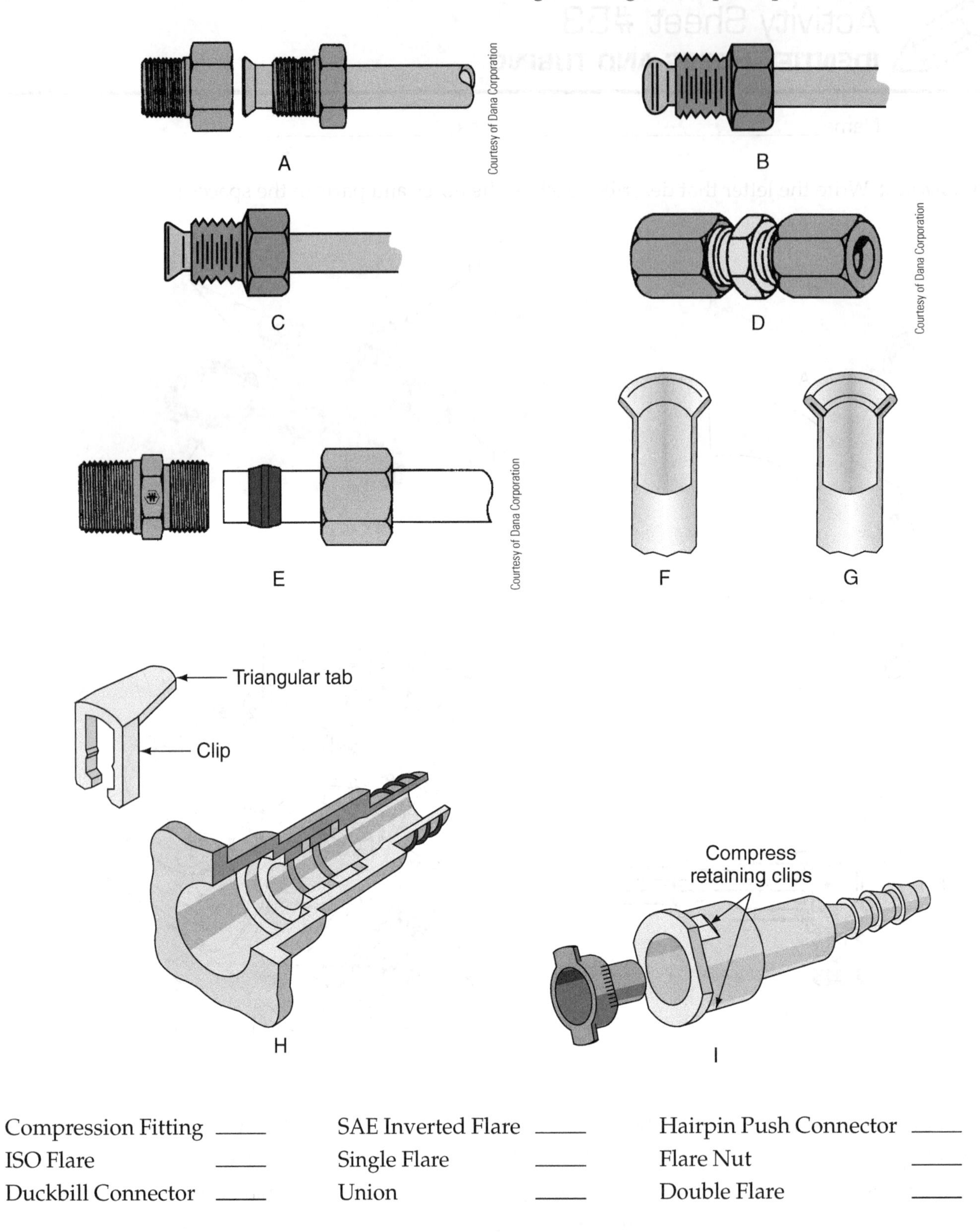

Compression Fitting	____	SAE Inverted Flare	____	Hairpin Push Connector	____
ISO Flare	____	Single Flare	____	Flare Nut	____
Duckbill Connector	____	Union	____	Double Flare	____

STOP

Part I

Activity Sheets

Section Six

Electrical System Theory and Service

Instructor OK ______________ Score ___________

Activity Sheet #54

ELECTRICAL THEORY

Name_____________________________ Class_____________________

Directions: Match the words on the left to the descriptions on the right. Write the letter for the correct word on the line provided. For the terms that you are not certain of, use the glossary in your textbook.

Term		Description
A. Atoms	_____	Also called a capacitor
B. Electricity	_____	Electrical flow
C. Circuit	_____	Composed of protons, neutrons, and electrons
D. Switch	_____	Unit of measurement for electrical resistance
E. Fuse	_____	Stores electricity
F. Stepped resistor	_____	An obstruction to electrical flow
G. Current	_____	Varies the voltage in a circuit
H. Amp	_____	Path for electrical flow
I. Capacitor	_____	Type of meter that has a dial
J. Resistance	_____	Magnetically controlled switch
K. Inductive pickup	_____	Circuit protection device
L. Ohm	_____	Solid state material that can act as either an insulator or a conductor
M. Current draw	_____	Unit of measurement for electrical pressure
N. Rheostat	_____	Unit of measurement for electrical current flow
O. Potentiometer	_____	Loss of voltage caused by current flow through a resistance
P. Series circuit	_____	Flow of electrons between atoms
Q. Parallel circuit	_____	The law governing the relationship between volts, ohms, and amps
R. Relay	_____	Oscillating electrical current
S. Alternating current	_____	Amount of current required to operate a load
T. Condenser	_____	A meter used only on a circuit that has no electrical power
U. Semiconductor	_____	Digital multimeter
V. Diode	_____	Used to turn a circuit on or off
W. Transistor	_____	Varies current flow through the circuit
X. Analog meter	_____	As resistance increases, current
Y. DMM	_____	Electrical pressure
Z. Voltage drop	_____	An electrical circuit where current must flow through all parts

TURN

AA.	Open circuit	_____	Circuit with different branches that current can flow through
AB.	Grounded circuit	_____	Name for a complete path provided for electrical flow
AC.	Closed circuit	_____	When current goes directly to ground
AD.	Decreases	_____	When there is a break in the path of electrical flow in a circuit
AE.	Ohmmeter	_____	An electronic relay
AF.	Voltage	_____	An electronic one-way check valve
AG.	Volt	_____	Clamps around wire to measure current
AH.	Ohm's law	_____	Used to control the speed of a heater motor
AI.	DVOM	_____	The force that moves the current in a circuit
AJ.	Electromotive force	_____	Another acronym for a DMM

STOP

Activity Sheet #55

OHM'S LAW

Name______________________________ Class______________________

ASE Education Foundation Correlation

This activity sheet addresses the following **MLR/AST/MAST** task:

VI.A.2 Demonstrate knowledge of electrical/electronic series, parallel, and series-parallel circuits using principles of electricity (Ohm's Law). **(P-1)**

We Support Education Foundation

Directions: Use Ohm's law to calculate the missing values.

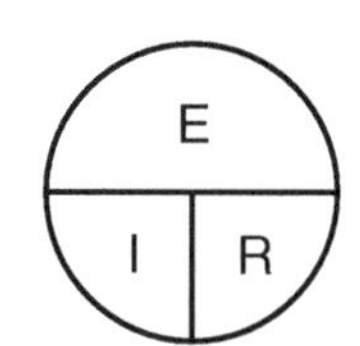

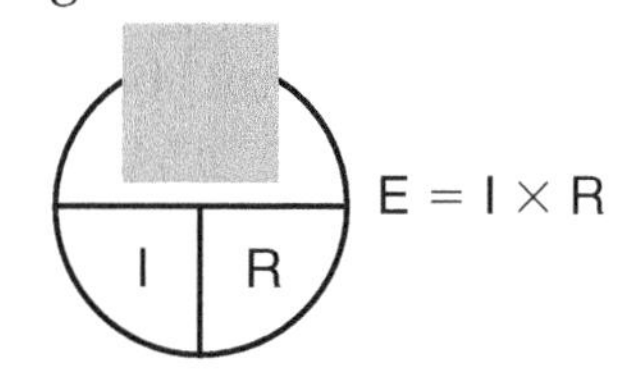

I = amperes (current)
R = ohms (resistance)
E = volts (electromotive force)

1.	4 ohms	12 volts	_____ amps
2.	_____ ohms	12 volts	2 amps
3.	2 ohms	_____ volts	6 amps
4.	3 ohms	12 volts	_____ amps
5.	6 ohms	_____ volts	3 amps
6.	_____ ohms	48 volts	12 amps
7.	6 ohms	48 volts	_____ amps
8.	_____ ohms	24 volts	6 amps
9.	4 ohms	_____ volts	3 amps

10. Calculate the current of the following and answer the question.

A.	2 ohms	12 volts	_____ amps
B.	3 ohms	12 volts	_____ amps
C.	4 ohms	12 volts	_____ amps

D. As the resistance increased, the current went ______________________. (up or down)

11. Calculate the current of the following and answer the question.

A.	2 ohms	6 volts	_____ amps
B.	2 ohms	12 volts	_____ amps
C.	2 ohms	24 volts	_____ amps

D. As the voltage increased, the current went ______________________. (up or down)

TURN

12. Complete the following Ohm's law relationships.
    A. When voltage goes up, the current will go ________. (up or down)
    B. When voltage goes down, the current will go ________. (up or down)
    C. When the resistance increases, the current will __________. (increase or decrease)
    D. When the resistance decreases, the current will __________. (increase or decrease)

Note: Wiring diagrams are drawn with the switch in its at-rest position. The first step in analyzing a circuit is to close the switch.

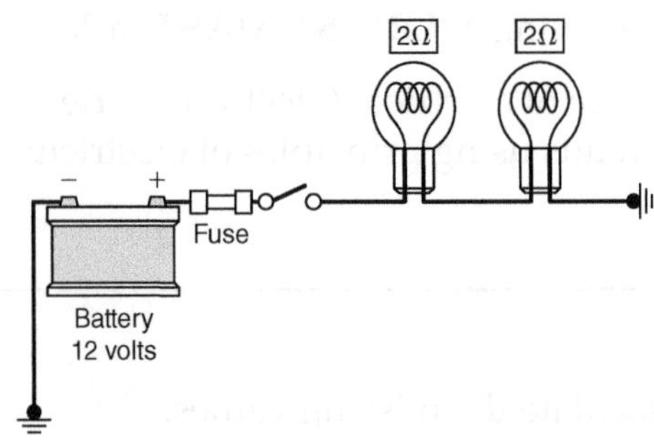

13. What is the current flow in the circuit above? _______ amps

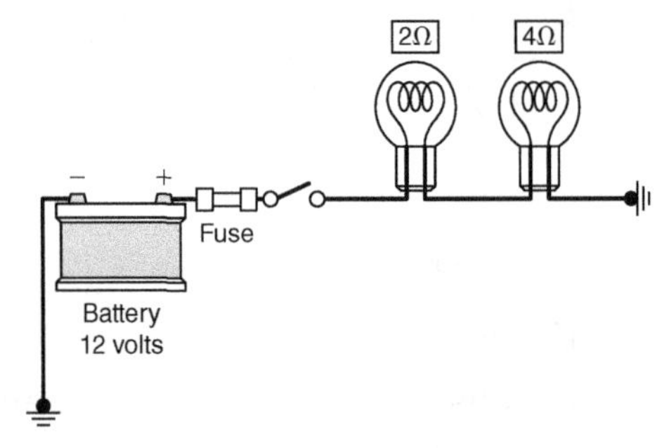

14. What is the current flow in the circuit above? _______ amps

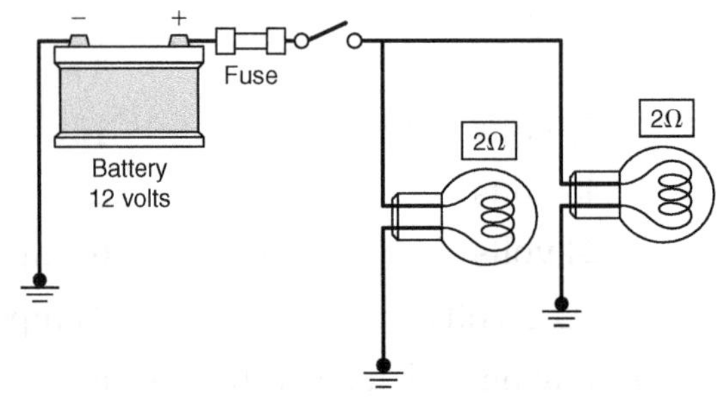

15. What is the current flow in the circuit above? _______ amps

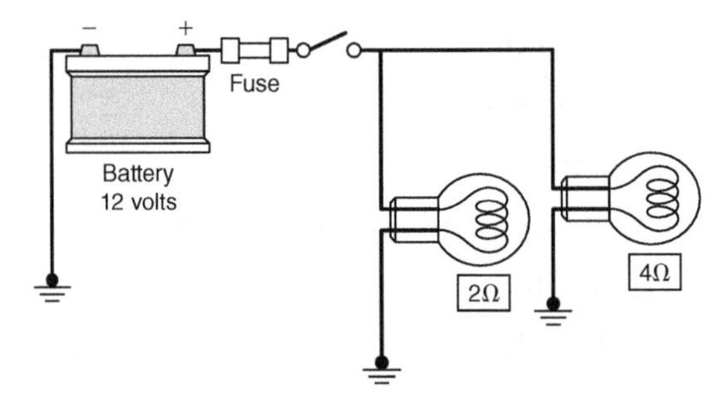

16. What is the current flow in the circuit above? _______ amps

Instructor OK ______________ Score __________

Activity Sheet #56

IDENTIFY ELECTRICAL CIRCUITS

Name ____________________ Class ______________

1. Draw lines to connect the light bulbs in series.

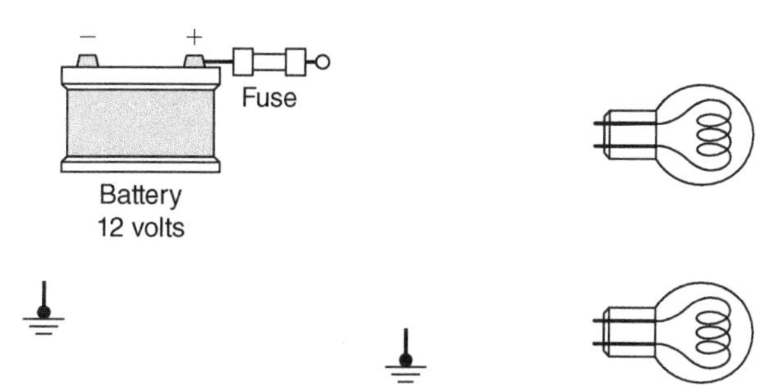

2. Draw lines to connect the light bulbs in parallel.

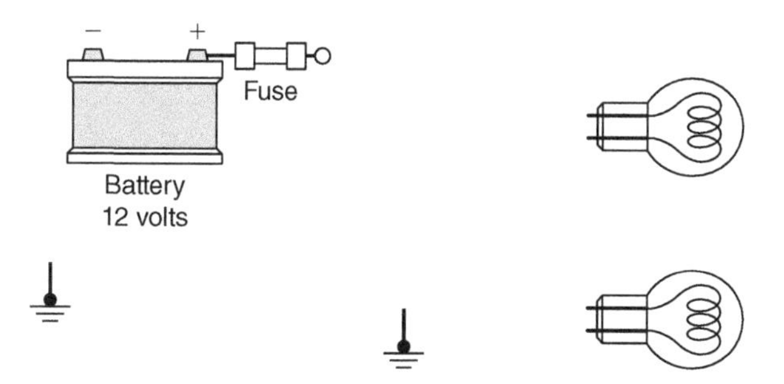

3. Draw lines to connect the light bulbs in series-parallel.

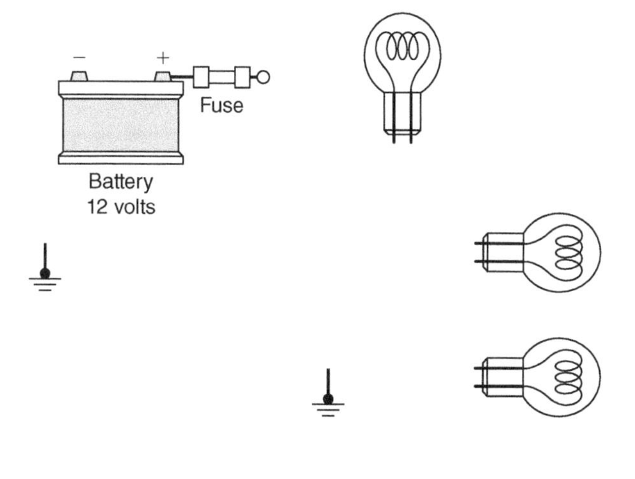

STOP

Instructor OK ______________ Score __________

Activity Sheet #57

BATTERY

Name______________________________ Class______________________

Directions: Match the words on the left to the descriptions on the right. Write the letter for the correct word on the line provided. For the terms that you are not certain of, use the glossary in your textbook.

A.	Electrolyte	_____	Unit of measurement for electrical power
B.	Element	_____	Low-maintenance battery
C.	Nickel metal hydride (NiMH)	_____	When the battery is allowed to run almost completely dead and then is recharged
D.	BCI	_____	Maintenance-free battery
E.	CCA	_____	Largest vehicle load on a car battery
F.	Reserve capacity	_____	Fully charged cell voltage
G.	Deep-cycle	_____	Gas given off as a battery is charged
H.	Nickel-cadmium (NiCd)	_____	Measurement of the battery's ability to provide current when there is no electricity from the charging system
I.	Cannot add water	_____	Battery Council International
J.	May need water	_____	Voltage of a fully charged battery
K.	Starter draw	_____	Cold cranking amps
L.	Lithium-ion (Li-ion) battery	_____	Largest battery post
M.	2.1 volts	_____	Mixture of sulfuric acid and water used in automotive batteries
N.	Hydrogen	_____	Connection on the side or top of a battery
O.	12.6 volts	_____	Group of battery plates connected in parallel
P.	Positive	_____	Type of electrical current produced by a battery
Q.	Battery terminal	_____	Highly flammable battery
R.	Watt	_____	Type of battery used in most hybrid vehicles
S.	DC	_____	Susceptible to memory effect

STOP

Instructor OK ________________ Score ____________

Activity Sheet #58

JUMP-STARTING

Name ______________________________ Class ____________________

ASE Education Foundation Correlation

This activity sheet addresses the following **MLR/AST/MAST** task:

VI.B.6 Jump-start vehicle using jumper cables and a booster battery or an auxiliary power supply. **(P-1)**

We Support ASE Education Foundation

Directions: Draw the jumper cables. Place a number next to each cable clamp to show the order in which each connection should be made.

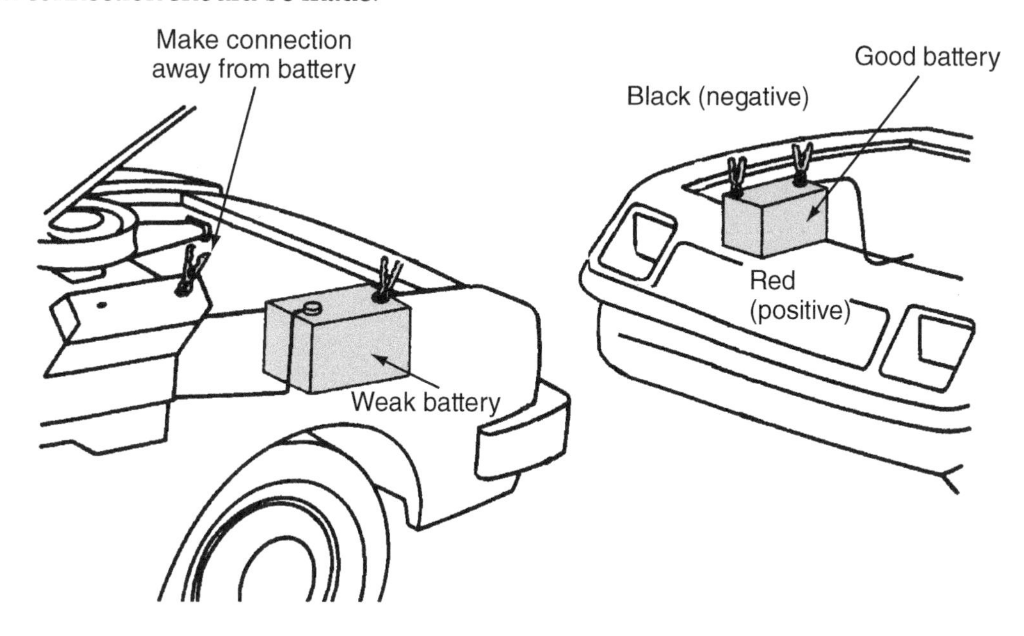

Note: The last connection should be lower than the battery and more than 12 inches away from the battery.

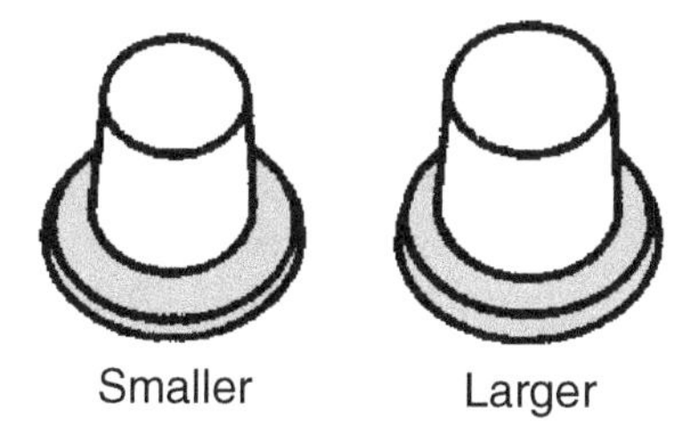

Top-terminal batteries

Identify the positive and negative battery terminals with a (+) and a (−).

Instructor OK ______________ Score ______________

Activity Sheet #59

IDENTIFY STARTING SYSTEM COMPONENTS

Name ______________________________ Class ______________________

ASE Education Foundation Correlation

This activity sheet addresses the following **MLR** task:

VI.A.11 Identify electrical/electronic system components and configuration. **(P-1)**

We Support ASE Education Foundation

Directions: Identify the starting system components in the drawing. Place the identifying letter next to the name of each component.

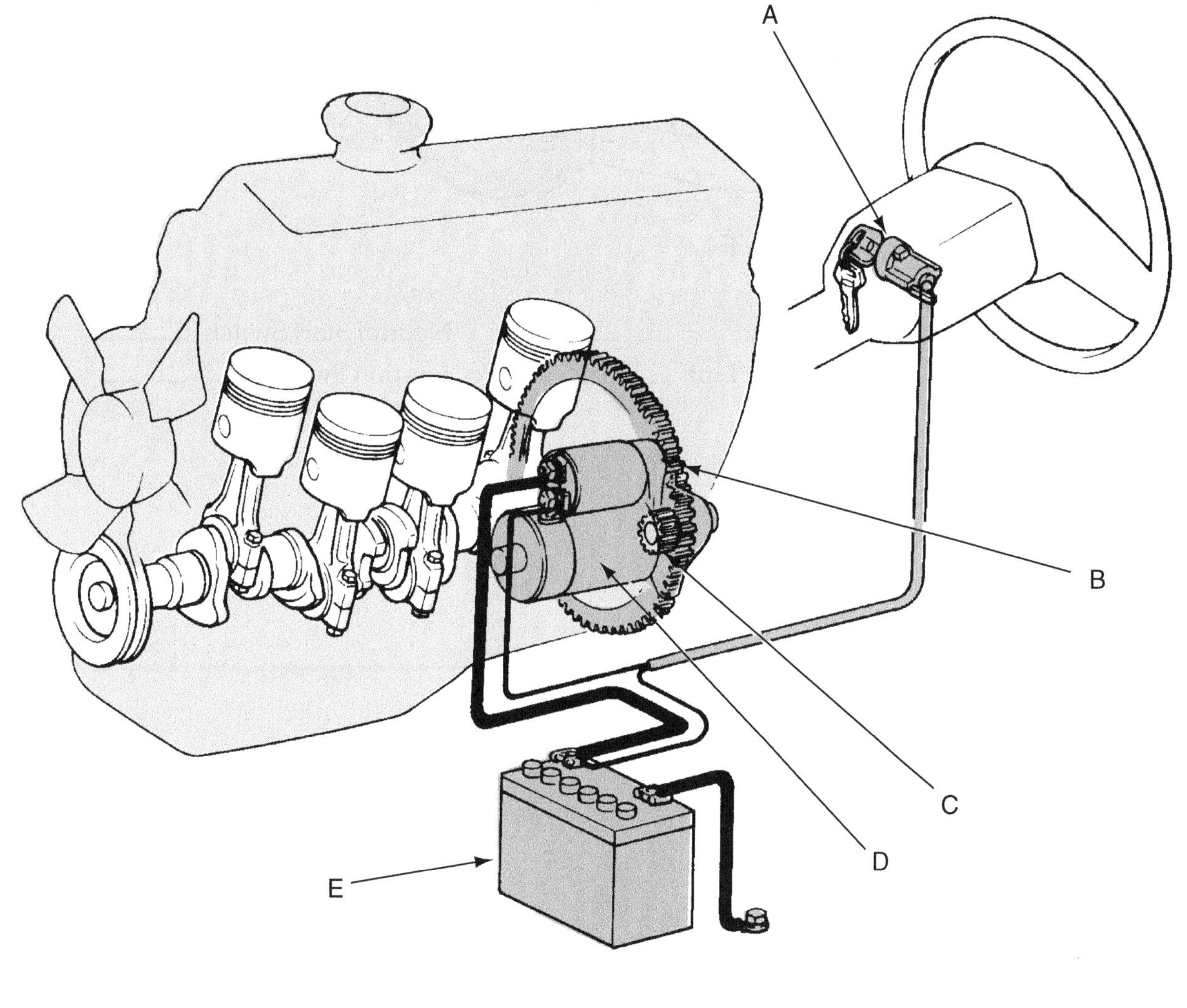

Pinion Drive Gear ______	Ring Gear ______	Battery ______
Ignition Switch ______	Starter Motor ______	

Directions: Identify the starting system components in the drawing. Place the identifying letter next to the name of each component.

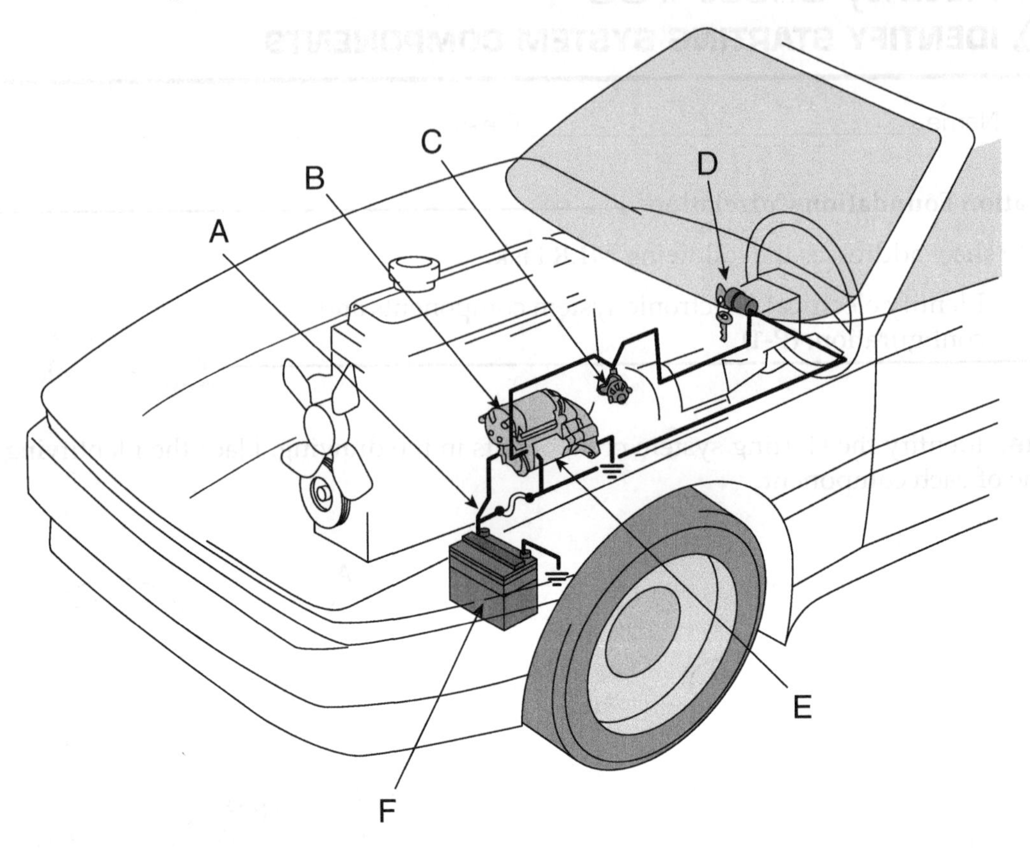

Solenoid	______	Battery	______	Neutral Start Switch	______
Starter Motor	______	Fusible Link	______	Ignition Switch	______

Instructor OK ______________ Score ______________

Activity Sheet #60

IDENTIFY STARTER MOTOR COMPONENTS

Name______________________________ Class______________________

ASE Education Foundation Correlation

This activity sheet addresses the following **MLR** task:

VI.A.11 Identify electrical/electronic system components and configuration. **(P-1)**

We Support ASE Education Foundation

Directions: Identify the starter motor components in the drawing. Place the identifying letter next to the name of each component.

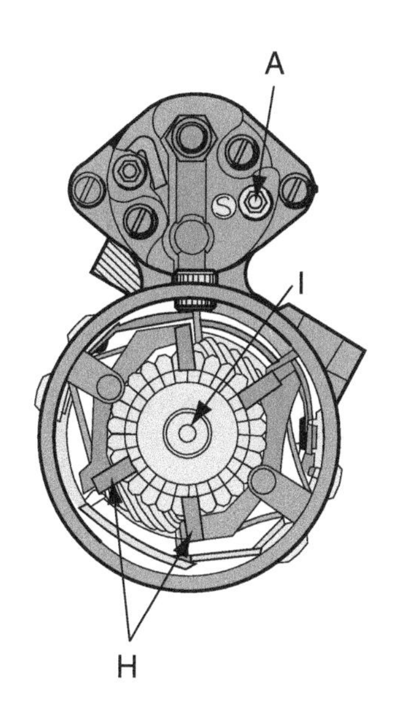

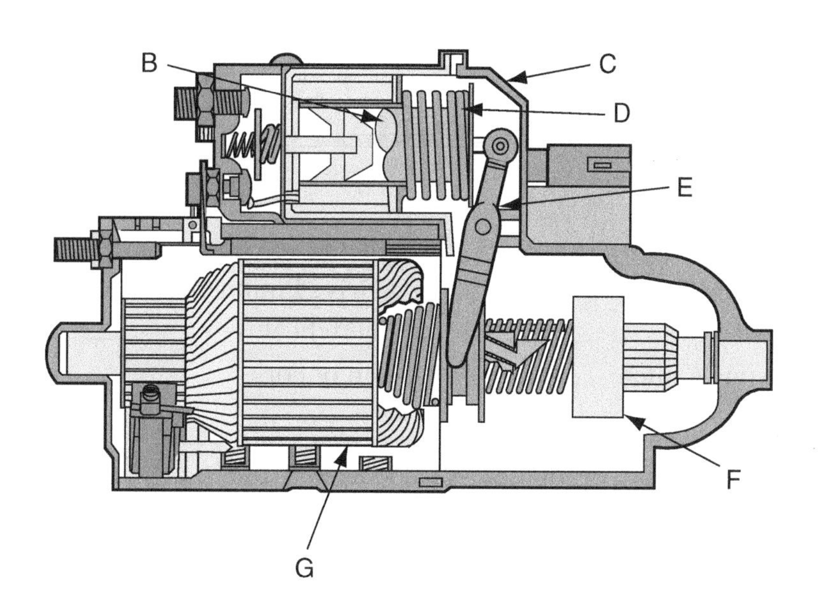

Energizing Terminal	______	Solenoid	______	Overrunning Clutch	______
Piston (Plunger)	______	Return Spring	______	Armature	______
Pivot Fork	______	Brushes	______	Commutator	______

Directions: Identify the starter motor components in the drawing. Place the identifying letter next to the name of each component.

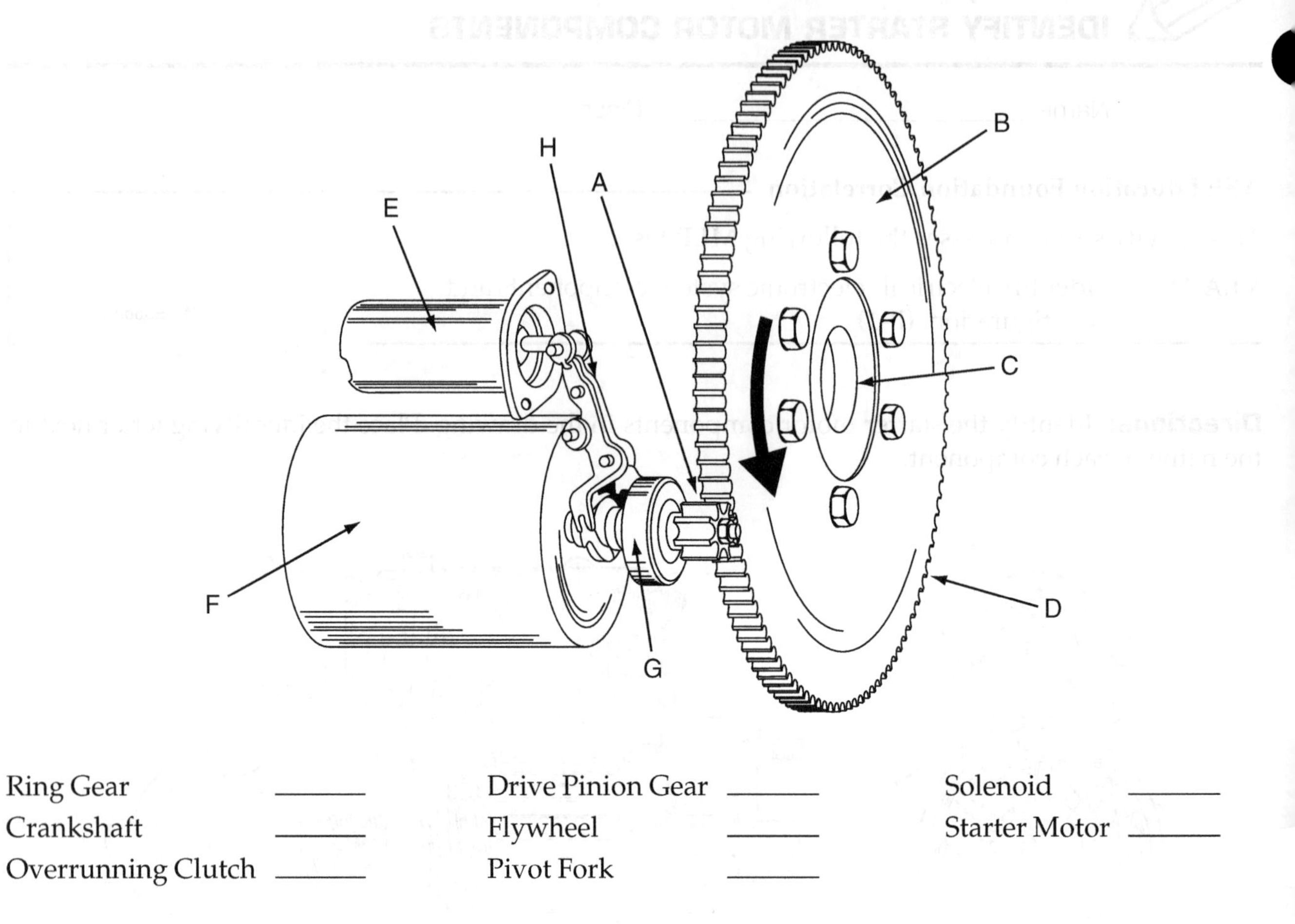

Ring Gear	______	Drive Pinion Gear	______	Solenoid	______
Crankshaft	______	Flywheel	______	Starter Motor	______
Overrunning Clutch	______	Pivot Fork	______		

What is the direction of rotation of the starter motor in the drawing above?

☐ Clockwise ☐ Counterclockwise

STOP

Instructor OK ______________ Score __________

Activity Sheet #61

IDENTIFY CHARGING SYSTEM COMPONENTS

Name ____________________________ Class ____________________

ASE Education Foundation Correlation

This activity sheet addresses the following **MLR** task:

VI.A.11 Identify electrical/electronic system components and configuration. **(P-1)**

We Support ASE Education Foundation

Directions: Identify the charging system components in the drawing. Place the identifying letter next to the name of each component.

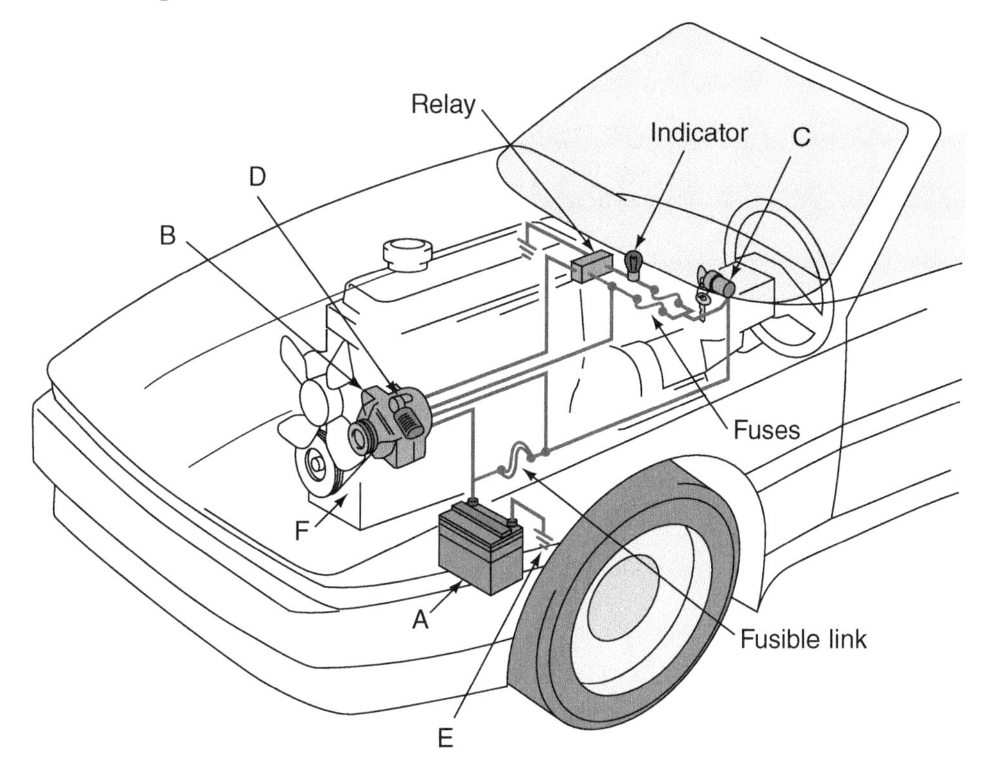

Regulator	______	Ignition Switch	______	Battery Ground	______
Alternator	______	Battery	______	Drive Belt	______

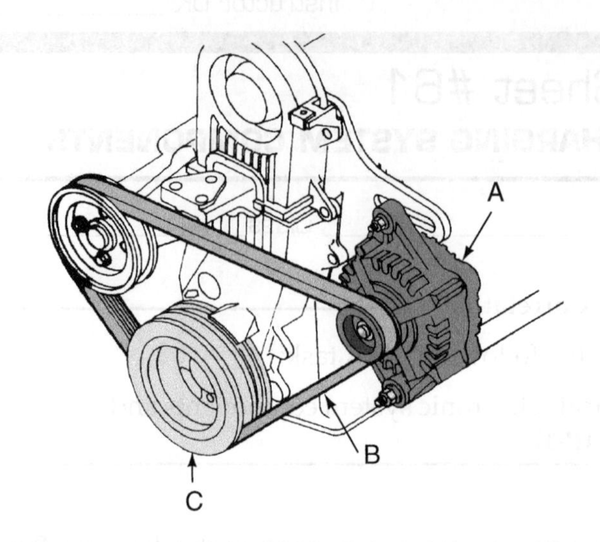

Crankshaft Pulley ______ Drive Belt ______ Alternator ______

STOP

Instructor OK _______________ Score __________

Activity Sheet #62

IDENTIFY ALTERNATOR COMPONENTS

Name ______________________________ Class ____________________

ASE Education Foundation Correlation

This activity sheet addresses the following **MLR** task:

VI.A.11 Identify electrical/electronic system components and configuration. **(P-1)**

We Support Education Foundation

Directions: Identify the alternator components in the drawing. Place the identifying letter next to the name of each component.

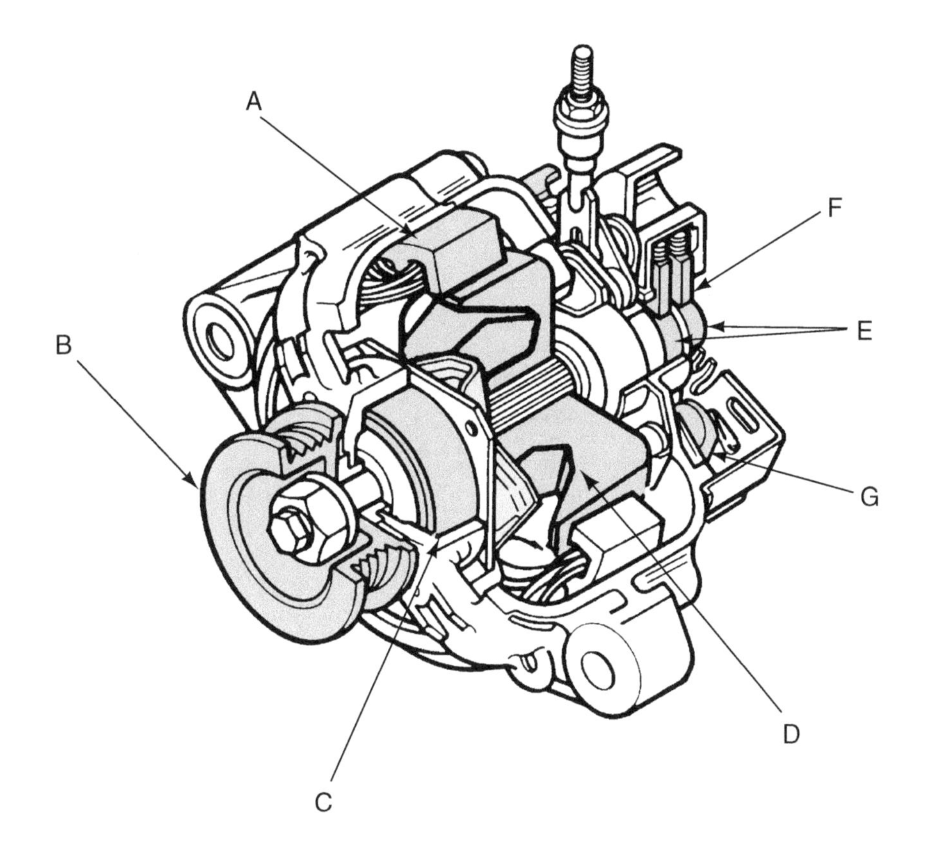

Rotor	______	Bearing	______	Pulley	______
Slip Rings	______	Brushes	______	Stator	______
Rectifier	______				

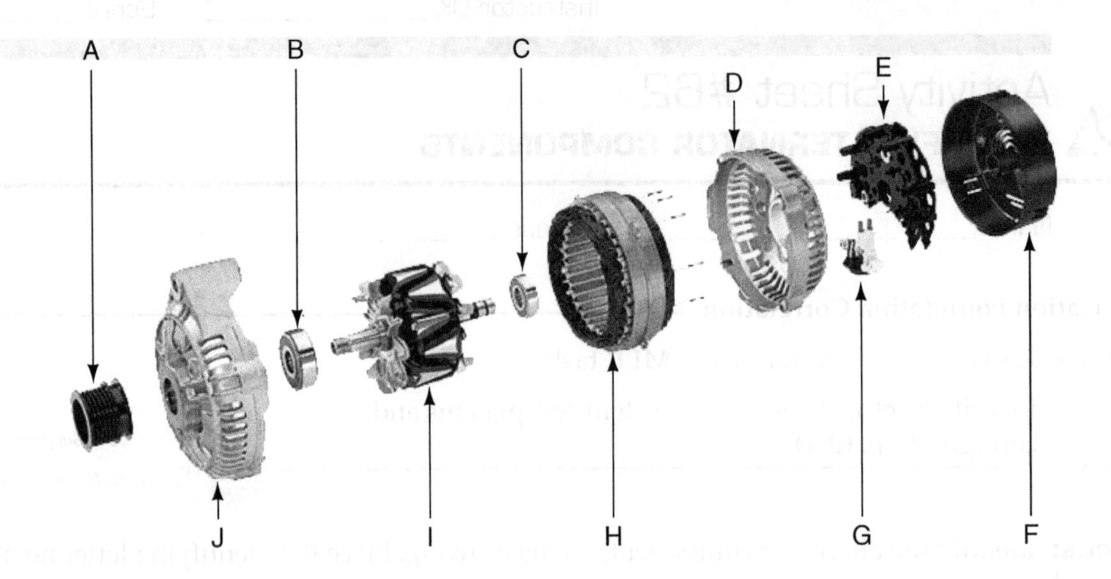

Pulley	______	Drive-End Shield	______	End Shield	______
Rotor	______	Front Bearing	______	Rectifier	______
Rear Bearing	______	Stator	______	Protective Cap	______
Regulator	______				

STOP

Instructor OK ______________ Score ______________

Activity Sheet #63

IDENTIFY LIGHTING AND WIRING COMPONENTS

Name ______________________________ Class ____________________

ASE Education Foundation Correlation

This activity sheet addresses the following **MLR** task:

VI.A.11 Identify electrical/electronic system components and configuration. **(P-1)**

We Support ASE | Education Foundation

Directions: Identify the wiring and lighting components in the drawing. Place the identifying letter next to the name of the component.

Contacts

Photo by Tim Gilles

A B C D

E F G

H

I

Halogen Headlamp	______	Single Filament Bulb	______	Dash Light Bulb	______
Blade Fuse	______	High Beam	______	Low Beam	______
Circuit Breaker	______	Cartridge Fuse	______	Dual Filament Bulb	______

TURN

Directions: Identify the vehicle lights in the drawing. Place the identifying letter next to the name of each light.

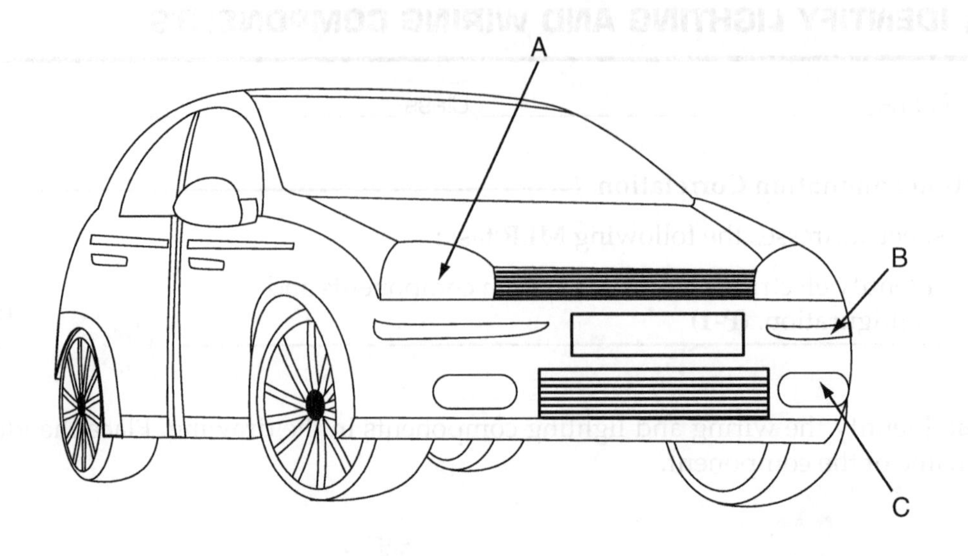

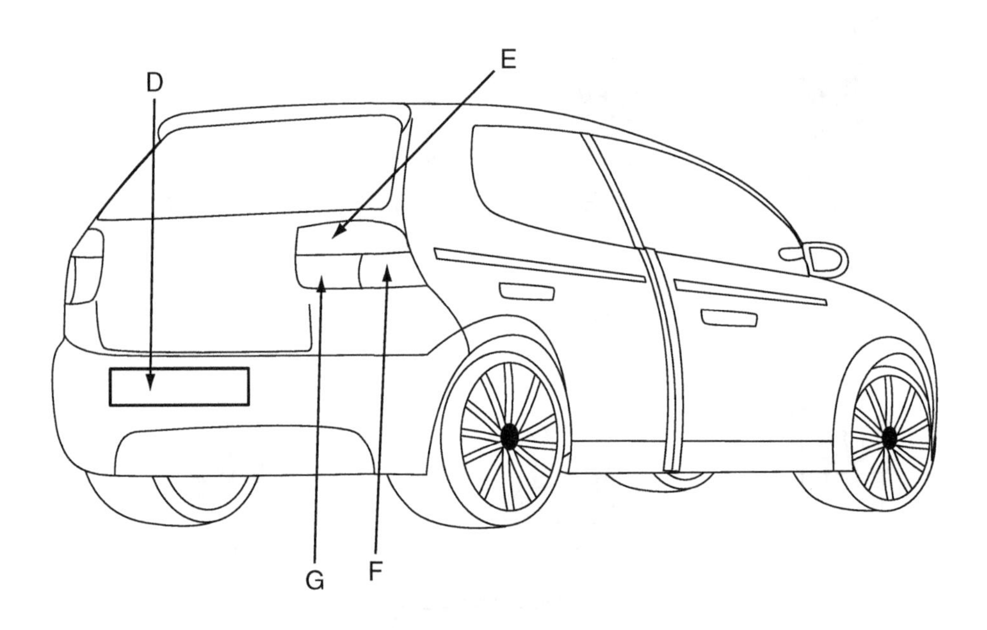

Tail and Brake Light ______	Fog Light ______	Front Turn Signal ______
Back-Up Light ______	Headlight ______	Rear Turn Signal ______
License Plate Light ______		

STOP

Instructor OK ______________ Score ____________

Activity Sheet #64

LIGHTING AND WIRING

Name ______________________________ Class ____________________

Directions: Match the words on the left to the descriptions on the right. Write the letter for the correct word on the line provided. For the terms that you are not certain of, use the glossary in your textbook.

Term		Description
A. Primary wiring	_____	A wire in a light bulb that provides a resistance to electron flow; when it heats up, it causes light
B. Secondary wiring	_____	The number that identifies a bulb for all manufacturers
C. Cables	_____	A rating for the intensity of a headlamp
D. AWG	_____	Two metal strips with different expansion rates
E. Fuse	_____	Has high beam only
F. SFE	_____	A device activated by the heat of the electricity that causes the turn signal bulbs to flash
G. Mini-fuse	_____	Larger wire means a AWG number
H. Fuse link	_____	American Wire Gauge
I. Circuit breaker	_____	A dimmer switch and turn signal lever together
J. Bimetal strip	_____	American National Standards Institute
K. Filament	_____	The Society of Fuse Engineers
L. Candlepower	_____	A type of fuse that allows for a heavier startup draw
M. Type I headlamp	_____	A circuit protection device designed to melt when the flow of current becomes too high for the wires or loads in the circuit
N. Type II headlamp	_____	Larger wires that allow more electrical current flow
O. Halogen headlamp	_____	The year rectangular headlamps were introduced
P. Composite headlamp	_____	The year hazard flashers were introduced
Q. ANSI	_____	A circuit protection device that resets automatically or can be manually reset after it trips
R. Bulb trade number	_____	Has both low and high beams
S. NA	_____	A smaller type of blade fuse; it has a fuse element cast into a clear plastic outer body
T. Signal flasher	_____	A circuit protection device that is a length of wire smaller in diameter than the wire it is connected to
U. Lower	_____	A brighter headlamp used on newer cars
V. Slow blow fuse	_____	A natural amber light bulb
W. HID lamp	_____	Low-voltage wiring
X. Multifunction switch	_____	A headlamp housing with a glass balloon that the halogen lamp fits inside
Y. 1975	_____	High-voltage ignition wiring
Z. 1967	_____	Headlight that is brighter, but uses less power

STOP

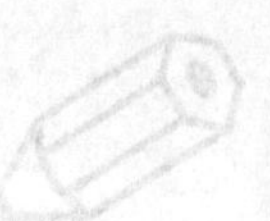

Instructor OK ______________ Score ____________

Activity Sheet #65

IDENTIFY THE SYMBOLS USED IN WIRING DIAGRAMS

Name ______________________________ Class ____________________

ASE Education Foundation Correlation

This activity sheet addresses the following **MLR** task:

VI.A.3 Use wiring diagrams to trace electrical/electronic circuits. **(P-1)**

This activity sheet addresses the following **AST/MAST** task:

VI.A.7 Use wiring diagrams during the diagnosis (troubleshooting) of electrical/electronic circuit problems. **(P-1)**

We Support ASE Education Foundation

Directions: Record the names of the parts that the symbols represent in the spaces provided.

SYMBOLS USED IN WIRING DIAGRAMS			
1		11	
2		12	
3		13	
4		14	
5		15	
6		16	
7		17	
8		18	
9		19	
10		20	

1. ____________________
2. ____________________
3. ____________________
4. ____________________
5. ____________________
6. ____________________
7. ____________________
8. ____________________
9. ____________________
10. ____________________
11. ____________________
12. ____________________
13. ____________________
14. ____________________
15. ____________________
16. ____________________
17. ____________________
18. ____________________
19. ____________________
20. ____________________

Directions: Use the schematic below to identify the components or answer the following questions.

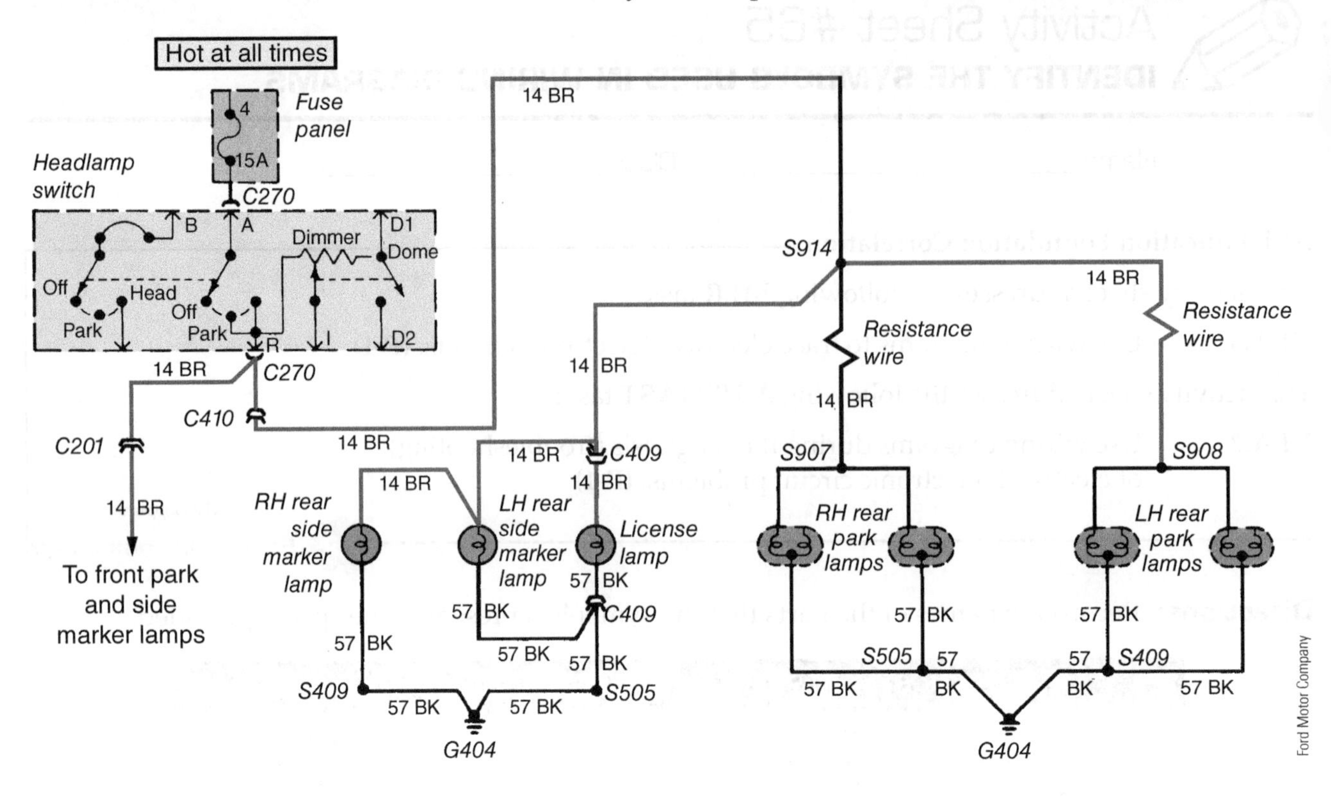

21. C410 ______________________________

22. S914 ______________________________

23. BK next to a wire ______________________________

24. G404 ______________________________

25. Is a single or double filament lamp used for the license lamp? ____________________

26. Is a single or double filament used for the RH rear parking lamps? ____________________

27. What does the symbol next to S409 represent? ______________________

28. What does the symbol next to C409 represent? ______________________

29. What does the symbol above G404 represent? ______________________

30. In which part of the vehicle is G409 located? ______________________

STOP

Instructor OK ______________ Score ___________

Activity Sheet #66
IDENTIFY ELECTRICAL PROBLEMS

Name______________________ Class______________

ASE Education Foundation Correlation

This activity sheet addresses the following **MLR** task:

VI.A.5 Demonstrate knowledge of the causes and effects from shorts, grounds, opens, and resistance problems in electrical/electronic circuits. **(P-1)**

This activity sheet addresses the following **AST/MAST** task:

VI.A.4 Demonstrate knowledge of the causes and effects from shorts, grounds, opens, and resistance problems in electrical/electronic circuits. **(P-1)**

We Support ASE | Education Foundation

Directions: Answer each of the questions with one of the following statements that would best identify the result of the problem: only bulb 1 is out, only bulb 2 is out, both bulbs are out, or there is no problem.

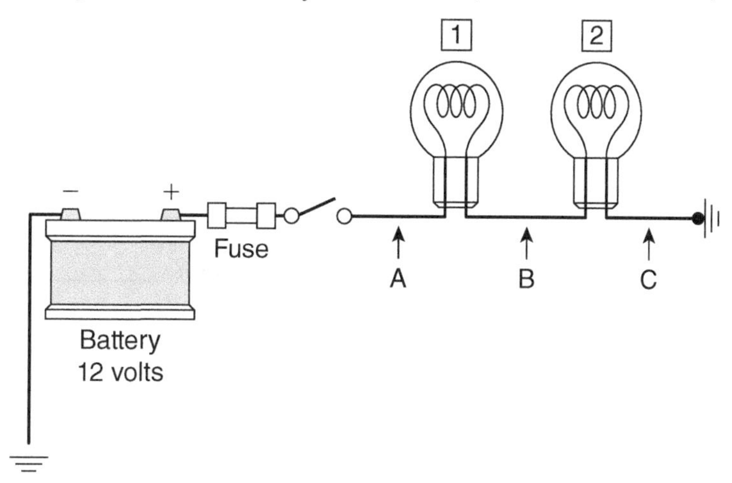

Note: Wiring diagrams are drawn with the switch in its at-rest position. The first step in analyzing a circuit is to close the switch.

1. What would be the result if the circuit was open at point A? ____________________
2. What would be the result if the circuit was shorted to ground at point A? ____________________
3. What would be the result if the circuit was open at point B? ____________________
4. What would be the result if the circuit was shorted to ground at point B? ____________________
5. What would be the result if the circuit was open at point C? ____________________
6. What would be the result if the circuit was shorted to ground at point C? ____________________

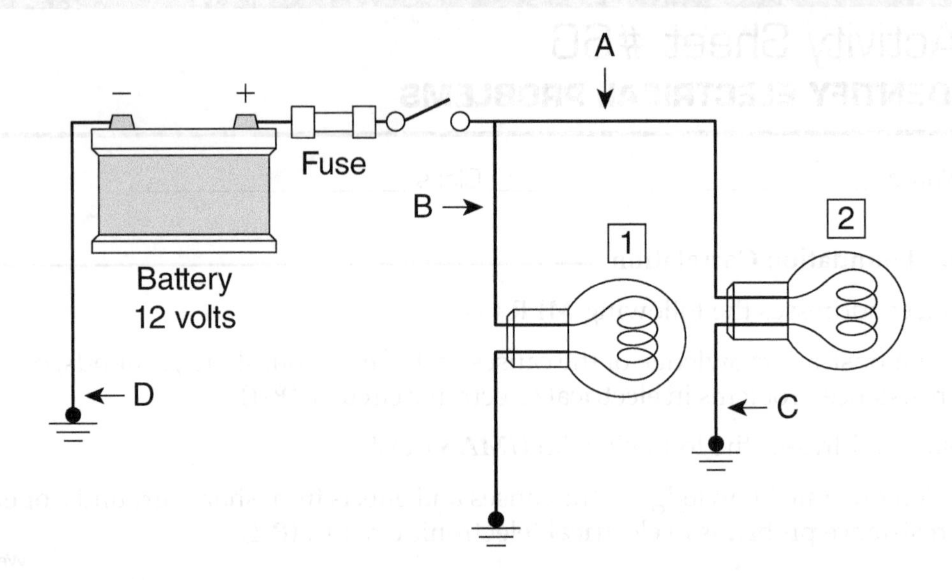

Directions: Use the drawing above to answer the following questions:

7. What would be the result if the circuit was open at point A? ______________________

8. What would be the result if the circuit was shorted to ground at point A? ______________________

9. What would be the result if the circuit was open at point B? ______________________

10. What would be the result if the circuit was shorted to ground at point B? ______________________

11. What would be the result if the circuit was open at point C? ______________________

12. What would be the result if the circuit was shorted to ground at point C? ______________________

13. What would be the result if the circuit was open at point D? ______________________

14. What would be the result if the circuit was shorted to ground at point D? ______________________

15. What problem would typically cause the following symptoms?

	High Voltage	Loose Connection	High Resistance	Open
Lights brighter than normal	☐	☐	☐	☐
Intermittent light operation	☐	☐	☐	☐
Dim lights	☐	☐	☐	☐
No light operation	☐	☐	☐	☐

STOP

Instructor OK ________ Score ________

Activity Sheet #67

USING AN ELECTRICAL WIRING DIAGRAM TO DETERMINE AVAILABLE VOLTAGE

Name________________ Class________________

ASE Education Foundation Correlation

This activity sheet addresses the following **MLR** task:

VI.A.3 Use wiring diagrams to trace electrical/electronic circuits. **(P-1)**

This activity sheet addresses the following **AST/MAST** task:

VI.A.7 Use wiring diagrams during the diagnosis (troubleshooting) of electrical/electronic circuit problems. **(P-1)**

We Support ASE Education Foundation

Directions: Use the wiring diagrams provided to identify the voltage in a circuit. Two colored highlighters will be needed to complete this assignment; one red and the other green.

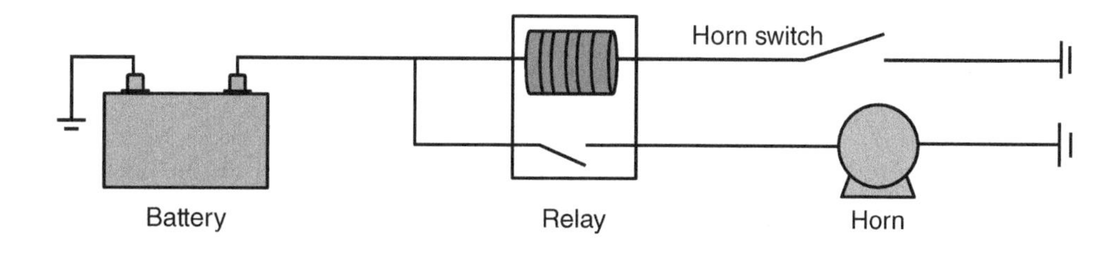

1. In the drawing above, complete the following:
   a. Trace in red where there is source voltage (12.6 volts) when the horn is *not* in operation.
   b. Trace in green where there is no voltage (0 volts) when the horn is *not* in operation.

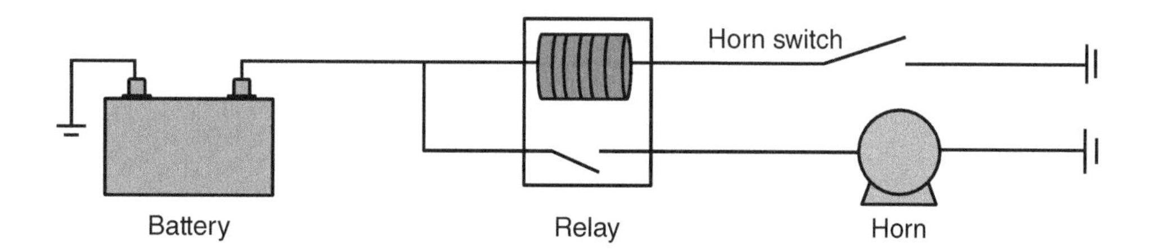

2. In the drawing above, complete the following: **Note:** The switches will need to be closed to complete the circuit.
   a. Trace in red where there is source voltage (12.6 volts) when the horn *is* in operation.
   b. Trace in green where there is no voltage (0 volts) when the horn *is* in operation.

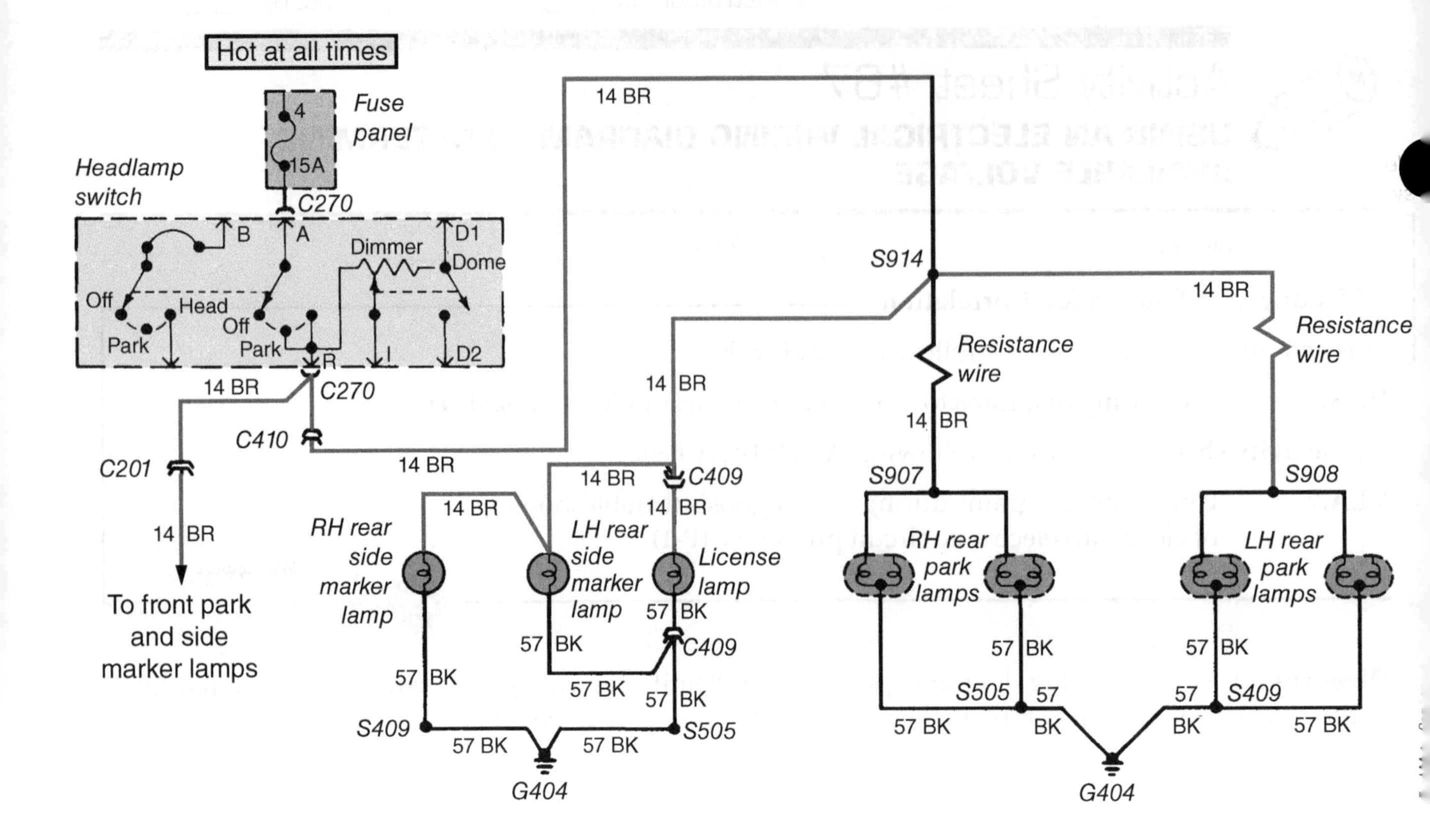

3. In the drawing above, complete the following:
   a. Trace in red where there is source voltage (12.6 volts) when all the rear lights are on.
   b. Trace in green where there is no voltage (0 volts) when all the rear lights are on.
4. What would be the result if connector C201 was disconnected (open)?

____

5. What would be the result if connector C409 was disconnected (open)?

____

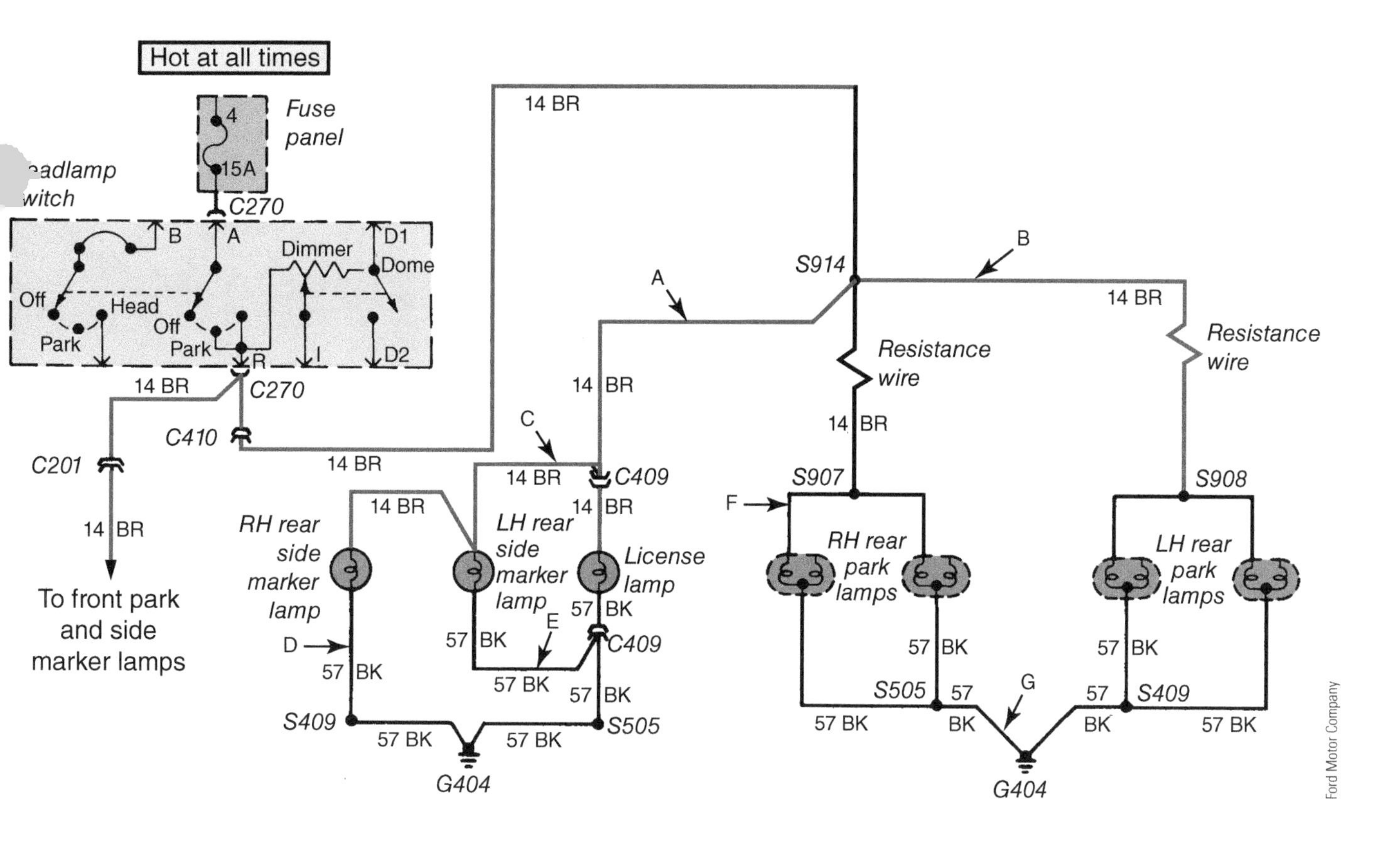

6. Refer to the drawing above and identify the available voltage at the points indicated, when all the lights are "on."

A. ______

B. ______

C. ______

D. ______

E. ______

F. ______

G. ______

STOP

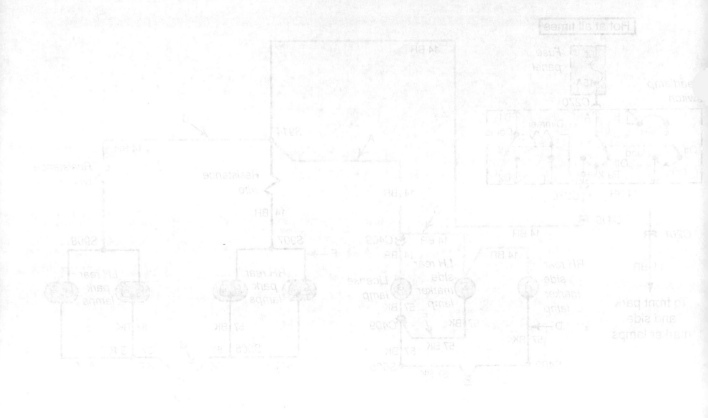

Instructor OK ______________ Score ______________

Activity Sheet #68

USING AN ELECTRICAL WIRING DIAGRAM TO DIAGNOSE AN ELECTRICAL PROBLEM

Name ______________________________ Class ____________________

ASE Education Foundation Correlation

This activity sheet addresses the following **MLR** task:

VI.A.3 Use wiring diagrams to trace electrical/electronic circuits. **(P-1)**

This activity sheet addresses the following **AST/MAST** task:

VI.A.7 Use wiring diagrams during the diagnosis (troubleshooting) of electrical/electronic circuit problems. **(P-1)**

We Support ASE Education Foundation

Directions: Use a wiring diagram to identify the voltage in a circuit.

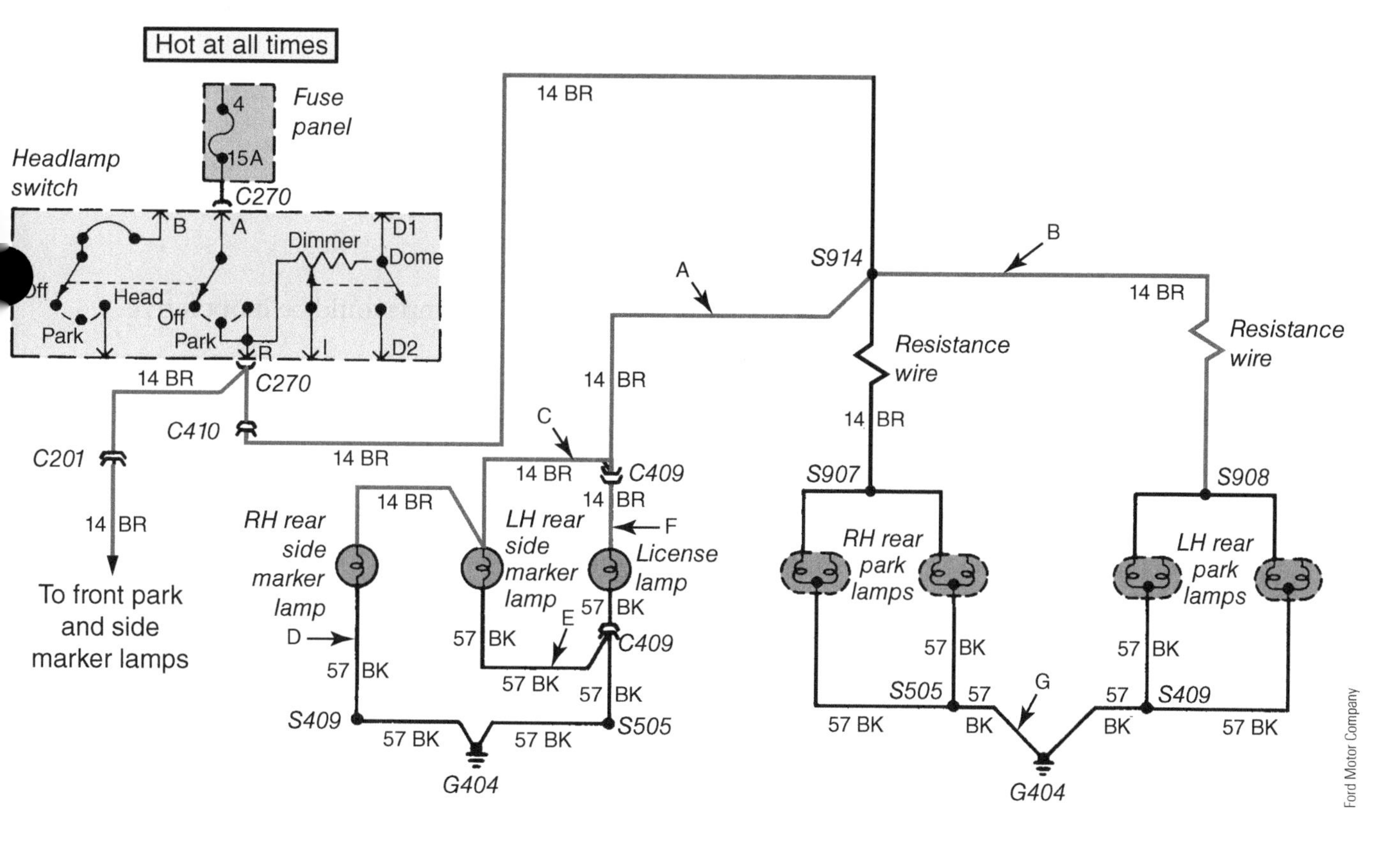

1. In the drawing above, complete the following:
   a. Trace in red where there is source voltage (12.6 volts) when the headlamp switch is in the "R" position.
   b. Trace in green where there is no voltage (0 volts) when the headlamp switch is in the "R" position.

2. What would be the result if the circuit was open at point "A" when the headlamp switch is in the "R" position?
   - ☐ None of the lights would work
   - ☐ RH rear side marker, LH rear side marker, and the license lamp would be inoperative
   - ☐ RH rear park lamp inoperative
   - ☐ LH rear park lamp inoperative
   - ☐ The 15A fuse would blow

3. What would be the result if there was a short at point "A"?
   - ☐ None of the lights would work
   - ☐ RH rear side marker, LH rear side marker, and the license lamp would be inoperative
   - ☐ RH rear park lamp inoperative
   - ☐ LH rear park lamp inoperative
   - ☐ The 15A fuse would blow

4. What would be the result if there was an open circuit at point "B"?
   - ☐ None of the lights would work
   - ☐ RH rear side marker, LH rear side marker, and the license lamp would be inoperative
   - ☐ RH rear park lamp would be inoperative
   - ☐ LH rear park lamp would be inoperative
   - ☐ The 15A fuse would blow

5. What would be the result if there was an open circuit at point "C"? Check all that apply.
   - ☐ RH rear side marker lamp would be inoperative
   - ☐ LH rear side marker lamp would be inoperative
   - ☐ License lamp would be inoperative
   - ☐ The 15A fuse would blow
   - ☐ RH rear park lamp would be inoperative
   - ☐ LH rear park lamp would be inoperative
   - ☐ The circuit would work normally

6. What would be the result if there was an open at point "D"? Check all that apply.
   - ☐ RH rear side marker lamp would be inoperative
   - ☐ LH rear side marker lamp would be inoperative
   - ☐ License lamp would be inoperative
   - ☐ The 15A fuse would blow
   - ☐ RH rear park lamp would be inoperative
   - ☐ LH rear park lamp would be inoperative
   - ☐ The circuit would work normally
7. What would be the result if there was a short at point "E"? Check all that apply.
   - ☐ RH rear side marker lamp would be inoperative
   - ☐ LH rear side marker lamp would be inoperative
   - ☐ License lamp would be inoperative
   - ☐ The 15A fuse would blow
   - ☐ RH rear park lamps would be inoperative
   - ☐ LH rear park lamps would be inoperative
   - ☐ The circuit would work normally
8. What would be the result if there was a short at point "F"? Check all that apply.
   - ☐ RH rear side marker lamp would be inoperative
   - ☐ LH rear side marker lamp would be inoperative
   - ☐ License lamp would be inoperative
   - ☐ The 15A fuse would blow
   - ☐ RH rear park lamp would be inoperative
   - ☐ LH rear park lamp would be inoperative
   - ☐ The circuit would work normally
9. What would be the result if there was an open at point "G"? Check all that apply.
   - ☐ RH rear side marker lamp would be inoperative
   - ☐ LH rear side marker lamp would be inoperative
   - ☐ License lamp would be inoperative
   - ☐ The 15A fuse would blow
   - ☐ RH rear park lamp would be inoperative
   - ☐ LH rear park lamp would be inoperative
   - ☐ The circuit would work normally

Instructor OK ____________ Score ________

Activity Sheet #69

IDENTIFY ELECTRICAL TEST INSTRUMENTS

Name ________________________ Class ________________

Directions: Identify the electrical test instruments in the drawing. Place the identifying letter next to the name of the instrument.

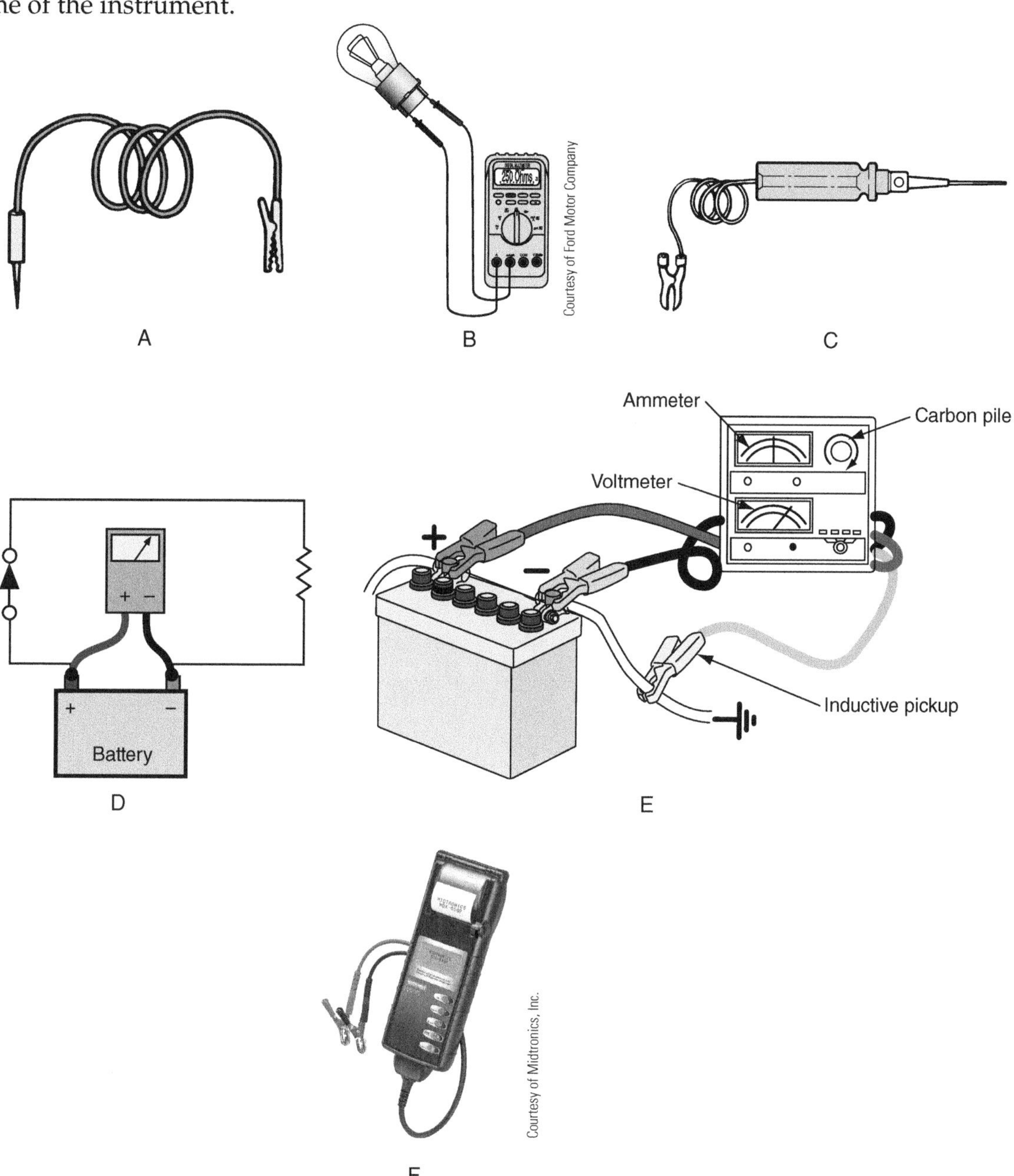

Analog Meter	______	Volt-Amp Tester	______	Jumper Wire	______
Digital Multimeter	______	Circuit Tester	______	Conductance Tester	______

Instructor OK ______________ Score ___________

Activity Sheet #70

SAFETY, SECURITY, COMFORT SYSTEMS, AND ELECTRICAL ACCESSORIES

Name ______________________________ Class ______________________

Directions: Match the words on the left to the descriptions on the right. Write the letter for the correct word on the line provided. For the terms that you are not certain of, use the glossary in your textbook.

Term		Description
A. Passive restraints	_____	Supplemental inflatable restraints
B. Active restraints	_____	Pendulum-like device that locks during sudden deceleration
C. Seat belts and air bags	_____	Initiator or igniter
D. Pretensioners	_____	Supplemental restraint system
E. LDW	_____	Located in the top of the dash
F. Seat belt warning system	_____	Manually buckled seat belt
G. SIR	_____	Global positioning system
H. SAR	_____	Automatic seat belts
I. SRS	_____	Produces high-frequency sounds
J. Air bag	_____	Heated gas inflators
K. Passenger side air bag	_____	Flexible nylon bag
L. Frequency modulation	_____	Used to inflate the air bag
M. Discriminating sensors	_____	Key fob transmitter
N. Safing sensors	_____	Dash light, with a bell or buzzer
O. Squib	_____	Resistance pellet is embedded in it
P. Deployment time	_____	Required on all cars and light trucks
Q. Nitrogen gas	_____	Used for power windows
R. HGI	_____	Automatically darken in response to sunlight
S. Resistance key	_____	Produces nondirectional sound
T. Amplitude modulation	_____	Found in the front of the vehicle
U. Keyless entry	_____	Supplemental air restraints
V. GPS	_____	Located in the steering wheel
W. A tweeter	_____	Controls the slack in seat belts
X. A woofer	_____	100 milliseconds
Y. Photochromatic mirrors	_____	Found in the center console
Z. Permanent magnet DC motors	_____	Receives a radio signal
AA. Smart key	_____	FM radio
AB. Mechanical pretensioners	_____	AM radio
AC. Driver side air bag	_____	Lane departure warning system
AD. Transponder key	_____	Replaces mechanical key

STOP

Instructor OK ______________ Score __________

Activity Sheet #71
IDENTIFY AIR BAG COMPONENTS

Name ____________________ Class ______________

ASE Education Foundation Correlation

This activity sheet addresses the following **Required Supplemental Task (RST):**

Shop and Personal Safety:

Task 13. Demonstrate awareness of the safety aspects of supplemental restraint systems (SRS), electronic brake control systems, and hybrid vehicle high voltage circuits.

This activity sheet addresses the following **AST** task:

VI.G.4 Describe operation of safety systems and related circuits (such as: horn, airbags, seat belt pretensioners, occupancy classification, wipers, washers, speed control/collision avoidance, heads-up display, parking assist, and back-up camera); determine needed repairs. **(P-3)**

This activity sheet addresses the following **MAST** task:

VI.G.4 Diagnose operation of safety systems and related circuits (such as: horn, airbags, seat belt pretensioners, occupancy classification, wipers, washers, speed control/collision avoidance, heads-up display, parking assist, and back-up camera); determine needed repairs. **(P-1)**

We Support ASE Education Foundation

TURN

Directions: Identify the air bag components in the drawing. Place the identifying letter next to the name of each component listed below.

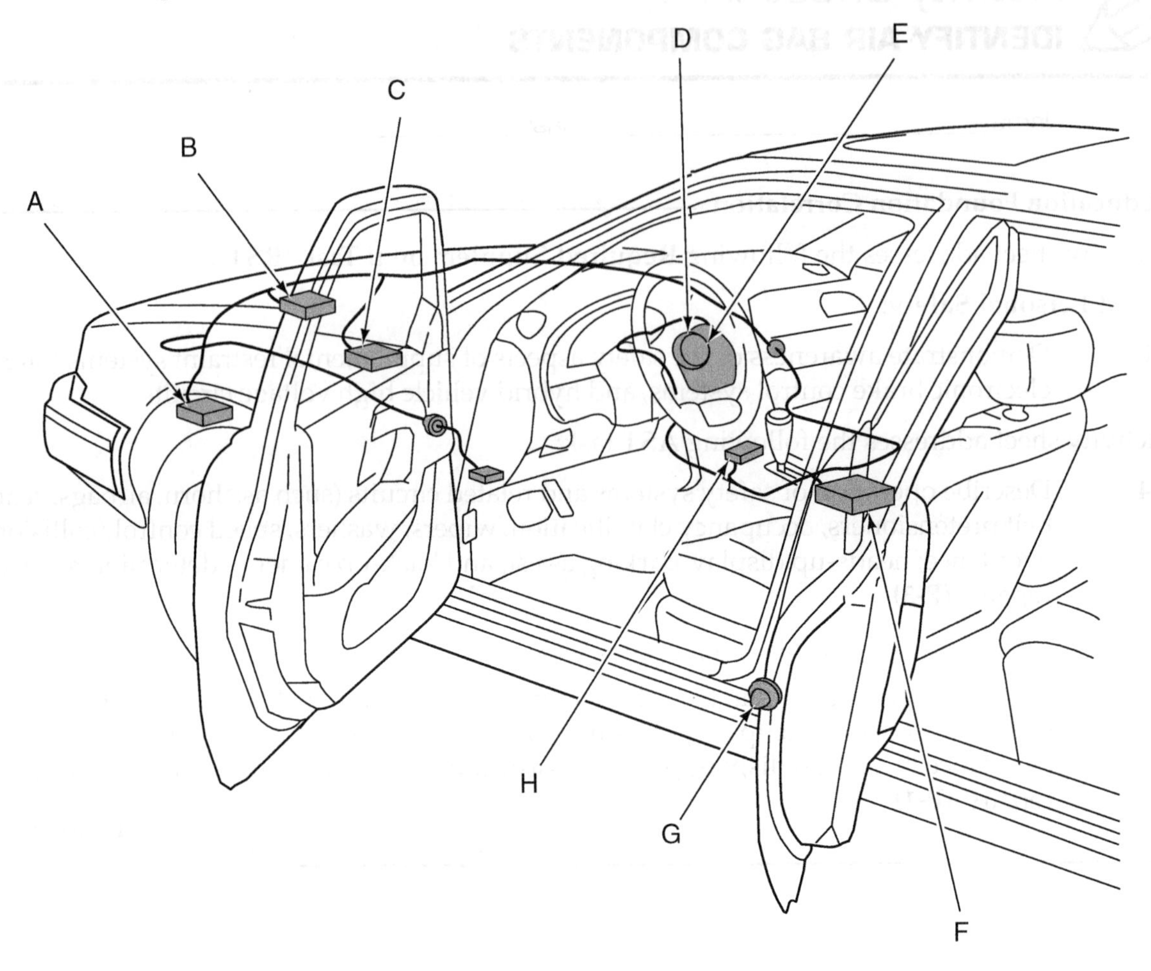

Driver Door Switch	______	Control Module	______	Right Impact Sensor	______
Center Impact Sensor	______	Clock Spring	______	Left Impact Sensor	______
Air Bag Module	______	Safing Sensor	______		

STOP

Instructor OK ____________ Score ____________

Activity Sheet #72

IDENTIFY SRS, ABS, AND HYBRID ELECTRICAL CIRCUITS

Name____________________ Class____________________

ASE Education Foundation Correlation

This activity sheet addresses the following **Required Supplemental Tasks (RST):**

Shop and Personal Safety:

Task 13. Demonstrate awareness of the safety aspects of supplemental restraint systems (SRS), electronic brake control systems, and hybrid vehicle high voltage circuits.

Shop and Personal Safety:

Task 14. Demonstrate awareness of the safety aspects of high voltage circuits (such as high intensity discharge (HID) lamps, ignition systems, injection systems, etc.).

This activity sheet addresses the following **MLR** task:

I.A.7 Identify service precautions related to service of the internal combustion engine of a hybrid vehicle. **(P-2)**

This activity sheet addresses the following **AST/MAST** task:

I.A.9 Identify service precautions related to service of the internal combustion engine of a hybrid vehicle. **(P-2)**

We Support
ASE | Education Foundation

There are certain safety precautions to be aware of when working on some vehicle systems. These include air bag (SRS) systems, anti-lock brakes (ABS), and hybrid electric vehicles (HEVs).

Directions: Match the words on the left to the descriptions on the right. Write the letter for the correct word on the line provided. If you are not certain of a term, use the glossary in your textbook.

A. Traction	_____	Continues to work normally when the ABS warning light is illuminated
B. Pulsate	_____	Disconnect before working on a hybrid vehicle
C. ABS warning light	_____	Color of high-voltage wiring, labels, and connectors
D. Brake system	_____	Location of SRS safety information
E. Amber	_____	During hard braking, ABS prevents tires from losing
F. Red	_____	Should be worn when working with high-voltage systems
G. Brake light	_____	Must use this when around a hybrid engine as it may start at any time

TURN

H.	Driver's sun visor	_____	A safety device that automatically shorts the circuit when an air bag module connector is disconnected
I.	Yellow	_____	During an ABS stop the brake pedal will
J.	144–650 volts	_____	The color of the ABS warning light is
K.	Orange	_____	When this is illuminated the ABS system will not work
L.	Protective gloves	_____	The color of the brake system warning light is
M.	Residual voltage	_____	Voltage range of hybrid drive systems
N.	Service information	_____	Color of SRS wiring and/or connectors
O.	Service plug	_____	Voltage needed to start HID headlights
P.	Caution	_____	Illuminates when the fluid level is low, emergency brake is on, or when there is a brake pressure problem
Q.	800 volts	_____	Available in SRS and hybrid systems when vehicle is off
R.	Shorting bar	_____	Consult before working on SRS or hybrid systems

STOP

Instructor OK ____________ Score ____________

Activity Sheet #73

AUTOMOTIVE SAFETY DEVICES

Name ____________________ Class ______________

Directions: Identify the safety devices by writing the letter that best describes each of the safety devices in the blank spaces below.

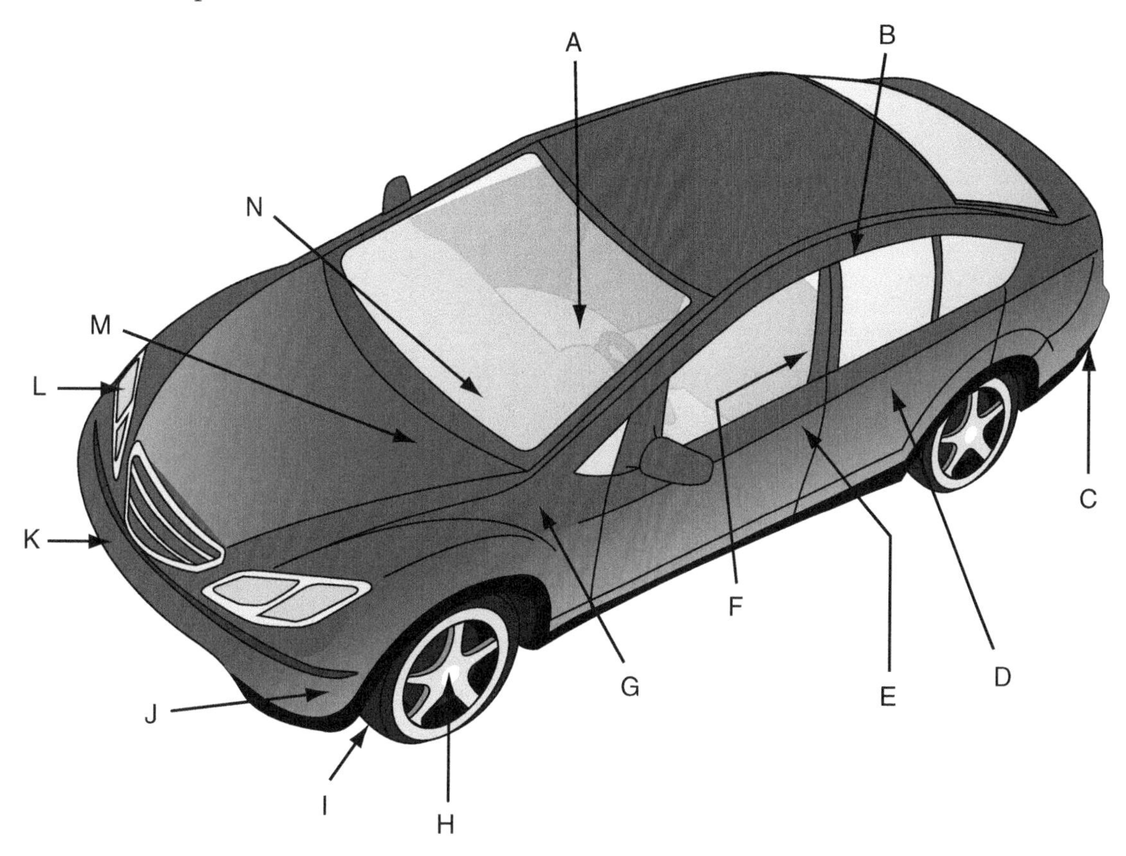

Adaptive Headlights _____
Adjustable Pedal _____
Antilock System _____
Avoidance Radar _____
Child Seat Anchor _____

Driver and Passenger Air Bags _____
Electronic Stability Program _____
Energy-Absorbing Steering
Column _____
Knee Blocker _____

Park Assist _____
Pretensioners _____
Side Curtain _____
Three-Point Belt _____
Tire Pressure Monitoring _____

STOP

Part I

Activity Sheets

Section Seven

Heating and Air Conditioning

Instructor OK ______________ Score ____________

Activity Sheet #74

IDENTIFY HEATING AND AIR-CONDITIONING SYSTEM COMPONENTS

Name ______________________________ Class ____________________

ASE Education Foundation Correlation

This activity sheet addresses the following **MLR** task:

VII.A.2 Identify heating, ventilation and air conditioning (HVAC) components and configuration. **(P-1)**

This activity addresses the following **AST/MAST** task:

VII.A.1 Identify and interpret heating and air conditioning problems; determine needed action. **(P-1)**

We Support ASE Education Foundation

Directions: Identify the cooling and heating system components in the drawing. Place the identifying letter next to the name of each component.

Fan and Blower Motor	______	Heater Core	______	Heater Valve	______
Heater Hoses	______	Thermostat	______	Radiator	______
Water Pump	______	Cooling Fan	______		

TURN

Directions: Identify the air-conditioning system components in the drawing below. Place the identifying letter next to the name of each component.

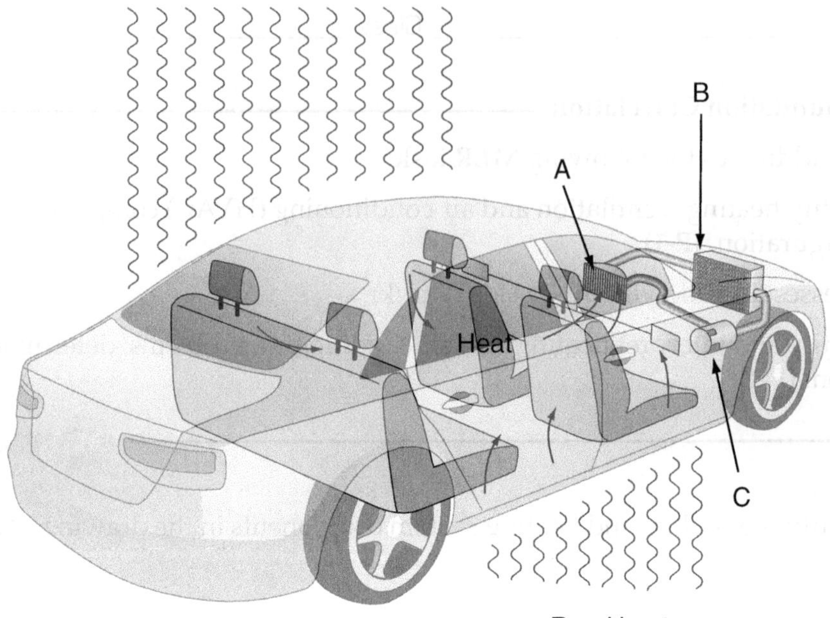

Condenser ______ Evaporator ______ Compressor ______

STOP

Instructor OK ______________ Score ____________

Activity Sheet #75

HEATING AND AIR CONDITIONING

Name ______________________________ Class ____________________

ASE Education Foundation Correlation

This activity sheet addresses the following **MLR** task:

VII.A.2 Identify heating, ventilation and air conditioning (HVAC) components and configuration. **(P-1)**

This activity addresses the following **AST/MAST** task:

VII.A.1 Identify and interpret heating and air conditioning problems; determine needed action. **(P-1)**

We Support ASE Education Foundation

Directions: Match the words on the left to the descriptions on the right. Write the letter for the correct word on the line provided. For the terms that you are not certain of, use the glossary in your textbook.

A. Air conditioning	______	This is also called freon (earlier refrigerant)
B. Blend door	______	System of measurement where water boils at 100 degrees
C. Convection	______	Protects the Earth's surface from ultraviolet rays
D. Radiation	______	Heat transfer where moisture is vaporized as it absorbs heat
E. Evaporation	______	System of measurement where water boils at 373 degrees
F. Humidity	______	When the air is totally saturated with moisture
G. Condensation	______	A term used to describe heat bouncing off a surface
H. Latent heat	______	The process in which air inside of the passenger compartment is cooled, dried, cleaned, and circulated
I. Montreal Protocol	______	System of measurement where water freezes at 32 degrees
J. R-12 refrigerant	______	When air becomes warmer and moves upward
K. Desiccant	______	When a vapor changes to a liquid
L. 100% humidity	______	The movement of heat when there is a difference in temperature between two objects

TURN

M. Ozone	______	Temperature of surrounding air
N. Kelvin	______	The moisture content of the air
O. Inches of mercury	______	Chlorofluorocarbon (abbreviation)
P. Fahrenheit	______	A refrigerant that is used in newer vehicles (hydrofluorocarbon)
Q. Ambient	______	Sets limits on the production of ozone-depleting chemicals
R. Heat transfer	______	Used to remove moisture
S. CFC	______	Used to measure vacuum or low pressure
T. R-134A refrigerant	______	Controls airflow in the heating and cooling systems
U. Celsius	______	Extra heat required for matter to change state

STOP

Instructor OK ______________ Score __________

Activity Sheet #76

IDENTIFY AIR-CONDITIONING SYSTEM COMPONENTS

Name ____________________________ Class ____________________

ASE Education Foundation Correlation

This activity sheet addresses the following **MLR** task:

VII.A.2 Identify heating, ventilation and air conditioning (HVAC) components and configuration. **(P-1)**

This activity addresses the following **AST/MAST** task:

VII.A.1 Identify and interpret heating and air conditioning problems; determine needed action. **(P-1)**

We Support ASE | Education Foundation

Directions: Identify the air-conditioning system components in the drawing. Place the identifying letter next to the name of each component.

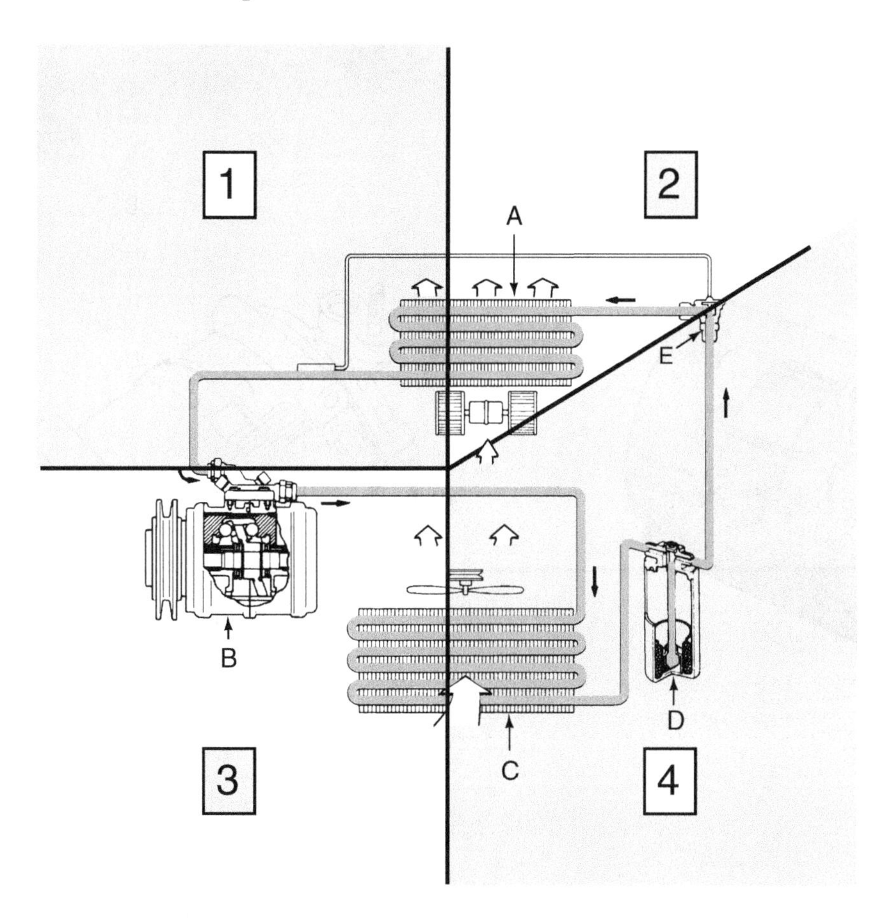

Receiver/Dryer ______ Expansion Valve ______ Evaporator ______

Compressor ______ Condenser ______

Refer to the drawing above and select the correct number for each of the following:

High-Pressure Liquid ______________________ Low-Pressure Gas ______________________

Low-Pressure Liquid ______________________ High-Pressure Gas ______________________

Directions: Identify the air-conditioning compressors in the drawings below. Place the identifying letter next to the name of the compressor.

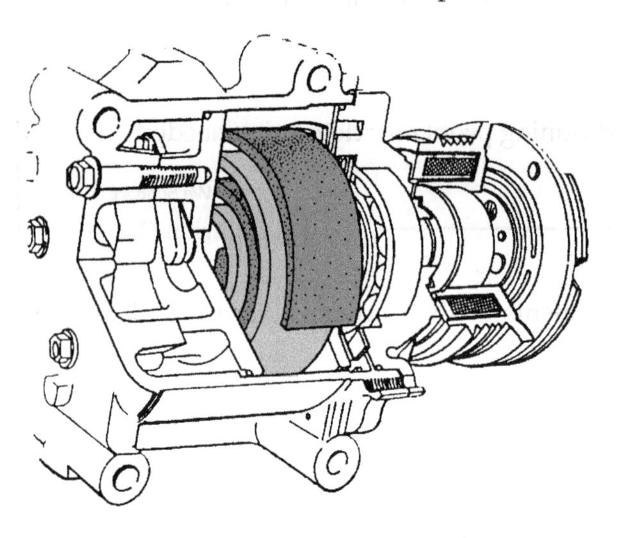

A

DENSO Corporation

B

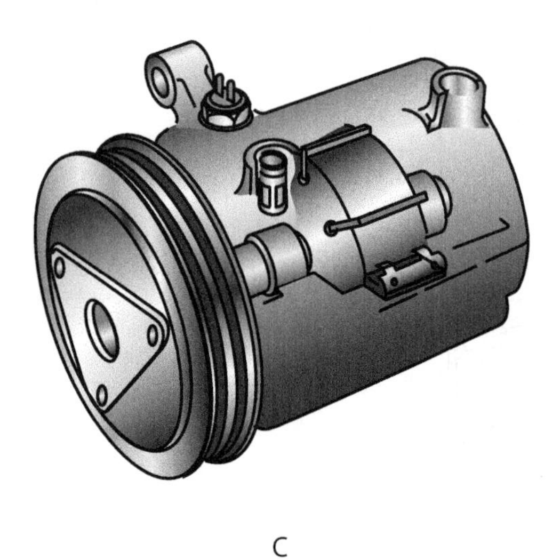

C

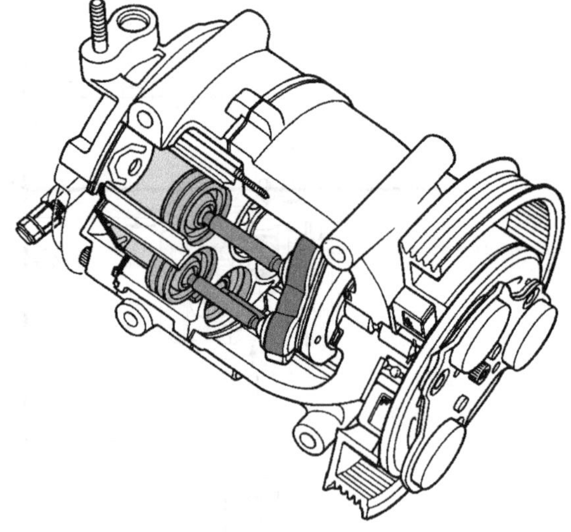

D

Scroll Compressor ______ Rotary Vane Compressor ______

Electric Compressor ______ Wobble Plate Compressor ______

STOP

Instructor OK ______________ Score ______________

Activity Sheet #77

IDENTIFY AIR-CONDITIONING PARTS

Name ______________________________ Class ____________________

ASE Education Foundation Correlation

This activity sheet addresses the following **MLR** task:

VII.A.2 Identify heating, ventilation and air conditioning (HVAC) components and configuration. **(P-1)**

This activity addresses the following **AST/MAST** task:

VII.A.1 Identify and interpret heating and air conditioning problems; determine needed action. **(P-1)**

We Support

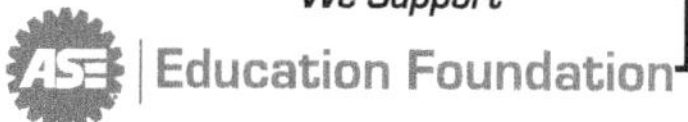

Directions: Identify the air-conditioning parts in the drawing. Place the identifying letter next to the name of the parts.

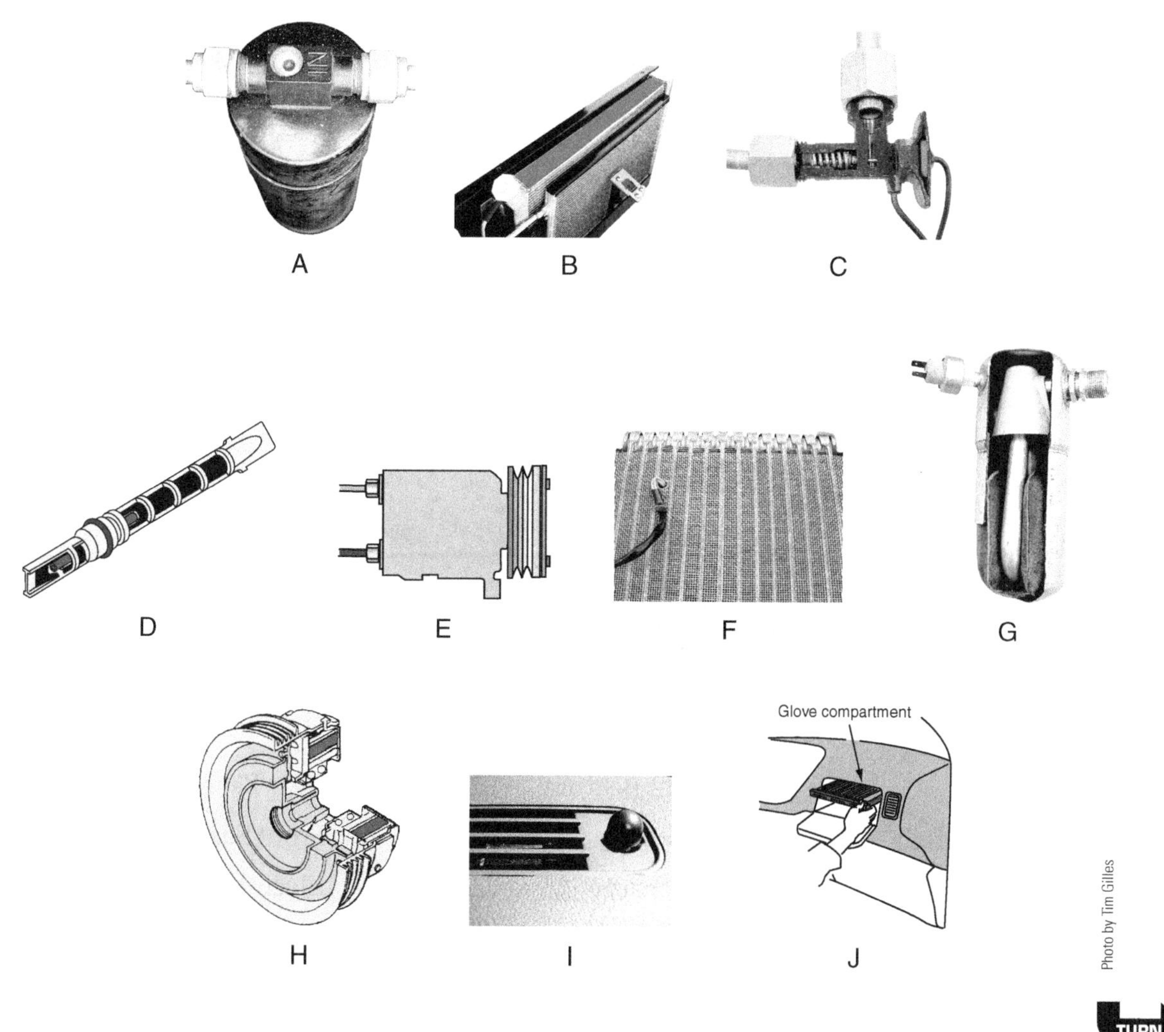

Photo by Tim Gilles

TURN

Condenser	______	Receiver/Dryer	______	Orifice Tube	______
Evaporator	______	Accumulator	______	Compressor	______
Expansion Valve	______	Sunload Sensor	______	Compressor Clutch	______
Cabin Filter	______				

Directions: Identify the parts in the drawing. Place the letter next to the name of each part.

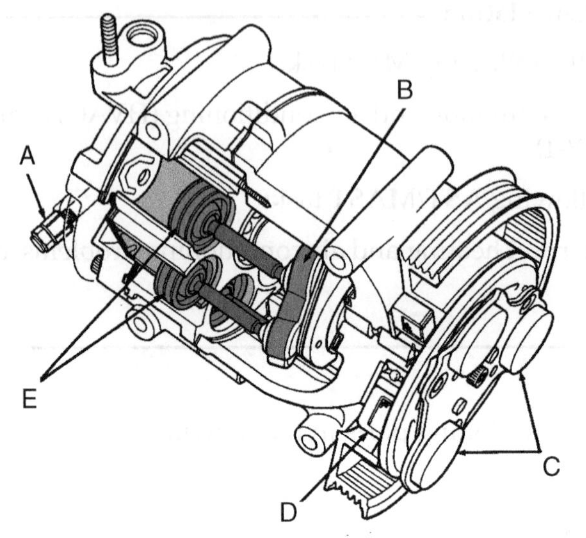

Relief Valve	______	Pressure Plate	______	Pistons	______
Wobble Plate	______	Field Coil	______		

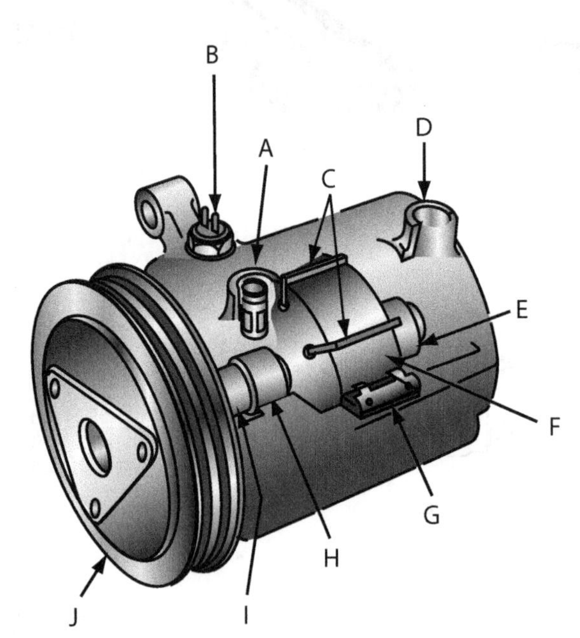

Discharge Port	______	Front Bearing	______	Rear Bearing	______
Case Temperature Switch	______	Rotor	______	Front Seal	______
Vane	______	Discharge Valve	______	Electromagnetic Clutch	______
Suction Port	______				

Part I

Activity Sheets

Section Eight

Engine Performance Diagnosis
Theory and Service

Instructor OK ______________ Score ____________

Activity Sheet #78

IDENTIFY CONVENTIONAL IGNITION SYSTEM COMPONENTS

Name ____________________________ Class ____________________

ASE Education Foundation Correlation

This activity sheet addresses the following **MLR** task:

VI.A.11 Identify electrical/electronic system components and configuration. **(P-1)**

We Support ASE | Education Foundation

Directions: Identify the ignition system components in the drawing. Place the identifying letter next to the name of each component.

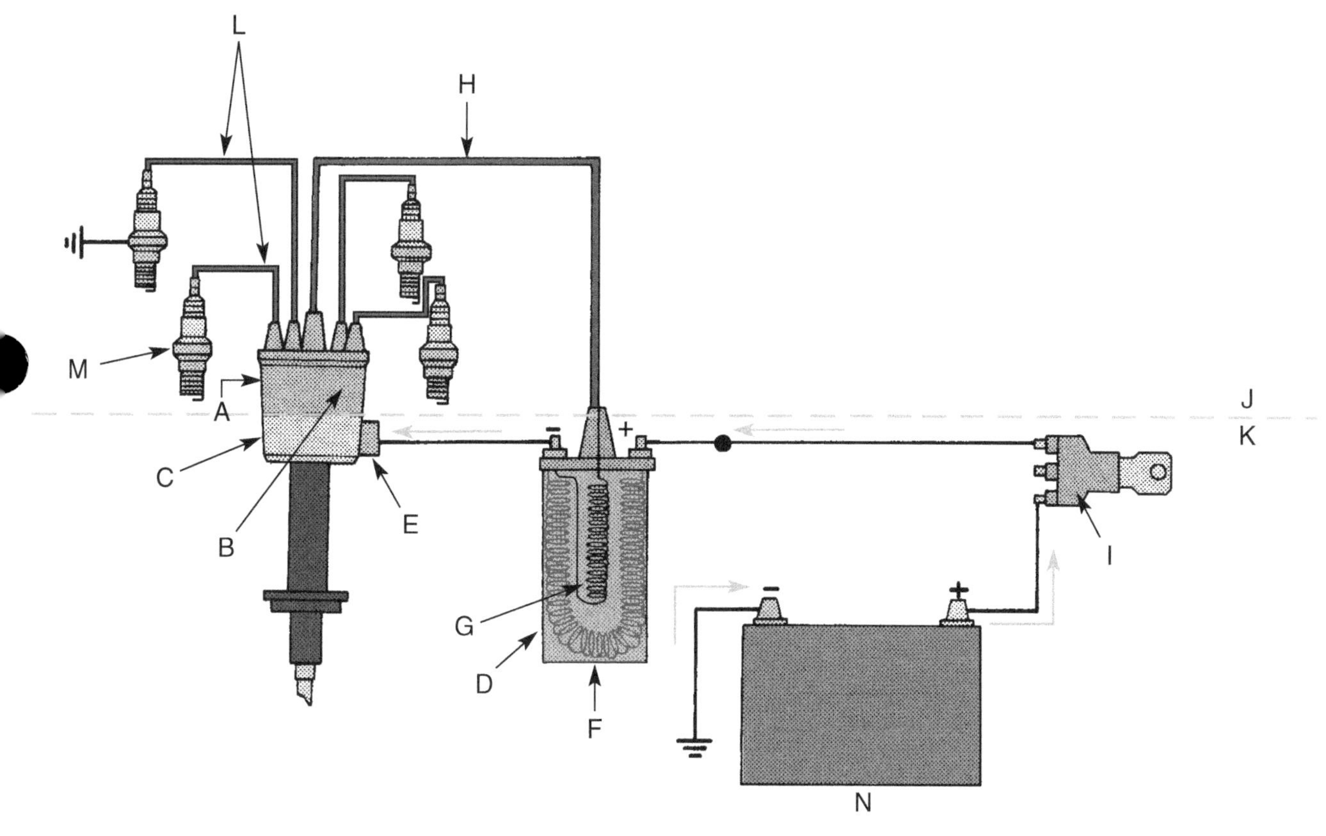

Distributor Cap	______	Secondary Winding	______	Control Module	______
Coil	______	Rotor	______	Coil Wire	______
Secondary Circuit	______	Primary Winding	______	Ignition Switch	______
Distributor	______	Primary Circuit	______	Spark Plug Cables	______
Battery	______	Spark Plug	______		

Instructor OK ____________ Score ____________

Activity Sheet #79

IDENTIFY IGNITION SYSTEM PARTS

Name ______________________ Class ______________________

ASE Education Foundation Correlation

This activity sheet addresses the following **MLR** task:

VI.A.11 Identify electrical/electronic system components and configuration. **(P-1)**

We Support ASE | Education Foundation

Directions: Identify the ignition system parts in the drawing. Place the identifying letter next to the name of each part.

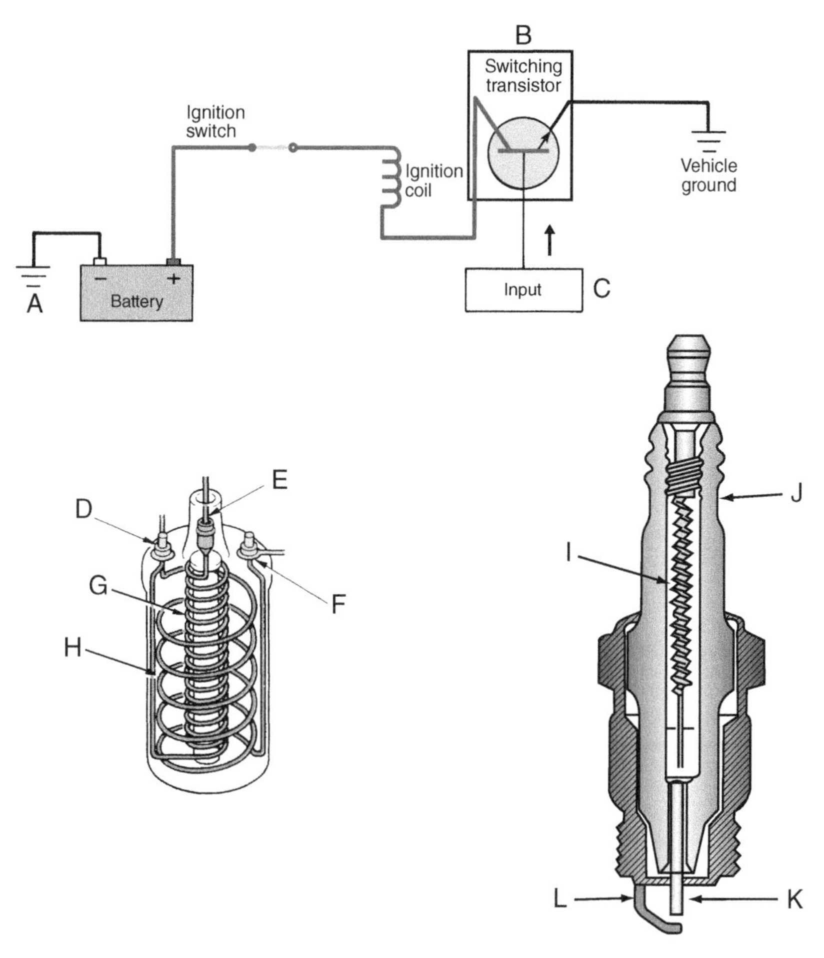

Secondary Winding	______	Coil Secondary Terminal	______	Primary Winding	______
Coil (−) Terminal	______	Crankshaft Position Sensor	______	Ignition Module	______
Battery Ground	______	Coil (+) Terminal	______	Ground Electrode	______
Spark Plug Insulator	______	Center Electrode	______	Spark Plug Resistor	______

Instructor OK ______________ Score ___________

Activity Sheet #80

IDENTIFY CONVENTIONAL IGNITION SYSTEM OPERATION

Name ______________________ Class ________________

ASE Education Foundation Correlation

This activity sheet addresses the following **MLR** task:

VI.A.11 Identify electrical/electronic system components and configuration. **(P-1)**

We Support ASE | Education Foundation

Directions: Identify the ignition system operation in the figure below by drawing the missing wiring.

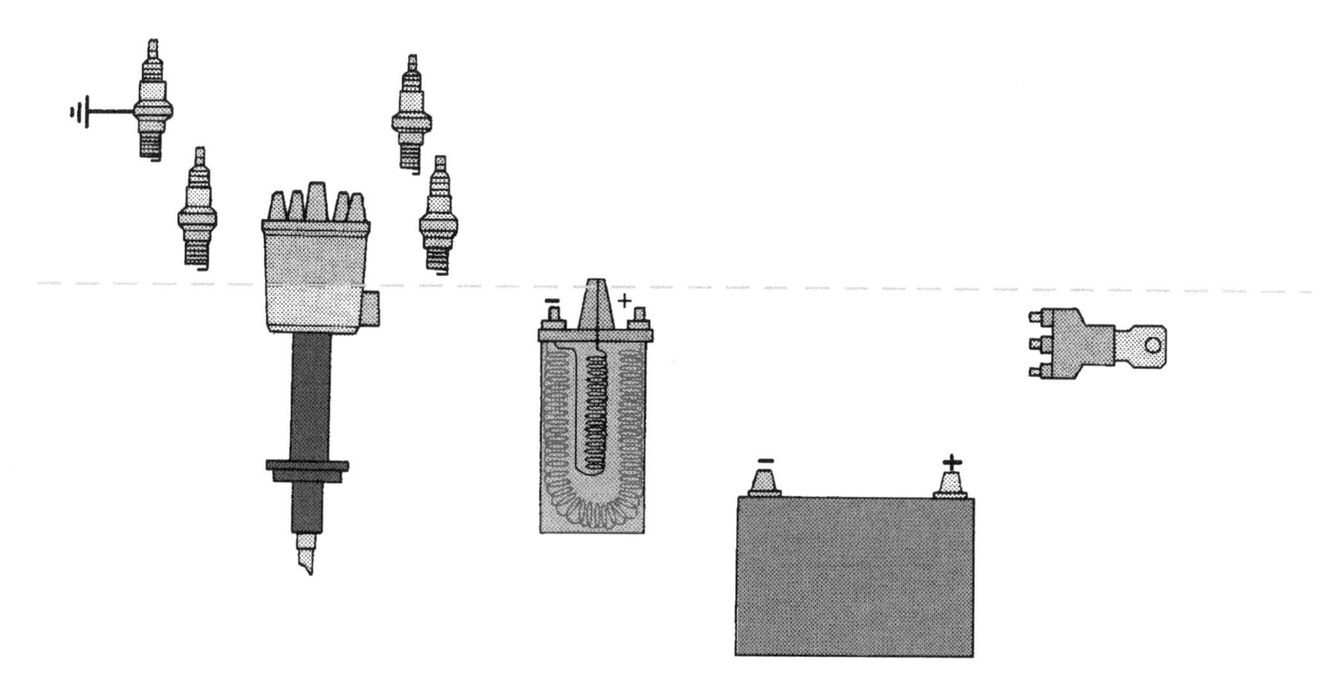

STOP

Instructor OK ______________ Score __________

Activity Sheet #81

IGNITION SYSTEMS

Name ______________________________ Class ______________________

Directions: Match the words on the left to the descriptions on the right. Write the letter for the correct word on the line provided. For the terms that you are not certain of, use the glossary in your textbook.

Term		Description
A. Coil and plug	______	Also called parade pattern, it displays all of the cylinders next to each other (side by side), so that the heights of the voltage spikes can be compared
B. Crossfire induction	______	Triggers the buildup and collapse of the magnetic field in the coil
C. Carbon trail	______	Causes a scope pattern to be upside down
D. Timing light	______	Also called stacked pattern, it displays all of the cylinders vertically, one above the next
E. Multi-strike	______	More compatible with computer systems
F. Firing line	______	Vertical movement on the oscilloscope screen
G. Spark line	______	A strobe light that is triggered by the voltage going through the number one spark plug cable
H. Display pattern	______	When one spark plug firing induces a spark in the one next to it, causing it to fire before its time
I. DIS	______	Buildup of carbon that shorts out a spark plug
J. Superimposed pattern	______	A horizontal line on the scope pattern that begins at the voltage level where electrons start to flow across the spark plug gap
K. MAP	______	Timing a distributor with the engine stopped
L. EMP	______	The upward line that starts the scope pattern
M. Square wave signal	______	A scope pattern used to compare all of the cylinders while their patterns are displayed one on top of the other
N. Waste spark	______	A type of sensor that triggers spark timing
O. Reversed coil polarity	______	A line of electrically conducting carbon that forms in a cracked distributor cap
P. Fouled spark plug	______	Used to sense engine load
Q. Static timing	______	Detect the frequency of spark knock
R. Raster pattern	______	Identifies when the cylinder is on the compression stroke
S. Keyless entry	______	Distributorless ignition system

TURN

T. Transistor	______	An ignition system that uses two spark plugs per cylinder
U. Hall switch	______	Each cylinder has a coil mounted on the spark plug
V. Voltage	______	When two cylinders are fired at the same time
W. Detonation sensor	______	Unlocks a vehicle without a key

Activity Sheet #82

FIRING ORDER

Name ____________________________ Class ____________________

Directions: Write the cylinder numbers in their correct positions in the distributor cap spark plug wire terminal holes and on the lines provided below the drawings.

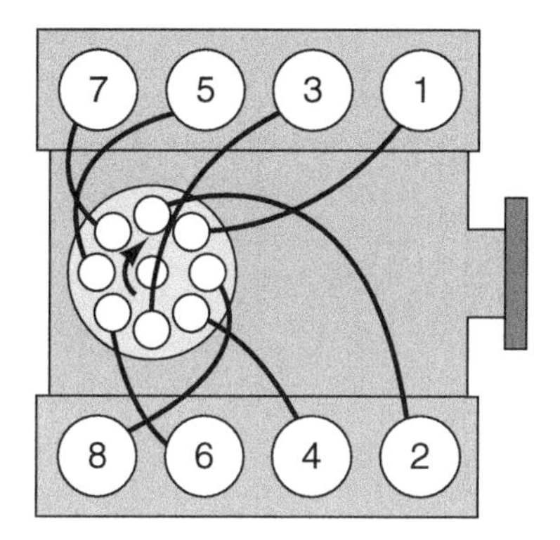

A. Firing Order: _ _ _ _ _ _ _ _

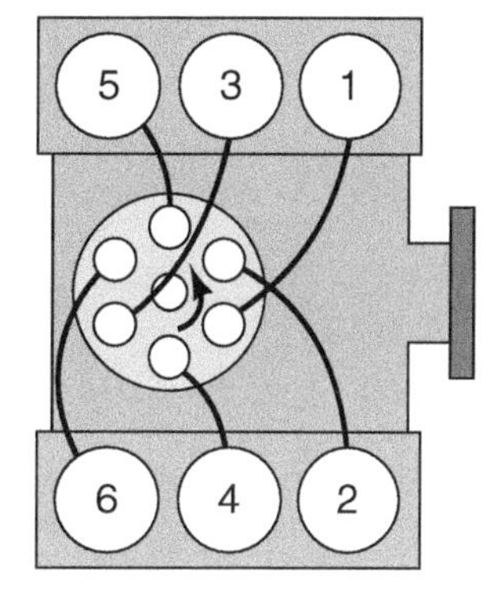

B. Firing Order: _ _ _ _ _ _

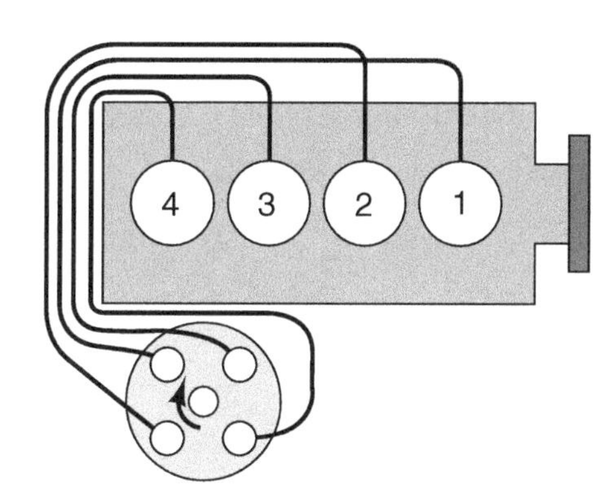

C. Firing Order: _ _ _ _

STOP

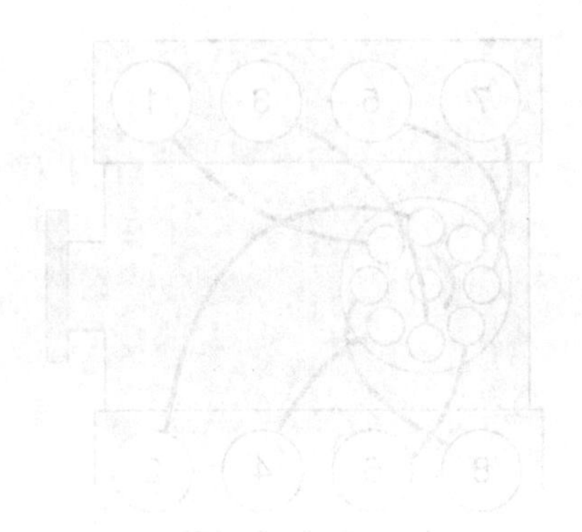

Instructor OK ______________ Score ____________

Activity Sheet #83

IDENTIFY DISTRIBUTORLESS IGNITION SYSTEM (DIS) COMPONENTS

Name ______________________________ Class ____________________

ASE Education Foundation Correlation

This activity sheet addresses the following **MLR** task:

VI.A.11 Identify electrical/electronic system components and configuration. **(P-1)**

We Support ASE Education Foundation

Directions: Identify the distributorless ignition system (DIS) components in the drawing. Place the identifying letter next to the name of each component.

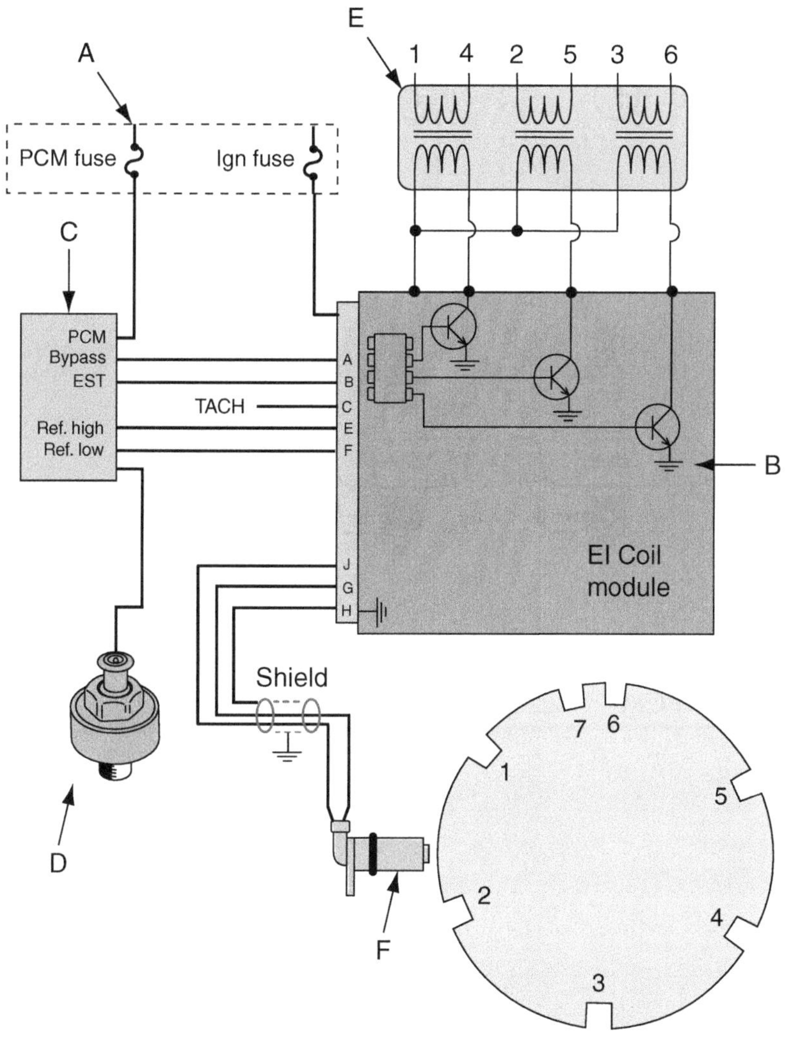

Crankshaft Position Sensor	______	Coil Module	______	Knock Sensor	______
Coil Pack	______	Fuse Block	______	PCM	______

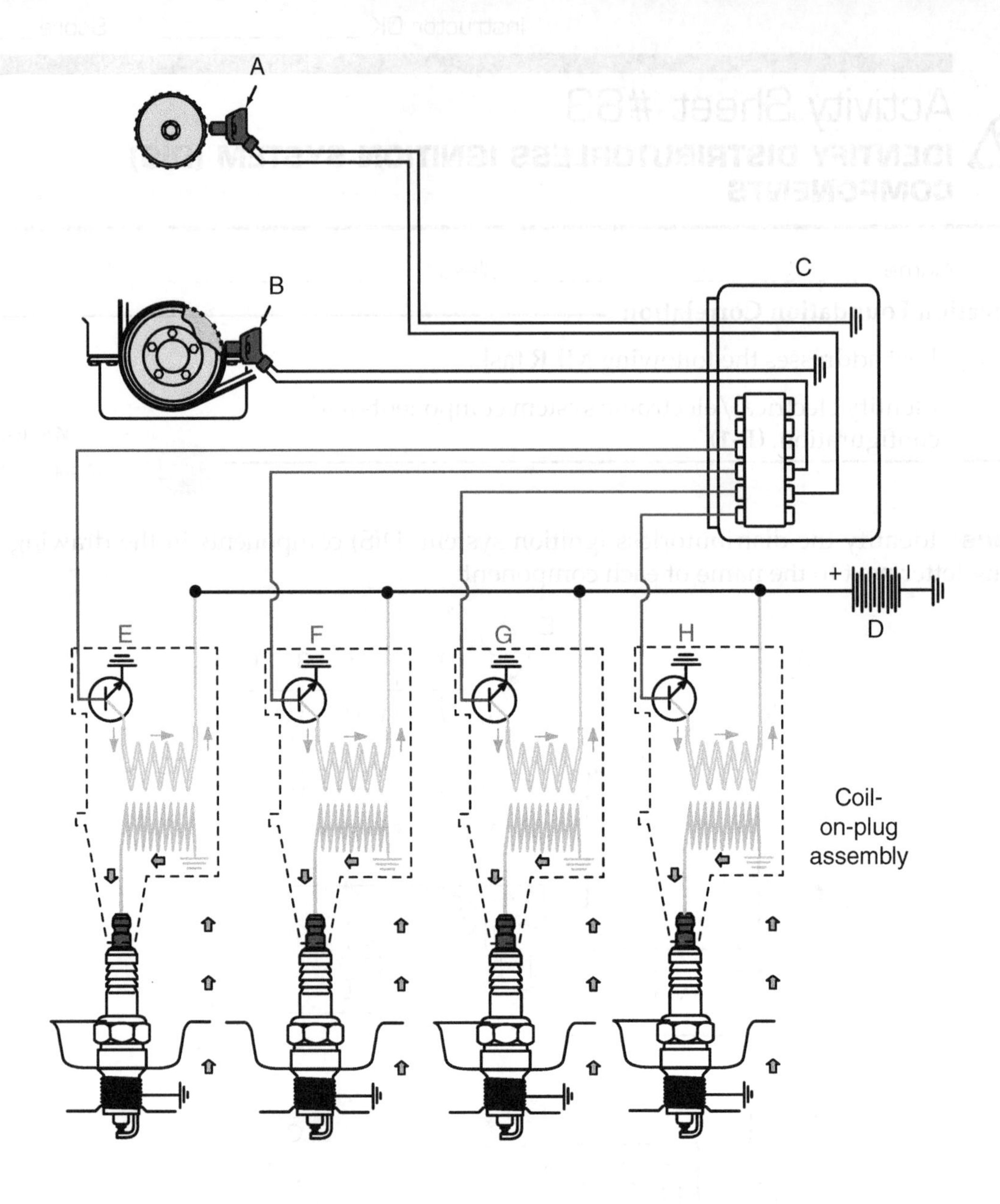

Coil 1	______	Coil 2	______	Coil 3	______	Coil 4	______
Battery	______	PCM	______	CKP	______	CMP	______

STOP

Instructor OK ______________ Score ____________

Activity Sheet #84

IDENTIFY THE BY-PRODUCTS OF COMBUSTION

Name ______________________________ Class ____________________

Directions: Identify the by-products of combustion in the drawing. Place the identifying letter next to the chemical produced.

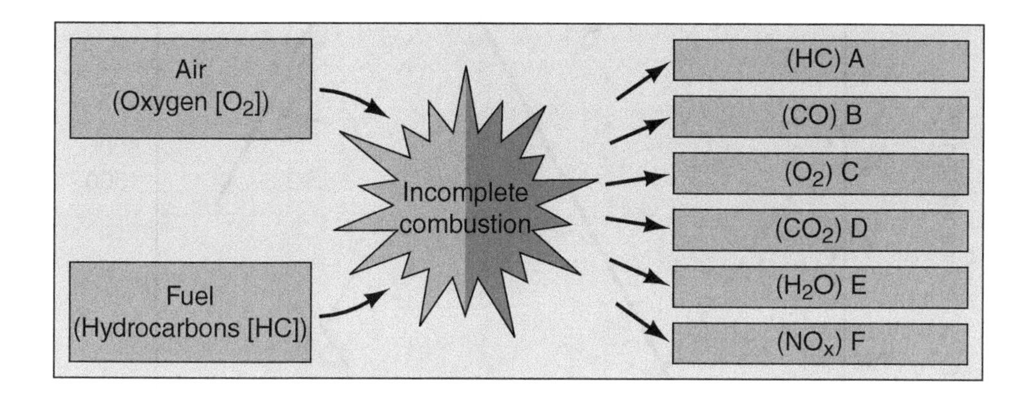

Carbon Monoxide (CO)	______	Oxides of Nitrogen (NO_x)	______	Hydrocarbons (HC)	______
Carbon Dioxide (CO_2)	______	Oxygen (O_2)	______	Water (H_2O)	______

Some of the by-products of combustion from an automobile engine are health hazards and others are not. List the hazardous and nonhazardous by-products of combustion below.

Hazardous	Nonhazardous
______________________	______________________
______________________	______________________
______________________	______________________
______________________	______________________
______________________	______________________

TURN

Directions: Identify the by-products of combustion relative to the air-fuel ratio. Place the identifying letter next to the by-product represented by each of the lines on the graph.

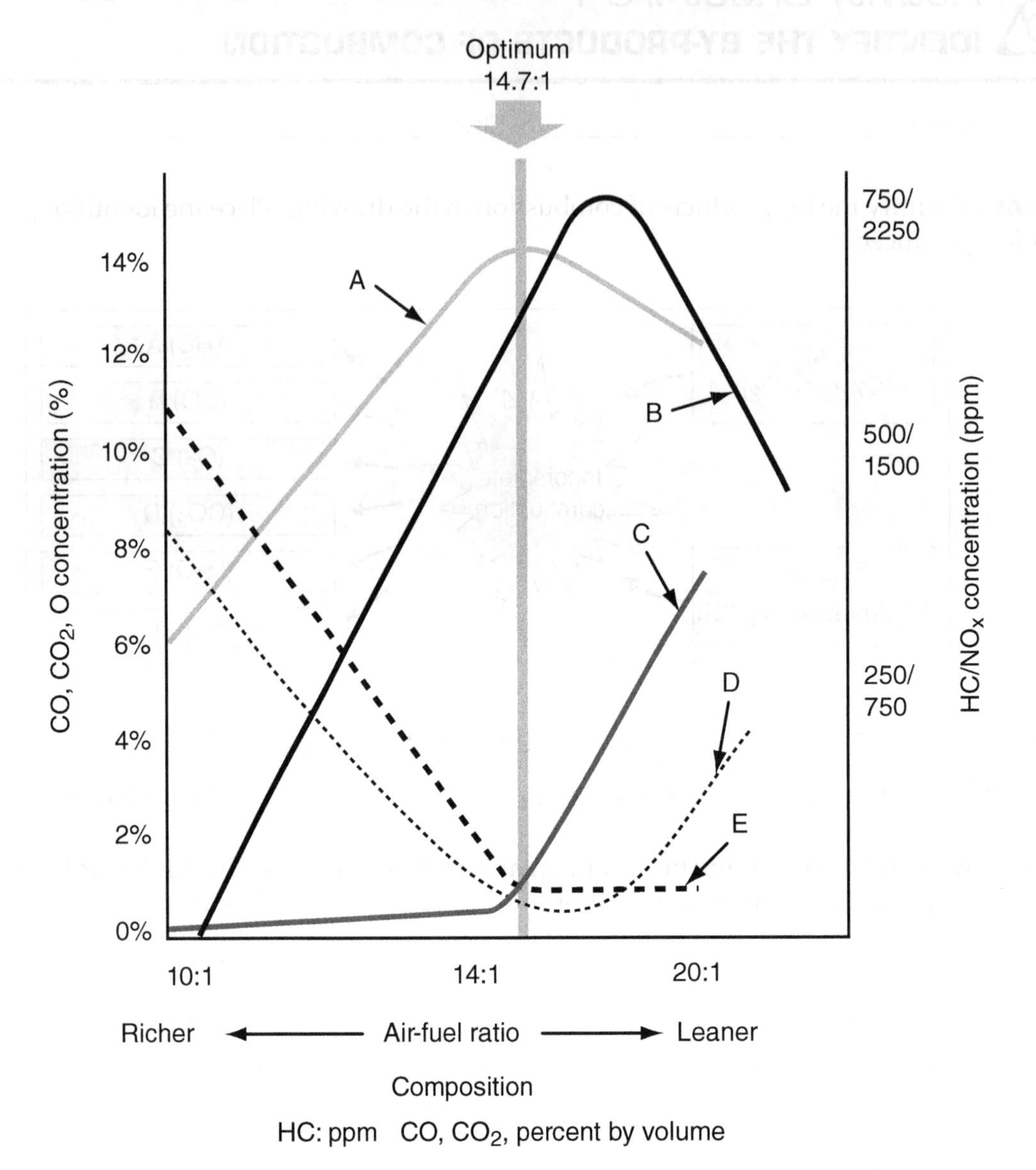

Carbon Dioxide (CO_2) ______ Hydrocarbons (HC) ______ Carbon Monoxide (CO) ______

Oxides of Nitrogen (NO_x) ______ Oxygen (O_2) ______

STOP

Instructor OK ______________ Score ____________

Activity Sheet #85

IDENTIFY EMISSION CONTROL SYSTEM PARTS

Name ______________________________ Class ______________________

ASE Education Foundation Correlation

This activity sheet addresses the following **MLR** task:

VIII.D.1 Inspect, test, and service positive crankcase ventilation (PCV) filter/breather, valve, tubes, orifices, and hoses; perform necessary action. **(P-2)**

This activity sheet addresses the following **AST** tasks:

VIII.E.2 Inspect, test, service, and/or replace positive crankcase ventilation (PCV) filter/breather, valve, tubes, orifices, and hoses; perform needed action. **(P-2)**

VIII.E.3 Diagnose emissions and driveability concerns caused by the exhaust gas recirculation (EGR) system; inspect, test, service and/or replace electrical/electronic sensors, controls, wiring, tubing, exhaust passages, vacuum/pressure controls, filters, and hoses of exhaust gas recirculation (EGR) system; determine needed action. **(P-3)**

VIII.E.4 Inspect and test electrical/electronically-operated components and circuits of secondary air injection systems; determine needed action. **(P-3)**

VIII.E.5 Diagnose emission and driveability concerns caused by the catalytic converter system; determine needed action. **(P-3)**

VIII.E.6 Inspect and test components and hoses of the evaporative emissions control (EVAP) system; determine needed action. **(P-1)**

This activity sheet addresses the following **MAST** tasks:

VIII.E.2 Inspect, test, service, and/or replace positive crankcase ventilation (PCV) filter/breather, valve, tubes, orifices, and hoses; perform needed action. **(P-2)**

VIII.E.3 Diagnose emissions and driveability concerns caused by the exhaust gas recirculation (EGR) system; inspect, test, service and/or replace electrical/electronic sensors, controls, wiring, tubing, exhaust passages, vacuum/pressure controls, filters, and hoses of exhaust gas recirculation (EGR) systems; determine needed action. **(P-2)**

VIII.E.4 Diagnose emissions and driveability concerns caused by the secondary air injection system; inspect, test, repair, and/or replace electrical/electronically-operated components and circuits of secondary air injection systems; determine needed action. **(P-2)**

VIII.E.5 Diagnose emissions and driveability concerns caused by the evaporative emissions control (EVAP) system; determine needed action. **(P-1)**

VIII.E.6 Diagnose emission and driveability concerns caused by catalytic converter system; determine needed action. **(P-2)**

We Support ASE Education Foundation

Directions: Identify the emission control system or component in the drawings. Place the identifying letter next to the correct name on the next page.

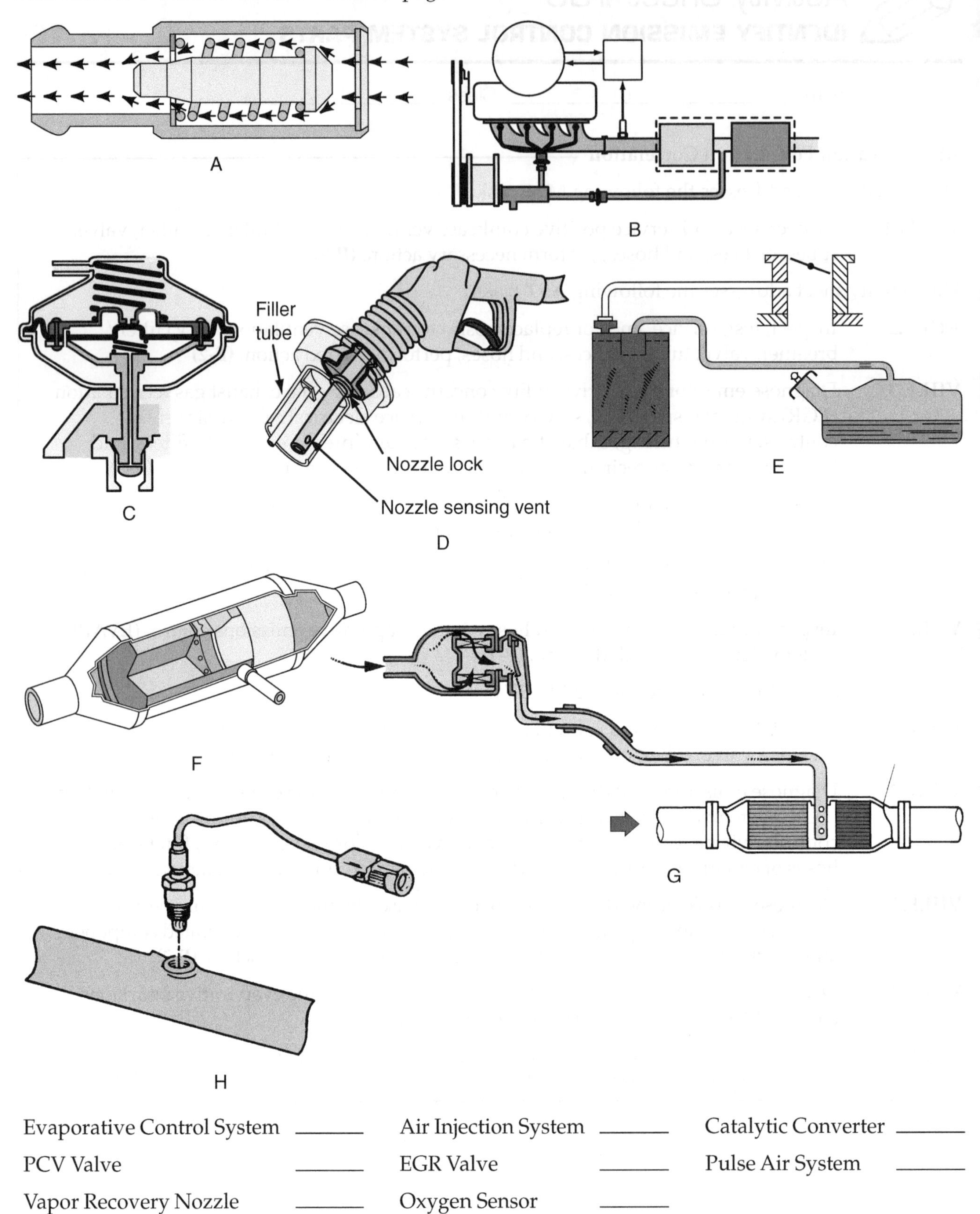

Evaporative Control System	______	Air Injection System	______	Catalytic Converter	______
PCV Valve	______	EGR Valve	______	Pulse Air System	______
Vapor Recovery Nozzle	______	Oxygen Sensor	______		

STOP

Instructor OK ______________ Score ____________

Activity Sheet #86

EMISSION CONTROL SYSTEMS

Name______________________ Class______________

ASE Education Foundation Correlation

This activity sheet addresses the following **MLR** task:

VIII.D.1 Inspect, test, and service positive crankcase ventilation (PCV) filter/breather, valve, tubes, orifices, and hoses; perform necessary action. **(P-2)**

This activity sheet addresses the following **AST** tasks:

VIII.E.2 Inspect, test, service, and/or replace positive crankcase ventilation (PCV) filter/breather, valve, tubes, orifices, and hoses; perform needed action. **(P-2)**

VIII.E.3 Diagnose emissions and driveability concerns caused by the exhaust gas recirculation (EGR) system; inspect, test, service and/or replace electrical/electronic sensors, controls, wiring, tubing, exhaust passages, vacuum/pressure controls, filters, and hoses of exhaust gas recirculation (EGR) system; determine needed action. **(P-3)**

VIII.E.4 Inspect and test electrical/electronically-operated components and circuits of secondary air injection systems; determine needed action. **(P-3)**

VIII.E.5 Diagnose emission and driveability concerns caused by the catalytic converter system; determine needed action. **(P-3)**

VIII.E.6 Inspect and test components and hoses of the evaporative emissions control (EVAP) system; determine needed action. **(P-1)**

This activity sheet addresses the following **MAST** tasks:

VIII.E.2 Inspect, test, service, and/or replace positive crankcase ventilation (PCV) filter/breather, valve, tubes, orifices, and hoses; perform needed action. **(P-2)**

VIII.E.3 Diagnose emissions and driveability concerns caused by the exhaust gas recirculation (EGR) system; inspect, test, service and/or replace electrical/electronic sensors, controls, wiring, tubing, exhaust passages, vacuum/pressure controls, filters, and hoses of exhaust gas recirculation (EGR) systems; determine needed action. **(P-2)**

VIII.E.4 Diagnose emissions and driveability concerns caused by the secondary air injection system; inspect, test, repair, and/or replace electrical/electronically-operated components and circuits of secondary air injection systems; determine needed action. **(P-2)**

VIII.E.5 Diagnose emissions and driveability concerns caused by the evaporative emissions control (EVAP) system; determine needed action. **(P-1)**;

VIII.E.6 Diagnose emission and driveability concerns caused by catalytic converter system; determine needed action. **(P-2)**

We Support ASE | Education Foundation

Directions: Match the words on the left to the descriptions on the right. Write the letter for the correct word on the line provided. For the terms that you are not certain of, use the glossary in your textbook.

A. Underhood label	______ Pollutant formed under high heat and pressure in the engine
B. Oxidizing	______ Uniting fuel with oxygen during combustion
C. Light off	______ The ideal air-fuel ratio, 14.7:1 by weight, in which all of the oxygen is consumed in the burning of the fuel
D. Pyrometer	______ A good rich indicator
E. Infrared thermometer	______ Term for carbon monoxide measurement
F. Combustion	______ Thermal vacuum switch
G. Flame front	______ A label found on cars since 1972 that describes such things as the size of the engine, ignition timing specifications, idle speed, valve lash clearance adjustment, and the emission devices that are included on the engine
H. Stoichiometric	______ Deceleration ___ can be due to a faulty air injection system valve
I. Oxides of nitrogen	______ Hydrocarbons are measured in ______
J. Idle should drop	______ Burning of fuel
K. TVS	______ CO_2 reading from an engine
L. 500°F	______ When the catalytic converter becomes hot enough and begins to oxidize pollutants
M. PPM	______ Condition under which NO_x is tested
N. CO_2	______ When heat ignites molecules next to already burning molecules and a chain reaction takes place, which results in a flame expanding evenly across the cylinder
O. 13% to 16%	______ A good lean indicator
P. Percentage	______ A temperature measuring device that is touched against a surface to obtain a reading
Q. Air pump	______ A thermometer that takes a temperature reading when it is aimed toward a surface
R. Hydrocarbons	______ Engine idle when the PCV valve is plugged
S. O_2 reading 2%–5%	______ The catalytic converter must be heated to approximately ______ °F before it starts to work
T. Backfire	______ Disabled during emission analyzer diagnosis
U. CO	______ Analyzer oxygen reading with exhaust dilution or smog pump working
V. O_2	______ If an engine that will not start is being cranked with an emission analyzer connected, the presence of what gas tells you that there is a fuel supply?
W. Under load	______ Good indicator of engine efficiency

Activity Sheet #87

IDENTIFY ENGINE PERFORMANCE TEST INSTRUMENTS

Name ______________________________ Class ____________________

Directions: Identify the engine performance test instruments given below. Place the identifying letter next to the name of each item on each next page.

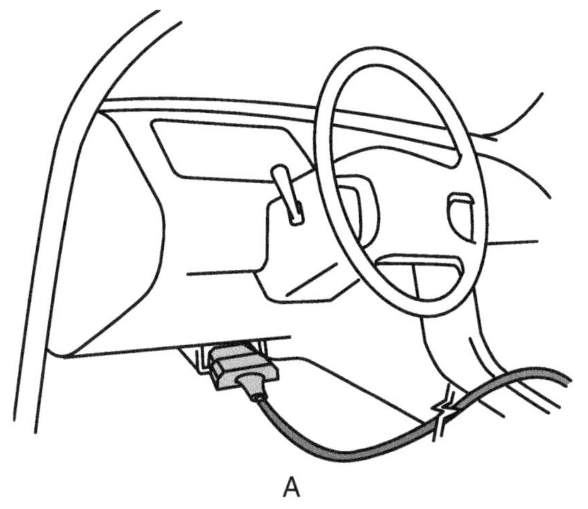

A

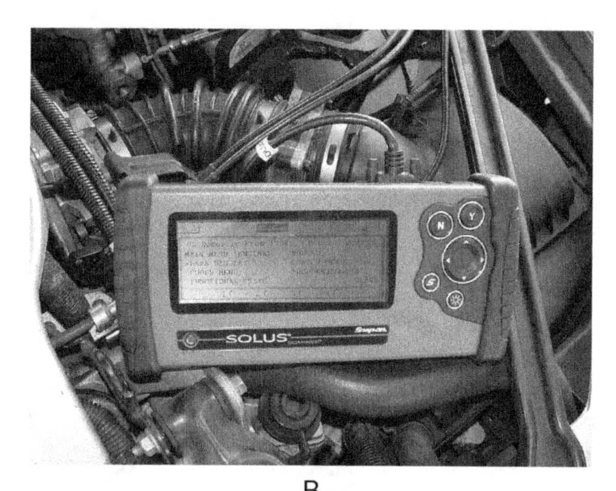

B

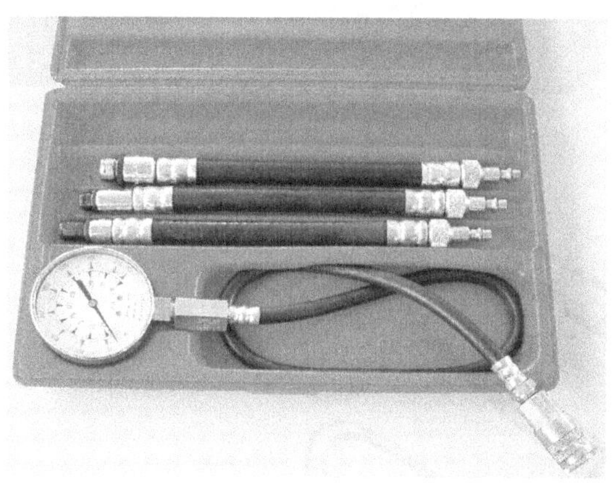

C

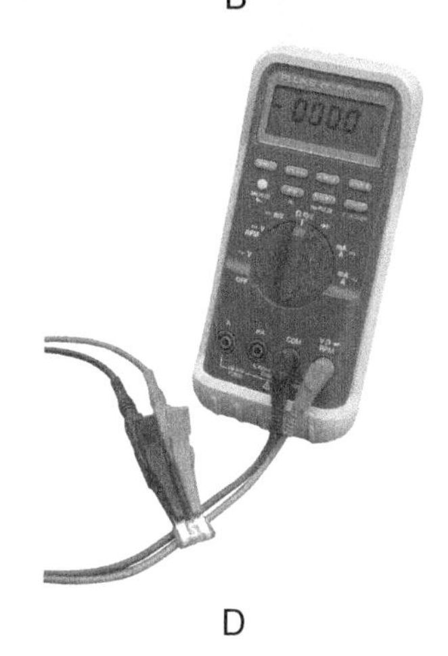

D

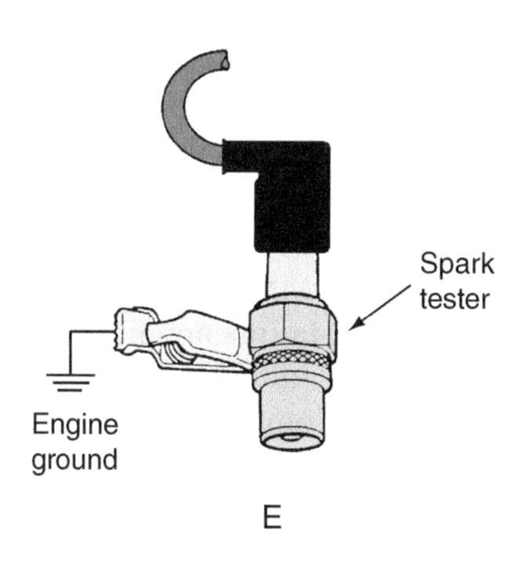

E

F

Photos by Tim Gilles

G

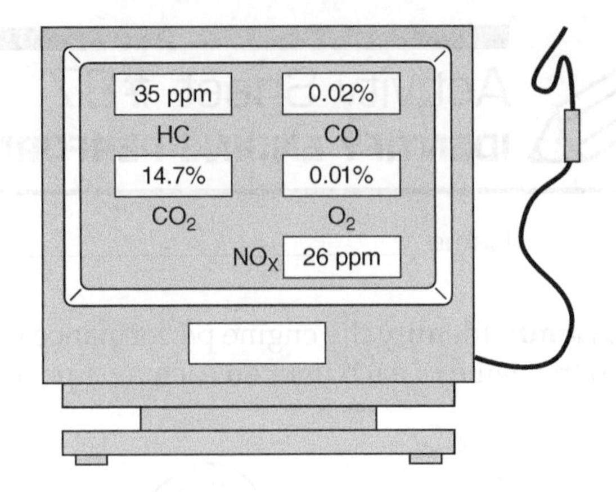

H

I

J

K

Freeze Frame Data	______	Spark Tester	______	DMM	______
Vacuum Gauge	______	Scan Tool	______	Malfunction Indicator Light	______
Cylinder Leakage Tester	______	Combustion Leak Tester	______	Compression Tester	______
Emission Analyzer	______	Logic Probe	______		

STOP

Instructor OK ________________ Score ____________

Activity Sheet #88

IDENTIFY ENGINE LEAKS

Name ______________________________ Class ____________________

ASE Education Foundation Correlation

This activity sheet addresses the following **MLR** task:

I.A.3 Inspect engine assembly for fuel, oil, coolant, and other leaks; determine necessary action. **(P-1)**

This activity sheet addresses the following **AST/MAST** task:

I.A.4 Inspect engine assembly for fuel, oil, coolant, and other leaks; determine needed action. **(P-1)**

Directions: Identify the places where an engine's gaskets can leak. Place the identifying letter next to the correct description.

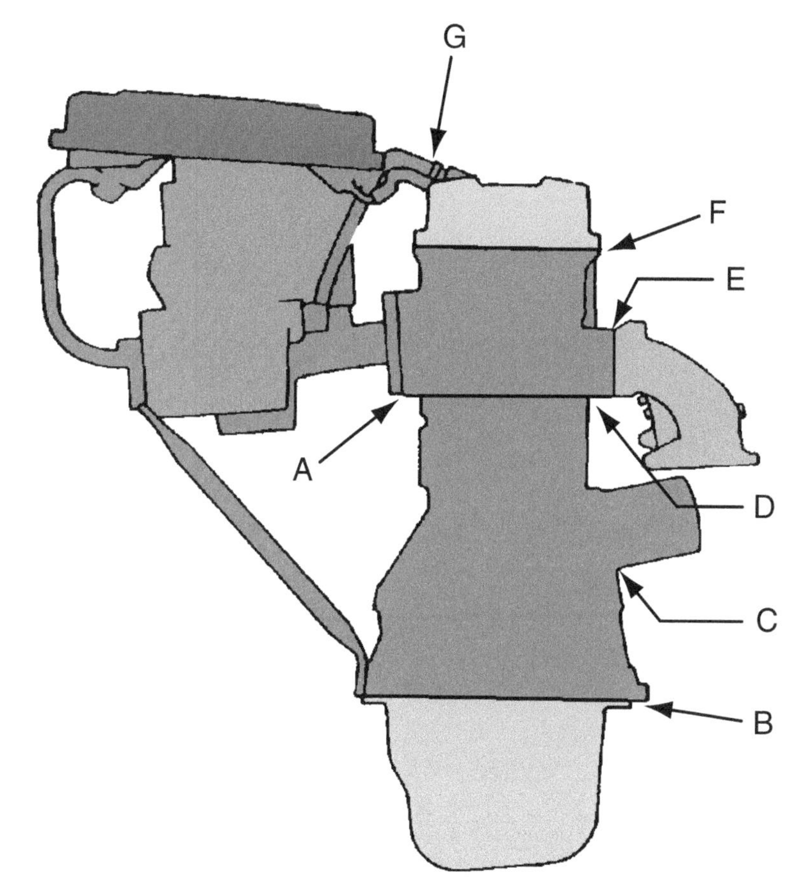

Oil Pan	______	Intake Manifold	______	Exhaust Manifold	______
Oil Filter	______	Valve Cover	______	Head Gasket	______
PCV	______				

STOP

Instructor OK ______________ Score ____________

Activity Sheet #89

IDENTIFY PARTS OF A COMPUTER SYSTEM

Name ______________________________ Class ______________________

Directions: Identify parts of a computer system in the drawing. Place the identifying letter next to the name of each part.

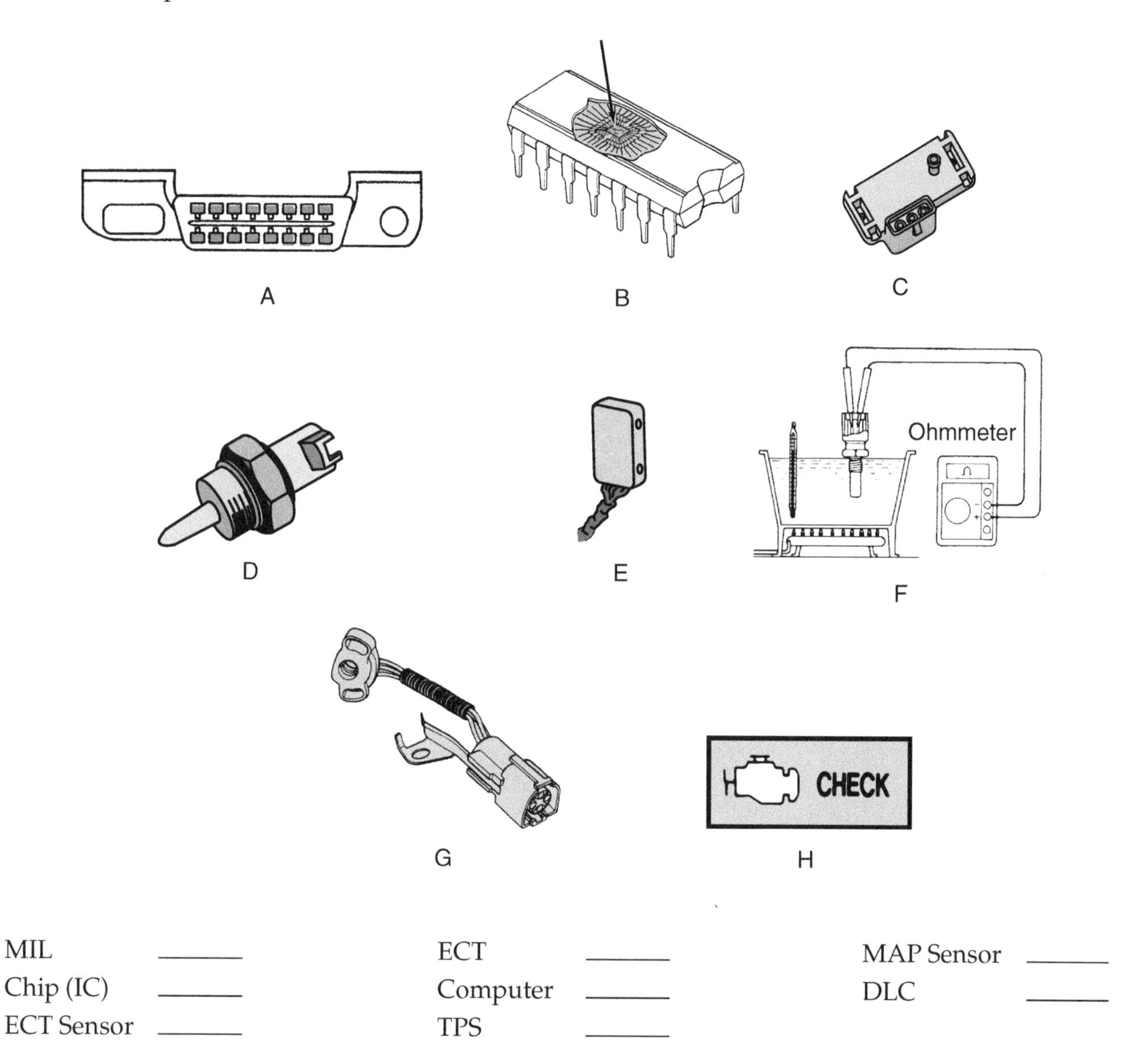

MIL	______	ECT	______	MAP Sensor	______
Chip (IC)	______	Computer	______	DLC	______
ECT Sensor	______	TPS	______		

STOP

Instructor OK ______________ Score ____________

Activity Sheet #90

IDENTIFY COMPUTER CONTROLS

Name ______________________________ Class ______________________

Directions: Identify computer *inputs* by checking the box next to items that require a sensor. Identify computer *outputs* by checking the actuator box.

	Sensor	Actuator
Crankshaft position	☐	☐
Torque converter clutch	☐	☐
Air temperature	☐	☐
Vehicle speed	☐	☐
Canister purge	☐	☐
Manifold pressure	☐	☐
Mass airflow	☐	☐
Spark timing	☐	☐
Idle air control	☐	☐
Exhaust oxygen	☐	☐
Air conditioning	☐	☐
Diagnostic data	☐	☐
Engine rpm	☐	☐
Engine cooling fan	☐	☐
Throttle position	☐	☐
Engine knock	☐	☐
Fuel injector	☐	☐
Coolant temperature	☐	☐
EGR valve	☐	☐
EGR position	☐	☐
Relay	☐	☐
Stepper motor	☐	☐
Camshaft position	☐	☐

STOP

Instructor OK ______________ Score __________

Activity Sheet #91

COMPUTER AND ELECTRONICS

Name ______________________ Class ______________

Directions: Match the words on the left to the descriptions on the right. Write the letter for the correct word on the line provided. For the terms that you are not certain of, use the glossary in your textbook.

Term		Description
A. Semiconductor	______	Information stored as electronic signals
B. Electron theory	______	The calculating and decision-making chip in the computer
C. Twisted pair	______	A variable resistor used to measure linear or rotary motion
D. Zener diode	______	Cycles per second
E. LED	______	Can be either a conductor or an insulator
F. Protocol	______	Malfunction indicator lamp
G. Microprocessor	______	Electronically erasable programmable read-only memory
H. Hardware	______	Electrons flow from − to +
I. Internet	______	Its resistance changes as its temperature changes
J. RAM	______	Diodes that give off light
K. ROM	______	Used as an electronic voltage regulator
L. LAN	______	Guidelines that provide standardization of terms
M. Actuator	______	The mechanical parts of an electronic system
N. Thermistor	______	A device that uses a piezoelectric element to sense vibration
O. Potentiometer	______	P-type and N-type crystals back to back
P. CAN	______	This is like a notepad that you can read from and write to
Q. Hertz	______	Permanently programmed information
R. Bluetooth	______	Sharing of electrical circuit conductors
S. OBD II	______	A device that is controlled by the computer
T. MIL	______	A complete miniaturized electric circuit
U. Multiplexing	______	Crystals that develop a voltage on their surfaces when pressure is applied
V. Diode	______	Uses a personal area network (PAN)

W. Integrated circuit ______ Type of network used in automobiles

X. Software ______ Most common network used in the home

Y. EEPROM ______ Wide area network (WAN) is the type of network used by the ___

Z. Piezoelectric ______ Language used for modules to communicate

AA. Knock sensor ______ A method of wiring to prevent radio interference

Instructor OK ______________ Score __________

Activity Sheet #92

EMISSIONS AND ON-BOARD DIAGNOSTICS (OBD II)

Name ______________________________ Class ______________________

Directions: Match the words on the left to the descriptions on the right. Write the letter for the correct word on the line provided. For the terms that you are not certain of, use the glossary in your textbook.

A. OBD	_______ Society of Automotive Engineers
B. OBD II	_______ Data link connector
C. Scan tool	_______ Measures emissions in grams per mile
D. Monitors	_______ Standardization of terms
E. FTP standard	_______ All monitors must operate to complete
F. FTP	_______ Powertrain control module
G. DLC	_______ Malfunction indicator lamp
H. Post CAT O_2 sensor	_______ Used to access stored codes
I. SAE	_______ Processed data used by the engine
J. SAE J1930	_______ Used primarily to improve air quality
K. Standard Communication Protocol	_______ Self-detects exhaust emission increase of over 50%
L. PCM	_______ Senses catalytic converter efficiency
M. DLC (OBD II)	_______ Used to look for malfunctions
N. VIN (OBD II)	_______ Requires various emission monitors to operate to complete
O. PID	_______ Used to indicate monitors are clear
P. Freeze frame data	_______ Requires manufacturers to use the same computer language
Q. MIL	_______ DTC that is always emissions related
R. Warm-up cycle	_______ Found under the left side of the dash
S. Trip	_______ Checks for leaks no larger than 0.040″
T. Drive cycle	_______ Automatically transmitted to the scan tool
U. Pending code	_______ Used to look for malfunctions
V. Type "A" code	_______ The speed of oxygen sensor oscillations
W. Monitor	_______ A type of automotive computer network
X. Readiness indicators	_______ Digital storage oscilloscope
Y. Comprehensive component	_______ Stored PIDs monitor
Z. Evaporative monitor	_______ Federal Test Procedure

AA. Switch ratio	______	Occurs every time the engine cools off and temperature rises to at least 40°F
AB. DSO	______	Set after first time a fault is identified
AC. CAN	______	Checks devices not tested by other OBD II monitors

STOP

Instructor OK ____________ Score __________

Activity Sheet #93

IDENTIFY ON-BOARD DIAGNOSTIC II COMPONENTS

Name ______________________ Class ________________

Directions: Identify the OBD II components in the drawings. Place the identifying letter next to the name of each component listed on the next page.

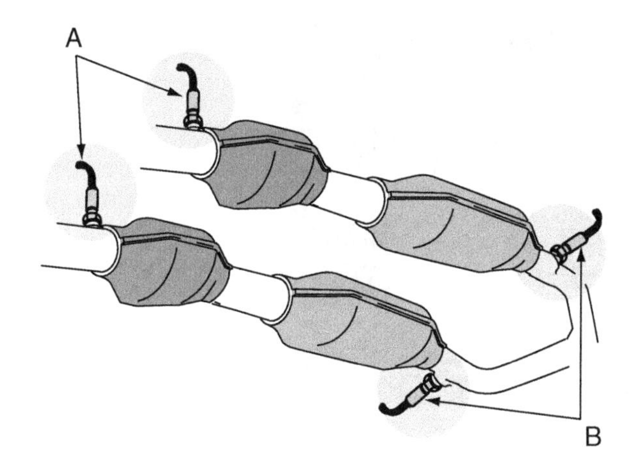

Injector	2.9 ms
Ign. Advance	7.0 des
Calc. Load	17%
Engine Speed	715 rpm
Coolant Temp.	190.4°F
CTP Switch	On
Vehicle Speed	0 mph
Starter Signal	Off
A/C Signal	Off
PNP Switch (new)	On

C

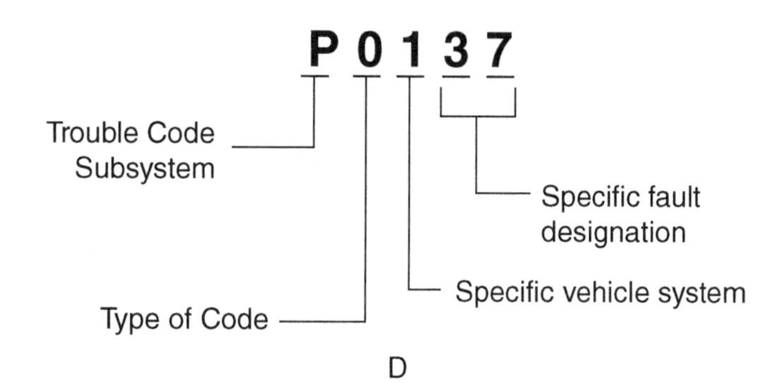

D

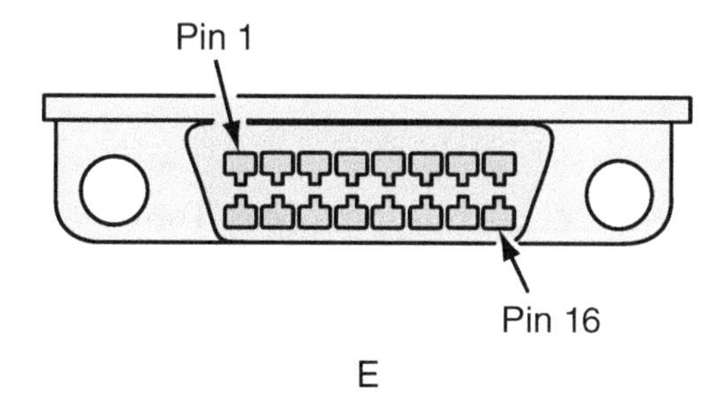

E

(Answer blanks are on next page)

- Misfire
- Comp component
- Heated catalyst
- AIR
- O_2 sensor
- EGR system
- Fuel system
- Catalyst
- EVAP
- A/C refrigerant
- O_2 sensor heater

F

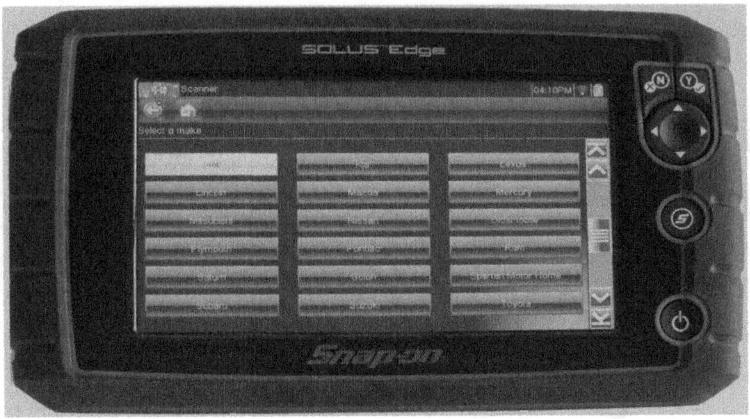

G

Photo by Tony Mollica

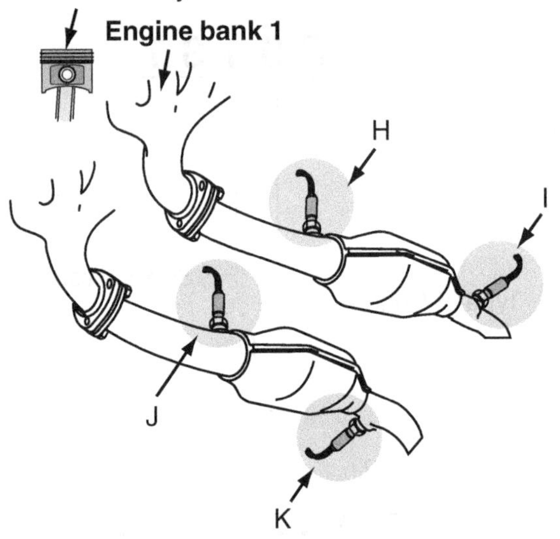

Monitors	______	Scan Tool	______
Upstream HO_2 Sensor	______	Downstream HO_2 Sensor	______
Freeze Frame Data	______	OBD II Trouble Codes	______
Bank 2 Sensor 2	______	Bank 1 Sensor 2	______
Diagnostic Link Connector	______	Bank 2 Sensor 1	______
Bank 1 Sensor 1	______		

Part I

Activity Sheets

Section Nine

Automotive Engine Service and Repair

Instructor OK ______________ Score ____________

Activity Sheet #94

ENGINE MECHANICAL PROBLEM DIAGNOSIS

Name ______________________________ Class ______________________

ASE Education Foundation Correlation

This activity sheet addresses the following **AST/MAST** task:

VIII.A.3 Diagnose abnormal engine noises or vibration concerns; determine needed action. (P-3)

We Support ASE | Education Foundation

Directions: Match the words on the left to the descriptions on the right. Write the letter for the correct word on the line provided. For the terms that you are not certain of, use the glossary in your textbook.

A. Valve train noise	_____	When a cam lobe wears out
B. Thrust bearing knock	_____	Noises that result from excessive clearance or abnormal combustion
C. Engine knocks	_____	Term used when a crankshaft will not turn
D. Piston slap	_____	Can occur at ½ engine rpm
E. Flat cam	_____	Carbon has built up in the neck area of a valve. What is the probable cause?
F. Rod knock	_____	Problem with an engine that has lower oil pressure at idle speed
G. Hydrolocked engine	_____	A test for oil leakage that uses fluorescent dye and an ultraviolet light source
H. Valve guide seal	_____	Bad valve guide seals could cause oil smoke from the exhaust during ___
I. Burned spark plug	_____	Noise that results from excessive clearance between the piston and the cylinder
J. Front main bearing knock	_____	Damage that results when an engine runs for a long period with an excessively lean air-fuel mixture
K. Deceleration	_____	When coolant or fuel in a cylinder prevents an engine from turning over
L. Black light	_____	Reduced when drive belts are removed
M. Seized engine	_____	Most noticeable when vehicle accelerates from a stop
N. Worn lower main bearings	_____	Sometimes accompanied by low oil pressure at idle

STOP

Part I

Activity Sheets

Section Ten

Brakes and Tires

Instructor OK ______________ Score __________

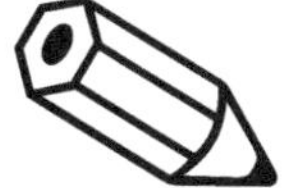

Activity Sheet #95

IDENTIFY BRAKE SYSTEM COMPONENTS

Name ______________________________ Class ____________________

ASE Education Foundation Correlation

This activity sheet addresses the following **MLR** task:

V.A.4 Identify brake system components and configuration. **(P-1)**

The activity sheet addresses the following **AST/MAST** task:

V.A.1 Identify and interpret brake system concerns; determine needed action. **(P-1)**

We Support ASE | Education Foundation

Directions: Identify the brake system components in the drawing. Place the identifying letter next to the name of each component.

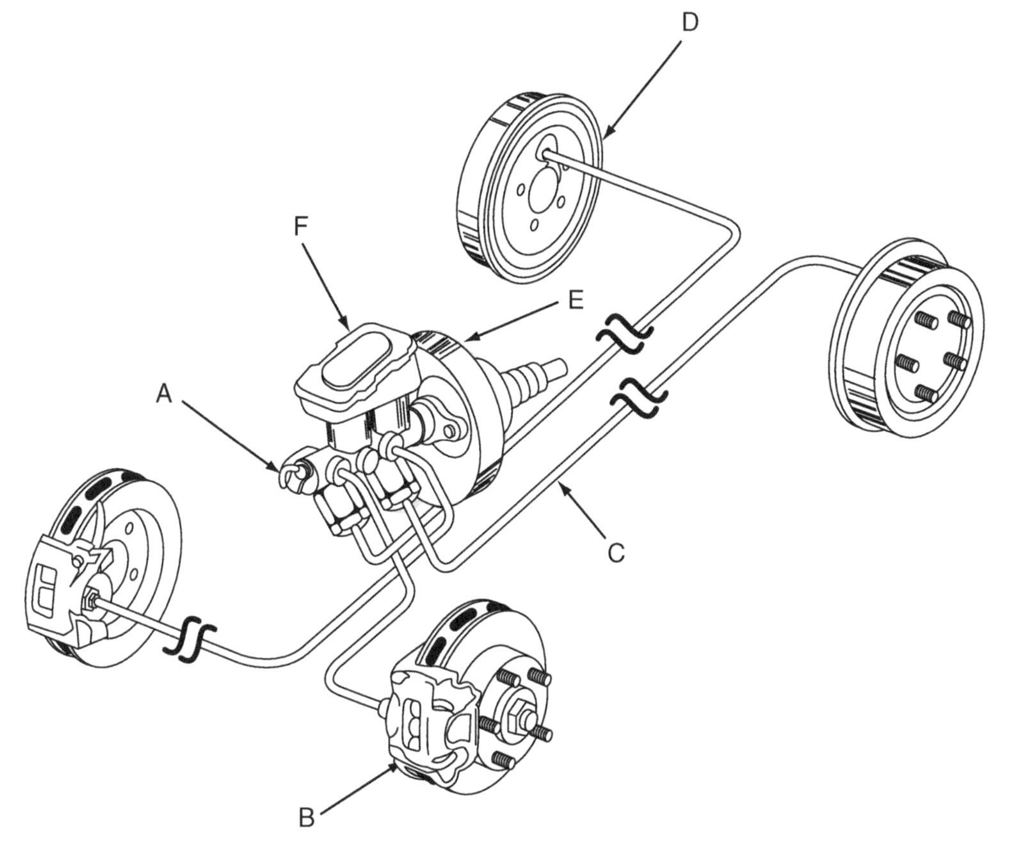

Drum Brakes	______	Brake Line	______	Master Cylinder	______
Disc Brakes	______	Brake Fluid Reservoir	______	Power Brake Booster	______

Instructor OK ______________ Score ______________

Activity Sheet #96

IDENTIFY MASTER CYLINDER COMPONENTS

Name ______________________________ Class ____________________

ASE Education Foundation Correlation

This activity sheet addresses the following **MLR** task:

V.A.4 Identify brake system components and configuration. **(P-1)**

The activity sheet addresses the following **AST/MAST** task:

V.A.1 Identify and interpret brake system concerns; determine needed action. **(P-1)**

We Support ASE Education Foundation

Directions: Identify the master cylinder components in the drawing. Place the identifying letter next to the name of each component.

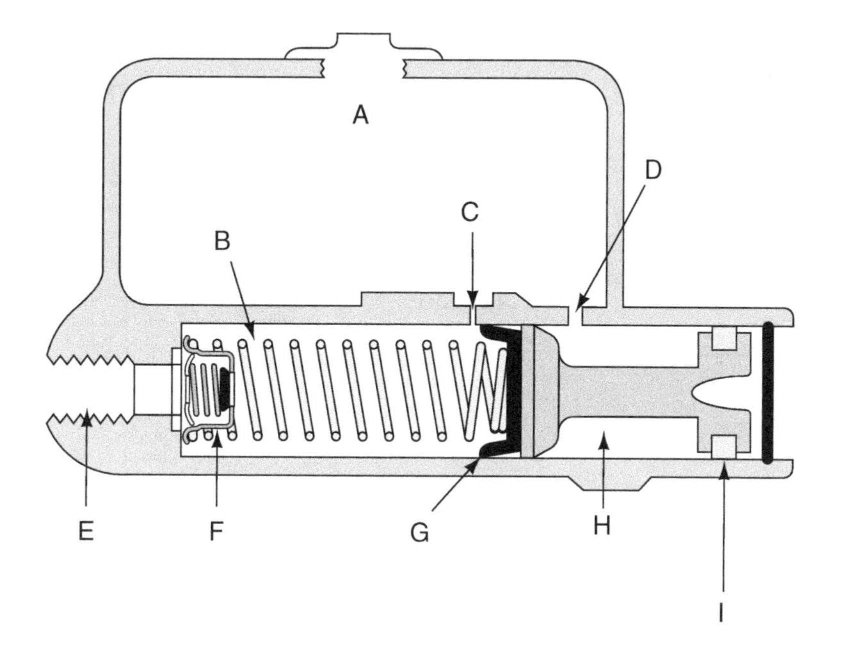

Pressure Chamber	______	Compensating Port	______	Secondary Reservoir	______
Fluid Outlet	______	Reservoir	______	Primary Cup	______
Replenishing Port	______	Secondary Seal	______	Check Valve	______

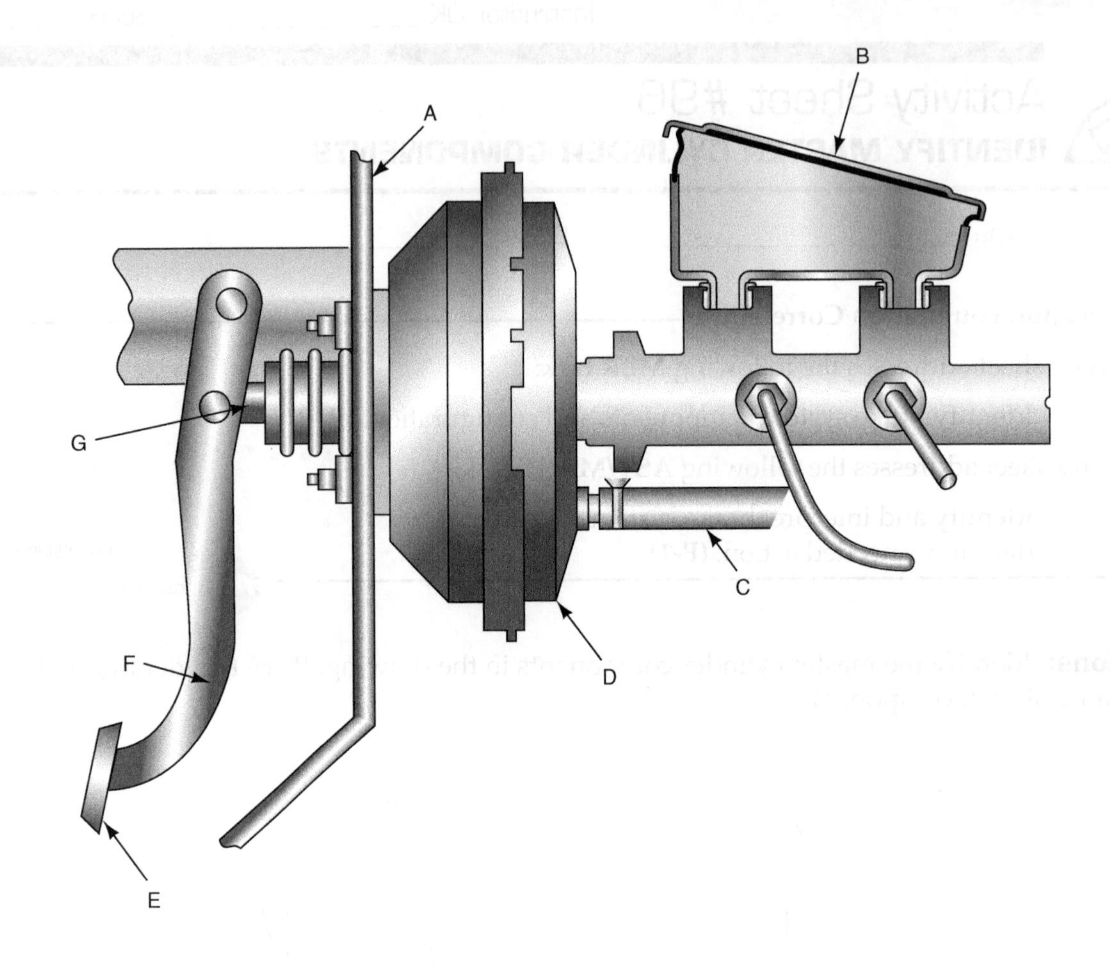

Master Cylinder	______	Power Booster	______	Pushrod	______
Brake Pedal	______	Bulkhead	______	Brake Lever Arm	______
Vacuum Hose	______				

STOP

Instructor OK ______________ Score __________

Activity Sheet #97

IDENTIFY TANDEM MASTER CYLINDER COMPONENTS

Name ______________________________ Class ____________________

ASE Education Foundation Correlation

This activity sheet addresses the following **MLR** task:

V.A.4 Identify brake system components and configuration. **(P-1)**

The activity sheet addresses the following **AST/MAST** task:

V.A.1 Identify and interpret brake system concerns; determine needed action. **(P-1)**

Directions: Identify the tandem master cylinder components in the drawing. Place the identifying letter next to the name of each component.

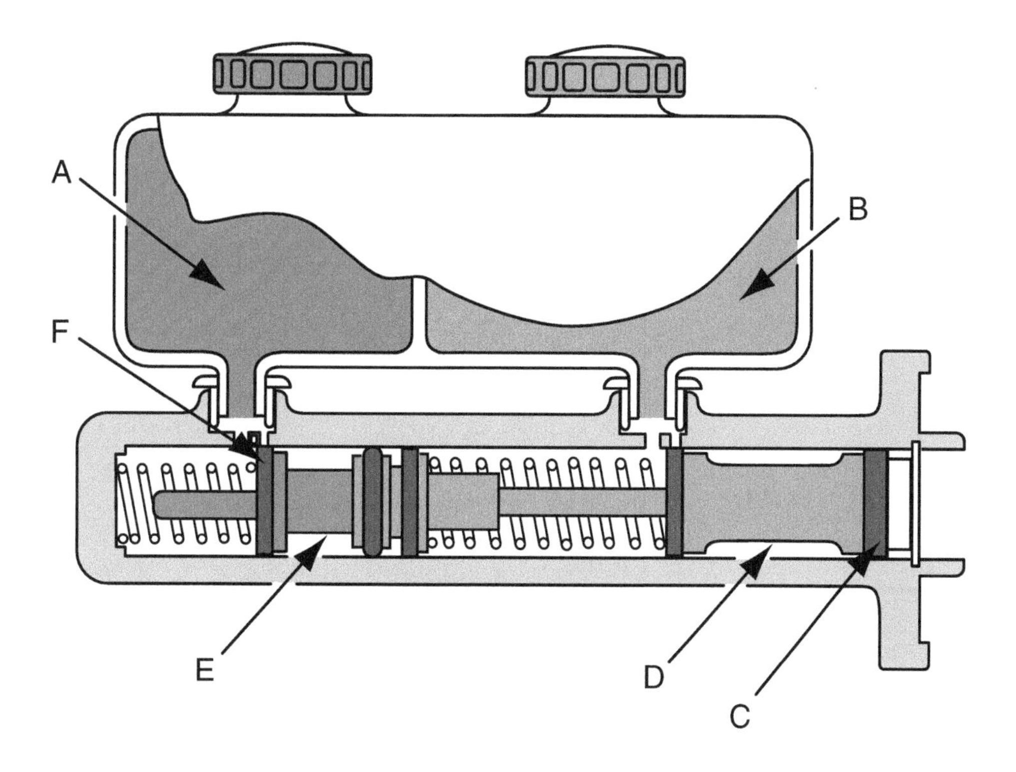

Primary Cup	______	Primary Reservoir	______	Primary Piston	______
Secondary Piston	______	Secondary Cup	______	Secondary Reservoir	______

STOP

Instructor OK ______________ Score ____________

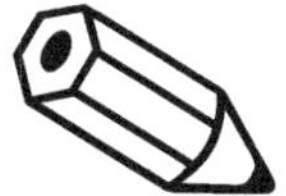

Activity Sheet #98

IDENTIFY WHEEL CYLINDER PARTS

Name ______________________________ Class ____________________

ASE Education Foundation Correlation

This activity sheet addresses the following **MLR** task:

V.A.4 Identify brake system components and configuration. **(P-1)**

The activity sheet addresses the following **AST/MAST** task:

V.A.1 Identify and interpret brake system concerns; determine needed action. **(P-1)**

Directions: Identify the wheel cylinder parts in the drawing. Place the identifying letter next to the name of each part.

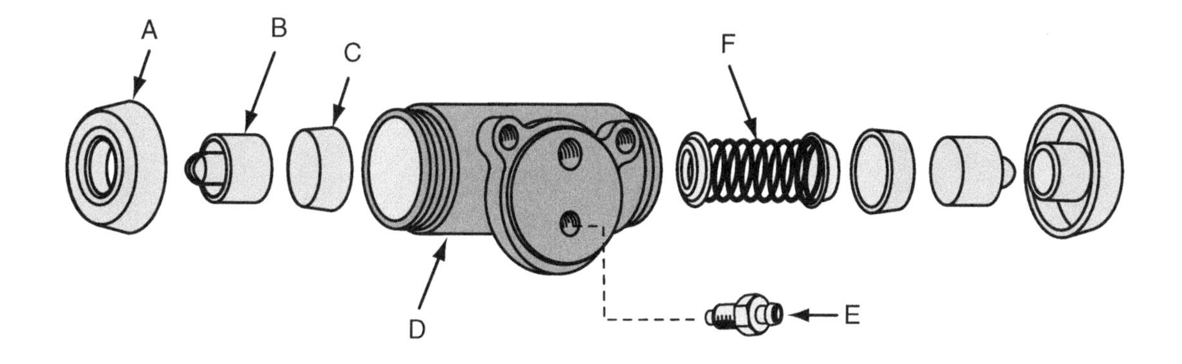

Cylinder	______	Cup	______	Piston	______
Bleed Screw	______	Boot	______	Return Spring	______

STOP

Instructor OK ______________ Score ______________

Activity Sheet #99

IDENTIFY DRUM BRAKE TERMS AND COMPONENTS

Name______________________________ Class____________________

ASE Education Foundation Correlation

This activity sheet addresses the following **MLR** task:

V.A.4 Identify brake system components and configuration. **(P-1)**

The activity sheet addresses the following **AST/MAST** task:

V.A.1 Identify and interpret brake system concerns; determine needed action. **(P-1)**

We Support ASE Education Foundation

Directions: Identify the drum brake components in the drawing. Place the identifying letter next to the name of each component on the next page.

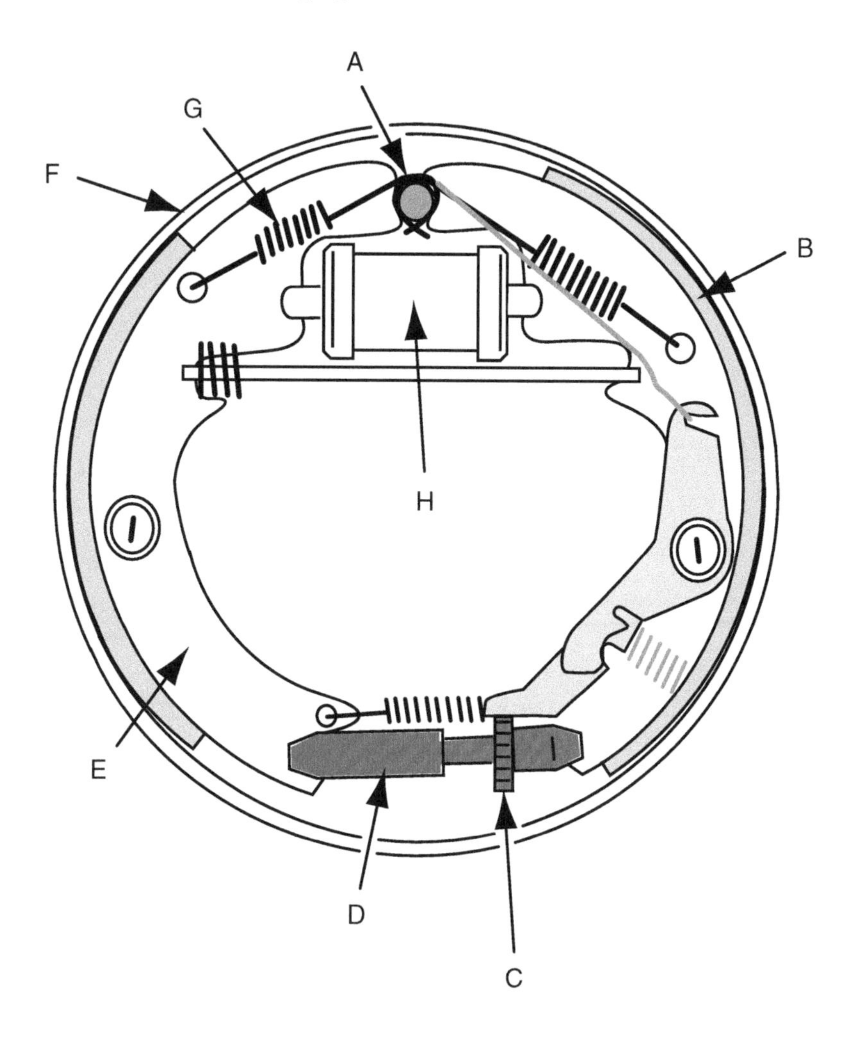

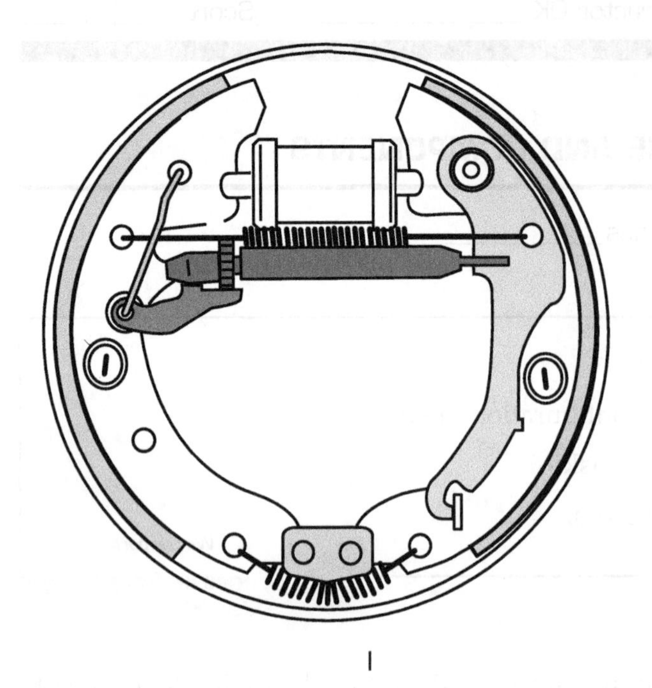

I

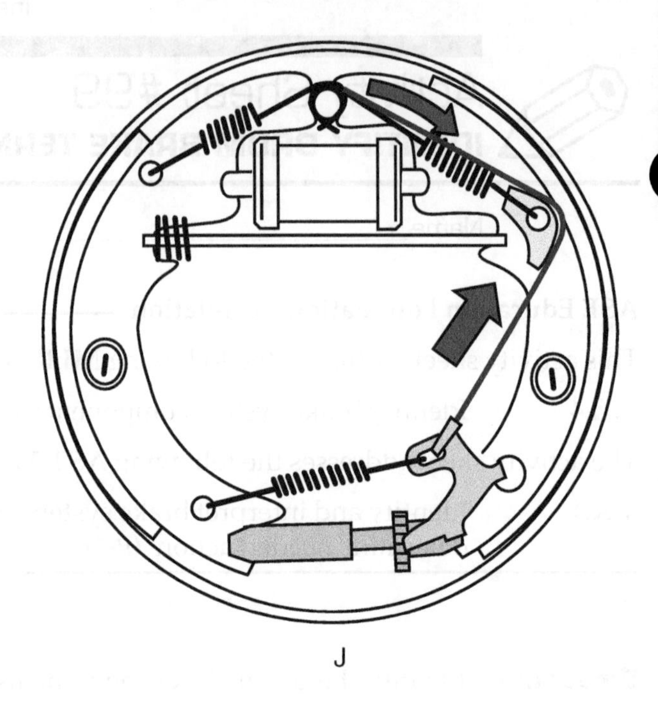

J

Shoe	______	Backing Plate	______	Wheel Cylinder	______
Leading/Trailing	______	Bendix (Dual Servo)	______	Lining	______
Anchor	______	Adjuster	______	Star Wheel	______
Return Spring	______				

STOP

Instructor OK ______________ Score __________

Activity Sheet #100

IDENTIFY DISC BRAKE COMPONENTS

Name______________________________ Class______________________

ASE Education Foundation Correlation

This activity sheet addresses the following **MLR** task:

V.A.4 Identify brake system components and configuration. **(P-1)**

The activity sheet addresses the following **AST/MAST** task:

V.A.1 Identify and interpret brake system concerns; determine needed action. **(P-1)**

We Support ASE Education Foundation

Directions: Identify the disc brake components in the drawing. Place the identifying letter next to the name of each component.

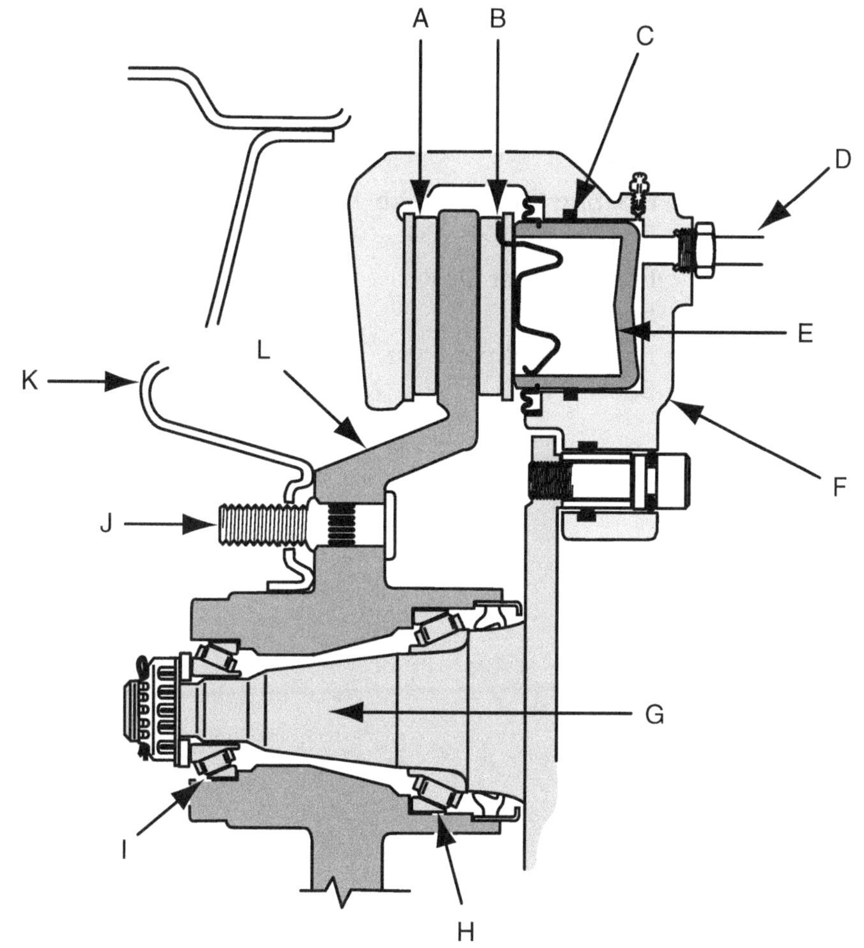

Inner Bearing	_____	Piston Seal	_____	Piston	_____
Wheel Stud	_____	Outboard Brake Pad	_____	Outer Bearing	_____
Inboard Brake Pad	_____	Wheel Rim	_____	Hydraulic Brake Hose	_____
Rotor	_____	Spindle	_____	Caliper	_____

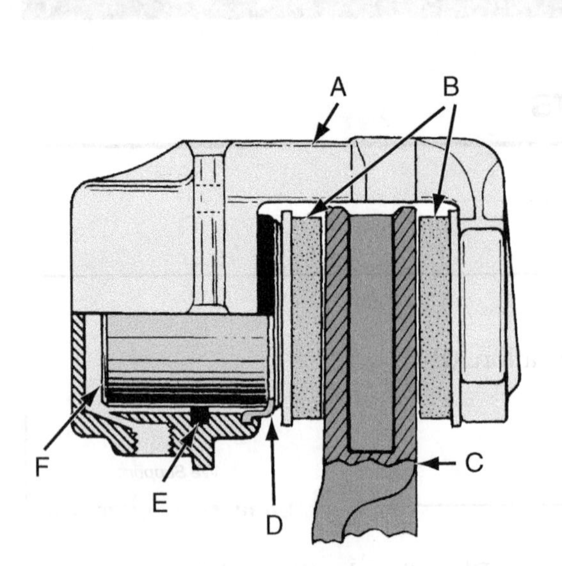

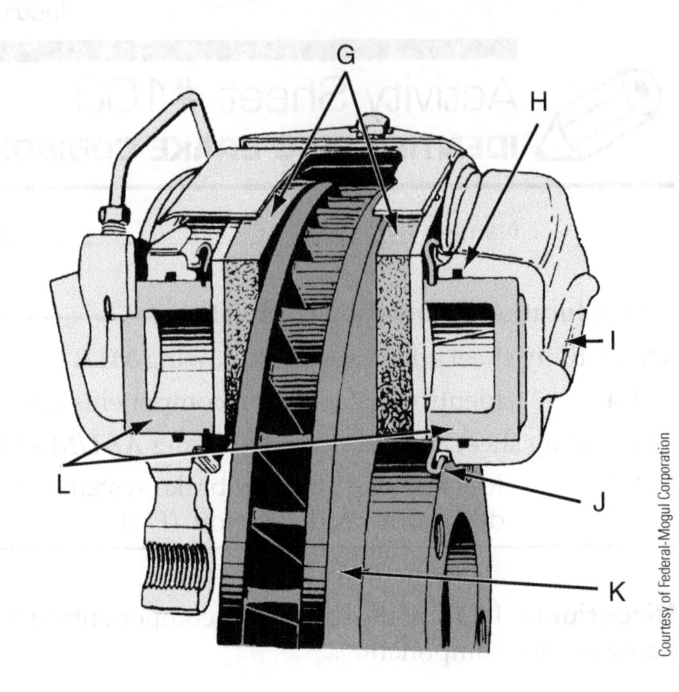

Courtesy of Federal-Mogul Corporation

Floating Caliper ______

Fixed Caliper Brake Pads ______

Floating Caliper Rotor ______

Fixed Caliper Rotor ______

Fixed Caliper Piston Seal ______

Floating Caliper Brake Pads ______

Fixed Caliper Dust Boot ______

Floating Caliper Dust Boot ______

Floating Caliper Piston ______

Fixed Caliper ______

Floating Caliper Piston Seal ______

Fixed Caliper Piston ______

STOP

Instructor OK ______________ Score ______________

Activity Sheet #101

IDENTIFY BRAKE HYDRAULIC COMPONENTS

Name ____________________________ Class ____________________

ASE Education Foundation Correlation

This activity sheet addresses the following **MLR** tasks:

V.A.4 Identify brake system components and configuration. **(P-1)**

V.B.5 Identify components of hydraulic brake warning light system. **(P-3)**

This activity sheet addresses the following **AST/MAST** tasks:

V.A.1 Identify and interpret brake system concerns; determine needed action. **(P-1)**

V.B.10 Inspect, test, and/or replace components of brake warning light system. **(P-3)**

V.B.11 Identify components of hydraulic brake warning light system. **(P-2)**

We Support ASE Education Foundation

Directions: Identify the brake hydraulic components in the drawings. Place the identifying letter next to the name of each component in the answer spaces on the next page.

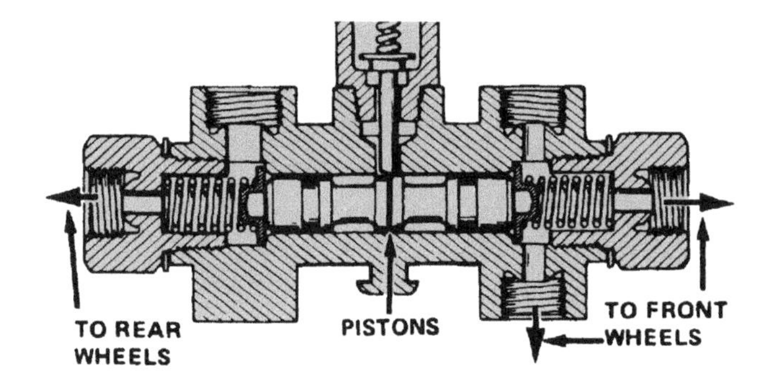

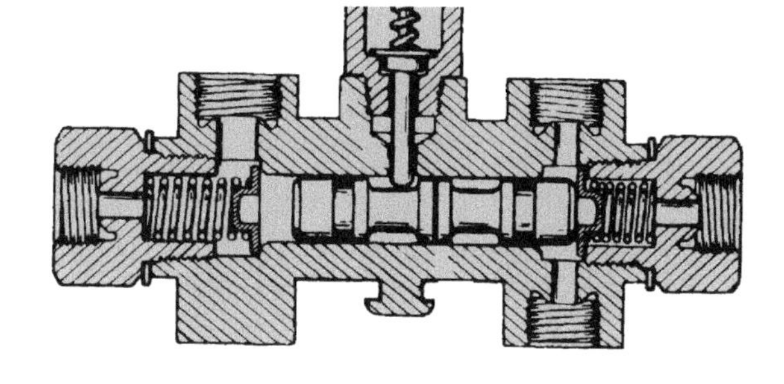

A

Related Questions:

1. Which of the above valves has a pressure loss? a. left ______ and b. right ______
2. Which side of the brake system has a pressure loss? a. front ______ and b. rear ______

TURN

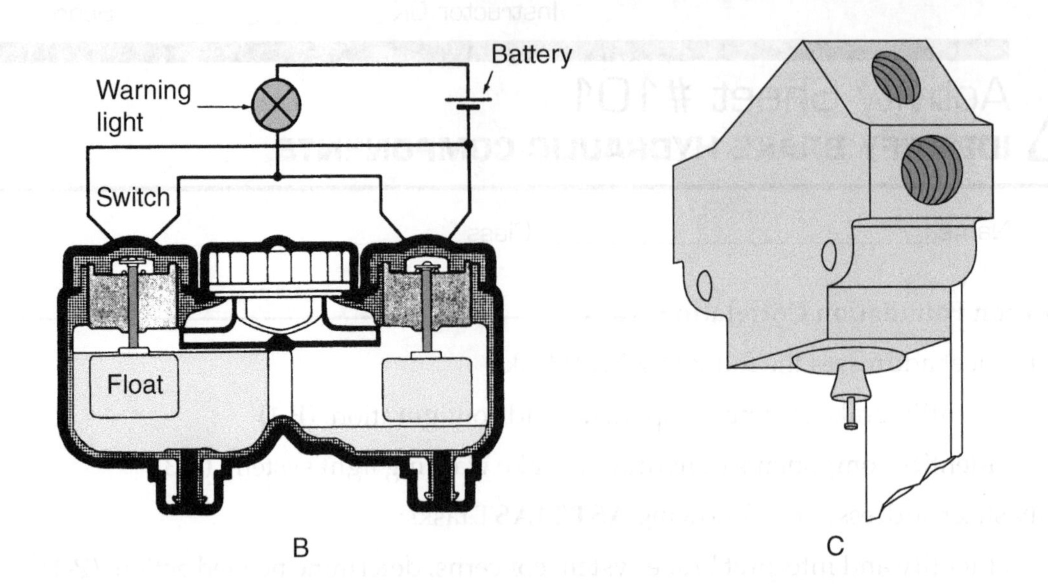

Pressure Differential Valve ______ Metering Valve ______ Fluid Level Switch ______

3. Identify three (3) conditions that could cause the brake warning light to illuminate.

A. ______________________________

B. ______________________________

C. ______________________________

STOP

Instructor OK ______________ Score ___________

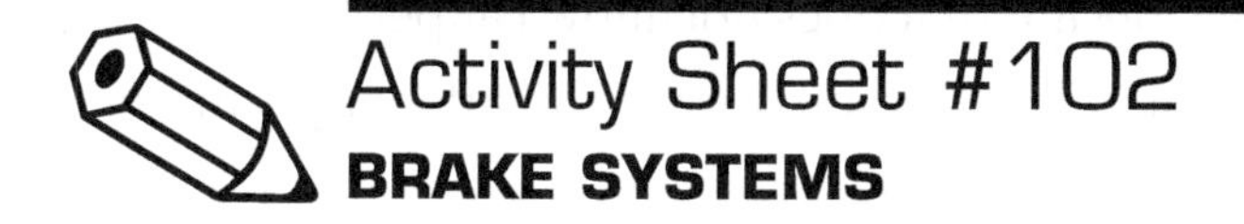

Activity Sheet #102

BRAKE SYSTEMS

Name__________________________ Class___________________

Directions: Match the words on the left to the descriptions on the right. Write the letter for the correct word on the line provided. For the terms that you are not certain of, use the glossary in your textbook.

Term		Description
A. Kinetic energy	______	Linings attached to their backings with fasteners
B. Coefficient of friction	______	When the leading shoe on a drum brake is forced into the brake drum
C. Bonded linings	______	When liquid is used to transfer motion or apply force
D. Riveted linings	______	Disc brake friction linings are sometimes called brake ___
E. Semimetallic linings	______	The ratio of the force required to slide one surface over another
F. Metallic linings	______	Fluid that absorbed 2% water
G. Brake shoes	______	Energy of motion
H. Brake pads	______	Operates the brakes on opposite corners of the vehicle
I. Hydraulics	______	Operates the front and rear brakes separately
J. Pascal's law	______	Organic linings with sponge iron and steel fibers mixed into them to add strength and temperature resistance
K. Hygroscopic	______	Linings made of metal that are used in heavy-duty conditions
L. DOT wet specification	______	On disc brakes the friction linings are called___
M. DOT dry specification	______	Holds drum brake friction material
N. Longitudinal braking	______	Loss of coefficient of friction in hot brakes
O. Diagonal braking system	______	The weight not supported by springs
P. Flapper valve	______	Specification for new fluid
Q. Self-energization	______	Linings that are glued to the brake shoe
R. Bleeding	______	The law of hydraulics
S. Brake fade	______	DOT number for synthetic brake fluid
T. Unsprung weight	______	Helps to prevent dangerous skids
U. Metering valve	______	Another name for bulkhead
V. Bulkhead	______	Allows fluid to flow in one direction only
W. Fire wall	______	Disc brakes do not require this
X. ABS	______	Removing air from a brake hydraulic system
Y. Pads	______	Material that absorbs water
Z. Equal	______	Separates engine and passenger compartments

TURN

AA. Regenerative braking	_____	Disc brake design that does not allow caliper to move
AB. DOT 5	_____	Caliper that is able to slide during and after application
AC. Ethylene glycol	_____	Antilock brake systems
AD. Trailing	_____	Energy source for power-assisted brakes
AE. Engine vacuum	_____	Pressure in an enclosed system is ___ and undiminished in all directions
AF. Generator	_____	Duo servo and the leading-___shoe are two types of drum brake designs
AG. Return springs	_____	Used to make brake fluid and automotive coolant
AH. Fixed caliper	_____	Captures energy lost during braking
AI. Conventional brakes	_____	Slows a vehicle when the hybrid motor changes to a generator
AJ. Floating caliper	_____	Their lifetime is increased by regenerative braking

STOP

Instructor OK ______________ Score ____________

Activity Sheet #103

IDENTIFY TAPERED WHEEL BEARING PARTS

Name ______________________________ Class ____________________

ASE Education Foundation Correlation

This activity sheet addresses the following **MLR** task:

V.A.4 Identify brake system components and configuration. **(P-1)**

This activity sheet addresses the following **AST/MAST** tasks:

V.F.1 Diagnose wheel bearing noises, wheel shimmy, and vibration concerns; determine needed action. **(P-1)**

V.F.2 Remove, clean, inspect, repack, and install wheel bearings; replace seals; install hub and adjust bearings. **(P-2)**

We Support ASE Education Foundation

Directions: Identify the tapered wheel bearing parts in the drawing. Place the identifying letter next to the name of each part.

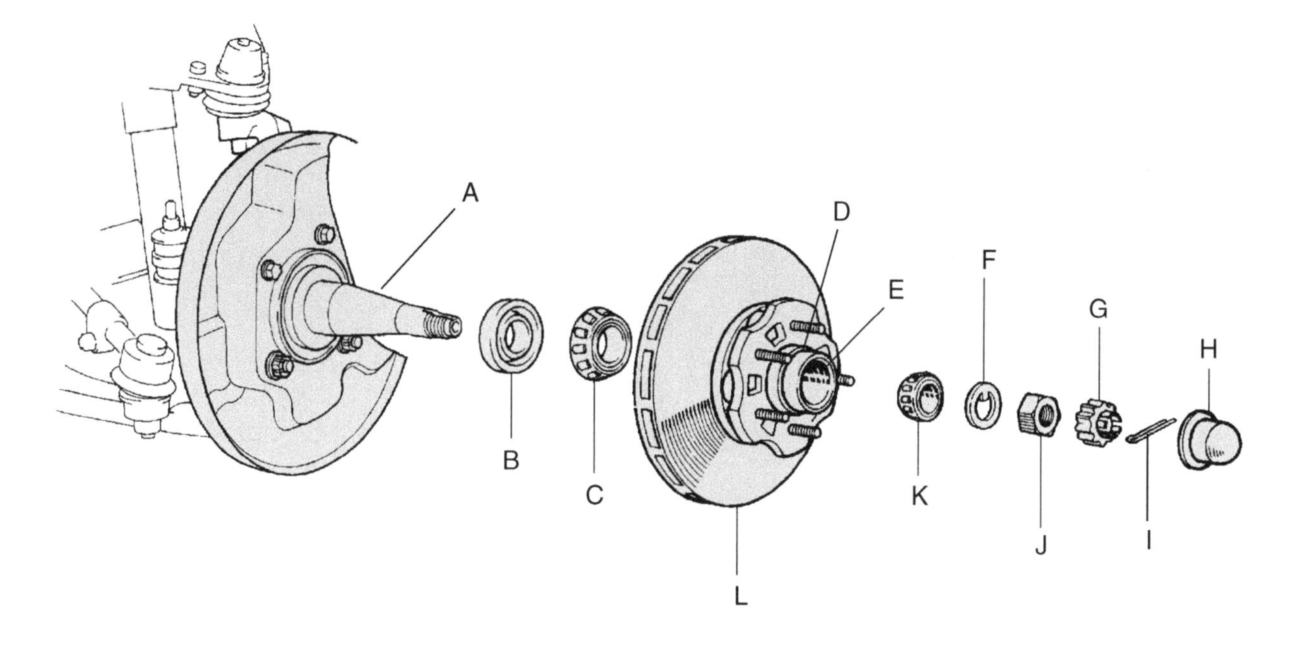

Outer Bearing	______	Nut Lock	______	Inner Bearing	______
Thrust Washer	______	Spindle	______	Oil Seal	______
Cap	______	Cotter Pin	______	Hub	______
Nut	______	Disc/Rotor	______	Bearing Cup	______

Instructor OK ______________ Score __________

Activity Sheet #104

IDENTIFY TAPERED AND DRIVE AXLE BEARING PARTS

Name ______________________________ Class ____________________

ASE Education Foundation Correlation

This activity sheet addresses the following **MLR** task:

V.A.4 Identify brake system components and configuration. **(P-1)**

This activity sheet addresses the following **AST/MAST** tasks:

V.F.1 Diagnose wheel bearing noises, wheel shimmy, and vibration concerns; determine needed action. **(P-1)**

V.F.2 Remove, clean, inspect, repack, and install wheel bearings; replace seals; install hub and adjust bearings. **(P-2)**

We Support ASE Education Foundation

Directions: Identify the tapered and drive axle bearing parts in the drawing. Place the identifying letter next to the name of each part.

D E F

A

G

B

H I J K

C

Chicago Rawhide

Enter the letter next to the correct bearing design.

Full-Floating ______ Tapered Roller ______ Semi-Floating ______

Enter the letter next to the correct part name.

Cage	______	Cup	______	Axle Shaft	______
Cone	______	Wheel Hub	______	Bearing	______
Roller	______	Axle Housing	______		

STOP

Instructor OK ______________ Score __________

Activity Sheet #105

BEARINGS AND GREASE

Name___________________________ Class____________________

Directions: Match the words on the left to the descriptions on the right. Write the letter for the correct word on the line provided. For the terms that you are not certain of, use the glossary in your textbook.

A. Bearing cage	_____ The term for non-drive front- and rear-wheel bearings
B. Radial load	_____ Indentations in the bearing or race from shock loads
C. End thrust	_____ The term for bearings that are on live axles (those that drive wheels)
D. Thrust bearing	_____ An axle design in which the bearings do not touch the axle but are located on the outside of the axle housing
E. Race	_____ A stamped steel or plastic insert that keeps bearing balls or rollers properly spaced around the bearing assembly
F. Needle bearing	_____ Side-to-side or front-to-rear force
G. Wheel bearings	_____ Moving
H. Axle bearings	_____ A grease of a consistency that allows it to be applied through a zerk fitting with a grease gun
I. Semi-floating axle	_____ A bearing cup
J. Full-floating axle	_____ When two parts have a pressed fit
K. Grease	_____ A bearing surface at 90 degrees to the load
L. NLGI	_____ A very small roller bearing used to control thrust or radial loads
M. Chassis lubricant	_____ Use one with the largest diameter that will fit into the hole
N. Dynamic	_____ A bearing design that tends to be self-aligning
O. Static	_____ An axle design in which the bearing rides on the axle
P. Brinelling	_____ A load in an up-and-down direction
Q. Interference fit	_____ A combination of oil and a thickening agent
R. Tapered roller bearings	_____ National Lubricating Grease Institute
S. Cotter pin	_____ At rest

STOP

Activity Sheet #106

IDENTIFY SECTIONS OF THE TIRE LABEL

Name ______________________________ Class ____________________

ASE Education Foundation Correlation

This activity sheet addresses the following **MLR** task:

IV.D.1 Inspect tire condition; identify tire wear patterns; check for correct tire size, application (load and speed ratings), and air pressure as listed on the tire information placard/label. **(P-1)**

This activity sheet addresses the following **AST/MAST** task:

IV.F.1 Inspect tire condition; identify tire wear patterns; check for correct tire size, application (load and speed ratings), and air pressure as listed on the tire information placard/label. **(P-1)**

We Support
ASE | Education Foundation

Directions: Identify the sections of the tire label in the drawing. Place the identifying letter next to the correct description.

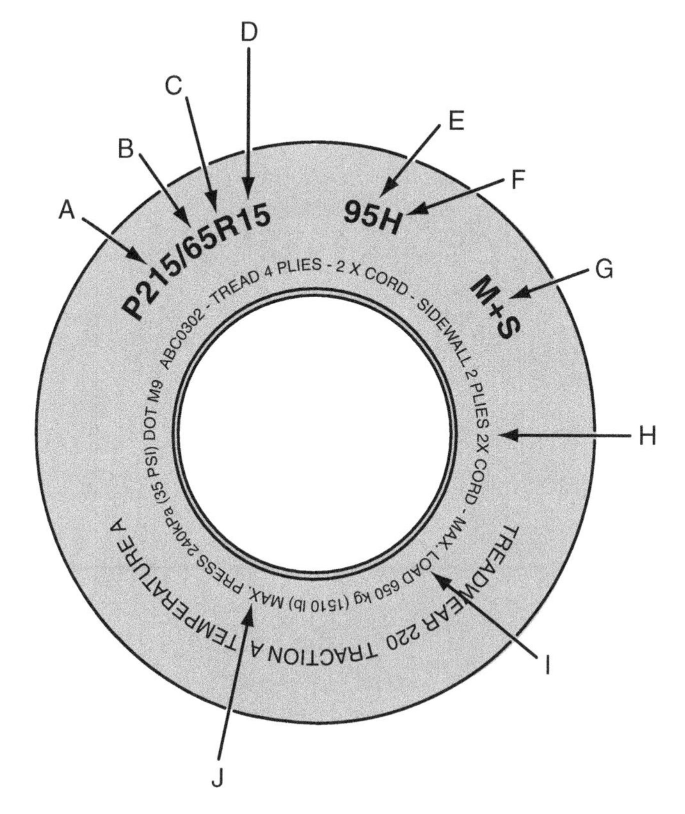

Rim Diameter	______	Ply Composition	______	Maximum Air Pressure	______
Maximum Load Rating	______	Speed Symbol	______	Snow Conditions	______
Construction	______	Aspect Ratio	______	Normal Width	______
Load Index	______				

TURN

Related Questions:

1. What is the meaning of treadwear 220? ______________________________
2. What is the meaning of traction A? ______________________________
3. What is the meaning of temperature A? ______________________________

STOP

Instructor OK ______________ Score __________

Activity Sheet #107
IDENTIFY PARTS OF TIRE SIZE DESIGNATION

Name ______________________ Class ______________

ASE Education Foundation Correlation

This activity sheet addresses the following **MLR** task:

IV.D.1 Inspect tire condition; identify tire wear patterns; check for correct tire size, application (load and speed ratings), and air pressure as listed on the tire information placard/label. **(P-1)**

This activity sheet addresses the following **AST/MAST** task:

IV.F.1 Inspect tire condition; identify tire wear patterns; check for correct tire size, application (load and speed ratings), and air pressure as listed on the tire information placard/label. **(P-1)**

We Support ASE Education Foundation

Directions: Identify the parts of this tire size designation. List the correct part of the listed size designation on the line next to its matching description.

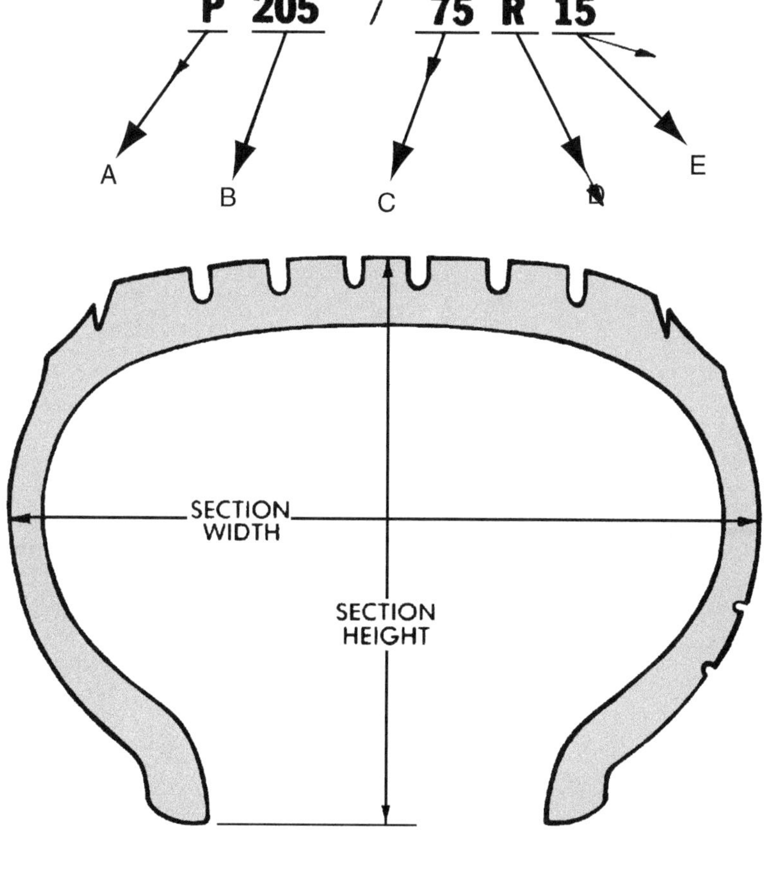

Tire Type	______	Section Width	______	Aspect Ratio	______
Construction Type	______	Rim Diameter	______		

Related Questions:

1. An aspect ratio of 50 means ____

 ____.

2. How wide is a 225 tire? ____

 ____.

Instructor OK ______________ Score __________

Activity Sheet #108

IDENTIFY PARTS OF THE TIRE

Name ____________________________ Class ____________________

ASE Education Foundation Correlation

This activity sheet addresses the following **MLR** tasks:

IV.A.3 Identify suspension and steering system components and configurations. **(P-1)**

IV.D.1 Inspect tire condition; identify tire wear patterns; check for correct tire size, application (load and speed ratings), and air pressure as listed on the tire information placard/label. **(P-1)**

This activity sheet addresses the following **AST/MAST** task:

IV.F.1 Inspect tire condition; identify tire wear patterns; check for correct tire size, application (load and speed ratings), and air pressure as listed on the tire information placard/label. **(P-1)**

We Support ASE Education Foundation

Directions: Identify the parts of the tire in the drawing. Place the identifying letter next to the name of each part.

Courtesy of Pirelli Tire of North America, LLC

Bead Filler	______	Rayon Plies	______	Tread Base	______
Tread	______	Nylon Plies	______	Bead Reinforcement	______
Steel Plies	______	Bead Core	______		

STOP

Instructor OK ______________ Score __________

Activity Sheet #109

TIRES

Name__________________________ Class____________________

Directions: Match the words on the left to the descriptions on the right. Write the letter for the correct word on the line provided. For the terms that you are not certain of, use the glossary in your textbook.

A. Aspect ratio	______ The speed at which the part of the tire that is contacting the ground is traveling
B. Lateral runout	______ The name for the up-and-down action of the tire that results in scalloped tire wear
C. Tire plug	______ A piece of rubber vulcanized to the inner liner of a tire to repair a leak
D. Sectional width	______ The combination of both static and couple imbalance
E. RMA	______ Material applied to the tire bead before installing a tire on a rim
F. Drop center	______ Wobble of a part in a side-to-side direction
G. 4 to 8 psi	______ Type of wheel balance measured with the wheel stationary
H. ZR	______ A piece of rubber vulcanized into a hole in a tire
I. Tire pressure monitor	______ A tire design that has a bulging sidewall when properly inflated
J. UTQG	______ Rubber Manufacturers Association
K. Negative offset	______ Approximate pressure increase in a tire as it warms up
L. 0 mph	______ Wobble of a part in an up-and-down direction
M. Dynamic imbalance	______ The tread depth at which wear bars show up around the tire tread
N. Radial runout	______ Term that describes tire height
O. Patch	______ Increases the tire track width
P. Radial	______ Rates traction, temperature, and treadwear
Q. 1/32"	______ Required on all vehicles after 2008
R. Wheel tramp	______ Highest speed rating
S. Rubber lube	______ Makes removing tire from a rim easier
T. Static	______ Measured at the widest part of a tire
U. Profile	______ Comparison of the height of a tire to its width
V. Run flat	______ A type of tire that can be driven when deflated

STOP

Instructor OK ______________ Score ____________

Activity Sheet #110

ANTILOCK BRAKES

Name ______________________________ Class ______________________

Directions: Match the words on the left to the descriptions on the right. Write the letter for the correct word on the line provided. For the terms that you are not certain of, use the glossary in your textbook.

Term		Description
A. Teves	______	Produces a digital signal
B. ABS	______	Part of
C. EBCM	______	What twisted pairs prevent
D. CAB	______	Called remote or add-on ABS
E. EBTCM	______	Synthetic brake fluid
F. PMV	______	ABS problems only
G. Radio interference	______	Separate
H. Hall effect	______	Rear-wheel antilock
I. Lateral acceleration sensor	______	Test for moisture
J. EHCU	______	Domestic antilock brake system
K. Integral	______	Typical hydraulic system warning light
L. Non-integral	______	Traction control system
M. Accumulator	______	Electronic brake control module
N. Non-integral ABS	______	Accumulates water
O. RWAL	______	Necessary to bleed some ABS
P. RABS	______	Controller antilock brake
Q. Four channel	______	Electronic brake and traction control module
R. BPMV	______	Measures force encountered while turning
S. TCS	______	Creates AC voltage
T. ASR	______	Electrohydraulic control unit
U. Amber light	______	Pressure modulator valves
V. Red light	______	Rear antilock brake system
W. Brake fluid test strips	______	Antilock brake system
X. DOT 5	______	Most effective ABS

TURN

Y.	Hygroscopic	______	Uses ABS, traction control, electronic brake-force distribution, and active yaw control computer inputs to determine if the car is actually traveling in the direction it was steered
Z.	Scan tool	______	Disables fuel injectors
AA.	Wheel speed sensor	______	Brake pressure modulator valve
AB.	Traction control	______	Stores brake fluid under very high pressure
AC.	ESC	______	Acceleration slip regulation

STOP

Instructor OK ______________ Score __________

Activity Sheet #111

IDENTIFY ANTILOCK BRAKE AND TRACTION CONTROL COMPONENTS

Name ______________________________ Class ______________________

ASE Education Foundation Correlation

This activity sheet addresses the following **MLR** task:

V.A.4 Identify brake system components and configuration. **(P-1)**

This task addresses the following **AST/MAST** task:

V.G.1 Identify and inspect electronic brake control system components (ABS, TCS, ESC); determine needed action. **(P-1)**

Directions: Identify the antilock brake components in the drawing. Place the identifying letter next to the name of each component identified in the diagram.

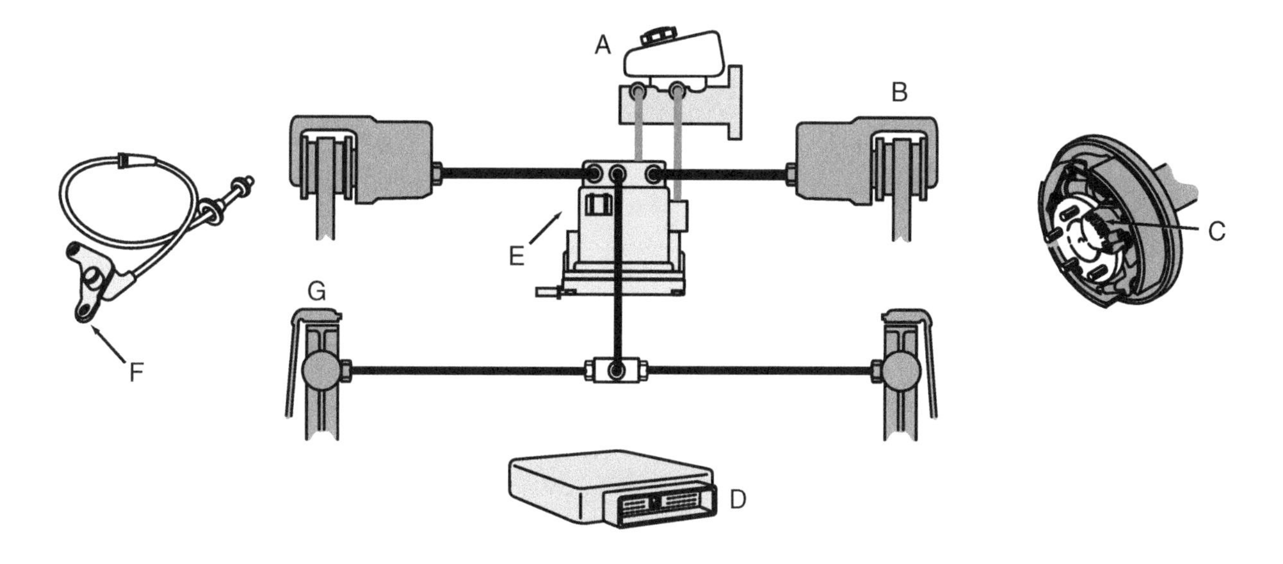

Wheel Sensor	______	Trigger Wheel	______	Drum Brake	______
ABS Modulator	______	ABS Control Unit	______	Disc Brake	______
Master Cylinder	______				

Related Questions:

What color is used by manufacturers to identify electrical wiring and connectors used in supplemental restraint systems?

_______ Red _______ Orange _______ Yellow _______ None

What color is used by manufacturers to identify electrical wiring and connectors used in traction control systems?

_______ Red _______ Orange _______ Yellow _______ None

Directions: In the list that follows, check off those components used by *both* the antilock brake system *and* the traction control system.

Left Rear-Wheel Sensor	❑	Driver's Fuse Box	❑	ABS Relay Box	❑
ABS Modulator	❑	Right Rear-Wheel Sensor	❑	Passenger's Fuse Box	❑
Right Front-Wheel Sensor	❑	ABS Control Unit	❑	Left Front-Wheel Sensor	❑
Underhood Fuse Box	❑	Data Link Connector	❑		

STOP

Part I

Activity Sheets

Section Eleven

Suspension, Steering, and Alignment

Instructor OK ______________ Score ____________

Activity Sheet #112

IDENTIFY SHORT-AND-LONG ARM SUSPENSION COMPONENTS

Name ______________________________ Class ____________________

ASE Education Foundation Correlation

This activity sheet addresses the following **MLR** task:

IV.A.3 Identify suspension and steering system components and configurations. **(P-1)**

This activity sheet addresses the following **AST** task:

IV.A.2 Identify and interpret suspension and steering system concerns; determine needed action. **(P-1)**

This activity sheet addresses the following **MAST** task:

IV.A.2 Identify and interpret suspension and steering system concerns; determine needed action. **(P-2)**

We Support ASE Education Foundation

Directions: Identify the short-and-long arm suspension components in the drawing. Place the identifying letter next to the name of each component.

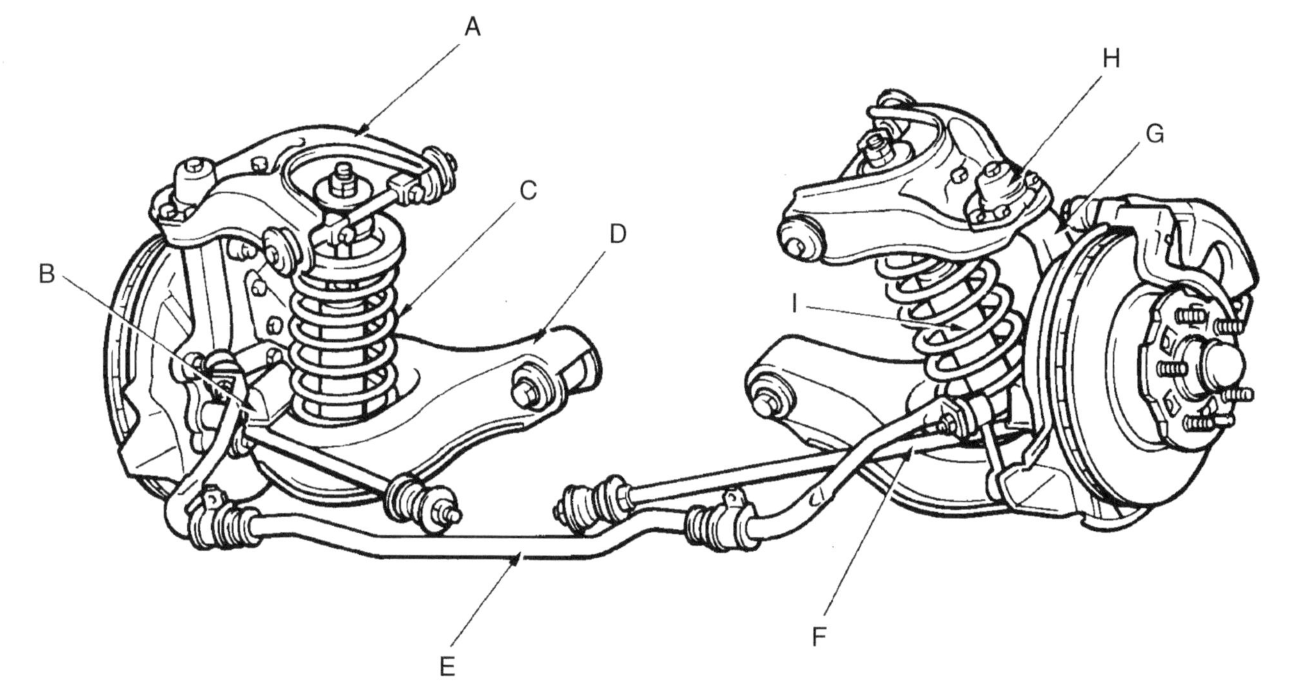

Steering Knuckle	______	Upper Control Arm	______	Upper Ball Joint	______
Lower Control Arm	______	Shock Absorber	______	Coil Spring	______
Bumper	______	Sway Bar	______	Strut Rod	______

STOP

Instructor OK ______________ Score ___________

Activity Sheet #113

IDENTIFY MACPHERSON STRUT SUSPENSION COMPONENTS

Name______________________________ Class____________________

ASE Education Foundation Correlation

This activity sheet addresses the following **MLR** task:

IV.A.3 Identify suspension and steering system components and configurations. **(P-1)**

This activity sheet addresses the following **AST** task:

IV.A.2 Identify and interpret suspension and steering system concerns; determine needed action. **(P-1)**

This activity sheet addresses the following **MAST** task:

IV.A.2 Identify and interpret suspension and steering system concerns; determine needed action. **(P-2)**

We Support ASE Education Foundation

Directions: Identify the MacPherson strut suspension parts in the drawing. Place the identifying letter next to the name of each part.

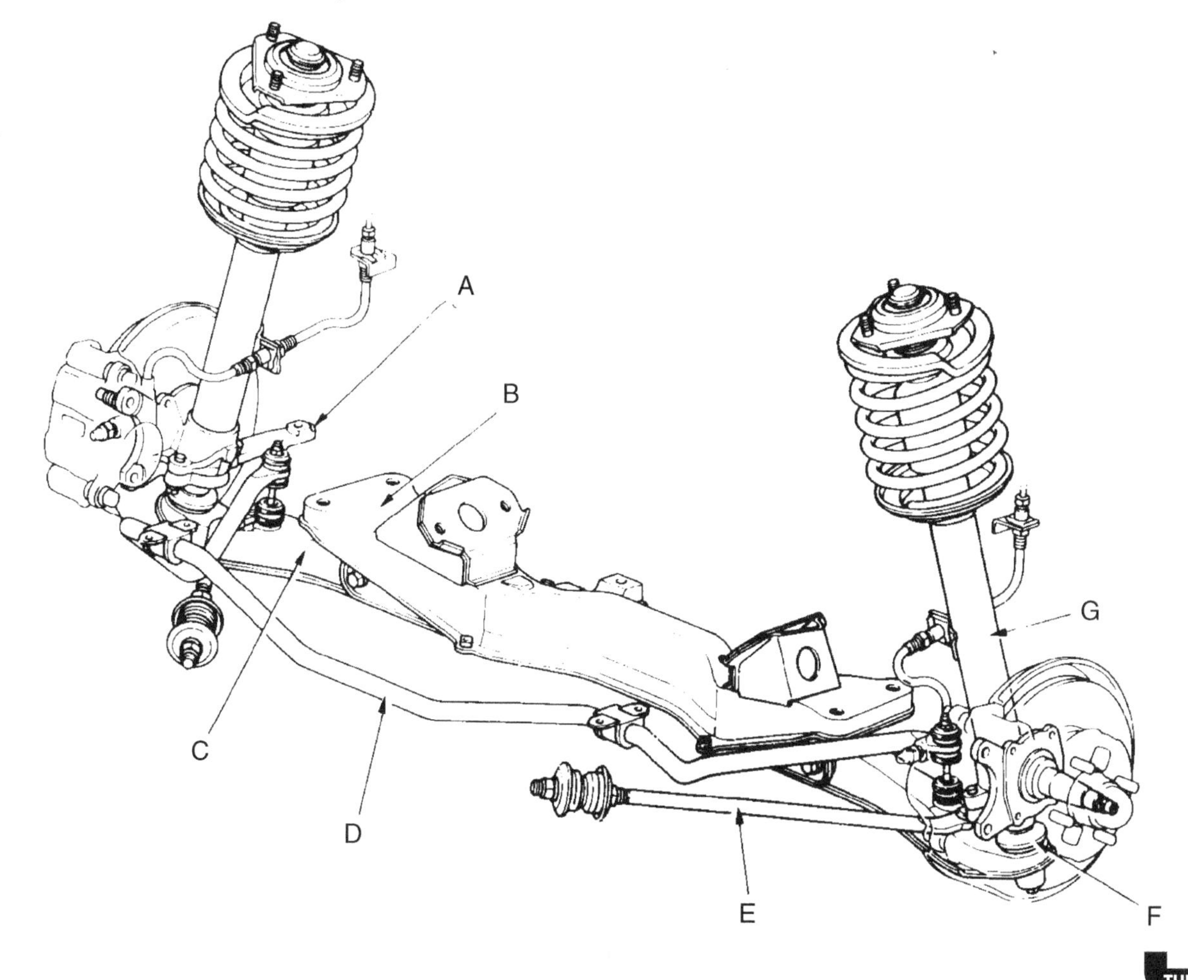

TURN

Steering Arm ______ Sway Bar ______ MacPherson Strut ______

Strut Rod ______ Control Arm ______ Ball Joint ______

Crossmember ______

STOP

Instructor OK ______________ Score __________

Activity Sheet #114

SUSPENSION SYSTEMS

Name ______________________________ Class ____________________

Directions: Match the words on the left to the descriptions on the right. Write the letter for the correct word on the line provided. For the terms that you are not certain of, use the glossary in your textbook.

Terms		Descriptions
A. Chassis	______	Steel rod wound into a coil
B. Suspension	______	Spring with a smooth ride that also allows for heavier carrying capacity
C. Rigid axle	______	Automatic suspension that keeps the car body level during all driving conditions
D. Independent suspension	______	Type of suspension system in which only one wheel will deflect
E. Coil spring	______	When the wheel moves up as the spring compresses
F. Variable rate spring	______	Weight not supported by springs
G. Torsion bar	______	The group of parts that includes the frame, shocks and springs, steering parts, tires, brakes, and wheels
H. Overload spring	______	A group of parts that supports the vehicle and cushions the ride
I. Sprung weight	______	A suspension design that incorporates the shock absorber into the front suspension
J. Unsprung weight	______	When hydraulic fluid becomes mixed with air
K. Compression or jounce	______	Also called aeration
L. Rebound	______	Straight rod that works as a spring
M. Short-and-long arm (SLA)	______	Found on heavy trucks
N. Shock absorber	______	An additional spring that only works under a heavy load
O. Shock ratio	______	Weight supported by springs
P. Aeration	______	Suspension leveling system that keeps the vehicle at the same height when weight is added to parts of the car
Q. Cavitation	______	When the wheel moves back down after compression
R. Gas shock	______	Suspension design that uses two control arms of unequal length

TURN

S.	MacPherson strut	______	Dampens spring oscillations
T.	Adaptive suspension system	______	The difference between the amount of control on compression and extension
U.	Active suspension	______	Pressurized to keep the bubbles from forming in the fluid

STOP

Instructor OK ______________ Score ______________

Activity Sheet #115

IDENTIFY STEERING COMPONENTS

Name ______________________ Class ______________

ASE Education Foundation Correlation

This activity sheet addresses the following **MLR** task:

IV.A.3 Identify suspension and steering system components and configurations. **(P-1)**

This activity sheet addresses the following **AST** task:

IV.A.2 Identify and interpret suspension and steering system concerns; determine needed action. **(P-1)**

This activity sheet addresses the following **MAST** task:

IV.A.2 Identify and interpret suspension and steering system concerns; determine needed action. **(P-2)**

We Support ASE Education Foundation

Directions: Identify the steering components in the drawing. Place the identifying letter next to the name of each component.

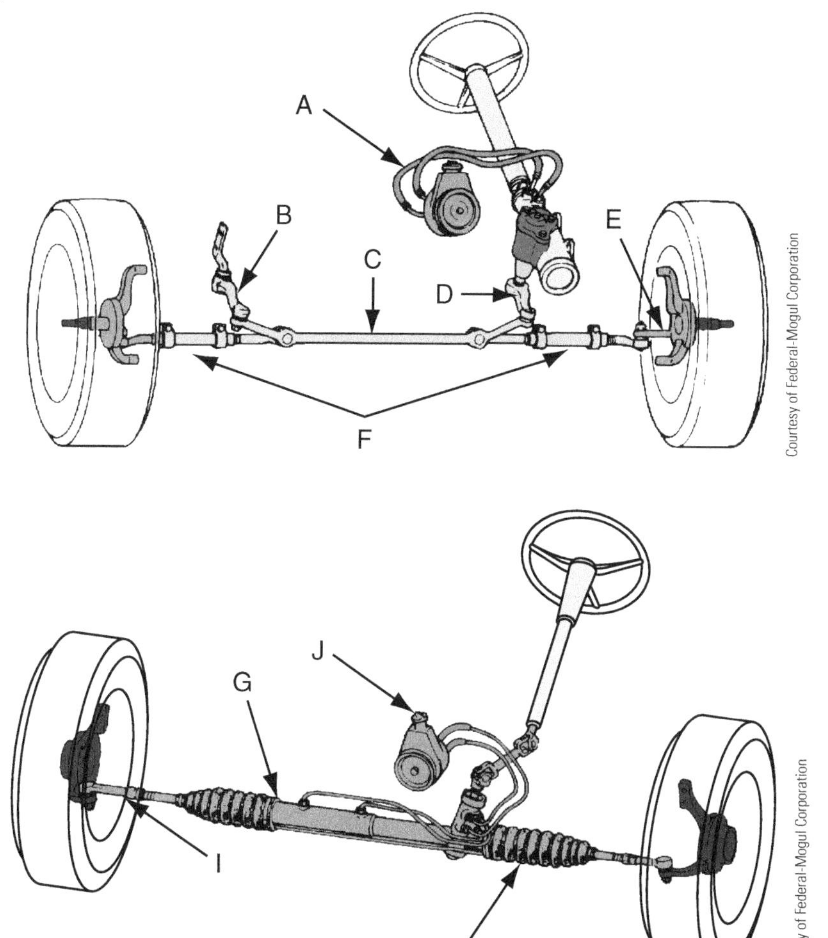

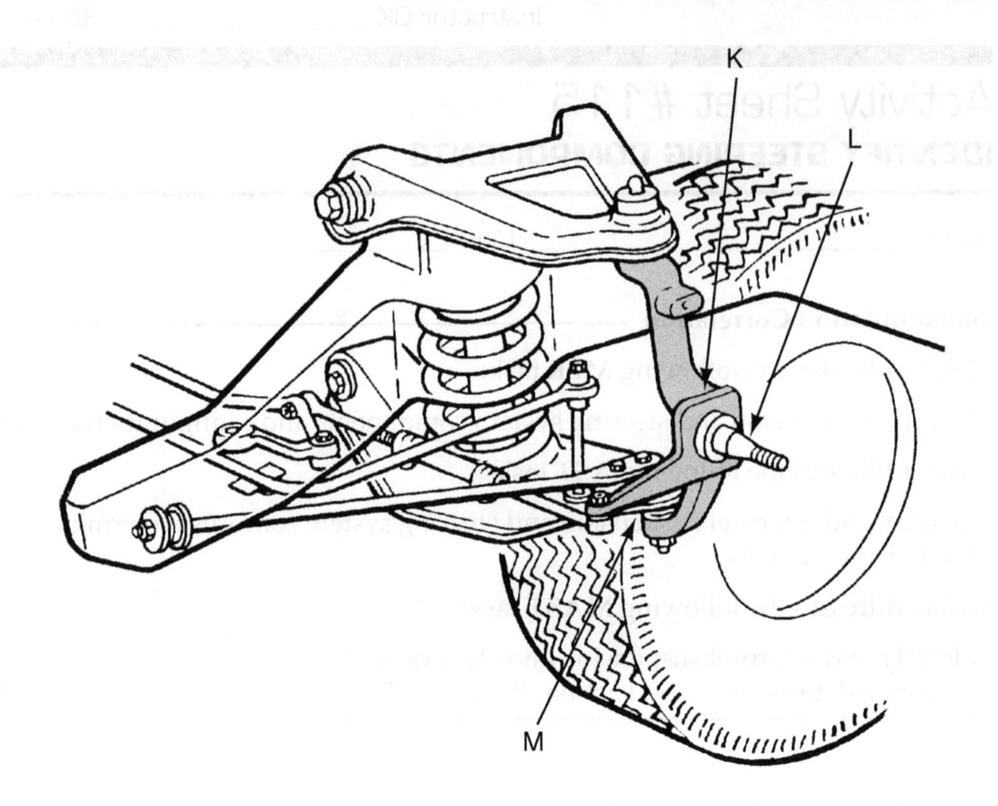

Tie-Rods	______	Steering Arm	______	Center Link	______
Pitman Arm	______	Idler Arm	______	Steering Knuckle	______
Spindle	______	Rack-and-Pinion Boot	______	Power Steering Pump	______
Rack-and-Pinion Assembly	______	Lower Ball Joint	______	Rack-and-Pinion Tie-Rod	______
Hydraulic Hoses	______				

STOP

Instructor OK ____________ Score ____________

Activity Sheet #116

IDENTIFY POWER STEERING COMPONENTS

Name ______________________ Class ______________

ASE Education Foundation Correlation

This activity sheet addresses the following **MLR** task:

IV.A.3 Identify suspension and steering system components and configurations. **(P-1)**

This activity sheet addresses the following **AST** task:

IV.A.2 Identify and interpret suspension and steering system concerns; determine needed action. **(P-1)**

This activity sheet addresses the following **MAST** task:

IV.A.2 Identify and interpret suspension and steering system concerns; determine needed action. **(P-2)**

We Support ASE | Education Foundation

Directions: Identify the power steering components in the drawing. Place the identifying letter next to the name of each component.

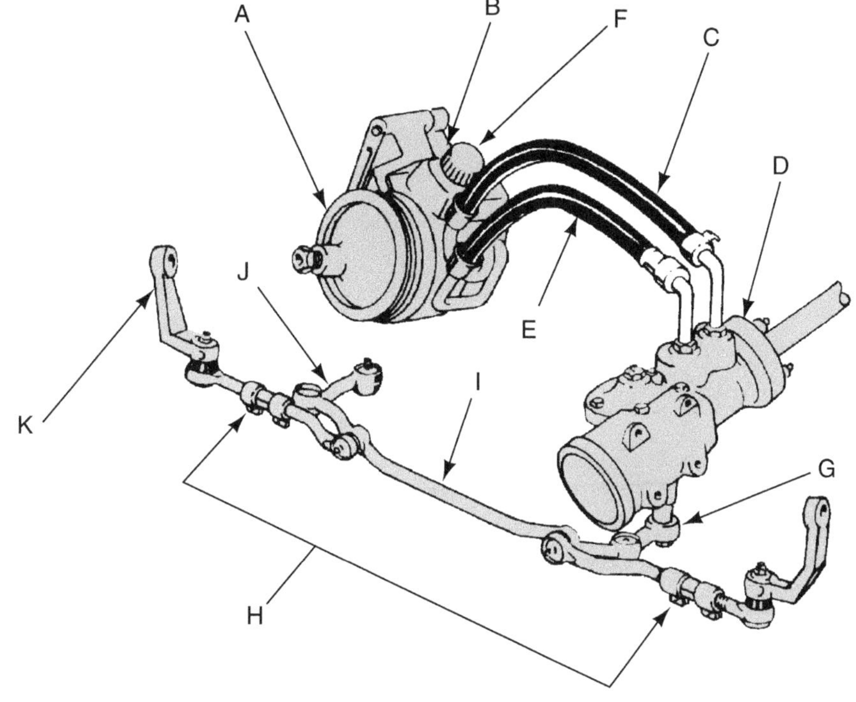

Pulley	______	Pressure Line	______	Pump	______
Return Line	______	Steering Gear	______	Reservoir	______
Pitman Arm	______	Steering Arm	______	Tie-Rods	______
Idler Arm	______	Center Link	______		

STOP

Instructor OK ______________ Score ____________

Activity Sheet #117

IDENTIFY STEERING GEAR COMPONENTS

Name ______________________________ Class ______________________

ASE Education Foundation Correlation

This activity sheet addresses the following **MLR** task:

IV.A.3 Identify suspension and steering system components and configurations. (P-1)

This activity sheet addresses the following **AST** task:

IV.A.2 Identify and interpret suspension and steering system concerns; determine needed action. **(P-1)**

This activity sheet addresses the following **MAST** task:

IV.A.2 Identify and interpret suspension and steering system concerns; determine needed action. **(P-2)**

We Support ASE Education Foundation

Directions: Identify the steering gear components in the drawing. Place the identifying letter next to the name of each component.

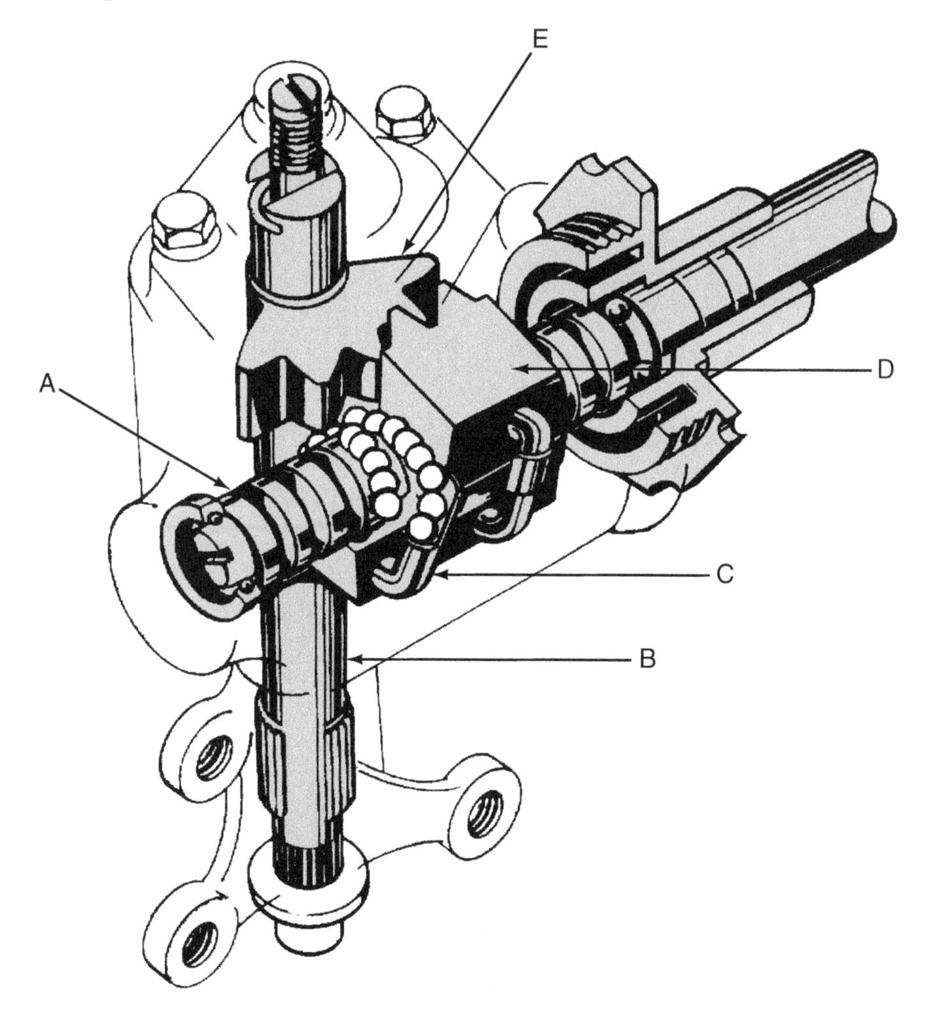

Ball Return Tubes	______	Sector Gear	______	Ball Nut	______
Sector Shaft	______	Worm Shaft	______		

STOP

Activity Sheet #118

IDENTIFY RACK-AND-PINION STEERING GEAR COMPONENTS

Name ______________________ Class ________________

ASE Education Foundation Correlation

This activity sheet addresses the following **MLR** task:

IV.A.3 Identify suspension and steering system components and configurations. **(P-1)**

This activity sheet addresses the following **AST** task:

IV.A.2 Identify and interpret suspension and steering system concerns; determine needed action. **(P-1)**

This activity sheet addresses the following **MAST** task:

IV.A.2 Identify and interpret suspension and steering system concerns; determine needed action. **(P-2)**

Directions: Identify the rack and pinion steering gear components in the drawing. Place the identifying letter next to the name of each component listed on the next page.

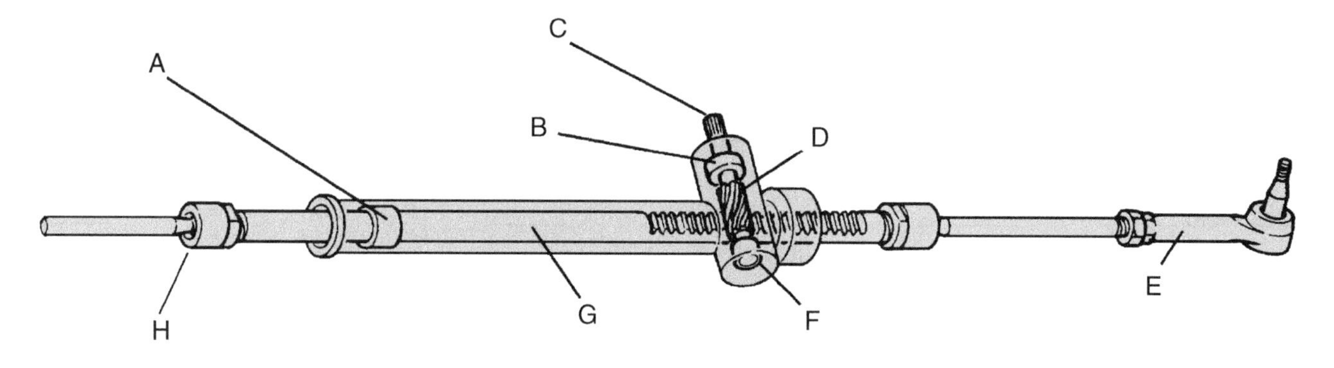

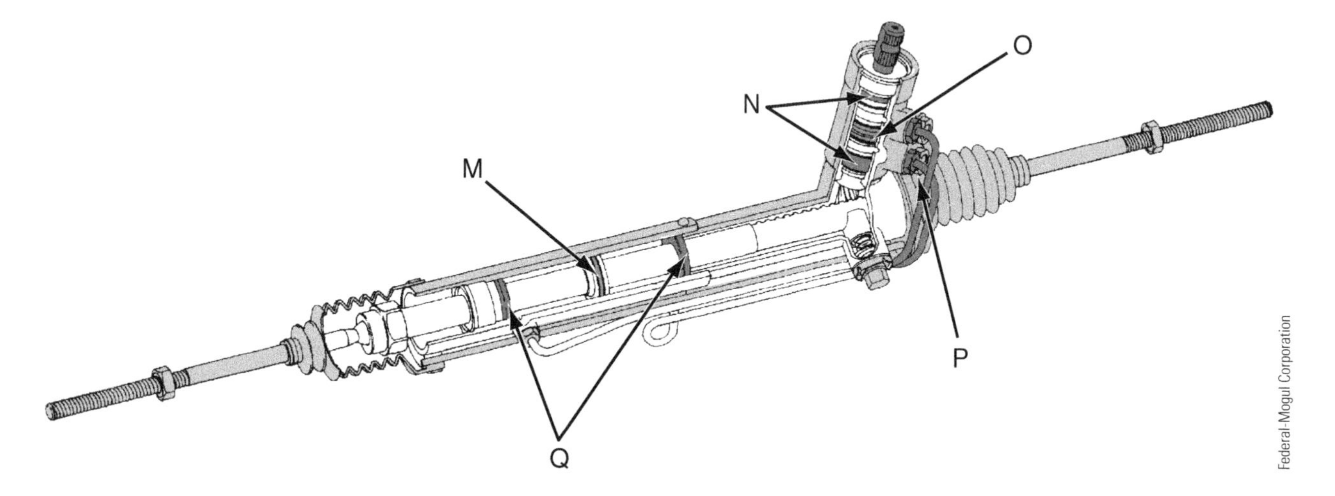

Federal-Mogul Corporation

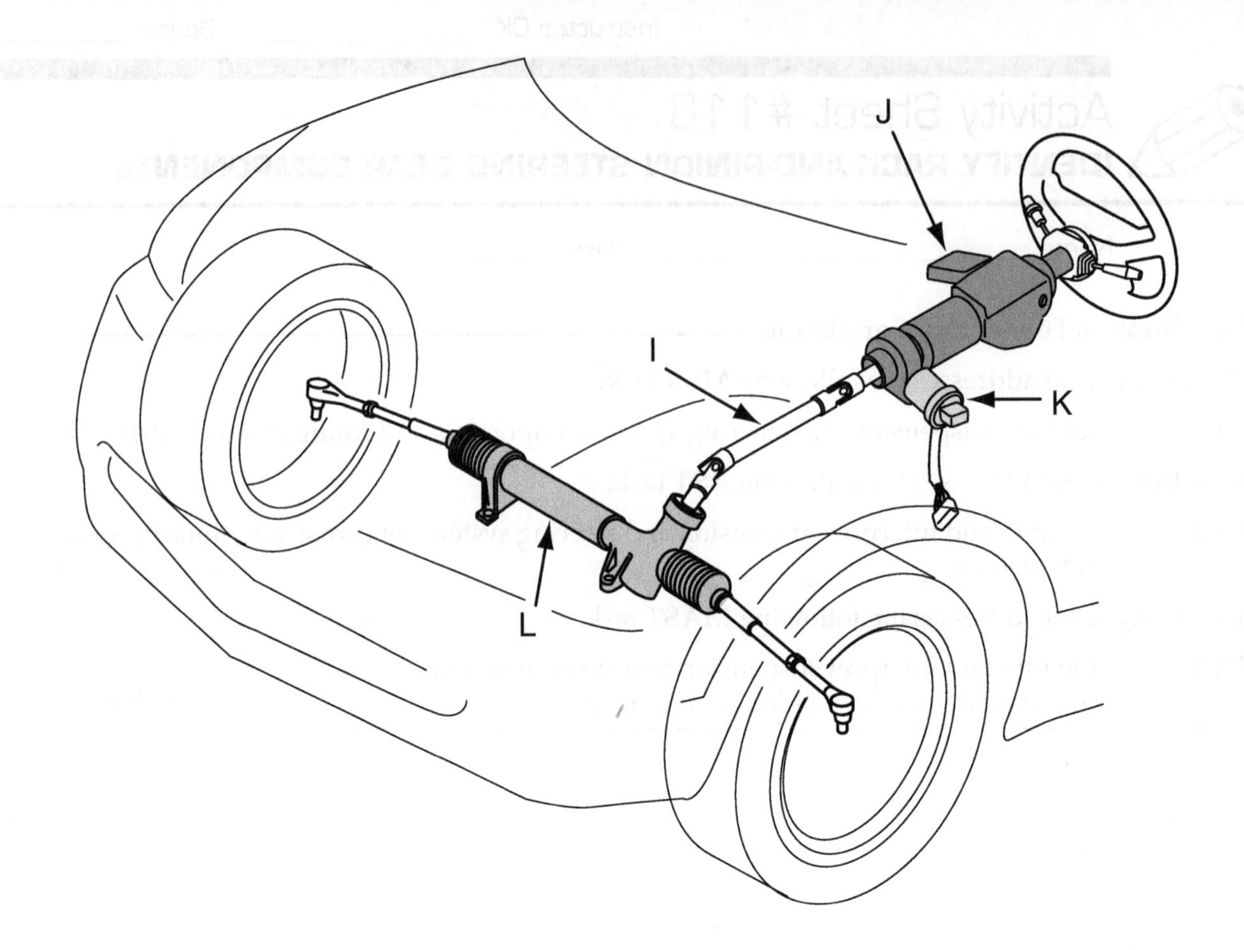

Rack Support Bushing	______	Lower Pinion Bearing	______	Upper Pinion Bearing	______
Inner Tie-Rod	______	Pinion Gear	______	Outer Tie-Rod	______
Rack Gear	______	Pinion Shaft	______	ECU	______
Steering Gear	______	Steering Shaft	______	Electric Assist	______
Power Piston Ring	______	Power Lines	______	Piston Seals	______
Spool Valve Seals	______	Spool Valve Rings	______		

STOP

Instructor OK ______________ Score __________

Activity Sheet #119

STEERING SYSTEMS

Name ____________________________ Class ____________________

Directions: Match the words on the left to the descriptions on the right. Write the letter for the correct word on the line provided. For the terms that you are not certain of, use the glossary in your textbook.

A. Lock to lock	______	Most common steering on new vehicles
B. Steering ratio	______	Used to shorten and lengthen shafts or rods
C. Recirculating ball	______	Power steering with a piston attached to the steering linkage
D. Toe-out-on-turns	______	Angled so that the front wheels toe out during a turn
E. Steering damper	______	Parts that connect the steering gear to the wheels
F. Steering linkage	______	Shock absorber on the steering linkage
G. Parallelogram steering	______	Number of teeth on the driving gear compared to the number of teeth on the driven gear
H. Flow control valve	______	Steering gear used with parallelogram steering
I. Turnbuckle	______	Most common type of power steering pump
J. Pressure relief valve	______	Type of steering shaft coupling
K. Steering arm	______	The faster the vehicle is driven, the ___ power assist is needed
L. Roller, vane, and slipper	______	Hydraulic valve that bleeds off excess pressure
M. Flex coupling	______	Steering linkage that uses a steering box
N. Spool valve	______	Wheel that turns sharper during a turn
O. Linkage power steering	______	Allow steering linkage parts to pivot
P. Integral power steering	______	When the steering wheel is turned all the way from one direction to the other
Q. Inner	______	Types of steering pumps
R. Variable assist	______	Power steering design that has the power steering components contained within the steering gear
S. Less	______	Limits maximum pressure
T. Rack-and-pinion steering	______	Used with rack-and-pinion to control power assist
U. Ball sockets	______	Ackerman angle or turning radius
V. Vane	______	Reduces power assist with increasing vehicle speed

STOP

Instructor OK ______________ Score ____________

Activity Sheet #120

IDENTIFY CAUSES OF TIRE WEAR

Name_________________________ Class_________________

ASE Education Foundation Correlation

This activity sheet addresses the following **MLR** task:

IV.D.1 Inspect tire condition; identify tire wear patterns; check for correct tire size, application (load and speed ratings), and air pressure as listed on the tire information placard/label. **(P-1)**

This activity sheet addresses the following **AST/MAST** task:

IV.F.1 Inspect tire condition; identify tire wear patterns; check for correct tire size, application (load and speed ratings), and air pressure as listed on the tire information placard/label. **(P-1)**

We Support
ASE | Education Foundation

Directions: Choose the most probable cause of the following types of tire wear from the list. Place the letter next to the cause.

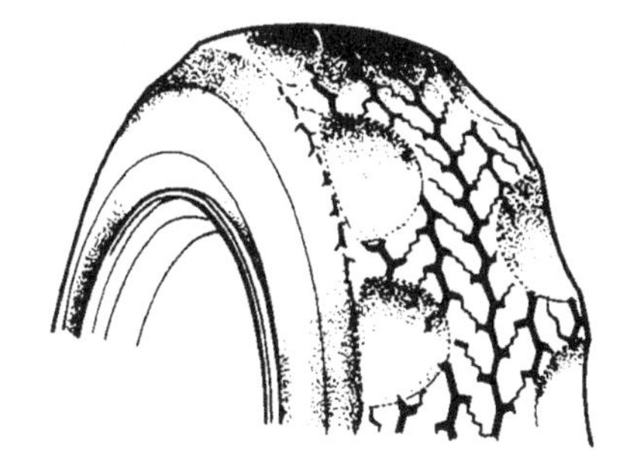

A

Photo by Tim Gilles

B

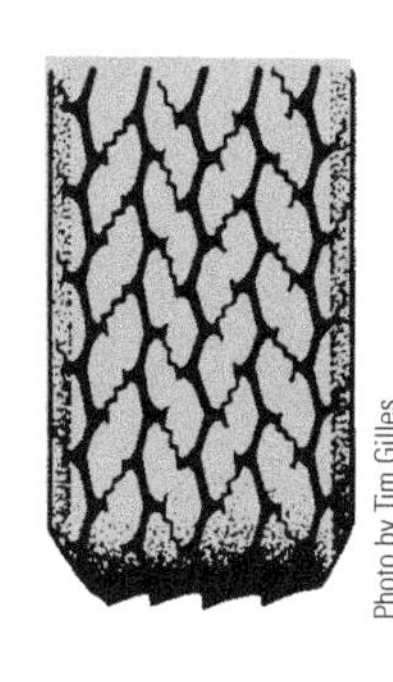

Photo by Tim Gilles

C

Toe Wear ______ Camber Wear ______ Loose Parts or Wheel Balance ______

D

E

1. What tire wear pattern indicates underinflation? ________
2. What tire wear pattern indicates that a tire is overinflated? ________

Instructor OK ______________ Score ______________

Activity Sheet #121

IDENTIFY WHEEL ALIGNMENT TERMS

Name ______________________________ Class ____________________

ASE Education Foundation Correlation

This activity sheet addresses the following **MLR** tasks:

IV.C.1 Perform prealignment inspection; measure vehicle ride height. **(P-1)**

IV.C.2 Describe alignment angles (camber, caster and toe). **(P-1)**

This activity sheet addresses the following **AST/MAST** tasks:

IV.E.2 Perform prealignment inspection; measure vehicle ride height; determine needed action. **(P-1)**

IV.E.3 Prepare vehicle for wheel alignment on alignment machine; perform four-wheel alignment by checking and adjusting front and rear wheel caster, camber and toe as required; center steering wheel. **(P-1)**

We Support ASE Education Foundation

Directions: Identify the correct wheel alignment term and place the letter that matches it on the line next to it on the next page.

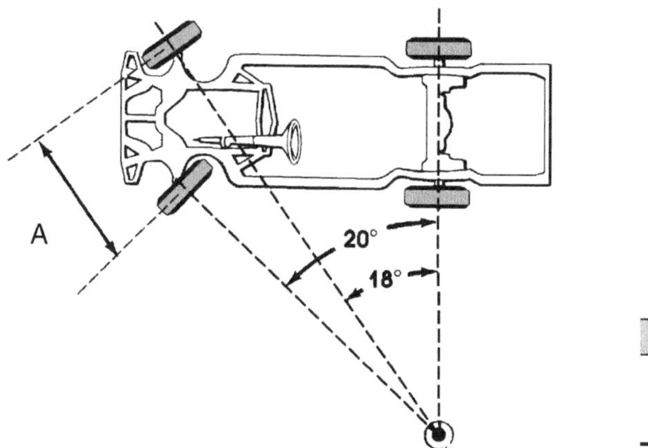

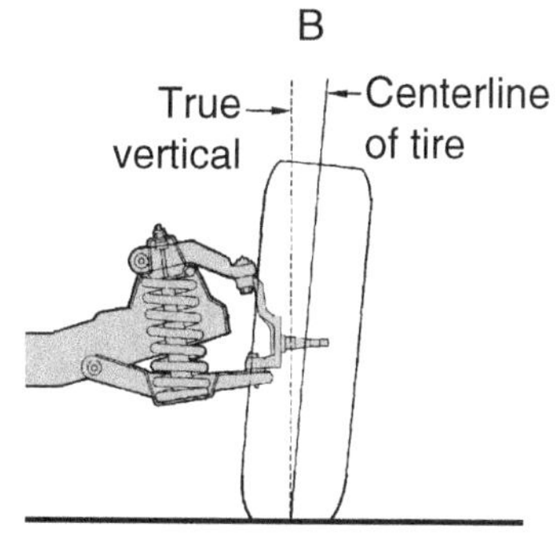

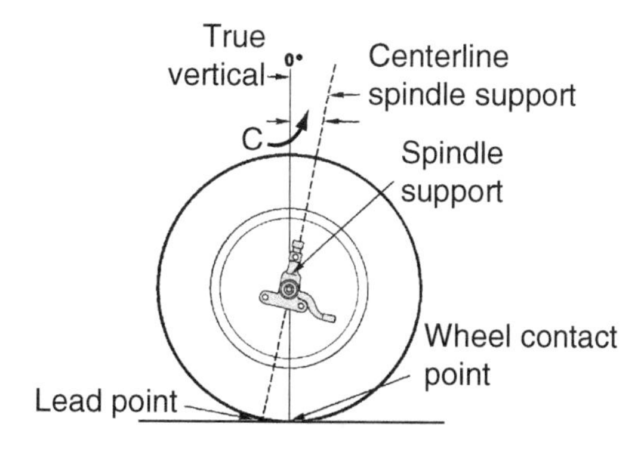

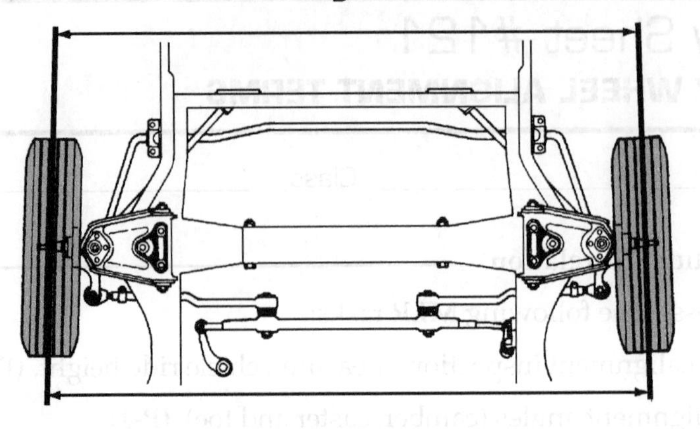

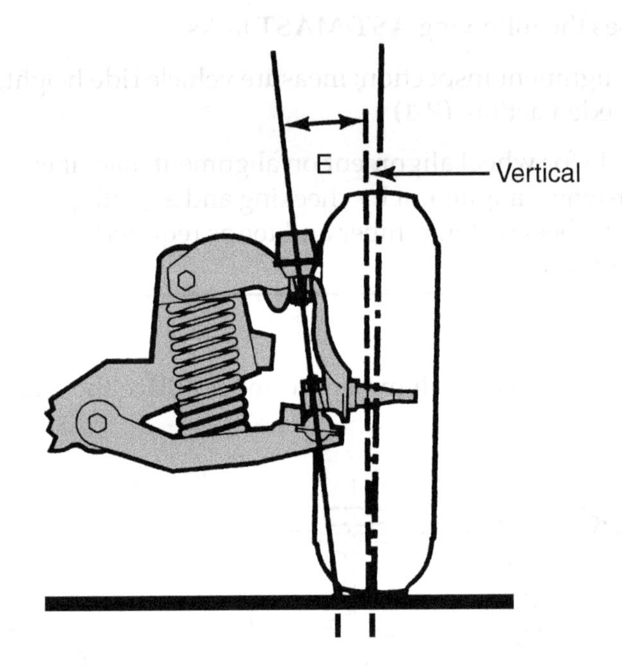

Caster	______	Turning Radius	______	Steering Axis Inclination	______
Camber	______	Toe	______		

STOP

Instructor OK ______________ Score __________

Activity Sheet #122

ALIGNMENT

Name ______________________ Class ______________

ASE Education Foundation Correlation

This activity sheet addresses the following **MLR** tasks:

IV.C.1 Perform prealignment inspection; measure vehicle ride height. **(P-1)**

IV.C.2 Describe alignment angles (camber, caster and toe). **(P-1)**

This activity sheet addresses the following **AST/MAST** tasks:

IV.E.2 Perform prealignment inspection; measure vehicle ride height; determine needed action. **(P-1)**

IV.E.3 Prepare vehicle for wheel alignment on alignment machine; perform four-wheel alignment by checking and adjusting front and rear wheel caster, camber and toe as required; center steering wheel. **(P-1)**

Directions: Match the words on the left to the descriptions on the right. Write the letter for the correct word on the line provided. For the terms that you are not certain of, use the glossary in your textbook.

Term		Description
A. Toe	______	Inward or outward tilt of a tire at the top
B. Toe-in	______	The distance between the front and the rear tires
C. Toe-out	______	The amount that the spindle support arm leans in at the top
D. Scuff	______	Also called turning radius or toe-out-on-turns
E. Camber	______	Forward or rearward tilt of the spindle
F. Positive camber	______	Also called the crossmember
G. Negative camber	______	When a car does not seem to respond to movement of the steering wheel during a hard turn
H. Camber roll	______	When a car turns too far in response to steering wheel movement
I. Caster	______	Comparison of the distances between the fronts and the rears of a pair of tires
J. Positive caster	______	The tendency during a turn for a tire to continue to go in the direction it was going before
K. Negative caster	______	When a tire is tilted out at the top
L. Included angle	______	When a tire is tilted in at the top

Term		Description
M. Scrub radius	______	The amount that one front wheel is behind the one on the other side of the car
N. Crossmember	______	A term that refers to the relationship between the average direction that the rear tires point and the average direction that the front tires point
O. Cradle	______	The forward tilt of the steering axis
P. Toe-out-on-turns	______	When the tires are closer together at the rear
Q. Ackerman angle	______	Tire wear resulting from incorrect toe adjustment
R. Wheel base	______	The steering axis pivot centerline
S. Track	______	The large steel part of the frame beneath the engine and between the front wheels
T. Tracking	______	Measurement to check for sagged springs
U. Setback	______	When the tires are closer together at the front
V. Slip angle	______	The side-to-side distance between an axle's tires
W. Understeer	______	The rearward tilt of the steering axis
X. Oversteer	______	SAI and camber together
Y. Steering axis inclination (SAI)	______	A term describing how a tire rolls in a circle like a cone
Z. Ride height	______	A term describing how tires toe out during a turn because the steering arms are bent at an angle

STOP

Instructor OK ______________ Score ____________

Activity Sheet #123

IDENTIFY ELECTRONIC STABILITY CONTROL COMPONENTS

Name ______________________________ Class ____________________

ASE Education Foundation Correlation

This activity sheet addresses the following **MLR** task:

IV.A.3 Identify suspension and steering system components and configurations. **(P-1)**

This activity sheet addresses the following **AST/MAST** task:

IV.D.3 Describe the function of suspension and steering control systems and components, (i.e. active suspension and stability control). **(P-3)**

We Support ASE Education Foundation

Directions: Identify the electronic stability control components in the drawing. Place the identifying letter next to the name of each component listed below the diagram.

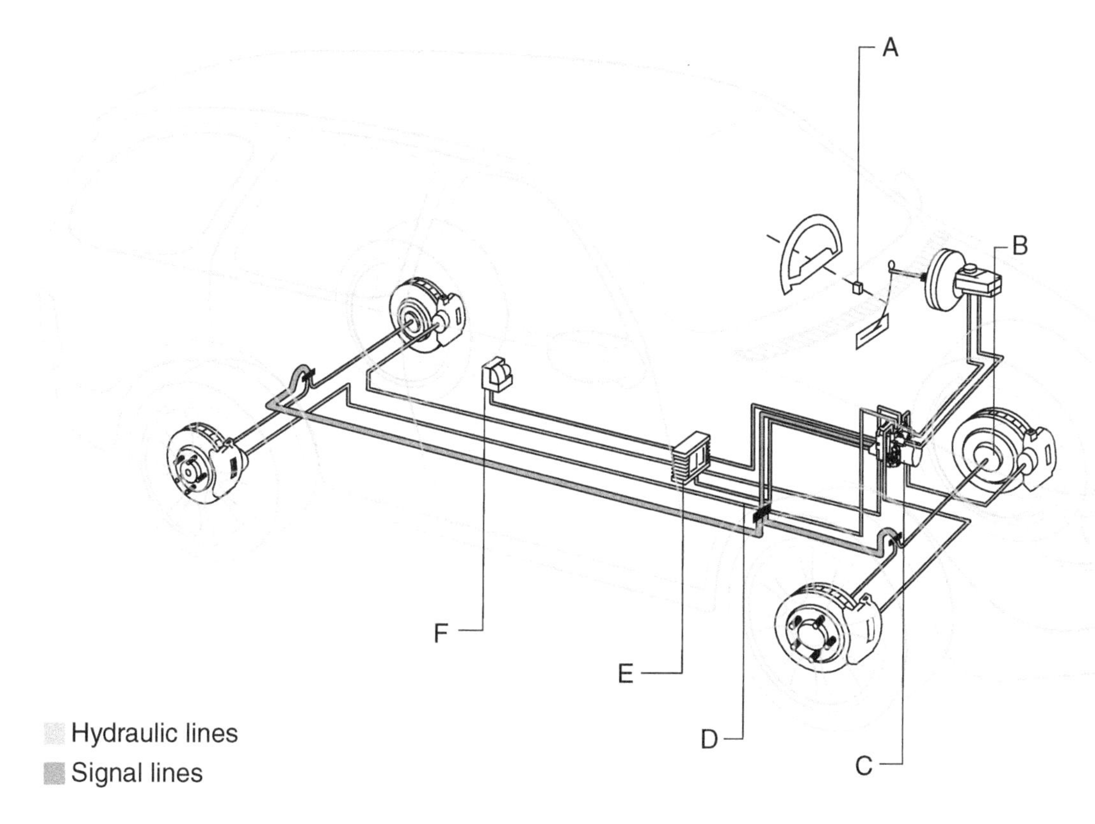

Yaw Rate Sensor	______	ABS/TCS Controller	______	Communication Link	______
Power Control Module	______	Brake Pedal Sensor	______	Wheel Speed Sensor	______

STOP

Part I

Activity Sheets

Section Twelve

Drivetrain

Instructor OK ________________ Score ____________

Activity Sheet #124

IDENTIFY CLUTCH COMPONENTS (ASSEMBLED)

Name ______________________________ Class ____________________

ASE Education Foundation Correlation

This activity sheet addresses the following **MLR** task:

III.A.4 Identify manual drivetrain and axle components and configuration. **(P-1)**

This activity sheet addresses the following **AST/MAST** task:

III.B.3 Inspect and/or replace clutch pressure plate assembly, clutch disc, release (throw-out) bearing, linkage, and pilot bearing/bushing (as applicable). **(P-1)**

We Support ASE Education Foundation

Directions: Identify the clutch components in the drawing. Place the identifying letter next to the name of each component.

Related Question: Is the clutch shown here engaged or disengaged? _______

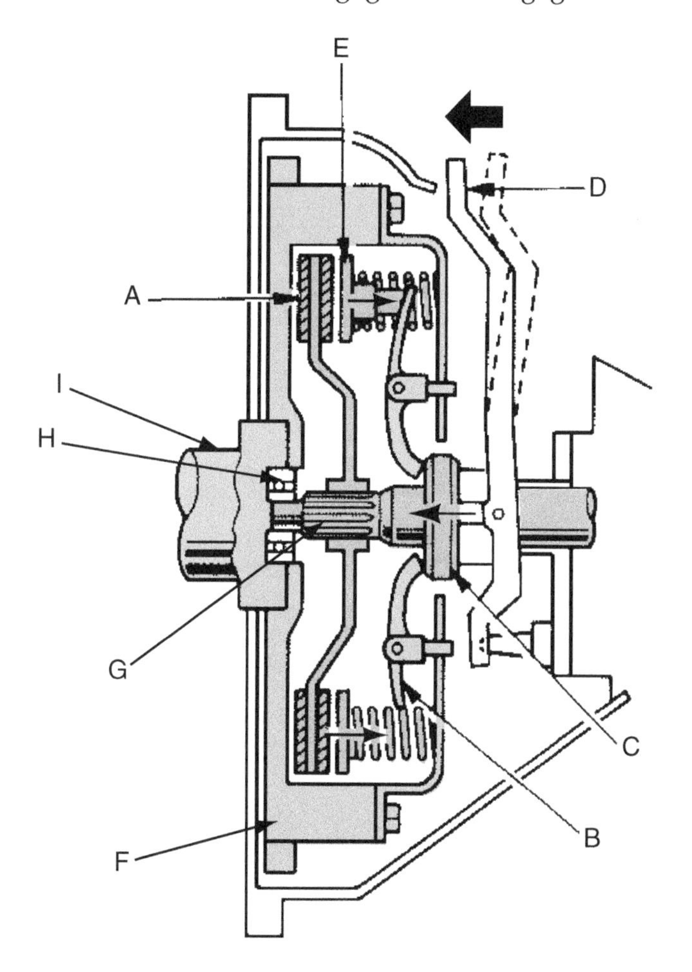

Clutch Disc	_____	Input Shaft	_____	Release Fork	_____
Release Bearing	_____	Flywheel	_____	Pressure Plate	_____
Release Lever	_____	Crankshaft	_____	Pilot Bearing	_____

STOP

Instructor OK _______________ Score __________

Activity Sheet #125

IDENTIFY CLUTCH COMPONENTS (EXPLODED VIEW)

Name ____________________________ Class ____________________

ASE Education Foundation Correlation

This activity sheet addresses the following **MLR** task:

III.A.4 Identify manual drivetrain and axle components and configuration. **(P-1)**

This activity sheet addresses the following **AST/MAST** task:

III.B.2 Inspect and/or replace clutch pressure plate assembly, clutch disc, release (throw-out) bearing, linkage, and pilot bearing/bushing (as applicable). **(P-1)**

Directions: Identify the clutch components in the drawing. Place the identifying letter next to the name of each component.

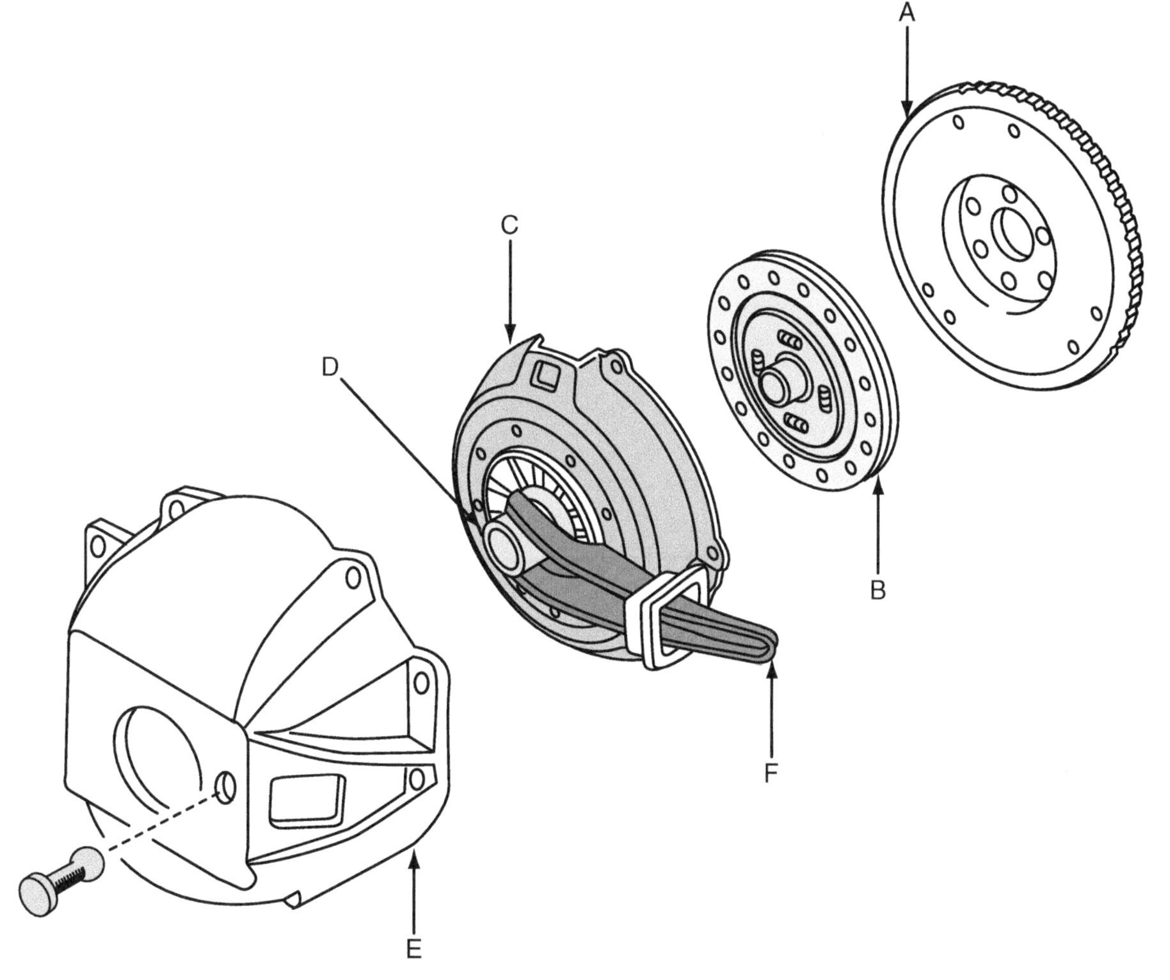

Release Bearing	______	Clutch Disc	______	Clutch Housing	______
Flywheel	______	Pressure Plate	______	Release Fork	______

STOP

Instructor OK ______________ Score ____________

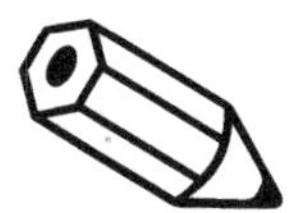

Activity Sheet #126

IDENTIFY CLUTCH RELEASE MECHANISM

Name ____________________________ Class ____________________

ASE Education Foundation Correlation

This activity sheet addresses the following **AST/MAST** task:

III.B.2 Inspect clutch pedal linkage, cables, automatic adjuster mechanisms, brackets, bushings, pivots, and springs; determine needed action. **(P-1)**

We Support ASE Education Foundation

Directions: Record the names of the parts and system type in the spaces provided.

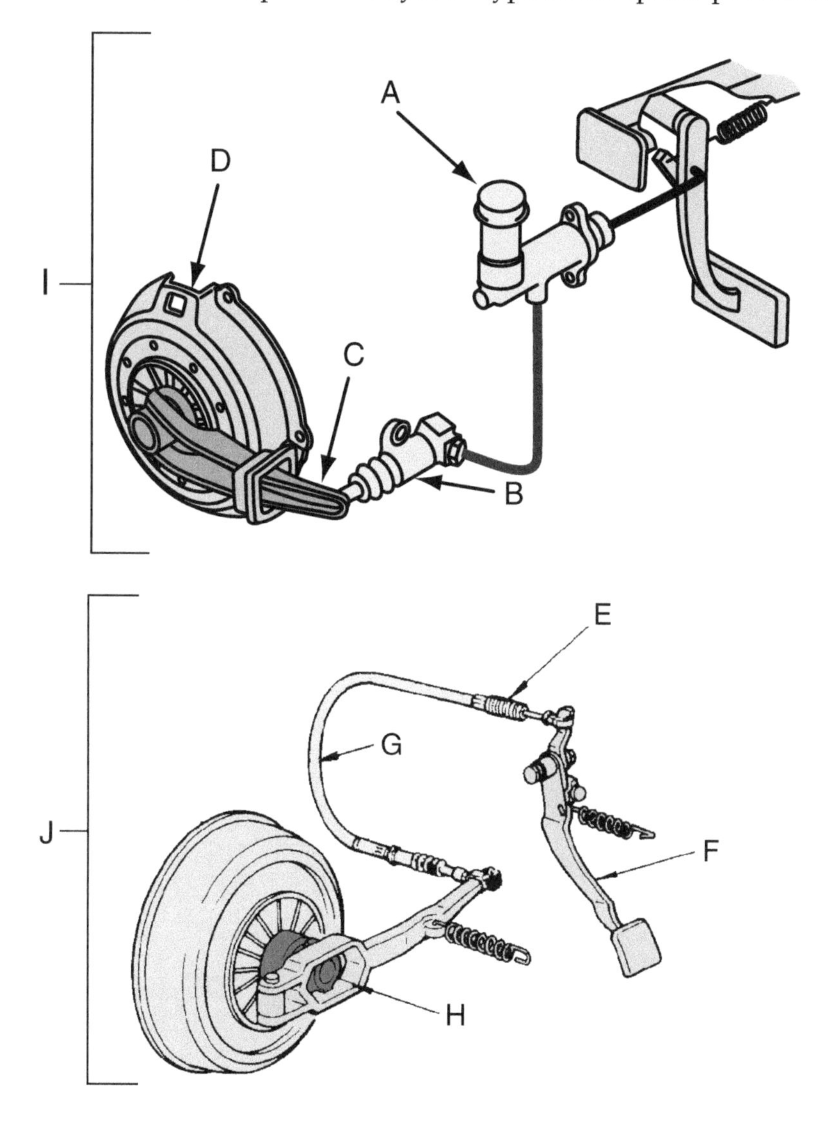

Master Cylinder	______	Release Fork	______	Clutch Pedal	______
Release Bearing	______	Hydraulically Operated Clutch	______	Slave Cylinder	______
Clutch Adjustment	______	Clutch Cover	______	Cable-Operated Clutch	______
Clutch Cable	______				

STOP

Instructor OK ______________ Score __________

Activity Sheet #127

CLUTCH

Name ______________________ Class ______________

Directions: Match the words on the left to the descriptions on the right. Write the letter for the correct word on the line provided. For the terms that you are not certain of, use the glossary in your textbook.

A. Friction disc	_____	It has splines and connects the clutch disc to the transmission
B. Clutch hub	_____	It connects the release bearing to the clutch cable or linkage
C. Dampened hub	_____	It contacts the rotating clutch to release the disc
D. Clutch facings	_____	A type of spring that replaces the release levers and coil springs in a diaphragm clutch
E. Clutch cushion plate	_____	Another name for the pressure plate assembly
F. Release levers	_____	The clutch pedal return spring
G. Diaphragm spring	_____	The output piston in a hydraulic clutch
H. Release bearing	_____	A bearing or bushing in the crankshaft that supports the transmission input shaft
I. Throwout bearing	_____	The part that presses the clutch disc against the flywheel
J. Clutch fork	_____	What the clutch does when you apply the clutch pedal
K. Overcenter spring	_____	The parts of a coil spring clutch that pull the pressure plate away from the flywheel
L. Slave cylinder	_____	The inner part of a clutch disc
M. Clutch freeplay	_____	Clutch part that absorbs shock during engagement
N. Input shaft	_____	Another name for a throwout bearing
O. Self centering	_____	A metal cushion that lets the clutch facings compress
P. Clutch cover	_____	Driven member of a clutch
Q. Pilot	_____	The friction material part of the clutch disc
R. Releases	_____	A term that describes movement measured at the clutch pedal
S. Pressure plate	_____	Type of release bearing used in FWD vehicles
T. Dual clutch transmission	_____	Operates alternating transmission input shafts

STOP

Instructor OK ______________ Score __________

Activity Sheet #128

IDENTIFY MANUAL TRANSMISSION COMPONENTS

Name ________________________ Class ________________

ASE Education Foundation Correlation

This activity sheet addresses the following **MLR** task:

III.A.4 Identify manual drivetrain and axle components and configuration. **(P-1)**

We Support Education Foundation

Directions: Identify the manual transmission components in the drawing. Place the identifying letter next to the name of each component.

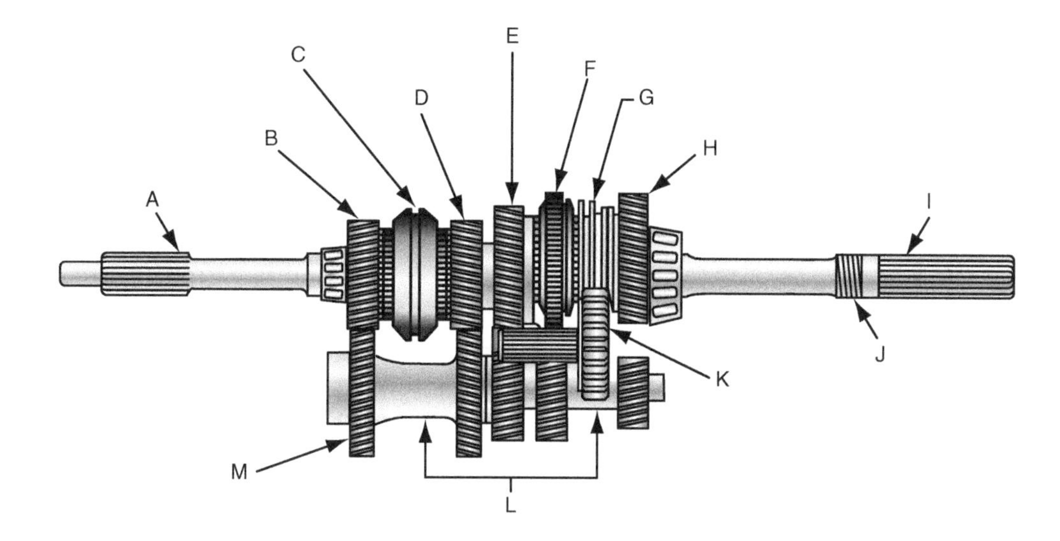

Second Gear	______	Input Shaft	______	Clutch Gear	______
Output Shaft	______	Reverse Idler Gear	______	Reverse Driven Gear	______
Speedometer Gear	______	First Gear	______	1st/2nd/Rev Synchronizer	______
Third Gear	______	3rd/4th Synchronizer	______	Gear Cluster Assembly	______
Counter Gear	______				

STOP

Instructor OK ________________ Score ____________

Activity Sheet #129

TRACE MANUAL TRANSMISSION POWER FLOW

Name ______________________________ Class ______________________

ASE Education Foundation Correlation

This activity sheet addresses the following **MAST** task:

III.C.3 Diagnose noise concerns through the application of transmission/transaxle powerflow principles. **(P-2)**

We Support ASE Education Foundation

Directions: Use a colored pen or pencil to trace the power flow through each of the gear ranges of the transmission in the drawings. On the next page, place the identifying letter next to the gear range indicated.

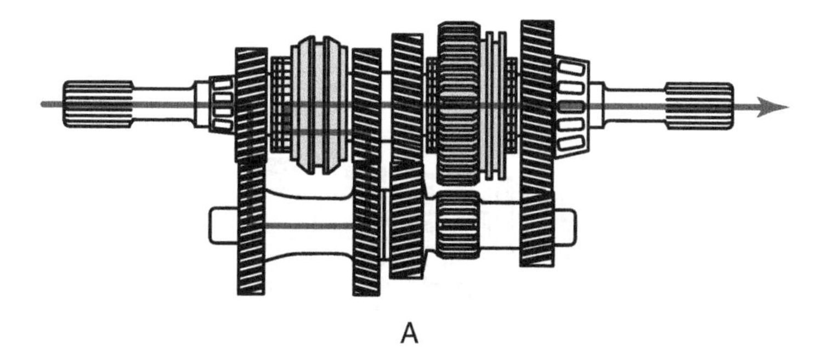

A

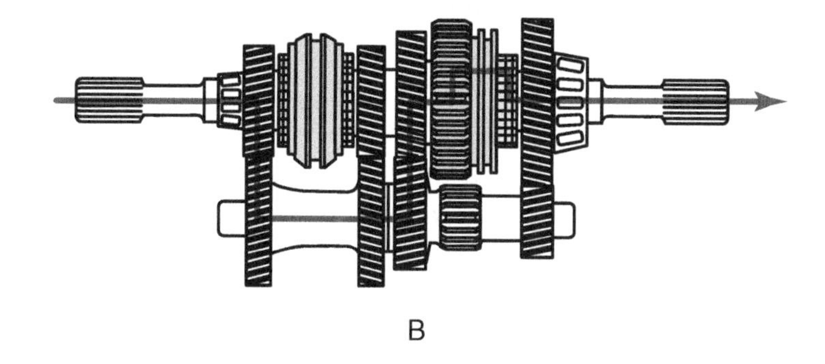

B

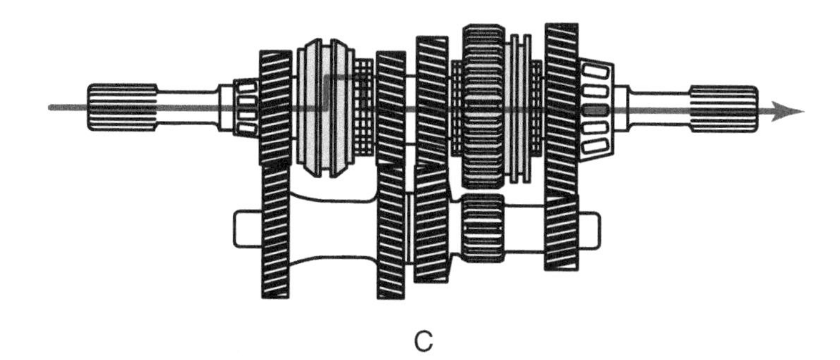

C

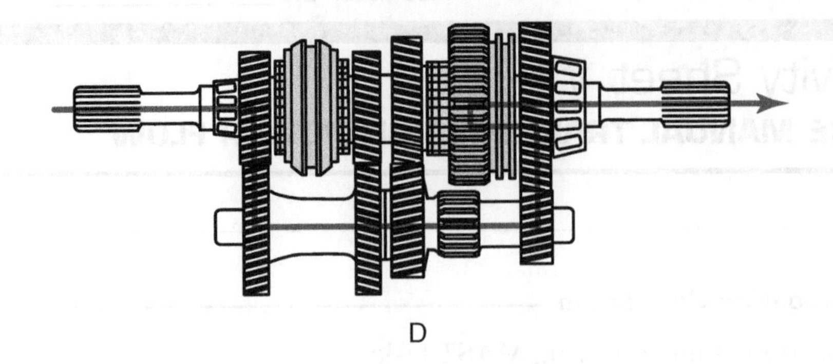

D

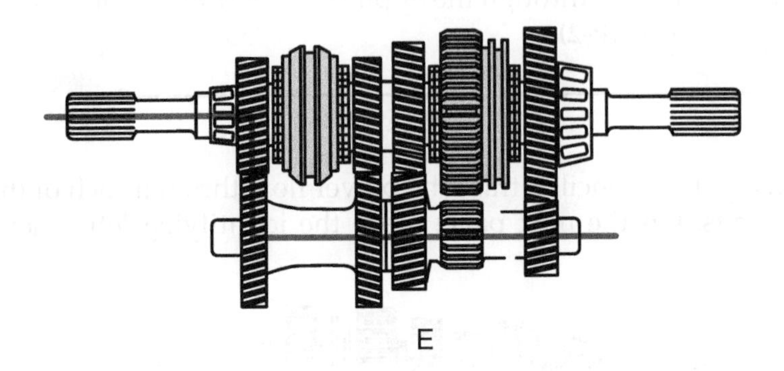

E

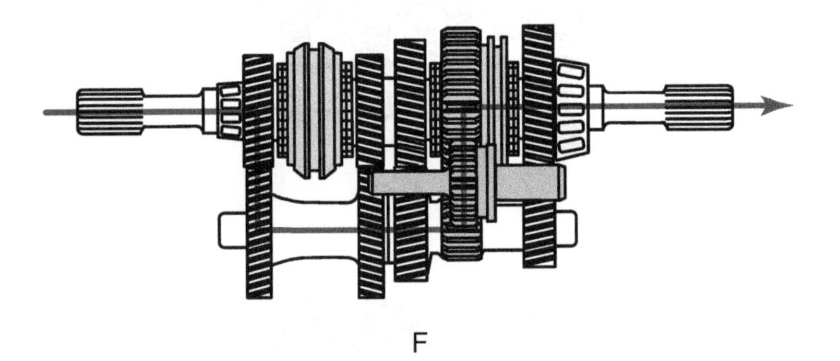

F

First Gear ______

Second Gear ______

Third Gear ______

Fourth Gear ______

Reverse ______

Neutral ______

STOP

Instructor OK ______________ Score __________

Activity Sheet #130

MANUAL TRANSMISSION

Name ______________________________ Class ____________________

Directions: Match the words on the left to the descriptions on the right. Write the letter for the correct word on the line provided. For the terms that you are not certain of, use the glossary in your textbook.

Term		Description
A. Lower gear ratio	___	The ratio between the transmission output shaft and the differential ring gear
B. Gear ratio	___	The clearance between meshing gear teeth
C. Dual mass flywheel	___	A gear used to change direction of rotation
D. Granny gear	___	Another name for the countergear
E. Final drive ratio	___	The most popular style of synchronizer
F. Meshed	___	When there is a large difference between the ratios of a transmission's forward gears
G. Spur gears	___	Gear ratio for high gear in a manual transmission
H. Backlash	___	When there is a small difference between the ratios of a transmission's forward gears
I. Helical gears	___	When a transmission has a very low first gear
J. Idler gear	___	Approximate ratio for first gear
K. Synchronizer	___	Calculated by dividing the number of teeth on the driven gear by the number of teeth on the driving gear
L. Blocker ring	___	Gears designed with straight-cut teeth
M. Countergear	___	When the output shaft turns faster than the input shaft
N. Dog teeth	___	A quieter operating gear design than spur gears
O. Final drive	___	A simple gear design with straight-cut teeth
P. Spur gear	___	Lubricant used in many manual transmissions
Q. 2:1	___	Keeps two meshing gears from clashing during a shift
R. Clutch shaft	___	Output speed is slower
S. Cluster	___	Little teeth around the circumference of a gear
T. 1:1	___	Output from the differential ring gear
U. Torque goes up	___	Another name for the input shaft
V. 3:1	___	One assembly made up of a series of gears (not cluster)
W. SAE 90	___	Result when a small gear drives a larger gear

TURN

X.	Reverse	___	Requires an idler gear
Y.	Wide ratio transmission	___	When two gears are engaged
Z.	Close ratio transmission	___	Common gear ratio for second gear
AA.	Overdrive	___	Reduces vibration and noise

STOP

Instructor OK ______________ Score ______________

Activity Sheet #131

IDENTIFY AUTOMATIC TRANSMISSION COMPONENTS

Name ______________________________ Class ______________________

ASE Education Foundation Correlation

This activity sheet addresses the following **MLR** task:

II.A.5 Identify drive train components and configuration. **(P-1)**

This activity sheet addresses the following **AST** task:

II.A.8 Diagnose transmission/transaxle gear reduction/multiplication concerns using driving, driven, and held member (power flow) principles. **(P-1)**

This activity sheet addresses the following **MAST** task:

II.A.10 Diagnose transmission/transaxle gear reduction/multiplication concerns using driving, driven, and held member (power flow) principles. **(P-1)**

We Support
ASE Education Foundation

Directions: Identify the automatic transmission components in the drawing. Place the identifying letter next to the name of each component.

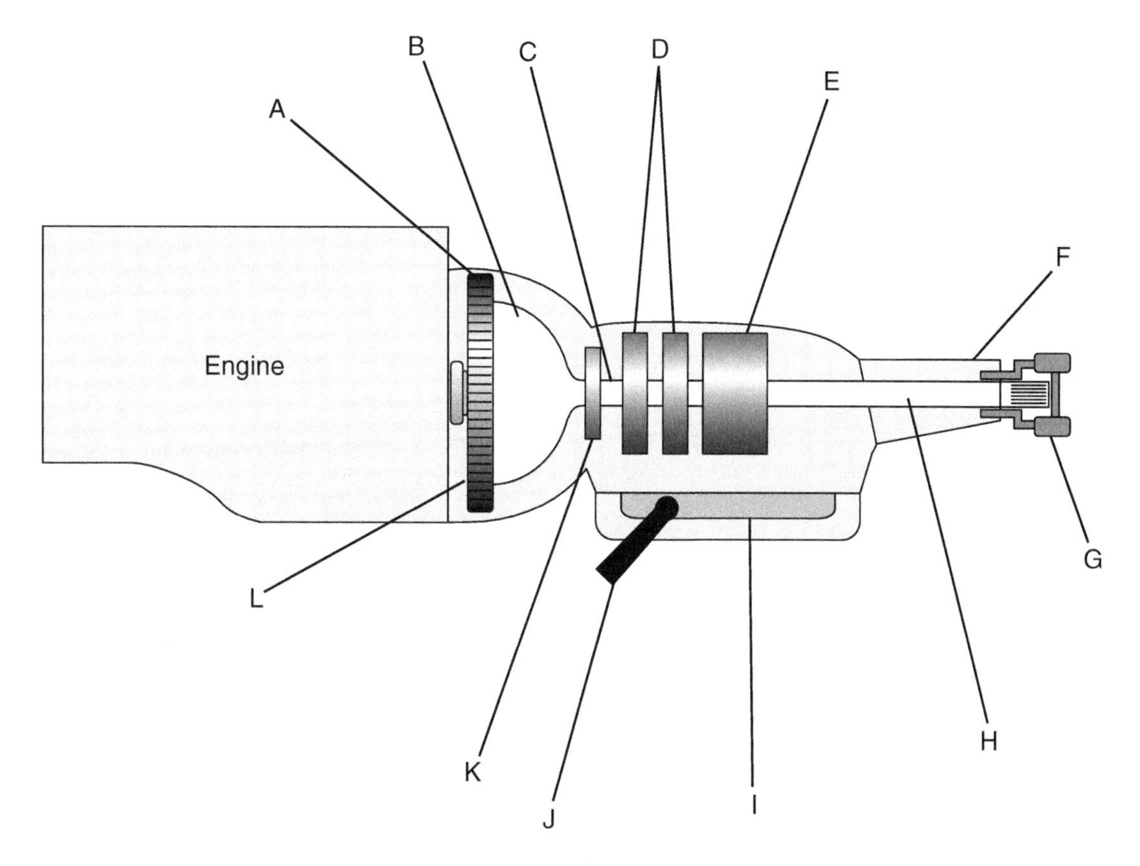

Extension Housing	______	Starter Ring Gear	______	Driveshaft Yoke	______
Clutch Packs	______	Pump	______	Input Shaft	______
Valve Body	______	Torque Converter	______	Planetary Gears	______
Flexplate	______	Output Shaft	______	Shift Lever	______

STOP

Instructor OK _______________ Score ___________

Activity Sheet #132

IDENTIFY AUTOMATIC TRANSMISSION PARTS

Name ______________________________ Class ____________________

ASE Education Foundation Correlation

This activity sheet addresses the following **MLR** task:

II.A.5 Identify drive train components and configuration. **(P-1)**

This activity sheet addresses the following **AST** task:

II.A.8 Diagnose transmission/transaxle gear reduction/multiplication concerns using driving, driven, and held member (power flow) principles. **(P-1)**

This activity sheet addresses the following **MAST** task:

II.A.10 Diagnose transmission/transaxle gear reduction/multiplication concerns using driving, driven, and held member (power flow) principles. **(P-1)**

We Support

Directions: Identify the automatic transmission parts in the drawing. Place the identifying letter next to the name of each part listed below the diagram.

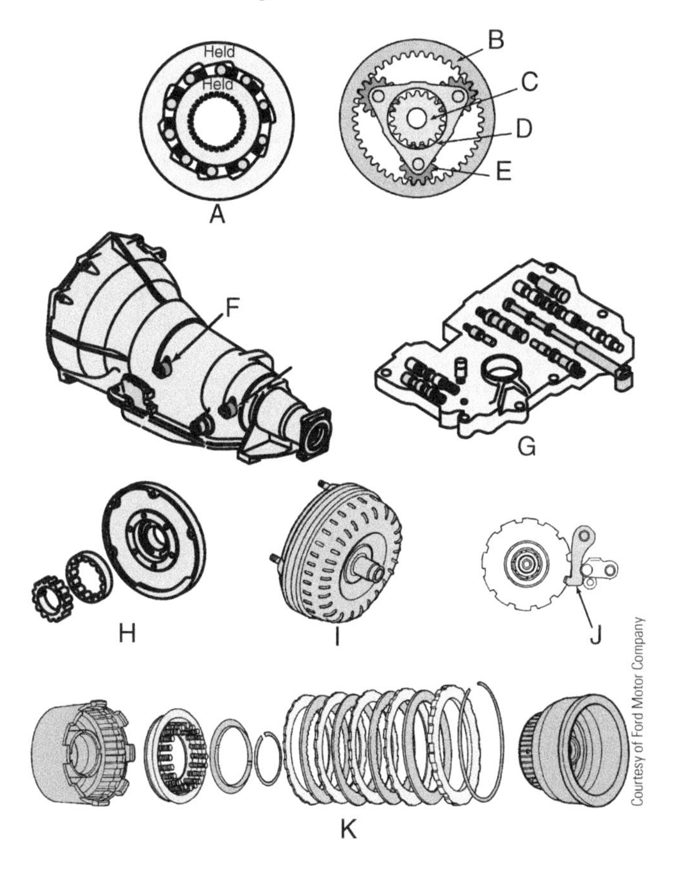

Clutch Pack	______	Carrier	______	Oil Pump	______
Pinion Gear	______	Torque Converter	______	Valve Body	______
Ring Gear	______	Park Pawl	______	Sun Gear	______
One-Way Clutch	______	Speed Sensors	______		

STOP

Instructor OK ______________ Score __________

Activity Sheet #133

AUTOMATIC TRANSMISSION

Name ______________________________ Class ______________________

Directions: Match the words on the left to the descriptions on the right. Write the letter for the correct word on the line provided. For the terms that you are not certain of, use the glossary in your textbook.

A. Fluid coupling
B. Torque converter
C. One-way clutch
D. Overrunning clutch
E. Coupling speed
F. Stall speed
G. Brake band
H. Accumulator
I. Spool valve
J. Orifice
K. Valve body
L. Shift quadrant
M. Upshift
N. Downshift
O. Throttle pressure
P. Governor pressure
Q. WOT
R. Detent
S. Park pawl
T. CVT

______ A restriction in a passage to slow down the flow of fluid

______ A valve that has lands, valleys, and faces

______ The device that locks the output shaft of the transmission when the shift lever is placed in park

______ Two planetary gearsets combined to provide more gear ratio possibilities

______ Forced kickdown

______ Another name for a fluid clutch

______ A torque converter with a friction disc that locks the impeller and turbine together

______ A continuously variable transmission

______ When the transmission shifts from a low gear to a higher gear; second to third, for instance

______ When the transmission shifts to a lower gear

______ The highest engine rpm that can be obtained when the vehicle is being prevented from moving while the engine is accelerated

______ A fluid coupling that multiplies torque

______ A compound planetary gear design that shares the same sun gear between the gearsets

______ A planetary gear design with long and short pinions

______ Another name for an overrunning clutch

______ The hydraulic control assembly of the transmission

______ The readout on the gear selector that selects what gear the transmission is in

______ The point at which the converter parts and ATF all turn as a unit

______

______ Pressure that results in response to engine load

______ A reservoir used in timing and cushioning gear shifts

TURN

U.	Impeller	_____	A device that locks in one direction and freewheels in the other
V.	Dual-clutch transmission	_____	Pressure that results from increases in vehicle speed
W.	Modulator	_____	An external brake planetary holding device
X.	Lock-up torque converter	_____	This part is actually part of the converter housing
Y.	Two clutches	_____	Wide-open throttle
Z.	Compound planetary gears	_____	A vacuum-operated diaphragm that controls shift points
AA.	Simpson geartrain	_____	Used instead of a torque converter in a DCT
AB.	Fifteen percent	_____	Uses two separate gear trains in one transmission
AC.	Ravigneaux geartrain	_____	Amount of fuel economy increase when a DCT is used

STOP

Instructor OK ______________ Score __________

Activity Sheet #134

IDENTIFY DIFFERENTIAL COMPONENTS

Name ____________________________ Class ____________________

ASE Education Foundation Correlation

This activity sheet addresses the following **MLR** task:

III.A.4 Identify manual drivetrain and axle components and configuration. **(P-1)**

We Support ASE | Education Foundation

Directions: Identify the differential components in the drawing. Place the identifying letter next to the name of each component.

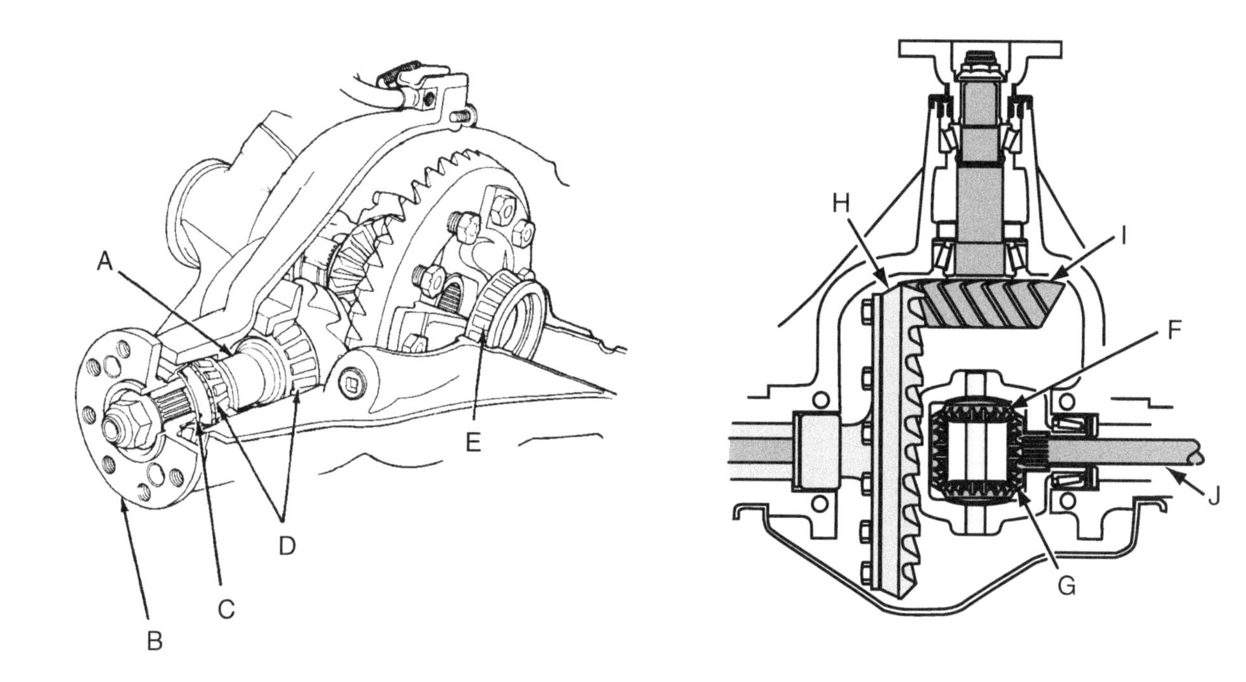

Drive Pinion Bearings	______	Companion Flange	______	Differential Bearing	______
Differential Pinion	______	Pinion Seal	______	Crush Sleeve	______
Drive Pinion	______	Ring Gear	______	Side Gear	______
Axle	______				

STOP

Instructor OK ______________ Score __________

Activity Sheet #135

DRIVELINE AND DIFFERENTIAL

Name ______________________________ Class ____________________

Directions: Match the words on the left to the descriptions on the right. Write the letter for the correct word on the line provided. For the terms that you are not certain of, use the glossary in your textbook.

Terms		Descriptions
A. Driveshaft	______	A universal joint whose output and input speed are constant
B. C-lock or C-clip axle	______	An axle with a groove on the inside that a clip fits into to keep it in place
C. Slip yoke	______	Another name for side gears and differential pinions
D. Constant velocity joint	______	A differential gearset where the pinion gear is lower than the centerline of the ring gear
E. Bearing retained axle	______	A pressed-fit axle with a bearing retainer ring
F. Cardan joint	______	A type of rear axle in which the differential is not removable as an assembly
G. EP additives	______	The part of the driveshaft assembly that slides in and out of the transmission
H. Salisbury axle	______	A universal joint used with RWD
I. Spider gears	______	It locks up the spider gears when one wheel starts to lose traction
J. Hypoid gears	______	The assembly that transfers power from the transmission to the rear wheels
K. Limited slip	______	Part of the lubricant package that prevents welding between metal surfaces

STOP

Instructor OK ______________ Score ____________

Activity Sheet #136

IDENTIFY FRONT-WHEEL-DRIVE (FWD) AXLE SHAFT COMPONENTS

Name______________________________ Class______________________

ASE Education Foundation Correlation

This activity sheet addresses the following **MLR** task:

III.A.4 Identify manual drivetrain and axle components and configuration. **(P-1)**

We Support ASE Education Foundation

Directions: Identify the front-wheel-drive (FWD) axle shaft components in the drawing. Place the identifying letter next to the name of each component.

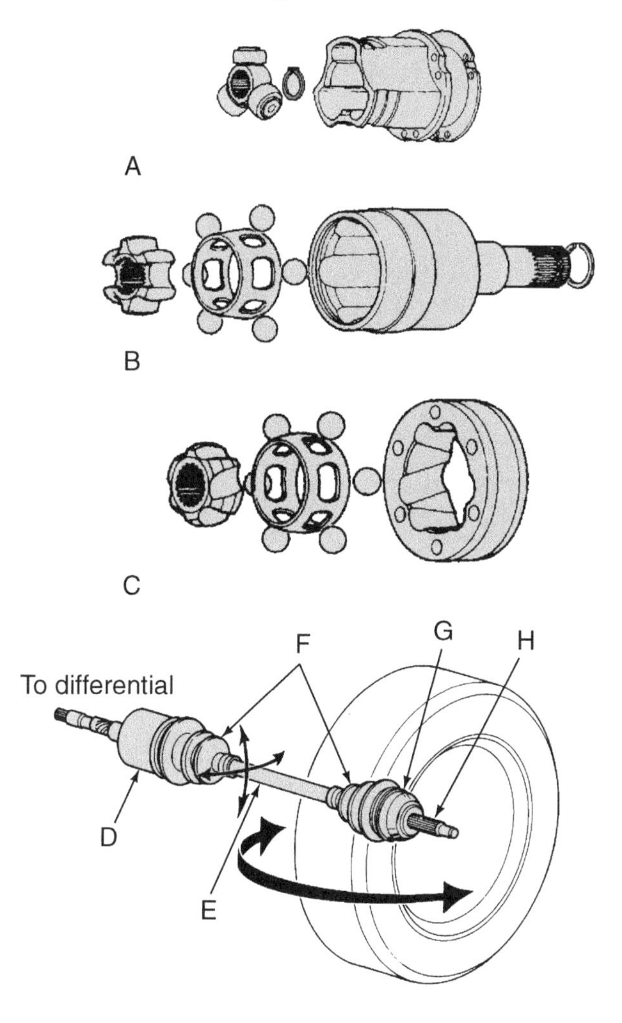

Cross Groove	______	Tripod Tulip	______	Double Offset	______
CV Joint Boots	______	Outboard Joint	______	Inboard Joint	______
Stub Axle	______	Drive Axle	______		

Instructor OK ______________ Score ___________

Activity Sheet #137

IDENTIFY FOUR-WHEEL-DRIVE (4WD) COMPONENTS

Name ______________________________ Class ______________________

ASE Education Foundation Correlation

This activity sheet addresses the following **MLR** task:

III.A.4 Identify manual drivetrain and axle components and configuration. **(P-1)**

We Support ASE | Education Foundation

Directions: Identify the four-wheel-drive components in the drawing. Place the identifying letter next to the name of each component listed below the illustration.

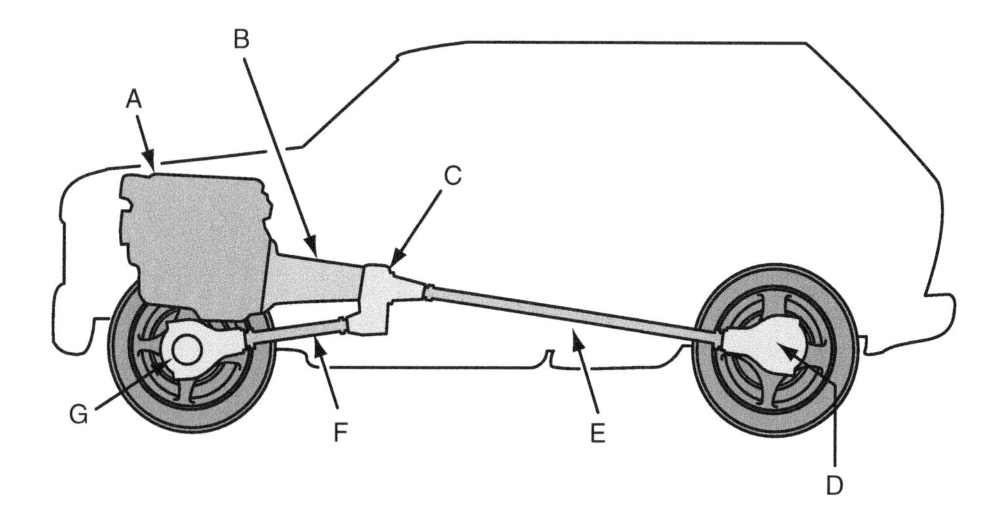

Front Driveshaft	______	Front Differential	______	Rear Driveshaft	______
Transfer Case	______	Engine	______	Transmission	______
Rear Differential	______				

STOP

Instructor OK ______________ Score __________

Activity Sheet #138

IDENTIFY TRANSAXLE COMPONENTS

Name ____________________________ Class ____________________

ASE Education Foundation Correlation

This activity sheet addresses the following **MLR** task:

III.A.4 Identify manual drivetrain and axle components and configuration. **(P-1)**

This activity sheet addresses the following **MAST** task:

III.3.C Diagnose noise concerns through the application of transmission/transaxle powerflow principles. **(P-2)**

We Support ASE | Education Foundation

Directions: Identify the transaxle components in the drawing. Place the identifying letter next to the name of each component.

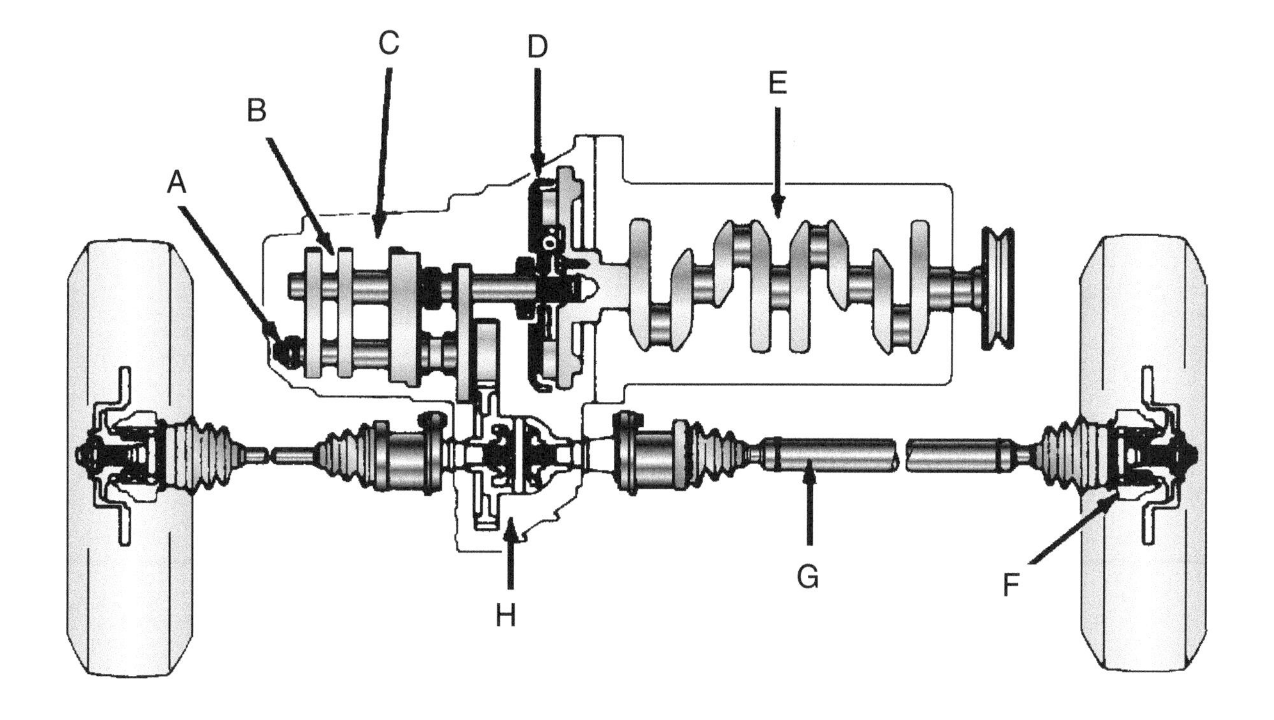

Drive Shaft	______	Transaxle	______	Input Shaft	______
Differential	______	Front Axle	______	Engine	______

Instructor OK ______________ Score __________

Activity Sheet #139

IDENTIFY MANUAL TRANSAXLE COMPONENTS

Name________________________ Class________________

Directions: Identify manual transaxle components in the drawing. Place the identifying letter next to the name of each component.

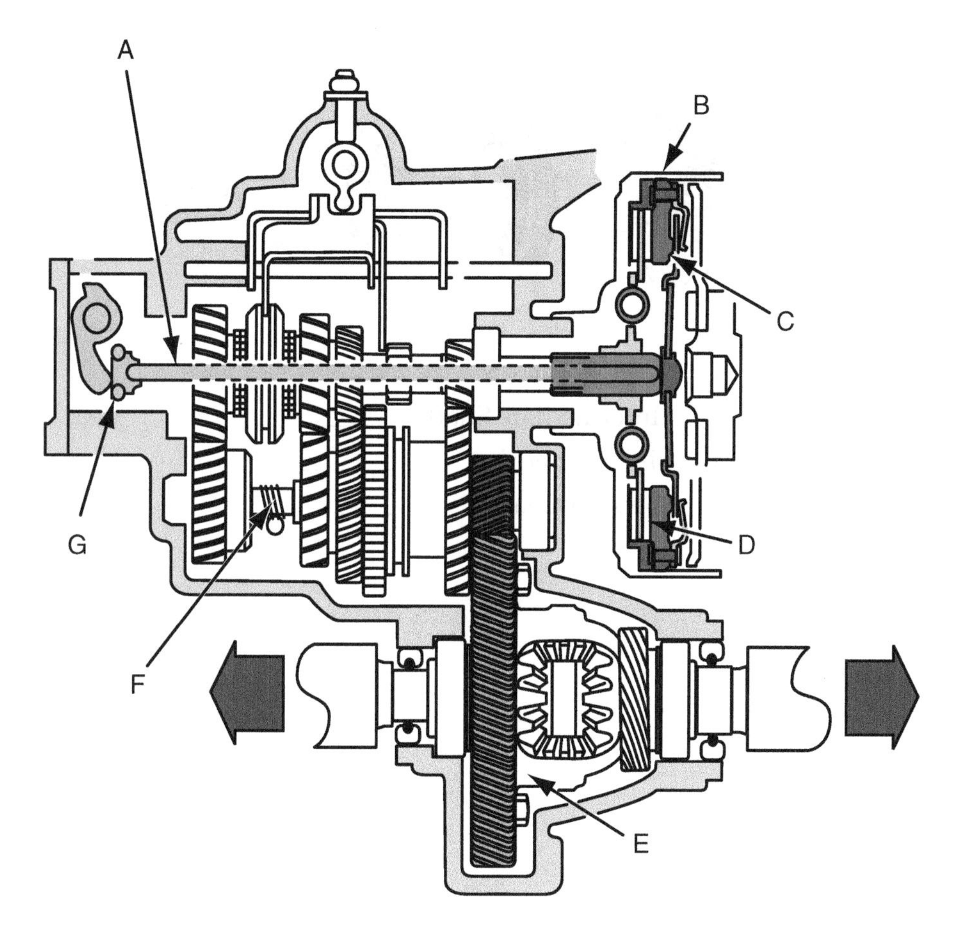

Clutch Disc	______	Clutch Apply Rod	______	Clutch Release Bearing	______
Flywheel	______	Differential Carrier	______	Clutch Pressure Plate	______
Intermediate Shaft	______				

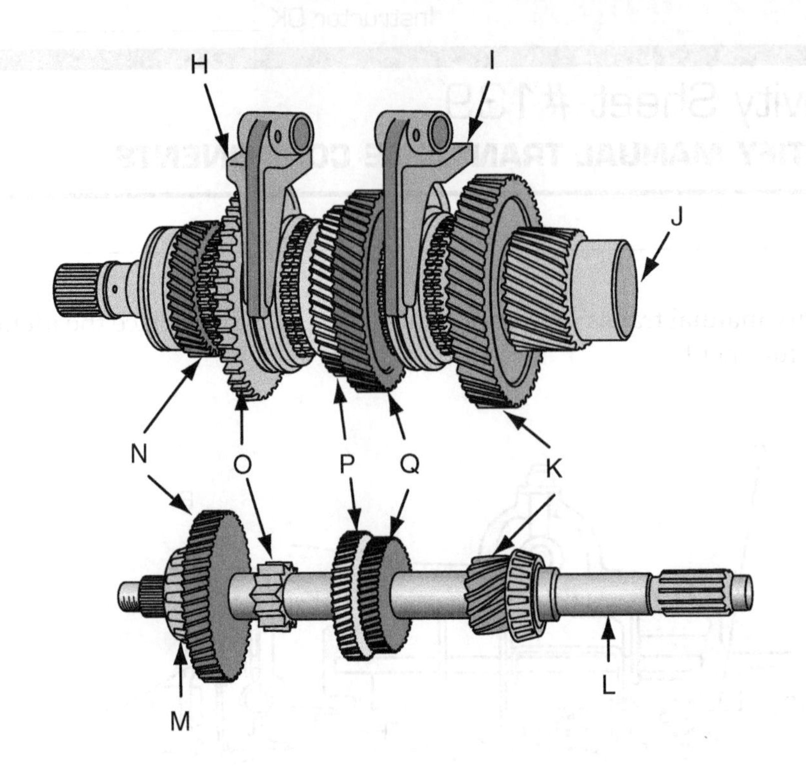

3-4 Shift Fork	______	Intermediate Shaft	______	Rear Bearing	______
Third Gear	______	1-2 Shift Fork	______	Input Shaft	______
Fourth Gear	______	Second Gear	______	First Gear	______
Reverse	______				

Activity Sheet #140

FRONT-WHEEL DRIVE

Name____________________________ Class____________________

Directions: Match the words on the left to the descriptions on the right. Write the letter for the correct word on the line next to the term. If you are uncertain of the meaning of a term, use the glossary in your textbook.

A. Transaxle	______	The most commonly replaced part of a front axle assembly
B. Halfshaft	______	The drive wheels on most cars currently produced
C. CV joint	______	A CV joint that allows for a change in the angle but not the length
D. CVT	______	This is done to the stub shaft nut after it is torqued
E. Inboard joint	______	The sound a worn outboard CV joint makes
F. Fixed joint	______	These allow the front wheels to turn at different speeds during a turn
G. Rzeppa CV joint	______	The name for a drive axle assembly on a front-wheel-drive vehicle
H. Better fuel economy	______	Constant velocity joint (abbreviation)
I. Front wheels	______	The most common type of outboard, fixed CV joint
J. Click	______	A combination transmission and differential
K. Clunk	______	The inside CV joint on a FWD car
L. FWD vehicles		The sound a worn inboard CV joint makes
M. Boot	______	When the car pulls to one side during hard acceleration
N. Staked	______	A transaxle that varies output speed by changing pulley diameter
O. Pushing chain		A CVT advantage because the engine can run at relatively constant RPM
P. Torque steer	______	A trait of the continuously variable transmission used in some small vehicles & snowmobiles
Q. Differential gears	______	Used in CVTs; it has less surface area than a flat one
R. Infinite gear ratios	______	The drivetrain design where a typical CVT is found

STOP

Instructor OK ______________ Score __________

Activity Sheet #141

TRACE MANUAL TRANSAXLE POWER FLOW

Name________________________ Class________________

ASE Education Foundation Correlation

This activity sheet addresses the following **MAST** task:

III.3.C Diagnose noise concerns through the application of transmission/transaxle powerflow principles. **(P-2)**

We Support ASE Education Foundation

Directions: Use a colored pen or pencil to trace the power flow through each of the gear ranges of the transaxle in the drawings. Place the identifying letter next to the indicated gear range.

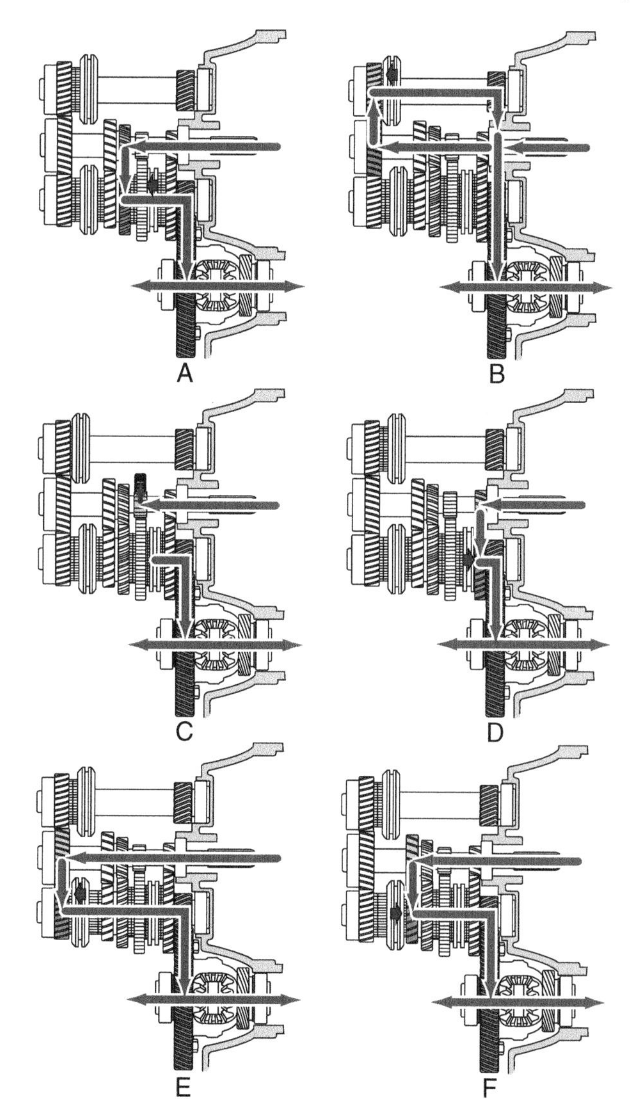

First Gear	______	Third Gear	______	Fifth Gear	______
Second Gear	______	Fourth Gear	______	Reverse	______

STOP

Instructor OK ______________ Score ________

Activity Sheet #142

IDENTIFY ALL-WHEEL-DRIVE (AWD) COMPONENTS

Name ______________________ Class ______________

Directions: Record the names of the components that the symbols represent in the spaces provided.

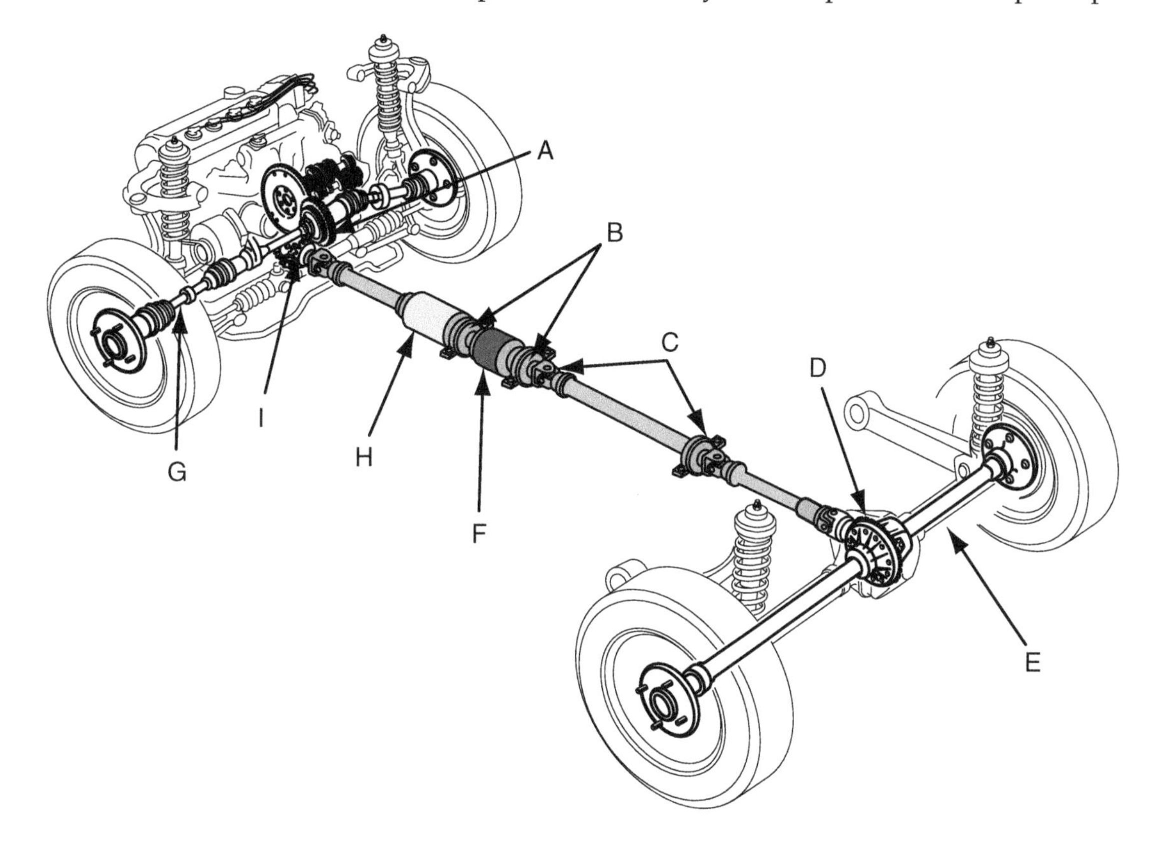

Viscous Coupling	______	Center Support Bearings	______	Tripod Joint	______
U-joints	______	Rear Differential	______	Front Driveshaft	______
Transfer Case	______	Rear Drive Axle	______	Front Differential	______

STOP

Instructor OK ______________ Score ____________

Activity Sheet #143

IDENTIFY HYBRID OPERATION

Name ______________________________ Class ______________________

ASE Education Foundation Correlation

This activity sheet addresses the following **MLR** task:

I.A.7 Identify service precautions related to service of the internal combustion engine of a hybrid vehicle. **(P-2)**

This activity sheet addresses the following **AST/MAST** task:

I.A.9 Identify service precautions related to service of the internal combustion engine of a hybrid vehicle. **(P-2)**

We Support ASE Education Foundation

Directions: Identify the electrical and mechanical energy flow in the drawing below in the following ways:

1. Identify the three operational modes of the hybrid system by placing the correct identifying letter in the blank.

   ______ Starting from a stop

   ______ Under full acceleration

   ______ Starting the engine

2. Identify the electrical energy flow during the three operating conditions by filling in the path with a blue pencil.

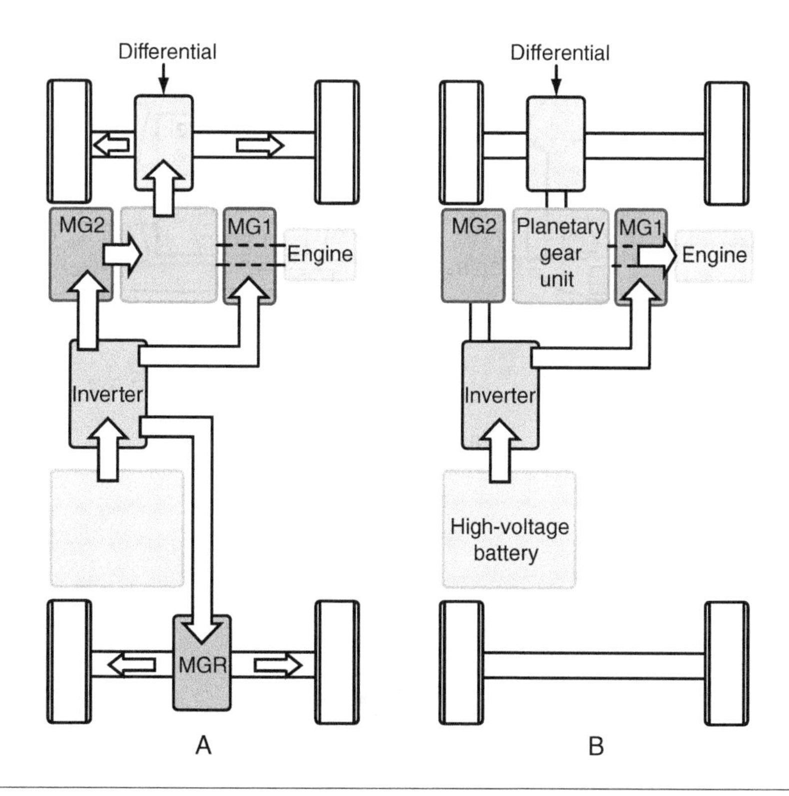

3. Identify the mechanical energy flow during the three operating conditions by filling in the path with a red pencil.

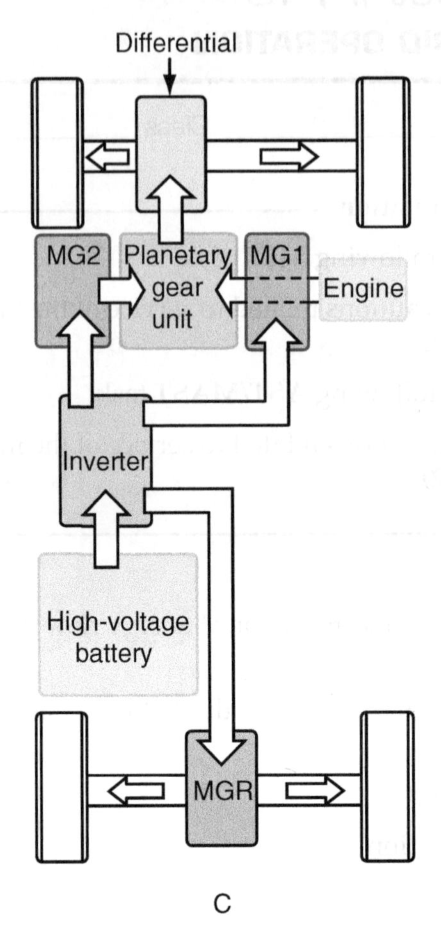

C

4. Identify a series or parallel hybrid system from the following illustrations.

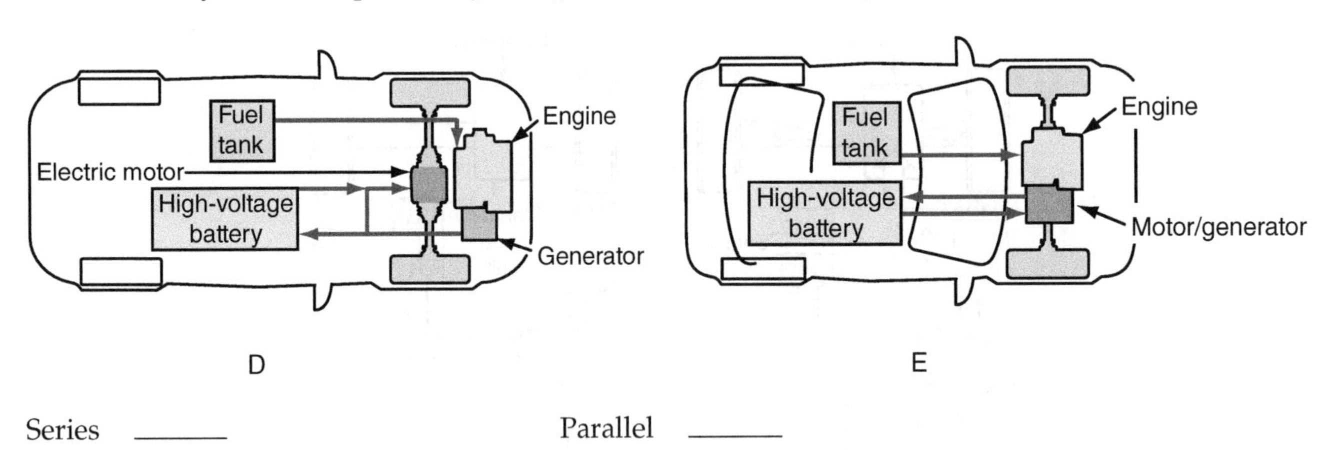

D

E

Series ______ Parallel ______

STOP

Part II

Lab Worksheets for ASE Maintenance and Light Repair

Introduction

ASE Education Foundation Worksheet #Intro-1

VEHICLE OWNER'S MANUAL

Name______________________________ Class____________________

Score: ☐ Excellent ☐ Good ☐ Needs Improvement **Instructor OK** ☐

Vehicle year __________ **Make** ____________________ **Model** ______________________

Objective: Upon completion of this assignment, you will be able to use a vehicle owner's manual to locate service specifications.

ASE Education Foundation Correlation

This worksheet addresses the following **Required Supplemental Task (RST):**

Preparing Vehicle for Service; Task 1:	Identify information needed and the service requested on a repair order.

This worksheet addresses the following **MLR** tasks:

I.A.1; II.A.1; III.A.1; IV.A.1; V.A.1; VI.A.1; VIII.A.1	Research vehicle service information, including fluid type, vehicle service history, service precautions, and technical service bulletins. **(P-1); (P-1); (P-1); (P-1); (P-1); (P-1); (P-1)**
VII.A.1	Research vehicle service information, including refrigerant/oil type, vehicle service history, service precautions, and technical service bulletins. **(P-1)**

This worksheet addresses the following **AST/MAST** tasks:

I.A.2	Research vehicle service information including fluid type, internal engine operation, vehicle service history, service precautions, and technical service bulletins. **(P-1)**
II.A.2; III.A.2; IV.A.1; V.A.2	Research vehicle service information including fluid type, vehicle service history, service precautions, and technical service bulletins. **(P-1); (P-1); (P-1); (P-1)**
VI.A.1; VIII.A.2	Research applicable vehicle and service information, vehicle service history, service precautions, and technical service bulletins. **(P-1); (P-1)**

We Support ASE | Education Foundation

Directions: Review this worksheet completely before starting. Use your vehicle owner's manual or one provided by your instructor. Record the requested information in the spaces provided. If you are completing this worksheet on your personal vehicle, you may want to save it for future reference.

Tools and Equipment Required: Vehicle owner's manual

Procedure:

Engine:	☐ 4-cylinder	☐ 6-cylinder	☐ 8-cylinder	
Transmission:	☐ Manual	☐ Automatic		
Drivetrain:	☐ RWD	☐ FWD	☐ AWD	☐ 4WD

What is the make and model of the owner's manual being used to complete this worksheet?

____

Capacities:

Battery	______ Cold cranking amps (CCA)	☐ N/A
Crankcase (with oil filter)	______ qt.	☐ N/A
Cooling system	______ qt.	☐ N/A
Differential capacity (RWD)	______ pt./qt.	☐ N/A
Transmission capacity (RWD)	______ pt./qt.	☐ N/A
Transaxle (FWD)	______ pt./qt.	☐ N/A
Fuel tank capacity	______ gallons	☐ N/A

Fluid Specifications:

SAE engine oil viscosity	______	☐ N/A
Oil service rating (grade)	______	☐ N/A
Transmission fluid type (RWD)	______	☐ N/A
Differential lubricant type (RWD)	______	☐ N/A
Transaxle fluid type (FWD)	______	☐ N/A
Brake fluid type	______	☐ N/A
Fuel type	______	☐ N/A
Fuel octane rating	______	☐ N/A

Fuel Specifications:

Octane Rating ______

Fuel Type Gasoline ______ Diesel ______ Other ______

Tires:

Rotation pattern (Draw here)

Wheel nut torque specification ______ ft.-lb

Belt Tension Specifications:

No belt tension—serpentine self-adjuster ______ ☐ N/A
Alternator ______ ☐ N/A
Power steering pump ______ ☐ N/A
Air-conditioning compressor ______ ☐ N/A
Smog pump ______ ☐ N/A

Maintenance Reminder Lights:

Is the vehicle equipped with a maintenance reminder light? ☐ Yes ☐ No

If so, how is it reset?

____

____

____

Notes: ____

STOP

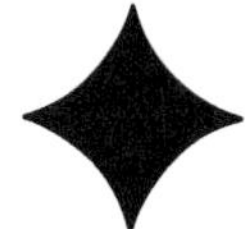

ASE Education Foundation Worksheet #Intro-2

VEHICLE IDENTIFICATION NUMBER (VIN)

Name________________________ Class__________________

Score: ☐ Excellent ☐ Good ☐ Needs Improvement **Instructor OK** ☐

Objective: Upon completion of this assignment, you will be able to identify the model year and country of manufacture of a vehicle.

ASE Education Foundation Correlation

This worksheet addresses the following **Required Supplemental Task (RST):**

Preparing Vehicle for Service, Task 1:	Identify information needed and the service requested on a repair order.

This worksheet addresses the following **MLR** tasks:

I.A.1; II.A.1; III.A.1; IV.A.1; V.A.1; VI.A.1; VIII.A.1	Research vehicle service information, including fluid type, vehicle service history, service precautions, and technical service bulletins. **(P-1); (P-1); (P-1); (P-1); (P-1); (P-1); (P-1)**
VII.A.1	Research vehicle service information, including refrigerant/oil type, vehicle service history, service precautions, and technical service bulletins. **(P-1)**

This worksheet addresses the following **AST/MAST** tasks:

I.A.2	Research vehicle service information including fluid type, internal engine operation, vehicle service history, service precautions, and technical service bulletins. **(P-1)**
II.A.2; III.A.2; IV.A.1; V.A.2	Research vehicle service information including fluid type, vehicle service history, service precautions, and technical service bulletins. **(P-1); (P-1); (P-1); (P-1)**
VI.A.1; VIII.A.2	Research applicable vehicle and service information, vehicle service history, service precautions, and technical service bulletins. **(P-1); (P-1)**

We Support ASE | Education Foundation

Directions: Before beginning this lab task, review the worksheet completely. Fill in the information in the spaces provided as you complete this assignment.

Related Information: Each manufacturer used its own sequence and meaning for the numbers and letters of the vehicle identification number (VIN) through the 1980 model year. After 1980 the VIN codes were standardized. VINs were required to include 17 characters. Each character identifies a characteristic of the vehicle. Information identified by the VIN includes the country of origin, model year, body style, engine, manufacturer, vehicle serial number, and more. The first and tenth characters are universal codes that all manufacturers must use. The first character of the VIN identifies the country where the vehicle was manufactured and the tenth identifies the model year of the vehicle. The chart on the next page identifies the meaning of the codes for these positions.

TURN

Procedure:

1. Use the code charts to identify the model year and country of manufacture for the following vehicle identification numbers.

VIN	Model Year	Country of Origin
a. 1G1CZ19H6HW135780	______	______________
b. J74VN13G9L5021104	______	______________
c. 1Y1SK5262TS005258	______	______________
d. 2MEPM6046KH637745	______	______________
e. 3T25V21E7WX055441	______	______________

2. Where is the VIN located on vehicles that are sold in the United States?

____

3. Locate the vehicle identification numbers on five vehicles manufactured after 1982. List the VIN, the model year, and the country of origin below.

VIN	Model Year	Country of Origin
a. ______________	______	______________
b. ______________	______	______________
c. ______________	______	______________
d. ______________	______	______________
e. ______________	______	______________

4. Which character (letter or digit) of the VIN of a domestic vehicle is used to identify the engine?

____

Code Charts:

	Country of Origin (1st character)		Model Year (10th character)		Model Year (10th character)
1	United States	A	1980/2010	S	1995/2025
2	Canada	B	1981/2011	T	1996/2026
3	Mexico	C	1982/2012	V	1997/2027
4	United States	D	1983/2013	W	1998/2028
5	United States	E	1984/2014	X	1999/2029
6	Australia	F	1985/2015	Y	2000/2030
9	Brazil	G	1986/2016	1	2001/2031
J	Japan	H	1987/2017	2	2002/2032
K	Korea	J	1988/2018	3	2003/2033
L	Taiwan	K	1989/2019	4	2004/2034
A	England	L	1990/2020	5	2005/2035
F	France	M	1991/2021	6	2006/2036
V	Europe	N	1992/2022	7	2007/2037
W	Germany	P	1993/2023	8	2008/2038
Y	Sweden	R	1994/2024	9	2009/2039
Z	Italy				

ASE Education Foundation Worksheet #Intro-3

IDENTIFYING SHOP EQUIPMENT

Name______________________________ Class______________________

Score: ☐ Excellent ☐ Good ☐ Needs Improvement **Instructor OK** ☐

Objective: Upon completion of this assignment, you will be able to identify the equipment commonly used in an automotive repair shop.

ASE Education Foundation Correlation

This worksheet addresses the following **Required Supplemental Task (RST)**:

Tools and Equipment, Task 1: Identify tools and their usage in automotive applications.

We Support ASE Education Foundation

Directions: Write your answers in the spaces provided. In column A, list ten pieces of automotive equipment that are available in your school shop. In column B, briefly describe the purpose of each piece of equipment.

	Column A	Column B
1.	____________________	______________________________
2.	____________________	______________________________
3.	____________________	______________________________
4.	____________________	______________________________
5.	____________________	______________________________
6.	____________________	______________________________
7.	____________________	______________________________
8.	____________________	______________________________
9.	____________________	______________________________
10.	____________________	______________________________

Use the following list to identify the equipment pictured on the next page. Write the answer in the space provided next to each picture.

1. Drill press	2. Bench vise	3. Bench grinder	4. Solvent tank
5. Tire changer	6. Battery charger	7. Tire balancer	8. Arbor press
9. Parts cleaner	10. Valve grinder	11. Oscilloscope	12. Grease gun
13. Jack stands/safety stands	14. Hydraulic jack	15. Bottle jack	16. Drill motor

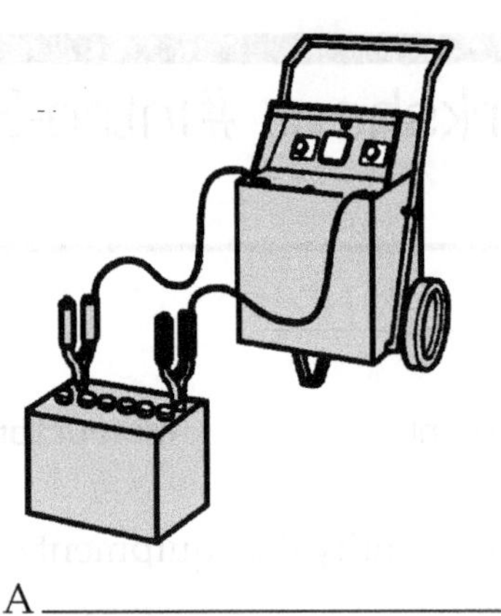

A ____________________

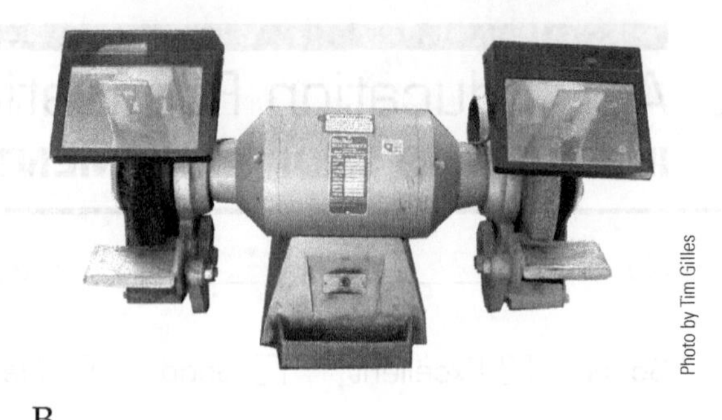

Photo by Tim Gilles

B ____________________

Photo by Tim Gilles

C ____________________

Courtesy of Hunter Engineering Company

D ____________________

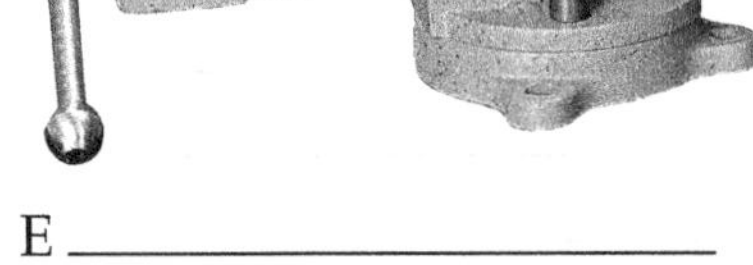

E ____________________

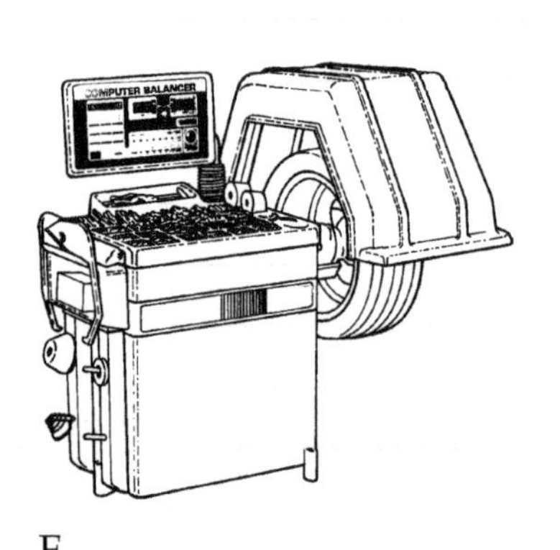

F ____________________

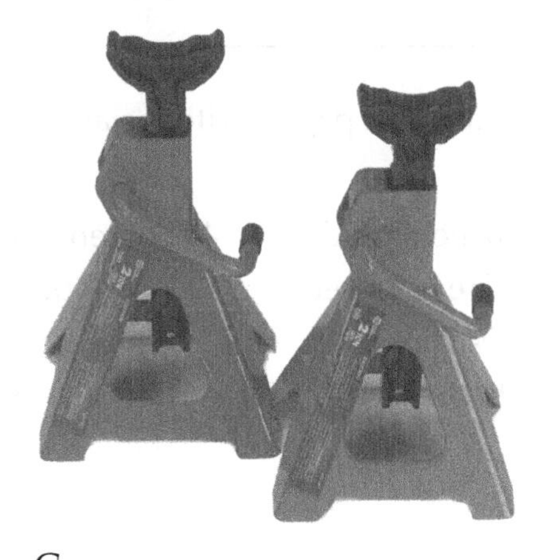

G ____________________

Photo by Tim Gilles

H ____________________

ASE Education Foundation Worksheet #Intro-4

SAFETY TEST

Name ______________________ Class ________________

Score: ☐ Excellent ☐ Good ☐ Needs Improvement **Instructor OK** ☐

Objective: Upon completion of this assignment, you will have an understanding of the hazards that are present in the automotive shop environment.

ASE Education Foundation Correlation

This worksheet addresses the following **Required Supplemental Tasks (RST)**:

Shop and Personal Safety, Task 1:	Identify general shop safety rules and procedures.
Shop and Personal Safety, Task 10:	Comply with the required use of safety glasses, ear protection, gloves, and shoes during lab/shop activities.
Shop and Personal Safety, Task 11:	Identify and wear appropriate clothing for lab/shop activities.
Shop and Personal Safety, Task 12:	Secure hair and jewelry for lab/shop activities.
Tools and Equipment, Task 3:	Demonstrate safe handling and use of appropriate tools.
Tools and Equipment, Task 4:	Demonstrate proper cleaning, storage, and maintenance of tools and equipment.

We Support
ASE Education Foundation

Most accidents are caused by impatience, carelessness, or poor judgment. The most common accidents in an automotive repair shop involve eye injuries and fires.

Directions: Choose the word from the list that best completes each statement. The words or phrases may be used more than once and not all words or phrases are used. Write your choice in the space provided.

vapor	radiator hose	rags	acid
clean sweep	clothes	black	blue
extinguisher	explosions	CO_2	eye protection
electrical	baking soda	water	one foot
CO	positive	hydraulic jack	jack
creeper	air	skin	tires
liquid	dust	fan belt	dressed
tool rest	side	ground	green
negative	battery	instructor	friend
gasoline	vehicle		

1. Gasoline in its ____________________ form is the most dangerous.
2. Place dirty____________________ in an approved receptacle.
3. Never use ____________________ to clean parts.
4. Wipe up or use ____________________ on all oil, brake fluid, or grease spills.
5. A fire ____________________ should be used on fuel fires.
6. Two types of common fire extinguishers are dry powder and ____________________.
7. Before an ____________________ fire can be extinguished, the electrical system must be disconnected.
8. Before opening a radiator, test for pressure in the system by squeezing the upper ____________________.
9. Battery ____________________ is a chemical combination of sulfuric acid and water.
10. Battery acid will cause holes in ____________________.
11. The most common cause of battery ____________________ is the battery charger.
12. ____________________ must be worn around batteries, air-conditioning machinery, compressed air, and other hazardous situations.
13. Battery acid may be neutralized with ____________________.
14. If acid gets on skin or eyes, immediately flush with ____________________ for at least 15 minutes.
15. Before raising a vehicle all the way on a hoist, raise it about ____________________ and shake the vehicle to be certain it is properly placed.
16. Use a ____________________ to raise and lower a vehicle only.
17. A jacked-up vehicle should be placed firmly on ____________________ stands.
18. When a ____________________ is not in use, it should be stored against a wall in a vertical position.
19. Compressed ____________________ is useful but dangerous.
20. Compressed air or grease can penetrate ____________________.
21. Exercise caution when inflating ____________________ that have been remounted on rims.

22. The keys should be out of the ignition any time a ________________ is being inspected or adjusted.
23. Mushroomed tools or chisels should be ________________ before use.
24. The ________________ must be positioned as close to the grinding wheel as possible.
25. Stand to the ________________ when starting a grinder.
26. The third terminal on electrical equipment is for ________________.
27. The color of the electrical ground wire on an extension cord is ________________.
28. When disconnecting a vehicle battery, disconnect the ________________ cable first.
29. Before removing the starter or alternator, disconnect the ________________.
30. If you should become injured while working in the shop, inform your ________________ immediately.

Student Signature ________________________________ Date ____________

ASE Education Foundation Worksheet #Intro-5

SHOP SAFETY

Name____________________________ Class____________________

Score: ☐ Excellent ☐ Good ☐ Needs Improvement **Instructor OK** ☐

Objective: Upon completion of this assignment, you will know the location of the shop emergency equipment and be aware of the emergency procedures.

ASE Education Foundation Correlation

This worksheet addresses the following **Required Supplemental Tasks (RST):**

Shop and Personal Safety, Task 5:	Utilize proper ventilation procedures for working within the lab/shop area.
Shop and Personal Safety, Task 6:	Identify marked safety areas.
Shop and Personal Safety, Task 7:	Identify the location and the types of fire extinguishers and other fire safety equipment; demonstrate knowledge of the procedures for using fire extinguishers and other fire safety equipment.
Shop and Personal Safety, Task 8:	Identify the location and use of eye wash stations.
Shop and Personal Safety, Task 10:	Comply with the required use of safety glasses, ear protection, gloves, and shoes during lab/shop activities.
Shop and Personal Safety, Task 11:	Identify and wear appropriate clothing for lab/shop activities.
Shop and Personal Safety, Task 12:	Secure hair and jewelry for lab/shop activities.
Shop and Personal Safety, Task 15:	Locate and demonstrate knowledge of material safety data sheets (MSDS).

We Support ASE | Education Foundation

Procedure:

1. Describe two major types of fires that may ignite in an automotive shop environment.
   a. ____________________
   b. ____________________
2. Check off the types of fire extinguishers in your school's shop.
   ☐ CO_2 ☐ Dry powder ☐ Foam ☐ Water
3. Place a check in the box following the hazardous wastes that may be encountered in your school's shop.
   ☐ Coolant ☐ Gasoline ☐ Solvent ☐ Motor oil
   ☐ Brake dust ☐ Freon (R-12) ☐ Dirty water

TURN

4. In the event that the building had to be evacuated, where outside the building would you meet with your instructor?

____

5. List the names of any equipment with a marked safety area on the floor.

a. ____________________

b. ____________________

c. ____________________

d. ____________________

6. Locate the main power panel. Does it have a marked safety area on the floor in front of it?

☐ Yes ☐ No

7. Which of the following are done when using a fire extinguisher? Check all that apply.

☐ Pull the pin.

☐ Aim at the base of the fire.

☐ Squeeze the handle.

8. All shops must have material safety data sheets (MSDS) for all chemicals used in the shop. Where are they located in your school shop? ____________________

9. On a blank piece of paper, sketch the layout of your school's automotive shop. Use the letters that precede each of the items listed to note their location in the shop. See example below.

A. All doors marked with an exit sign

B. Fire extinguishers/fire blanket

C. First-aid kits

D. Emergency telephone

E. Floor mops

F. Emergency eyewash

I. Push brooms

J. Water hose

K. Hazardous materials poster

L. Air hoses

M. Sink

N. Exhaust ventilation hoses

G. Hand brooms

H. Dust pans

O. Marked safety areas

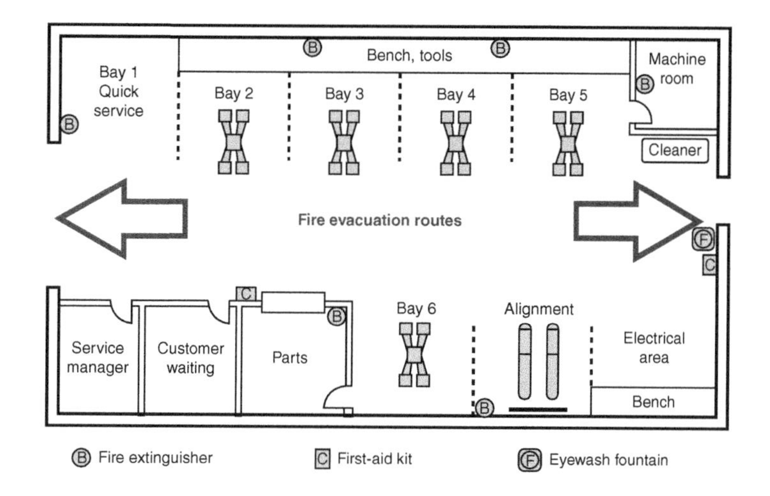

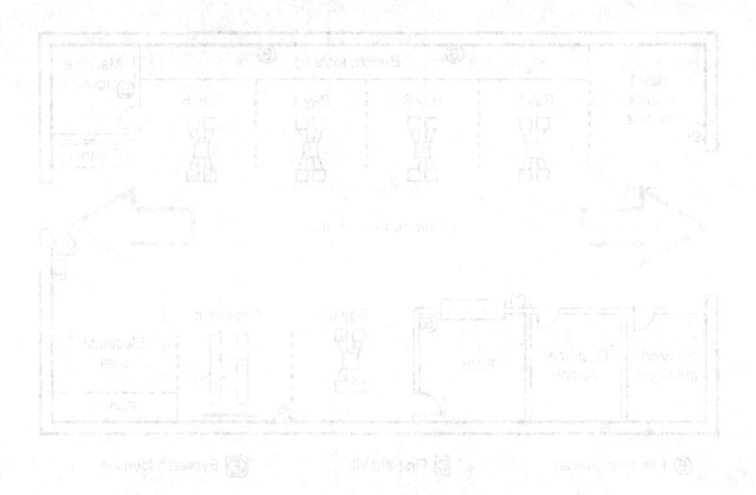

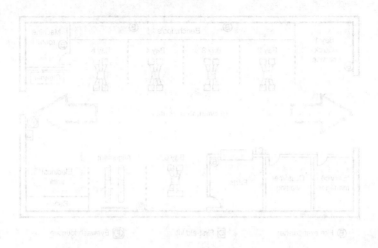

ASE Education Foundation Worksheet #Intro-6

LOCATE VEHICLE LIFT POINTS

Name ______________________ Class ______________

Score: ☐ Excellent ☐ Good ☐ Needs Improvement **Instructor OK** ☐

Vehicle year ________ **Make** ______________ **Model** ______________

Objective: Upon completion of this assignment, you should be able to locate the correct lift points to safely raise a vehicle.

ASE Education Foundation Correlation

This worksheet addresses the following **Required Supplemental Tasks (RST)**:

Shop and Personal Safety, Task 4: Identify and use proper procedures for safe lift operation.

We Support ASE Education Foundation

Directions: Before beginning this lab assignment, review the worksheet completely. Fill in the information in the spaces provided as you complete each task.

Procedure:

1. Before you start working, check the service information for proper lift points.

   Manual ______________________ Page # ______

   On the drawings below, place an X at each of the proper lifting points.

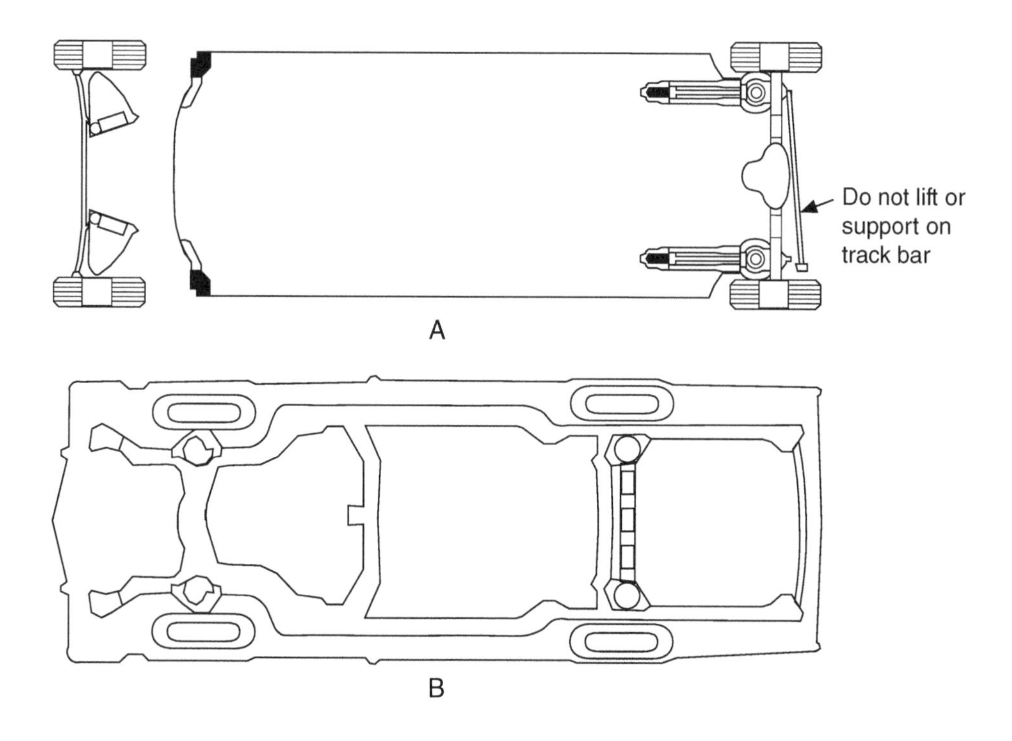
Do not lift or support on track bar
A
B

TURN

2. Place the letter in the space provided that best identifies the equipment in the drawings.

Single-post frame-contact lift ______

Two-post frame-contact lift ______

Surface-mount frame-contact lift ______

Surface-mount wheel-contact lift ______

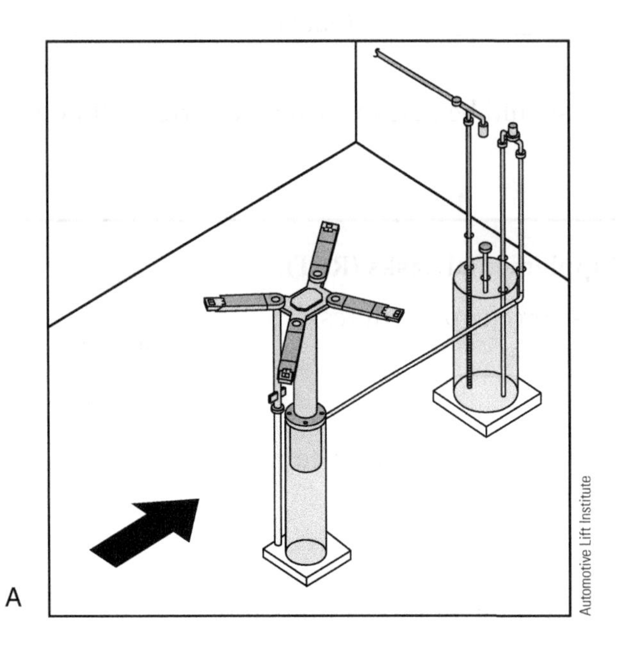

A

Automotive Lift Institute

B

Automotive Lift Institute

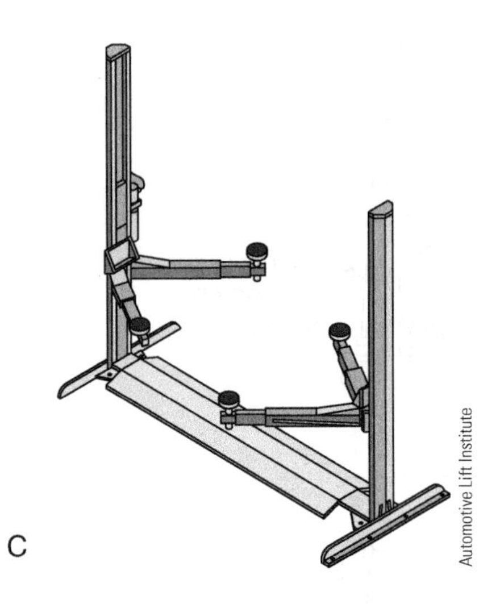

C

Automotive Lift Institute

D

Rotary Lift

ASE Education Foundation Worksheet #Intro-7

COMPLETING A REPAIR ORDER (R.O.)

Name______________________________ Class______________________

Score: ☐ Excellent ☐ Good ☐ Needs Improvement **Instructor OK** ☐

Objective: Upon completion of this assignment, you will be able to complete a repair order for required repairs.

ASE Education Foundation Correlation

This worksheet addresses the following **Required Supplemental Tasks (RST)**:

Preparing Vehicle for Service, Task 1: Identify information needed and the service requested on a repair order.

Preparing Vehicle for Service, Task 3: Demonstrate use of the three C's (concern, cause, and correction).

Preparing Vehicle for Service, Task 4: Review vehicle service history.

Preparing Vehicle for Service, Task 5: Complete work order to include customer information, vehicle identifying information, customer concern, related service history, cause, and correction.

This worksheet addresses the following **AST/MAST** task:

I.A.1 Complete work order to include customer information, vehicle identifying information, customer concern, related service history, cause, and correction. **(P-1)**

We Support ASE Education Foundation

Directions: Before beginning this assignment, review the worksheet completely. Fill in the information in the spaces provided as you complete each task.

Related Information: Repair orders are used to keep an accurate record of work completed and parts used during a vehicle repair. They serve as an agreement between the customer and the repair shop as to what repairs are to be completed and the cost of the parts and repairs. Agreeing to the specific repairs and their cost before work is started prevents many surprises and disagreements when the vehicle is picked up by the customer.

The repair order is filled out by the repair shop when the customer brings the vehicle in for repair. The customer's name, address, description of the vehicle, and the needed service or repairs are recorded. An estimate of the date and time that the job will be finished and of the cost of repairs is included.

Note: An estimate of the cost of repairs is required by law in several states.

Procedure:

Complete a repair order for the following customer:

Customer #1

Mrs. Jane Pollano	1989 Ford, Taurus (Blue)
456 Willard Drive	License # GHY 385
Susanwash, CA 93004	145,467 miles
Phone # 123-2145	VIN # 1G1CZ19H6KW135675
Service requested:	Sixty thousand mile service $225.00
Parts needed:	5 quarts oil $2.50 each 1 oil filter $8.50

Photo by Tim Gilles

1. Complete a repair order using the above information.
   a. Enter the customer's name, address, and telephone number in the spaces provided.
   b. Fill in the date and your name (the person writing the repair order).
   c. Insert the related information about the vehicle in the correct spaces.
   d. List the customer concerns in the section labeled *concern.*
   e. Estimate the total price and write it in the space marked *preliminary estimate.*
2. The repair order is now ready to be signed by the customer. Ask your instructor to review and sign the repair order in the space for customer authorization.
3. Review the vehicle history for any previous related repairs.
4. Check for any related Technical Service Bulletins (TSBs).
5. The repairs have been completed.
   a. Record the cause and correction for the customer concern.
   b. Record any parts used. The charges for parts and labor are then priced and totaled.
   c. Calculate the sales tax on the parts only. (Some states charge tax on the labor.) Multiply the parts total by the percentage of the tax (6%).

      Example: $22.00 (Parts)
      × .06 (% of tax)
      $1.320 Tax

   d. Total the parts, labor, and tax. The vehicle is now ready for delivery to the customer.

Complete a repair order for the following customer:

Customer #2

Mr. William Black comes in to your shop with his Silver Honda Accord. He is concerned because the engine speed increases but the vehicle does not maintain speed whenever he drives up the hill leading to his house at 1435 Hill Street, Oakhill, CA 93005. This problem has become more pronounced lately. Yesterday he had to call his wife to pick him up at the bottom of the hill. After hearing his concerns, it is explained that it would be necessary to fill out a repair order before diagnosing his problem. The service writer told Mr. Black that there would

be a half-hour labor charge at $50.00 per hour for the diagnosis. Mr. Black agreed to the charge and the 7% tax on the parts and labor. He said that he could be reached by phone at home at 876-4876 or at work at 456-9834. The vehicle identification number (1Y1SK52626R0076546), license number (YTR 746), and mileage (75,736) were obtained from the vehicle. Remember, the VIN can be used to identify the year of the vehicle.

Prepare the repair order (estimate) before proceeding any further. Have your instructor review it and sign in the space for the customer authorization.

The technician checked the Honda and found that the clutch was slipping and needed to be replaced. The technician checked the parts and labor guide and found that the job would require 6.5 hours to complete. He also made a list of the required parts and got prices from a local parts store.

Pressure plate	$124.98
Clutch disc	65.68
Throwout bearing	24.58

A call was made to Mr. Black at 2:45 PM and he approved the repairs. The repairs were completed. It is time for you to complete the repair order.

STOP

TERMS:
Cash ☐
Credit Card ☐
Prior Approval ☐

VEHICLE IDENTIFICATION NUMBER (VIN): | YEAR: | MAKE: MODEL: | LICENSE NO: | REPAIR ORDER NO:

CUSTOMER NAME/ADDRESS: | COLOR: | MILEAGE: | R.O. DATE:

CALL WHEN READY ☐ YES ☐ NO | RESIDENCE PHONE: | CELL PHONE: | PRELIMINARY ESTIMATE: | **CUSTOMER SIGNATURE** | ADVISOR: | HAT NO:

SAVE REMOVED PARTS FOR CUSTOMER ☐ YES ☐ NO | TIME RECEIVED: | DATE/TIME PROMISED: | REVISED ESTIMATE: | REASON FOR ADDITIONAL REPAIRS: | ADDITIONAL COST:

CUSTOMER PAY ☐
WARRANTY ☐

VEHICLE HISTORY ATTACHED ☐
TECHNICAL/SERVICE BULLETINS ☐

ADDITIONAL REPAIRS AUTHORIZED BY: DATE: TIME:
☐ IN PERSON ☐ PHONE #

WE USE NEW PARTS UNLESS OTHERWISE SPECIFIED. | TEARDOWN ESTIMATE. IF THE CUSTOMER CHOOSES NOT TO AUTHORIZE THE SERVICES RECOMMENDED, THE VEHICLE WILL BE REASSEMBLED WITHIN ____ DAYS OF THE DATE OF THIS REPAIR ORDER. | ADDITIONAL COST (TEARDOWN ESTIMATE):

LABOR INSTRUCTIONS

PARTS				
NAME	PRICE	LABOR		CUSTOMER STATES: **Concern**
		PARTS		CHECK AND ADVISE: **Cause**
		SUBLET		
		TAX		REPAIR(S) PERFORMED: **Correction**
TOTAL		TOTAL		

WE DO NOT ASSUME RESPONSIBILITY FOR LOSS OR DAMAGE OF ARTICLES LEFT IN YOUR VEHICLE. PLEASE REMOVE ALL PERSONAL PROPERTY.
____ CUSTOMER RENTAL
____ COURTESY VEHICLE
____ SHUTTLE

NOTES (specs, procedures, additional service, or repair information):

ADDITIONAL RECOMMENDATIONS FOR SERVICE OR REPAIRS:

TERMS: Cash ☐ Credit Card ☐ Prior Approval ☐	VEHICLE IDENTIFICATION NUMBER (VIN):	YEAR:	MAKE: MODEL:	LICENSE NO:	REPAIR ORDER NO:
	CUSTOMER NAME/ADDRESS:		COLOR:	MILEAGE:	R.O. DATE:

CALL WHEN READY ☐ YES ☐ NO	RESIDENCE PHONE:	CELL PHONE:	PRELIMINARY ESTIMATE:	CUSTOMER SIGNATURE	ADVISOR: HAT NO:
SAVE REMOVED PARTS FOR CUSTOMER ☐ YES ☐ NO	TIME RECEIVED:	DATE/TIME PROMISED:	REVISED ESTIMATE:	REASON:	ADDITIONAL COST:

CUSTOMER PAY ☐ WARRANTY ☐	VEHICLE HISTORY ATTACHED ☐ TECHNICAL/SERVICE BULLETINS ☐	AUTHORIZED BY: ☐ IN PERSON ☐ PHONE #	DATE:	TIME:

WE USE NEW PARTS UNLESS OTHERWISE SPECIFIED.	TEARDOWN ESTIMATE. IF THE CUSTOMER CHOOSES NOT TO AUTHORIZE THE SERVICES RECOMMENDED, THE VEHICLE WILL BE REASSEMBLED WITHIN ____ DAYS OF THE DATE OF THIS REPAIR ORDER.	ADDITIONAL COST (TEARDOWN ESTIMATE):

LABOR INSTRUCTIONS

PARTS		LABOR		CUSTOMER STATES: **Concern**
NAME	PRICE			
		PARTS		CHECK AND ADVISE: **Cause**
		SUBLET		
		TAX		REPAIR(S) PERFORMED: **Correction**
TOTAL		TOTAL		

WE DO NOT ASSUME RESPONSIBILITY FOR LOSS OR DAMAGE OF ARTICLES LEFT IN YOUR VEHICLE. PLEASE REMOVE ALL PERSONAL PROPERTY.

____ CUSTOMER RENTAL
____ COURTESY VEHICLE
____ SHUTTLE

NOTES (specs, procedures, additional service, or repair information):

ADDITIONAL RECOMMENDATIONS FOR SERVICE OR REPAIRS:

STOP

NOTES (specs, procedures, additional service, or repair information):

ADDITIONAL RECOMMENDATIONS FOR SERVICE OR REPAIRS:

Part II

Lab Worksheets for ASE Maintenance and Light Repair

Service Area 1

Oil Change Service

ASE Education Foundation Worksheet #1-1

MAINTENANCE SPECIFICATIONS

Name____________________________ Class____________________

Score: ☐ Excellent ☐ Good ☐ Needs Improvement **Instructor OK** ☐

Vehicle year __________ **Make** ___________________ **Model** ____________________

Objective: Upon completion of this assignment, you will be able to use a vehicle maintenance guide to locate maintenance and service specifications.

ASE Education Foundation Correlation

This worksheet addresses the following **MLR** tasks:

I.A.1; II.A.1; III.A.1; IV.A.1; V.A.1; VI.A.1; VIII.A.1	Research vehicle service information, including fluid type, vehicle service history, service precautions, and technical service bulletins. **(P-1); (P-1); (P-1); (P-1); (P-1); (P-1); (P-1)**
VII.A.1	Research vehicle service information, including refrigerant/oil type, vehicle service history, service precautions, and technical service bulletins. **(P-1)**

This worksheet addresses the following **AST/MAST** tasks:

I.A.2	Research vehicle service information including fluid type, internal engine operation, vehicle service history, service precautions, and technical service bulletins. **(P-1)**
II.A.2; III.A.2; IV.A.1; V.A.2	Research vehicle service information including fluid type, vehicle service history, service precautions, and technical service bulletins. **(P-1); (P-1); (P-1); (P-1)**
VI.A.1; VIII.A.2	Research applicable vehicle and service information, vehicle service history, service precautions, and technical service bulletins. **(P-1); (P-1)**

We Support ASE | Education Foundation

Directions: Review this worksheet completely before starting. Use your own vehicle or one provided by your instructor. Use a Car Care Guide, service manual, or computer program to locate the requested information. Record the information in the spaces provided. If the information is not available or does not apply, write N/A in the answer space.

> **Note:** If you are completing this worksheet on your personal vehicle, you may want to save it for future reference.

Tools and Equipment Required: Car Care Guide, vehicle service manual, or computer program

Procedure:

Engine: ☐ 4-cylinder ☐ 6-cylinder ☐ 8-cylinder

Transmission: ☐ Manual ☐ Automatic

Is there an underhood label? ☐ Yes ☐ No

Is there an underhood vacuum diagram? ☐ Yes ☐ No

What is the name of the service manual or program being used to complete this worksheet?

____

Capacities:

Battery _____ Cold cranking amps (CCA) ☐ N/A

Crankcase (without oil filter) _____ qt. ☐ N/A

Oil filter capacity _____ qt. ☐ N/A

Cooling system (without AC) _____ qt. ☐ N/A

Cooling system (with AC) _____ qt. ☐ N/A

Differential capacity (RWD) _____ pt./qt. ☐ N/A

Transmission (RWD) _____ pt./qt. ☐ N/A

Transaxle (FWD) _____ pt./qt. ☐ N/A

Fuel tank capacity _____ gallons ☐ N/A

Diesel exhaust fluid capacity _____ gallons ☐ N/A

Fluid Specifications:

SAE engine oil viscosity _______________ ☐ N/A

Service rating (grade) _______________ ☐ N/A

Transmission fluid type (RWD) _______________ ☐ N/A

Differential lubricant type (RWD) _______________ ☐ N/A

Transaxle (FWD) _______________ ☐ N/A

Brake fluid type _______________ ☐ N/A

Fuel type _______________ ☐ N/A

Fuel octane rating _______________ ☐ N/A

Tires:

Pressures Front __________ psi

Rear __________ psi

Lug nut torque specification __________ ft.-lb

Belt Tension Specifications:

		Self-Adjusting	
Self adjusting belts	______________	☐	☐ N/A
Alternator	______________	☐	☐ N/A
Power steering pump	______________	☐	☐ N/A
Air-conditioning compressor	______________	☐	☐ N/A
Smog pump	______________	☐	☐ N/A

Notes:

____

____

____

____

STOP

ASE Education Foundation Worksheet #1-2

RAISE AND SUPPORT A VEHICLE (JACK STANDS)

Name_______________________ Class_________________

Score: ☐ Excellent ☐ Good ☐ Needs Improvement **Instructor OK** ☐

Vehicle year ________ **Make** ________________ **Model** __________________

Objective: Upon completion of this assignment, you will be able to safely raise a vehicle with a hydraulic floor jack and support it on jack stands.

ASE Education Foundation Correlation

This worksheet addresses the following **Required Supplemental Task (RST):**

Shop and Personal Safety, Task 3: Identify and use proper placement of floor jacks and jack stands.

We Support ASE Education Foundation

Directions: Before beginning this lab assignment, review the worksheet completely. Fill in the information in the spaces provided as you complete each task.

Related Information: Before starting this worksheet, complete Worksheet #Intro-6 (Locate Vehicle Lift Points)

Procedure:

Raise the front of the vehicle:

a. Center the jack under the front crossmember or frame. *Do not jack on the radiator, oil pan, or the steering linkage!*

Jack centered? ☐ Yes ☐ No

b. Raise the vehicle until both front wheels are about 6″ off the ground. Are both wheels leaving the ground equally?

☐ Yes ☐ No

c. Place the jack stand in the recommended position.

d. Lower the vehicle onto jack stands.

e. The front wheels should still be off ground after lowering the vehicle onto the jack stands.

f. After instructor signs off, remove the jack stands and lower the vehicle.

Instructor OK ____________________

Raise all four wheels:

Note: When all four wheels are to be raised off the ground, raise the rear and support it first. Then the front can be raised and supported.

Note: When raising the rear wheels, be careful not to damage the fuel tank.

TURN

Raise Rear Wheels

a. Center the jack on the rear axle or crossmember so that both sides of the vehicle are raised equally.

b. Raise the vehicle until the tires are about a foot off the ground.

c. In front of the rear wheels, there is a bend in the frame. Place the jack stands there. If the car has leaf springs, place them in front of the spring eye.

d. Ask your instructor to check your work.

Instructor OK ____________________

Lower and remove the jack.

Photo by Tim Gilles

Note: When rear wheels are to be removed from a vehicle, the vehicle should be supported by the frame with the suspension system hanging free. Otherwise, there may not be enough clearance for the wheels to be removed.

Raise Front Wheels

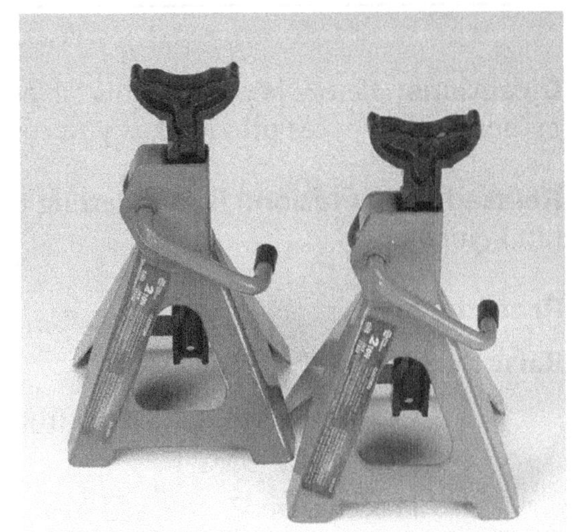

a. Center the jack under the front crossmember or frame. *Do not jack on the radiator, oil pan, or the steering linkage!*

b. Raise the vehicle until both front wheels are about 6″ off the ground. Be sure both wheels are leaving the ground equally.

c. Place the jack stands on the frame, just behind the front wheels.

d. Ask your instructor to check your work.

Instructor OK ____________________

e. Lower the jack and return it so other students can use it.

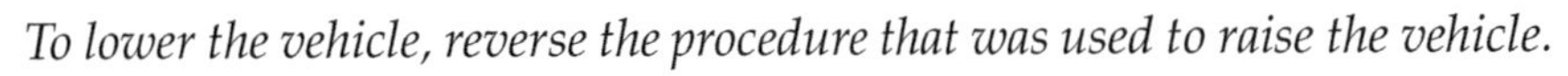

To lower the vehicle, reverse the procedure that was used to raise the vehicle.

When you are finished, clean your work area and put the tools in their proper places.

ASE Education Foundation Worksheet #1-3

RAISE A VEHICLE USING A FRAME-CONTACT LIFT

Name_______________________________ Class_______________________

Score: ☐ Excellent ☐ Good ☐ Needs Improvement **Instructor OK** ☐

Vehicle year __________ **Make** ____________________ **Model** ______________________

Objective: Upon completion of this assignment, you will be able to safely raise a vehicle using a frame-contact lift.

ASE Education Foundation Correlation

This worksheet addresses the following **Required Supplemental Task (RST)**:

Shop and Personal Safety, Task 4: Identify and use proper procedures for safe lift operation.

We Support ASE Education Foundation

Directions: Before beginning this lab assignment, review the worksheet completely. Fill in the information in the spaces provided as you complete each task.

Procedure:

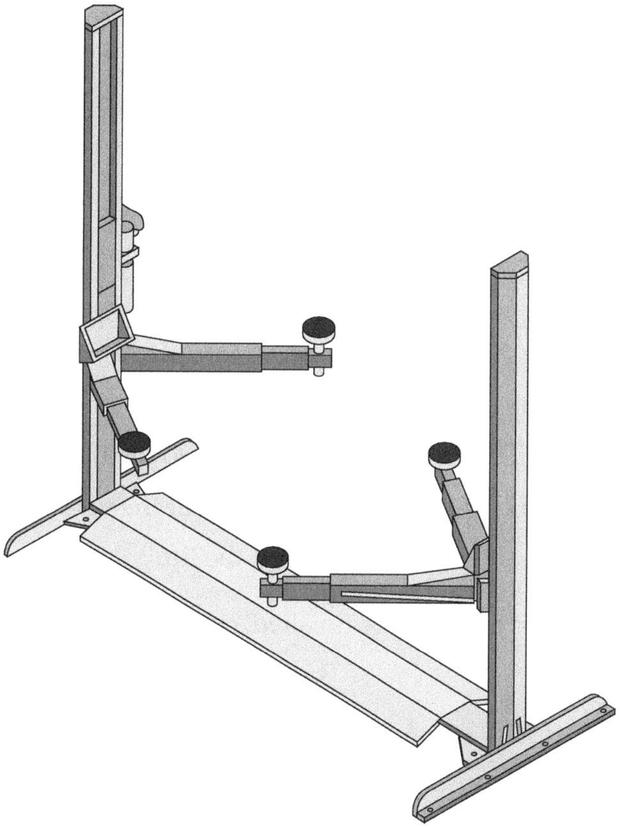

Note: Get assistance and approval from your instructor before lifting a vehicle more than 6″ off the ground.

1. Before you start working, check the service manual for proper lift points.

   Service Information ____________________

2. Prepare to lift the vehicle:

   a. Center the vehicle over the lift.

   b. Turn off the engine.

   c. Put the shift lever in neutral position.

   d. Did you apply the parking brake?

      ☐ Yes ☐ No

   e. Adjust the lift pads to contact the appropriate lift points on the vehicle. Be careful that the vehicle's center of gravity is over the posts of the lift.

      Vehicle centered? ☐ Yes ☐ No

Note: The vehicle's center of gravity is not always the center of the vehicle. The center of gravity is dependent on the vehicle's weight distribution.

TURN

f. The center of gravity of a FWD vehicle is:

_____Below the driver seat

_____In front of the driver seat, below the steering wheel

_____ Under the dashboard

g. The center of gravity of a RWD vehicle is:

_____Below the driver seat

_____In front of the driver seat, below the steering wheel

_____ Under the dashboard

3. Lifting the vehicle:

a. Raise the lift slowly until the pads contact the lift points. Double-check to see that they are centered under the lift points. Also check that the lift is not going to contact the exhaust system or any lines or cables.

Are the lift points centered?	☐ Yes	☐ No
Is the lift clear of the exhaust system?	☐ Yes	☐ No
Is the lift clear of any cables or lines?	☐ Yes	☐ No

b. Raise the vehicle until all tires leave the ground.

CAUTION Be certain that the lift arms or contact pads do not contact the vehicle's tires when the vehicle is raised.

c. Shake the vehicle to be sure it will not fall off the lift when raised further.

d. Before proceeding further, have the instructor check your work.

Instructor OK ____

e. Raise the vehicle to the desired height and set any safety devices that apply.

4. Lowering the vehicle:

a. Be certain all toolboxes, air or electrical hoses, or lubrication equipment are removed from under the vehicle before lowering it.

b. Release any safety devices (if the lift is so equipped).

c. Lower the lift *all the way* to the floor.

d. Move the lift arms to clear the vehicle.

e. Back the vehicle off the lift.

5. When you are finished, clean the work area and put the tools in their proper places.

ASE Education Foundation Worksheet #1-4

CHECK ENGINE OIL LEVEL

Name ______________________ Class ________________

Score: ☐ Excellent ☐ Good ☐ Needs Improvement **Instructor OK** ☐

Vehicle year ________ **Make** ______________ **Model** ________________

Objective: Upon completion of this assignment, you will be able to check and adjust a vehicle's engine oil level.

ASE Education Foundation Correlation

This worksheet addresses the following **MLR** task:

I.A.3 Inspect engine assembly for fuel, oil, coolant, and other leaks; determine needed action. **(P-1)**

This worksheet addresses the following **AST/MAST** task:

I.A.4 Inspect engine assembly for fuel, oil, coolant, and other leaks; determine needed action. **(P-1)**

We Support ASE Education Foundation

Directions: Before beginning this lab assignment, review the worksheet completely. Fill in the information in the spaces provided as you complete each task.

Tools and Equipment Required: Safety glasses, shop towel

Procedure:

Engine size ____________ # of cylinders _____

Odometer reading ____________ miles

1. Locate the following oil specifications for the vehicle:

   Recommended viscosity ________ Recommended oil service rating (grade) ________

2. Engine oil level is checked with the engine off and at normal operating temperature.

   Is the engine at normal operating temperature? ☐ Yes ☐ No

   Is the engine off? ☐ Yes ☐ No

3. When possible, the oil should be checked after the engine has been off for about 5 minutes. Oil remaining in other parts of the engine will have a chance to return to the oil pan.

   Has the engine been off for 5 minutes? ☐ Yes ☐ No

4. Remove the dipstick and wipe it clean with a shop towel.

TURN

5. Push the dipstick *all the way* into the dipstick tube and then pull it back out. Hold a towel under it so that oil does not accidentally drip onto the vehicle's fender.

6. Read the oil level on the dipstick. If the reading is unclear, flip the dipstick over and repeat the test (read the back side of the dipstick).

   Dipstick reading: ☐ Clear ☐ Unclear

7. The correct oil level is between the "add" and the "full" lines. When the level is below the "add" line, 1 quart of oil is added.

   Note: It is not necessary to add oil when the level is between the "add" and "full" marks unless you are doing an oil change.
   What is the oil level?

   ☐ Full ☐ Oil needed

8. If the oil level is low, check the service sticker to see if the vehicle is due for servicing.

   Is there a service sticker? ☐ Yes ☐ No

   What is the vehicle's current mileage? __________

   Mileage at last oil change __________________ Service required? ☐ Yes ☐ No

9. Reset the maintenance reminder light if necessary.

   Maintenance light reset?

   ☐ Yes ☐ No

   When is it necessary to reset the maintenance reminder light?

   ☐ After completing a repair

   ☐ After rotating the tires

   ☐ When doing an oil change service

   ☐ Whenever the light is on

Photo by Tim Gilles

   Note: Be sure that the dipstick is correctly seated on the dipstick tube. If it is not, the crankcase ventilation system can draw in dirty air. This can lead to premature engine failure. Also, some fuel-injected vehicles will not run properly unless the crankcase is sealed.

10. Close the hood and store the shop towel in a flammable storage container.

ASE Education Foundation Worksheet #1-5

OIL AND FILTER CHANGE

Name_______________________________ Class______________________

Score: ☐ Excellent ☐ Good ☐ Needs Improvement **Instructor OK** ☐

Vehicle year __________ **Make** ____________________ **Model** ______________________

Objective: Upon completion of this assignment, you should be able to change a vehicle's engine oil and filter.

ASE Education Foundation Correlation

This worksheet addresses the following

Required Supplemental Task (RST):

Preparing Vehicle for Service, Task 2: Identify purpose and demonstrate proper use of fender covers, mats.

This worksheet addresses the following **MLR** tasks:

I.C.5 Perform engine oil and filter change; use proper fluid type per manufacturer specification; reset maintenance reminder as required. **(P-1)**

VI.E.7 Verify operation of instrument panel gauges and warning/indicator lights; reset maintenance indicators. **(P-1)**

This worksheet addresses the following **AST/MAST** tasks:

I.D.10 Perform engine oil and filter change; use proper fluid type per manufacturer specification. **(P-1)**

VI.F.3 Reset maintenance indicators as required. **(P-2)**

We Support ASE Education Foundation

Directions: Before beginning this lab assignment, review the worksheet completely. Fill in the information in the spaces provided as you complete each task.

Tools and Equipment Required: Safety glasses, fender covers, jack and jack stands or vehicle lift, shop towel, drain receptacle, combination wrench, and oil filter wrench

Procedure:

Before you begin this procedure, be sure you have the correct number of quarts of oil and the correct oil filter for the vehicle. List the type of oil and filter below.

Oil and Filter Information:

Oil brand __________ API/ILSAC oil service rating __________ Oil quantity _______ quarts

Filter brand __________ Filter part number ________

Before draining the oil, confirm with your instructor that you have the correct oil and filter.

Instructor OK ______________

TURN

Drain the Oil

Note: For best results, drain the crankcase when the engine is warm.

☐ Engine warm ☐ Engine cold

Note: Never allow the soft side of the fender cover to come into contact with dirt or grease.

1. Remove the oil filler cap and put it on the top of the air cleaner or in another conspicuous place. This will remind you (or someone teaming with you) to refill the crankcase before you start the engine after changing the oil.
2. Raise the vehicle with a jack or lift.
3. Position a drain pan under the oil pan drain plug.

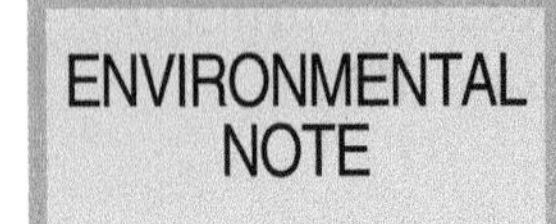

Be environmentally aware! Dispose used oil in a responsible manner. More than 220 million gallons of oil are disposed of improperly each year. Used oil can damage the water supply and kill plants and wildlife. One gallon of improperly disposed oil can pollute 1 million gallons of drinking water.

4. Loosen the drain plug using a wrench of the correct size. Which direction did you turn to loosen?

   ☐ Clockwise ☐ Counterclockwise

5. Unscrew the drain plug by hand. As it becomes loose and begins to leak oil, hold it against the drain opening. When the threads are no longer holding the plug, quickly pull it away so it does not fall into the drain oil. Commercial oil drain tanks include a screen to catch an accidentally dropped drain plug.
6. Allow the oil to drain until it no longer drips from the opening in the oil pan. This can sometimes take a few minutes. You can change the oil filter while you wait.
7. Check the condition of the plastic, aluminum, or copper drain plug gasket. Replace it if necessary.

   Condition of drain plug gasket

   ☐ Good ☐ Requires replacement

8. Thread the drain plug by hand into the threads in the oil pan.

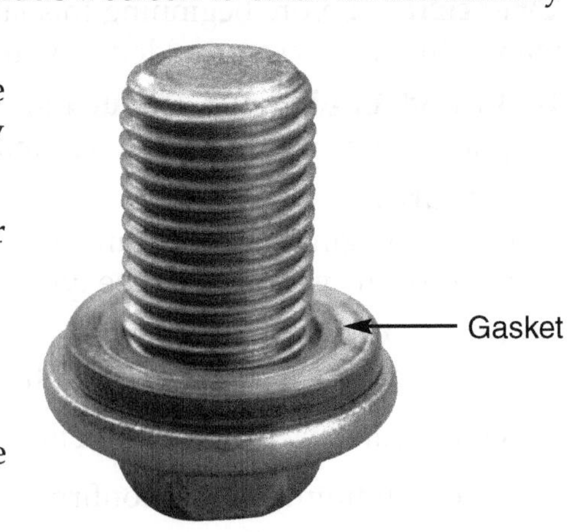

SHOP TIP It is often easier to start threading by first turning counterclockwise until the first threads of the plug drop into alignment with the threads in the oil pan. Then turn clockwise to tighten. Under no circumstances should the plug be tightened with a wrench unless it has first been finger-tightened all the way. Repairing a stripped oil pan thread can be a costly job.

9. Tighten the drain plug.

Remove and Replace the Oil Filter:

1. Use a filter wrench installed all the way against the base of the oil filter, which is the strongest area of the filter. This will prevent the filter housing from being crushed during disassembly.

Photo by Tim Gilles

2. Position a drain pan under the filter and remove it. Oil can spill from the filter when removed. Be careful not to drop it into the drain pan.
3. Clean the filter mounting surface on the engine. Check to see that the filter base O-ring is not stuck to the engine block.
4. Compare the oil and new oil filters to be certain they are identical. The new filter may not be exactly the same size as the oil filter, but the sealing surfaces must be identical.
5. Put a few drops of oil on the filter's rubber O-ring and install the filter by hand until it contacts the filter mounting surface on the engine.
6. For tightening specifications, refer to the instructions on the filter housing or its box. The typical recommendation is to tighten the filter an additional ⅔ turn after its O-ring seal contacts the block.
7. Locate the crankcase oil capacity from service information.

   Capacity: with filter ___ qt. Without filter ______qt.
8. Fill the crankcase and replace the oil filler cap.
9. Start the engine and allow it to idle. Stay in the car until the gauge light goes out or until pressure is indicated on the gauge. Then check for leaks.
10. Double-check the crankcase oil level and top off as needed.
11. How much oil did you add to the crankcase to bring the oil level to full? _______ qt.

12. Fill out a service reminder label and install it. Some stickers are transparent and are installed on the inside of the top of the windshield. Others are installed on the door jamb.

OIL CHANGE	
DATE	MILES
FILTER CHANGE	
☐ OIL ☐ AIR ☐ FUEL	
DATE	MILES
LUBRICATION	
DATE	MILES
OTHER SERVICE	
DATE	MILES
DESCRIPTION	
OIL CHANGE RECORD (USE UNDER HOOD)	
CUSTOMER'S NAME	BRAND-WEIGHT
DATE	MILES

TYPICAL DOOR SERVICE STICKER

13. Reset the maintenance reminder light if necessary. Maintenance light reset? ☐ Yes ☐ No
14. Before completing your paperwork, clean your work area, clean and return tools to their proper places, and wash your hands.
15. Record your recommendations for needed service or additional repairs and complete the repair order. ____

____

Part II

Lab Worksheets for ASE Maintenance and Light Repair

Service Area 2

Underhood Inspection

ASE Education Foundation Worksheet #2-1

CHECK THE BRAKE MASTER CYLINDER FLUID LEVEL

Name_______________ Class_______________

Score: ☐ Excellent ☐ Good ☐ Needs Improvement **Instructor OK** ☐

Vehicle year ________ **Make** ______________ **Model** ______________

Objective: Upon completion of this assignment, you will be able to check and refill a brake master cylinder to the proper level.

ASE Education Foundation Correlation

This worksheet addresses the following **MLR** tasks:

V.B.2 Check master cylinder for external leaks and proper operation. **(P-1)**

V.B.4 Select, handle, store, and fill brake fluids to proper level; use proper fluid type per manufacturer specification. **(P-1)**

V.B.7 Test brake fluid for contamination. **(P-1)**

This worksheet addresses the following **AST/MAST** tasks:

V.B.3 Check master cylinder for internal/external leaks and proper operation; determine needed action. **(P-1)**

V.B.9 Select, handle, store, and fill brake fluids to proper level; use proper fluid type per manufacturer specification. **(P-1)**

V.B.13 Test brake fluid for contamination. **(P-1)**

We Support ASE | Education Foundation

Directions: Before beginning this lab assignment, review the worksheet completely. Fill in the information in the spaces provided as you complete each task.

Tools and Equipment Required: Safety glasses, fender covers, shop towel, brake fluid test strips, refractometer

Note: Brake fluid rapidly absorbs moisture when exposed to air. Do not leave the lid off the brake fluid container, or the cover off the master cylinder. Also, remember that brake fluid can damage the vehicle's paint.

Procedure:

1. Install fender covers on the vehicle.
2. Remove the master cylinder cover.

   Note: Before removing the cover from the master cylinder, clean around it to prevent dirt from entering the system.

   Note: It is not necessary to remove the reservoir cap if the fluid level can be seen through the side of a translucent reservoir.

   Is the reservoir translucent? ☐ Yes ☐ No

   Type of cover: ☐ Screw-type cap
   ☐ Pry-off bale-type
   ☐ Plug-type cap

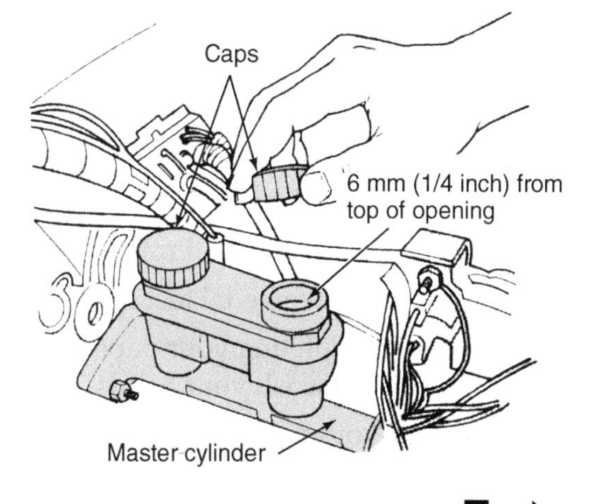

3. Inspect the fluid level.

☐ OK (between full and add lines) ☐ Low

4. Inspect the fluid condition.

☐ Clear ☐ Cloudy ☐ Dark

Note: Most vehicles have had two chambers (tandem) since 1967 for safety reasons. Many vehicles incorporate a dual reservoir as well.

5. Test the brake fluid condition (moisture content) in one of the following ways:

A. Insert a brake fluid test strip into the brake fluid.

Test results: ☐ Good ☐ Needs attention ☐ NA

B. Use a brake fluid tester to test the moisture content of the brake fluid.

Test results: ☐ Good ☐ Needs attention ☐ NA

C. Use a brake fluid refractometer to test the moisture content of the brake fluid.

Test results: ☐ Good ☐ Needs attention ☐ NA

6. Does the vehicle have a single- or dual-chamber master cylinder? ☐ Single ☐ Dual

How many reservoirs are used on the master cylinder? ☐ One ☐ Two

Note: If there are two reservoirs and one is larger, the larger one is for disc brakes. As disc brakes wear, the level drops as more fluid is required. If the large chamber is low, check the disc brake linings for wear. Refer to Worksheet #9-7, Inspect Front Disc Brakes.

The larger reservoir is: ☐ Full ☐ Low ☐ N/A

Note: Low fluid level can be an indication of worn disc brake pads.

7. Add the proper brake fluid as necessary to fill the brake master cylinder reservoir to the proper level. What type of brake fluid does this system require?

☐ DOT 3 ☐ DOT 4 ☐ DOT 5

Note: Clean any brake fluid spills immediately. Remember, brake fluid will damage the vehicle's paint. Work carefully and dilute accidental spills with water.

8. Reinstall the reservoir cover.

9. Look for any leakage or dampness around the master cylinder. Dampness can indicate that the cylinder might need to be rebuilt or replaced.

Are there any signs of leakage or dampness? ☐ Yes ☐ No

If so, where on or near the master cylinder is the leakage or dampness located?

☐ At the cover ☐ Front of the cylinder

☐ Rear of the cylinder ☐ Lines or hoses Other______

10. Was it necessary to add any brake fluid to the system? ☐ Yes ☐ No

11. Check the brake pedal height. ☐ Normal ☐ Low

12. Check the brake pedal travel. ☐ Normal ☐ Excessive (Low pedal)

13. Does the brake pedal feel spongy when depressed? ☐ Yes ☐ No

14. Before completing your paperwork, clean your work area, clean and return tools to their proper places, and wash your hands.

15. Record your recommendations for needed service or additional repairs and complete the repair order.

STOP

ASE Education Foundation Worksheet #2-2

CHECK CLUTCH MASTER CYLINDER FLUID LEVEL

Name_______________________________ Class_______________________

Score: ☐ Excellent ☐ Good ☐ Needs Improvement **Instructor OK** ☐

Vehicle year __________ **Make** ____________________ **Model** ______________________

Objective: Upon completion of this assignment, you will be able to check and refill a clutch master cylinder to the proper level and inspect the system.

ASE Education Foundation Correlation

This worksheet addresses the following **MLR** task:

III.B.1 Check and adjust clutch master cylinder fluid level; use proper fluid type per manufacturer specification. **(P-1)**

This worksheet addresses the following **AST/MAST** task:

III.B.5 Check and adjust clutch master cylinder fluid level; check for leaks; use proper fluid type per manufacturer specification. **(P-1)**

We Support ASE Education Foundation

Directions: Before beginning this lab assignment, review the worksheet completely. Fill in the information in the spaces provided as you complete each task.

Tools and Equipment Required: Safety glasses, fender covers, shop towel

Procedure:

Note: The brake fluid used in clutch master cylinders rapidly absorbs moisture when exposed to air. Do not leave the lid off the fluid container or the cover off the master cylinder. Also, remember that brake fluid can damage the vehicle's paint. Clean spills by diluting them with water.

1. What type of transmission/transaxle is this vehicle equipped with?

   ☐ Manual ☐ Automatic

2. Install fender covers on the vehicle.
3. Remove the clutch cylinder reservoir cover.

   Note: Before removing the cover, clean around the clutch master cylinder cover to prevent dirt from entering the system.

   Type of cover:

   ☐ Screw-type cap ☐ Plug-type cap

TURN

4. Inspect the fluid level.

   ☐ OK (within ⅜″ of top) ☐ Low

5. Inspect the fluid condition.

   ☐ Clear ☐ Cloudy ☐ Dark

6. Add brake fluid if necessary.

   ☐ Yes ☐ No

7. What type of brake fluid does this system require?

   ☐ DOT 3 ☐ DOT 4 ☐ DOT 5

8. Reinstall the reservoir cover.
9. Look for any leakage or dampness around the master cylinder. Dampness can indicate that the cylinder needs to be rebuilt or replaced.

   Is there any leakage or dampness? ☐ Yes ☐ No

   If so, where on or near the clutch master cylinder is the leakage or dampness located?

   ____

10. Before completing your paperwork, clean your work area, clean and return tools to their proper places, and wash your hands.
11. Record your recommendations for needed service or additional repairs and complete the repair order.

   ____

   ____

STOP

ASE Education Foundation Worksheet #2-3

CHECK POWER STEERING

Name______________________________ Class______________________

Score: ☐ Excellent ☐ Good ☐ Needs Improvement **Instructor OK** ☐

Vehicle year __________ **Make** ____________________ **Model** ______________________

Objective: Upon completion of this assignment, you will be able to check the power steering fluid and refill it to the proper level.

ASE Education Foundation Correlation

This worksheet addresses the following **MLR** tasks:

IV.B.2 Inspect power steering fluid level and condition. **(P-1)**

IV.B.21 Inspect electric power-assisted steering. **(P-2)**

This worksheet addresses the following **AST** tasks:

IV.B.9 Inspect power steering fluid level and condition. **(P-1)**

IV.B.19 Inspect electric power steering assist system. **(P-3)**

This worksheet addresses the following **MAST** tasks:

IV.B.9 Determine proper power steering fluid type; inspect fluid level and condition. **(P-1)**

IV.B.18 Inspect, test and diagnose electrically- assisted power steering systems (including using a scan tool); determine needed action. **(P-2)**

We Support ASE Education Foundation

Directions: Before beginning this lab assignment, review the worksheet completely. Fill in the information in the spaces provided as you complete each task. Only complete the inspection that applies to the vehicle being inspected.

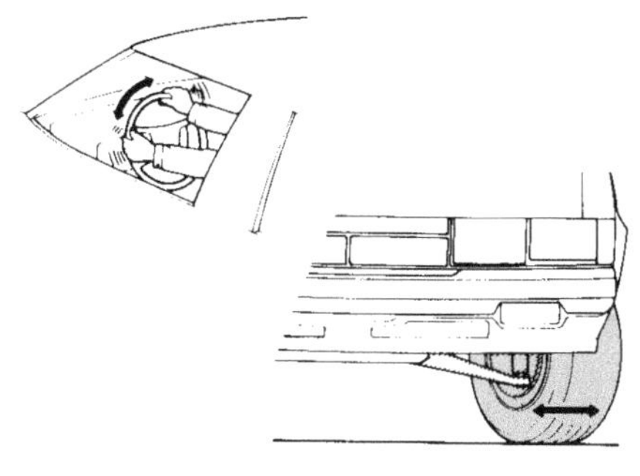

Tools and Equipment Required: Safety glasses, fender covers, shop towel

Procedure:

1. Perform this procedure when the fluid is warm. Cycle the system by turning the steering wheel through its complete range of travel from left to right. This will increase the temperature of the power steering fluid.

   Note: Do not hold the steering wheel at full right or left for more than 10 seconds. Damage to the system could result.

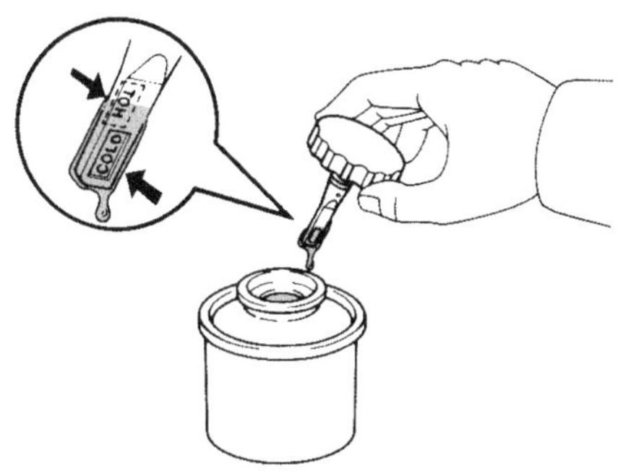

2. Does the vehicle have hydraulic or electric power steering?

   ☐ Hydraulic ☐ Electric

Hydraulic Power Steering Inspection:

3. Shut off the engine and check the fluid level.

   ☐ OK ☐ Low

   Fluid temperature:

   ☐ Hot ☐ Cold

   Fluid condition:

   ☐ Good ☐ Bad

4. Add specified fluid as needed.

   Type of fluid specified: __________

5. Was additional fluid required?

   ☐ Yes ☐ No

6. Is there evidence of a leak?

   ☐ Yes ☐ No

   Note: When looking for a power steering fluid leak, it will be necessary to inspect the entire power steering system.

   If a leak is noticed, where is it located?

   ☐ Pressure hose

   ☐ Return hose

   ☐ Reservoir

   ☐ Steering gear

   ☐ Power steering pump

Electric Power Steering Inspection:

7. Where is the power steering motor located? ______________________________

8. Inspect the condition of the electric power steering wiring and connectors.

   ☐ Good ☐ Needs attention ☐ N/A

9. Rotate the steering wheel to the left and right.

   a. Does it turn easily? ☐ Yes ☐ No

   b. Do you notice any unusual noise while turning the steering wheel? ☐ Yes ☐ No

10. If there is a problem with the electric power steering, describe the problem.

    ____

    ____

STOP

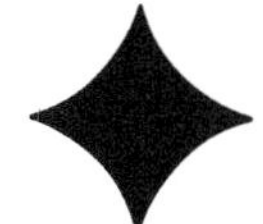

ASE Education Foundation Worksheet #2-4

CHECK AND CORRECT COOLANT LEVEL

Name______________________________ Class______________________

Score: ☐ Excellent ☐ Good ☐ Needs Improvement **Instructor OK** ☐

Vehicle year __________ **Make** ____________________ **Model** ______________________

Objective: Upon completion of this assignment, you will be able to check and correct the radiator coolant level and add coolant to the proper level.

ASE Education Foundation Correlation

This worksheet addresses the following **MLR** task:

I.C.1 Perform cooling system pressure and dye tests to identify leaks; check coolant condition and level; inspect and test radiator, pressure cap, coolant recovery tank, heater core, and galley plugs; determine necessary action. **(P-1)**

This worksheet addresses the following **AST/MAST** task:

I.D.1 Perform cooling system pressure and dye tests to identify leaks; check coolant condition and level; inspect and test radiator, pressure cap, coolant recovery tank, heater core, and galley plugs; determine needed action. **(P-1)**

We Support ASE | Education Foundation

Directions: Before beginning this lab assignment, review the worksheet completely. Fill in the information in the spaces provided as you complete each task.

Tools and Equipment Required: Safety glasses, fender covers, shop towel

Procedure:

1. Open the hood and place fender covers over the fenders.
2. Inspect coolant level in the coolant recovery (overflow) tank. It should be filled to the "cold" line if the coolant is cold.

   Coolant temperature:

   ☐ Cold ☐ Warm

Atmospheric-type pressure cap

Overflow tube

Overflow tank

3. What is the coolant condition/color?

   ☐ Clear ☐ Green ☐ Orange

   ☐ Yellow ☐ Rusty ☐ Red

4. Fill the recovery tank as needed.

   ☐ Coolant level OK ☐ Water added ☐ Coolant added

   Note: If the recovery tank was empty, it will be necessary to check the coolant level in the radiator. Normally, it is not necessary to remove the radiator pressure cap to check the coolant level if the recovery tank has coolant in it.

TURN

5. Before removing the radiator pressure cap check the following:

   a. Check the top of the radiator cap. What is the pressure rating of the cap? ________ psi

   b. What radiator cap pressure is specified for the vehicle? ________ psi

   Is this the correct radiator cap for the vehicle? ☐ Yes ☐ No

   c. What is the temperature of the cooling system? ☐ Hot ☐ Cold

SAFETY NOTE The radiator should not be opened when there is pressure in the system. Before opening the radiator cap, squeeze the upper radiator hose to be sure the system is not under pressure.

   d. Check the cooling system pressure. Is the upper radiator hose hard or soft when you squeeze it?

   ☐ Hard ☐ Soft

Photo by Tim Gilles

6. To remove the radiator pressure cap:

   a. If the top hose is soft, fold a shop towel and place it over the radiator cap.

   b. Hold down firmly on the cap and turn it counterclockwise ¼ turn until the cap is opened to the safety catch.

SAFETY NOTE Let up slowly on the pressure you are exerting on the cap. If coolant escapes, press the cap back down. (On most cars, cap pressure will be no more than 17 psi.) If pressure is allowed to escape from a hot system, the coolant boiling point will be lowered and it may boil.

   c. If coolant escapes, retighten the cap.

   ☐ Coolant escapes ☐ No coolant escapes

   d. If coolant escaped while attempting to open the radiator cap it will be necessary to wait for the vehicle to cool, or consult with your instructor before proceeding.

   e. If no coolant escapes, press down on the cap while turning it counterclockwise to remove it.

Cover cap with shop towel

Turn slowly counterclockwise

CAUTION After removing the radiator cap, DO NOT look into the radiator for at least 30 seconds. Sometimes it takes the coolant several seconds before it begins to boil.

7. Observe the coolant level. ☐ Full ☐ Low

8. Inspect the coolant condition and color.

   ☐ Clear ☐ Green Orange

   ☐ Yellow ☐ Rusty ☐ Red

9. Fill the radiator/cooling system as needed.

10. Replace the radiator cap, making sure that it is fully locked in place.

11. Before completing your paperwork, clean your work area, clean and return tools to their proper places, and wash your hands.

12. Record your recommendations for needed service or additional repairs and complete the repair order.

STOP

ASE Education Foundation Worksheet #2-5

SERPENTINE V-RIBBED BELT INSPECTION

Name________________________ Class________________

Score: ☐ Excellent ☐ Good ☐ Needs Improvement **Instructor OK** ☐

Vehicle year ________ **Make** ______________ **Model** ________________

Objective: Upon completion of this assignment, you will be able to inspect and replace a serpentine V-ribbed belt.

ASE Education Foundation Correlation

This worksheet addresses the following **MLR** tasks:

I.C.2 Inspect, replace, and adjust drive belts, tensioners, and pulleys; check pulley and belt alignment. **(P-1)**

IV.B.5 Remove, inspect, replace, and/or adjust power steering pump drive belt. **(P-1)**

VI.D.2 Inspect, adjust, and/or replace generator (alternator) drive belts; check pulleys and tensioners for wear; check pulley and belt alignment. **(P-1)**

VII.B.1 Inspect and replace A/C compressor drive belts, pulleys, and tensioners; visually inspect A/C components for signs of leaks; determine necessary action. **(P-1)**

This worksheet addresses the following **AST/MAST** tasks:

I.D.3 Inspect, replace, and adjust drive belts, tensioners, and pulleys; check pulley and belt alignment. **(P-1)**

IV.B.12 Remove, inspect, replace, and/or adjust power steering pump drive belt. **(P-1)**

VI.D.3 Inspect, adjust, and/or replace generator (alternator) drive belts; check pulleys and tensioners for wear; check pulley and belt alignment. **(P-1)**

VII.B.1 Inspect, remove, and/or replace A/C compressor drive belts, pulleys, tensioners and visually inspect A/C components for signs of leaks; determine needed action. **(P-1)**

We Support ASE Education Foundation

Directions: Before beginning this lab assignment, review the worksheet completely. Fill in the information in the spaces provided as you complete each task.

Tools and Equipment Required: Safety glasses, fender covers, shop towel, hand tools, belt tension gauge

Procedure:

1. Locate the belt routing diagram under the hood.

   Is there a diagram? ☐ Yes ☐ No

   Does it match the current installation of the belt?

   ☐ Yes ☐ No
2. Draw a sketch of the diagram in the box to the right.
3. Locate the belt tensioning roller. What type of adjustment does it have? Check one.

   ☐ Automatic spring loaded ☐ Locked center

   ☐ Jack screw
4. Loosen the belt tension, remove the belt, and inspect it. Check all that apply.

   ☐ Good condition ☐ Cracks ☐ Missing chunks

   ☐ Edges damaged ☐ Other ____________________
5. Inspect the idler pulley and bearing.

   Did it feel rough?

   ☐ Yes ☐ No

   Did the spring-loaded part move smoothly during unloading of the belt tension?

   ☐ Yes ☐ No ☐ N/A
6. Inspect the condition of the pulley grooves. Check below.

   ☐ Good condition

   ☐ Rust ☐ Damaged
7. Does the belt drive the coolant pump? If so, which side of the belt drives the pump?

   ☐ The flat side

   ☐ The grooved side
8. Reinstall the belt and set tensioner.
9. Run the engine and watch the belt for correct alignment.

   ☐ It runs true. ☐ It is out of alignment.
10. Before completing your paperwork, clean your work area, clean and return tools to their proper places, and wash your hands.
11. Record your recommendations for needed service or additional repairs and complete the repair order.

____

____

Good belt

Worn belt

Photo by Tim Gilles

STOP

ASE Education Foundation Worksheet #2-6

BATTERY VISUAL INSPECTION

Name ______________________ Class ________________

Score: ☐ Excellent ☐ Good ☐ Needs Improvement **Instructor OK** ☐

Vehicle year ________ **Make** ______________ **Model** ______________

Objective: Upon completion of this assignment, you will be able to inspect a battery for condition and electrolyte level.

ASE Education Foundation Correlation

This worksheet addresses the following **MLR/AST/MAST** task:

VI.B.4 Inspect and clean battery; fill battery cells; check battery cables, connectors, clamps, and hold-downs. **(P-1)**

We Support ASE Education Foundation

Directions: Before beginning this lab assignment, review the worksheet completely. Fill in the information in the spaces provided as you complete each task.

Tools and Equipment Required: Safety glasses, fender covers, shop towel

Procedure:

1. Open the hood and install fender covers on the vehicle.

SAFETY NOTE

- In addition to being dangerous to skin and eyes, battery acid can damage clothing and the car's paint. Work carefully and dilute accidental spills immediately with water.
- Batteries give off hydrogen gas when charging. Be careful to avoid an accidental spark.

2. Inspect the alternator belt. ☐ OK ☐ Loose
3. Check the condition of the battery terminal clamps/posts.

   ☐ Tight ☐ Loose ☐ Clean ☐ Corroded

Post or top terminal

Side terminal

"L" terminal

4. What type of terminals does the battery have?

   ☐ Top ☐ Side ☐ "L"
5. Battery cable condition:

   ☐ Clean ☐ Frayed ☐ Corroded

   ☐ Well-insulated ☐ Worn insulation

Cable end and terminal

Frayed insulation

POS

Corrosion

TURN

6. Check the battery's external condition.

   ☐ Clean ☐ Dirty

   Note: A mixture of baking soda and water may be used to clean the top of a battery. Battery acid may be neutralized with baking soda. Do not allow the baking soda mixture to enter the battery.

7. Check the condition of the battery holddown.

   ☐ OK ☐ Needs service

8. Does the battery have a built-in hydrometer?

   ☐ Yes ☐ No

   If it has a built-in hydrometer:

   a. What color is it? ____________________

   b. What does this color indicate? ☐ Charged ☐ Low charge ☐ Electrolyte low

9. Does the battery have removable cell caps? ☐ Yes ☐ No

10. Check the level of the battery electrolyte by looking through the translucent case or by looking into the cells. The electrolyte should be at least ½" above the separator plates.

    ☐ OK ☐ Low ☐ N/A

    If you selected N/A, describe why you did so? ________________________________

11. If the electrolyte level is low, fill the battery with water to just below the "split ring" full indicator or to the maximum level mark on the side of the battery.

    Note: Low electrolyte often indicates a problem with the charging system.

12. Use a paper towel to clean up any water or electrolyte that has spilled. Battery acid will ruin shop towels and damage the vehicle's paint.

13. Check yes or no if any of the following services or repairs are required.

	Yes	No
Replace battery clamps	☐	☐
Replace battery cables	☐	☐
Replace battery holddown	☐	☐
Adjust AC generator belt	☐	☐
Check charging system	☐	☐
Other	____________________	

14. Before completing your paperwork, clean your work area, clean and return tools to their proper places, and wash your hands.

15. Record your recommendations for needed service or additional repairs and complete the repair order.

    ____

    ____

STOP

ASE Education Foundation Worksheet #2-7

INSPECT AND REPLACE AN AIR FILTER

Name____________________________ Class____________________

Score: ☐ Excellent ☐ Good ☐ Needs Improvement **Instructor OK** ☐

Vehicle year __________ **Make** ____________________ **Model** ______________________

Objective: Upon completion of this assignment, you will be able to inspect and replace an air filter.

ASE Education Foundation Correlation

This worksheet addresses the following **MLR** task:

VIII.C.2 Inspect, service, or replace air filters, filter housings, and intake duct work. **(P-1)**

This worksheet addresses the following **AST** task:

VIII.D.4 Inspect, service, or replace air filters, filter housings, and intake duct work. **(P-1)**

This worksheet addresses the following **MAST** task:

VIII.D.5 Inspect, service, or replace air filters, filter housings, and intake duct work. **(P-1)**

We Support ASE Education Foundation

Directions: Before beginning this lab assignment, review the worksheet completely. Fill in the information in the spaces provided as you complete each task.

Procedure:

1. Raise the hood and install fender covers on the vehicle.
2. Remove the cover from the air filter housing.
3. Remove the filter from the housing and check the housing for dirt, debris, and oil.

   ☐ Dirt ☐ Debris ☐ Oil ☐ Clean
4. Inspect the filter for excessive dirt.

   ☐ Dirt ☐ Clean
5. Shine a light through the filter to identify any small holes that would allow dirt to pass through the filter.

   Are there any holes? ☐ Yes ☐ No
6. Does the air filter need to be replaced? ☐ Yes ☐ No
7. Reinstall the filter and its housing.
8. Before completing your paperwork, clean your work area, clean and return tools to their proper places, and wash your hands.
9. Record your recommendations for needed service or additional repairs and complete the repair order.

Lid

Filter element

Air cleaner housing

STOP

ASE Education Foundation Worksheet #2-8

INSPECT OPERATION OF THE LIGHTING SYSTEM

Name________________________ Class__________________

Score: ☐ Excellent ☐ Good ☐ Needs Improvement **Instructor OK** ☐

Vehicle year ________ **Make** ______________ **Model** ________________

Objective: Upon completion of this assignment, you will be able to inspect the operation of a vehicle's lighting system.

ASE Education Foundation Correlation

This worksheet addresses the following **MLR** task:

VI.E.1 Inspect interior and exterior lamps and sockets including headlights and auxiliary lights (fog lights/driving lights); replace as needed. **(P-1)**

This worksheet addresses the following **AST/MAST** task:

VI.E.2 Inspect interior and exterior lamps and sockets including headlights and auxiliary lights (fog lights/driving lights); replace as needed. **(P-1)**

We Support ASE Education Foundation

Directions: Before beginning this lab assignment, review the worksheet completely. Fill in the information in the spaces provided as you complete each task.

Procedure:

Lighting System			OK	Problem	N/A
1. Headlights	Low beam	Left	☐	☐	☐
		Right	☐	☐	☐
	High beam	Left	☐	☐	☐
		Right	☐	☐	☐
	Aim		☐	☐	☐

TURN

Note: The following systems are usually checked with the *key on* and *engine off* (KOEO).

			OK	Problem	N/A
2.	License plate lights		☐	☐	☐
3.	Turn signals	Front	☐	☐	☐
		Rear	☐	☐	☐
4.	Emergency flashers	Front	☐	☐	☐
		Rear	☐	☐	☐
5.	Back-up lights		☐	☐	☐
6.	Brake lights	Right	☐	☐	☐
		Left	☐	☐	☐
		Center	☐	☐	☐
7.	Running lights (side marker lights)	Left	☐	☐	☐
		Right	☐	☐	☐
8.	Interior courtesy lights		☐	☐	☐
9.	Interior dome light		☐	☐	☐
10.	Dash indicator lights				
	a. Turn signal dash indicators		☐	☐	☐
	b. Oil pressure indicator		☐	☐	☐
	c. Water temperature indicator		☐	☐	☐
	d. Brake warning light (emergency brake applied to check bulb)		☐	☐	☐
	e. Malfunction indicator lamp (MIL)		☐	☐	☐
	f. Maintenance reminder light		☐	☐	☐
	g. Air bag (SLR)		☐	☐	☐
	h. Antilock brake system (ABS)		☐	☐	☐
	i. Other		☐	☐	☐
11.	Other lighting systems (describe)				
	a. ____________		☐	☐	☐
	b. ____________		☐	☐	☐
	c. ____________		☐	☐	☐
	d. ____________		☐	☐	☐

12. Before completing your paperwork, clean your work area, clean and return tools to their proper places, and wash your hands.
13. Record your recommendations for needed service or additional repairs and complete the repair order. ____

____

STOP

ASE Education Foundation Worksheet #2-9

VISIBILITY CHECKLIST

Name______________________________ Class____________________

Score: ☐ Excellent ☐ Good ☐ Needs Improvement **Instructor OK** ☐

Vehicle year __________ **Make** ____________________ **Model** ______________________

Objective: Upon completion of this assignment, you should be able to inspect the windows, windshield wipers, and mirrors to ensure safe vehicle operation.

ASE Education Foundation Correlation

This worksheet addresses the following **Required Supplemental Task (RST):**

Preparing Vehicle for Customer, Task 1: Ensure vehicle is prepared to return to customer per school/company policy (floor mats, steering wheel cover, etc.).

We Support ASE | Education Foundation

Directions: Before beginning this lab assignment, review the worksheet completely. Fill in the information in the spaces provided as you complete each task.

Procedure:

1. Inspect the condition of the following:
   a. Windshield glass: ☐ Fogged ☐ Cracked ☐ Chipped ☐ Pitted ☐ Good
   b. Windshield rubber molding: ☐ Good ☐ Cracked ☐ Evidence of leaks
   c. Rear window glass: ☐ Good ☐ Fogged ☐ Cracked ☐ Chipped ☐ Pitted
   d. Rear window rubber molding: ☐ Good ☐ Cracked ☐ Evidence of leaks
   e. How are the side windows operated? ☐ Hand crank ☐ Power
2. Check the operation and condition of the wiper blades.
   a. Do they operate properly? ☐ Yes ☐ No
   b. Do they operate in the proper range (without going off the window or hitting the trim)? ☐ Yes ☐ No
   c. Is the blade rubber soft or torn? ☐ Yes ☐ No
   d. Is the tension spring good? ☐ Yes ☐ No
3. Check the condition and operation of the windshield washer.
   a. Check the reservoir liquid level. ☐ OK ☐ Low
   b. Are the washer nozzles aimed correctly? ☐ Yes ☐ No
   c. Are the washer nozzles plugged? ☐ Yes ☐ No
   d. Is there a sufficient volume of fluid? ☐ Yes ☐ No

4. Check the condition of the side window glass.
   a. Right front window:
      Is the glass broken? ☐ Yes ☐ No
      Will the window roll up and down? ☐ Yes ☐ No
      Condition of the window molding: ☐ Good ☐ Cracked ☐ Evidence of leaks
   b. Left front window:
      Is the glass broken? ☐ Yes ☐ No
      Will the window roll up and down? ☐ Yes ☐ No
      Condition of the window molding: ☐ Good ☐ Cracked ☐ Evidence of leaks
   c. Right rear window:
      Is the glass broken? ☐ Yes ☐ No ☐ N/A
      Will the window roll up and down? ☐ Yes ☐ No
      Condition of the window molding: ☐ Good ☐ Cracked ☐ Evidence of leaks
   d. Left rear window:
      Is the glass broken? ☐ Yes ☐ No N/A
      Will the window roll up and down? ☐ Yes ☐ No
      Condition of the window molding: ☐ Good ☐ Cracked ☐ Evidence of leaks
5. Check the condition of the mirrors.
   a. Interior mirror: ☐ Tight ☐ Glass clear
   b. Driver side exterior mirror: ☐ Loose ☐ Missing ☐ Cracked ☐ Good
   c. Passenger side mirror: ☐ Loose ☐ Missing ☐ Cracked ☐ N/A ☐ Good
   d. Are the outside mirrors power operated? ☐ Yes ☐ No
      If so, do they both move up and down, as well as right to left? ☐ Yes ☐ No
6. Check the front window defroster.
   a. Does the blower motor work?
      ☐ At all speeds ☐ Only middle speed ☐ Not working
      ☐ Only high speed ☐ Only slow speed
   b. Do the heater controls operate smoothly without binding? ☐ Yes ☐ No
   c. Does air blow from the ducts? ☐ Yes ☐ No
   d. Do the windows fog when the defroster is turned on? ☐ Yes ☐ No
   Note: This could be due to a leaking heater core.
7. Check the rear window defroster.
   a. Does the vehicle have a rear window defroster? ☐ Yes ☐ No
   b. If it has a defroster, check the electrical strips in the window. Are any of them scratched, torn, or obviously damaged? ☐ Yes ☐ No ☐ N/A
8. Before completing your paperwork, clean your work area, clean and return tools to their proper places, and wash your hands.
9. Record your recommendations for needed service or additional repairs and complete the repair order.

____

____

STOP

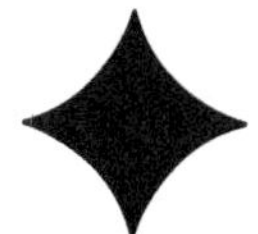

ASE Education Foundation Worksheet #2-10

REPLACE A WIPER BLADE

Name______________________________ Class______________________

Score: ☐ Excellent ☐ Good ☐ Needs Improvement **Instructor OK** ☐

Vehicle year __________ **Make** ____________________ **Model** ______________________

Objective: Upon completion of this assignment, you will be able to check and replace a vehicle's wiper blades.

ASE Education Foundation Correlation

This worksheet addresses the following **MLR** task:

VI.E.8 Verify windshield wiper and washer operation; replace wiper blades. **(P-1)**

This worksheet addresses the following **AST** task:

VI.G.4 Describe operation of safety systems and related circuits (such as: horn, airbags, seat belt pretensioners, occupancy classification, wipers, washers, speed control/collision avoidance, heads-up display, park assist, and back-up camera); determine needed repairs. **(P-3)**

This worksheet addresses the following **MAST** task:

VI.G.4 Diagnose operation of safety systems and related circuits (such as: horn, airbags, seat belt pretensioners, occupancy classification, wipers, washers, speed control/collision avoidance, heads-up display, park assist, and back-up camera); determine needed repairs. **(P-1)**

We Support ASE | Education Foundation

Directions: Before beginning this lab assignment, review the worksheet completely. Fill in the information in the spaces provided as you complete each task.

Tools and Equipment Required: Safety glasses, fender covers

Parts and Supplies: Wiper blades

Procedure:

1. Measure the length of the old wiper blade.

   ☐ 12″ ☐ 13″ ☐ 14″ ☐ 15″

   ☐ Other (list here) __________

2. The wiper blade assembly is:

   ☐ Refillable ☐ Nonrefillable

3. If the wiper blade is of the refillable type, what type of locking mechanism does it use?

   ☐ Plastic button ☐ Metal end clip ☐ Notched

Push Button Refill

End Clip Refill

Notched Flexor Refill Coin Removal

TURN

4. Obtain the two new replacement wiper blade assemblies or refills.
5. Compare the new parts with the old wiper blades. Do they look like they are the right replacement parts?

   ☐ Yes ☐ No
6. If there is any doubt, consult your instructor.
7. Carefully read and follow the instructions that come with the replacement parts. Did you read the instructions before you started to replace the wiper blades?

   ☐ Yes ☐ No
8. Be especially careful not to scratch the windshield glass or paint. Remove the old blade assembly from the wiper arm.

   Note: Some shops place a piece of cardboard over the windshield to avoid damaging it during a wiper blade replacement.
9. Install the new wiper blade assemblies or refills. Be certain that they fit properly.

   Note: Failure to properly install the wiper blade can result in a scratched windshield, which is costly to replace.
10. Check the operation of the windshield wipers. Do the windshield wipers work?

    ☐ Yes ☐ No
11. If the wipers do not work properly, what is the problem?
12. Before completing your paperwork, clean your work area, clean and return tools to their proper places, and wash your hands.
13. Record your recommendations for needed service or additional repairs and complete the repair order. ____________________

    ____

STOP

ASE Education Foundation Worksheet #2-11

ON-THE-GROUND SAFETY CHECKLIST

Name______________________________ Class______________________

Score: ☐ Excellent ☐ Good ☐ Needs Improvement **Instructor OK** ☐

Vehicle year __________ **Make** ____________________ **Model** ______________________

Objective: Upon completion of this assignment, you should be able to inspect a vehicle's safety features that are accessible without raising the vehicle.

ASE Education Foundation Correlation

This worksheet addresses the following **Required Supplemental Task (RST):**

Preparing Vehicle for Customer, Task 1: Ensure vehicle is prepared to return to customer per school/company policy (floor mats, steering wheel cover, etc.).

We Support ASE | Education Foundation

Directions: Before beginning this lab assignment, review the worksheet completely. Fill in the information in the spaces provided as you complete each task.

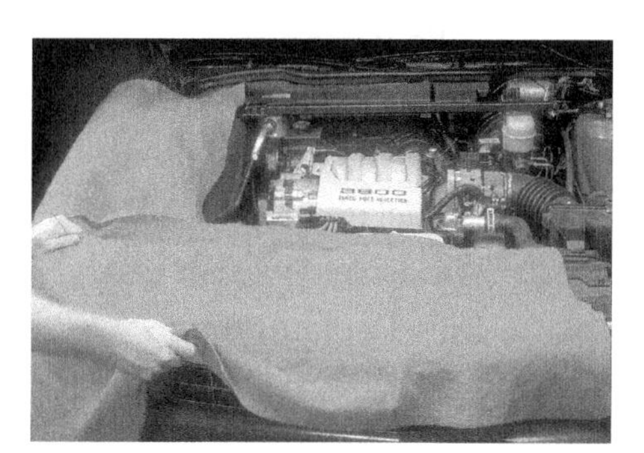

Tools and Equipment Required: Safety glasses, fender covers, shop towel

Procedure:

1. Open the hood and place fender covers on the fenders and over the front body parts.
2. Inspect the condition of the following:

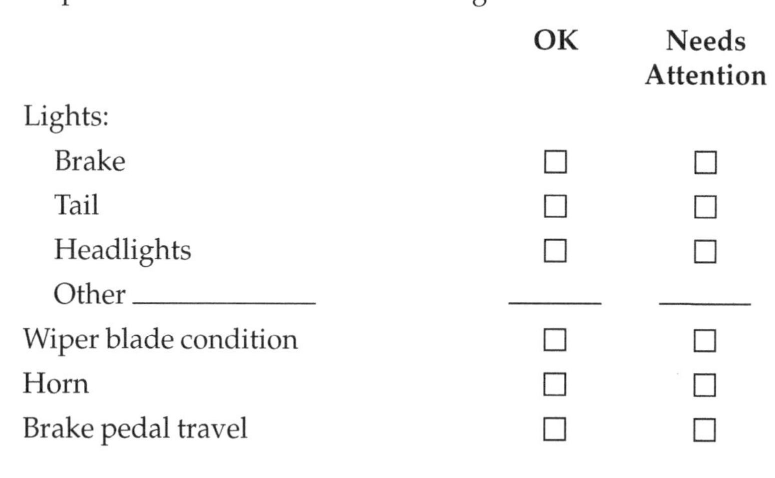

	OK	Needs Attention
Lights:		
Brake	☐	☐
Tail	☐	☐
Headlights	☐	☐
Other ____________	______	______
Wiper blade condition	☐	☐
Horn	☐	☐
Brake pedal travel	☐	☐

TURN

	OK	Needs Attention	
Emergency brake adjustment	☐	☐	
Seat belts (not loose or damaged)	☐	☐	
Windows (cracked or broken)	☐	☐	
Mirrors (cracked, loose, or broken)	☐	☐	
Door latches (lube)	☐	☐	
Exhaust leaks (listen)	☐	☐	
Tire pressure check (cold)	☐	☐	
Check tires for nails	☐	☐	
Tire condition	☐	☐	
Lug nut torque check (Spec. ___ ft.-lb)	☐	☐	
V-belt tension and condition	☐	☐	
Hood latch	☐	☐	
Brake fluid level	☐	☐	
Battery water check and fill	☐	☐	
Electrical wiring (inspect)	☐	☐	
Hose condition:	☐	☐	
Radiator hoses	☐	☐	
Heater hoses	☐	☐	
Power steering hoses	☐	☐	
Fuel hoses	☐	☐	
Windshield washer reservoir	☐	☐	
Diesel exhaust fluid	☐	☐	☐N/A

Describe any other problems noticed during the vehicle inspection.

____

____

3. Before completing your paperwork, clean your work area, clean and return tools to their proper places, and wash your hands.

ASE Education Foundation Worksheet #2-12

CHECK AUTOMATIC TRANSMISSION FLUID (ATF) LEVEL

Name______________________ Class______________

Score: ☐ Excellent ☐ Good ☐ Needs Improvement **Instructor OK** ☐

Vehicle year ________ **Make** ______________ **Model** ______________

Objective: Upon completion of this assignment, you will be able to check an automatic transmission for the correct fluid level.

ASE Education Foundation Correlation

This worksheet addresses the following **MLR** tasks:

II.A.2 Check fluid level in a transmission or a transaxle equipped with a dipstick. **(P-1)**

II.A.3 Check fluid level in a transmission or a transaxle not equipped with a dipstick. **(P-1)**

This worksheet addresses the following **AST/MAST** tasks:

II.A.4 Check fluid level in a transmission or a transaxle equipped with a dipstick. **(P-1)**

II.A.5 Check fluid level in a transmission or a transaxle not equipped with a dipstick. **(P-1)**

We Support ASE Education Foundation

Directions: Before beginning this lab assignment, review the worksheet completely. Fill in the information in the spaces provided as you complete each task. Complete only the sections that apply to the vehicle being inspected.

Tools and Equipment Required: Safety glasses, fender covers

Procedure:

Note: Some automatic transmissions do not have a dipstick. The fluid level is checked by removing a plug and observing if any fluid overflows. If the transmission does not have a dipstick, it will be necessary to refer service information to learn the correct method of checking the fluid level.

1. What type of ATF is specified for this vehicle?

   ☐ Dexron ☐ Dexron II ☐ Type F ☐ CJ ☐ Mercon ☐ Type T

   ☐ Other (describe) __

2. Does the transmission have a dipstick? ☐ Yes ☐ No

To check the fluid level on a dipstick-type automatic transmission.

a. To obtain an accurate fluid level reading, the fluid must be at normal operating temperature.

   Normal operating temperature? ☐ Yes ☐ No

   Note: Cold fluid will be approximately 1 pint lower than warm fluid.

b. The vehicle must be on level ground.

c. Locate and remove the transmission dipstick.

d. Wipe the dipstick with a clean shop towel.

TURN

e. Does the fluid wipe off easily?

☐ Yes ☐ No

Note: If the fluid has become excessively hot, it can become sticky. The transmission will require rebuilding.

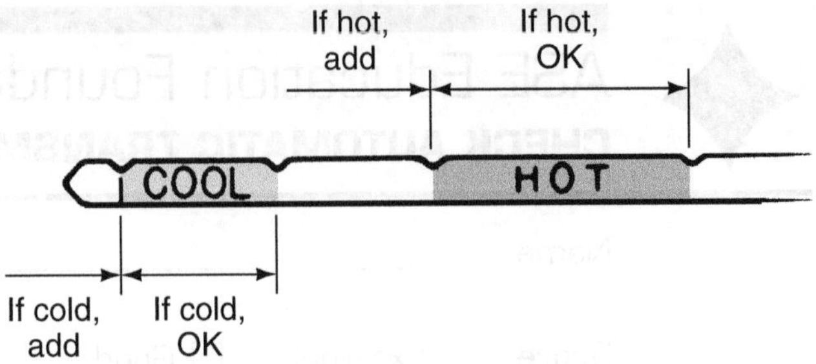

f. Are fluid checking instructions stamped on the dipstick?

☐ Yes ☐ No

g. What is the specified gearshift selector position during the fluid level check?

☐ P ☐ R ☐ N ☐ D ☐ L

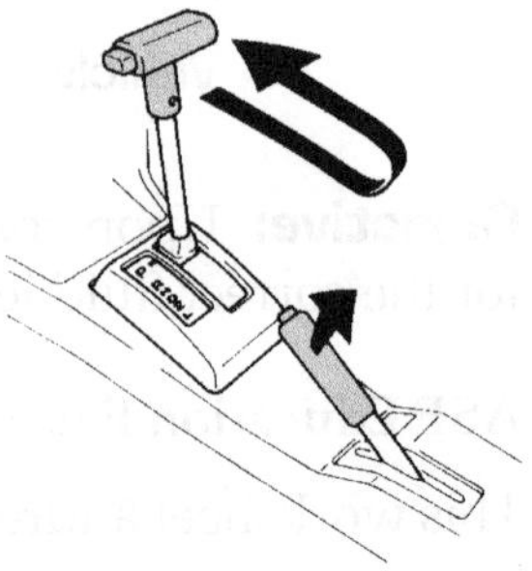

h. Be certain the parking brake is set.

i. Start the engine. Insert the dipstick until it seats firmly on the dipstick tube.

j. Remove the dipstick and note the reading. ☐ Full ☐ Low

k. What color is the fluid? ☐ Red ☐ Brown ☐ Pink

l. Does the fluid smell burnt? ☐ Yes ☐ No

m. Does any fluid need to be added to the transmission?

☐ Yes ☐ No

n. Was any ATF added to the transmission?

☐ Yes ☐ No

What type?

☐ Dexron ☐ Dexron II ☐ Type F CJ

☐ Mercon ☐ Type T ☐ Other ________

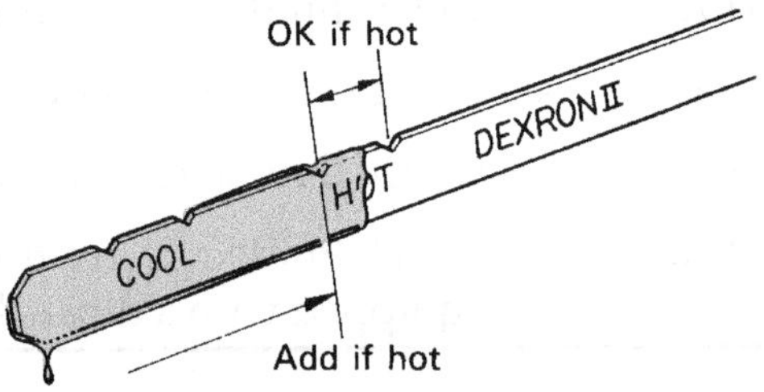

Checking automatic transmission fluid level on a vehicle that does not have a dipstick.

a. Refer to the service information for the suggested method for checking the fluid level. Briefly describe how to check the fluid level.

____

____

b. Are any special tools needed to check the fluid level? ☐ Yes ☐ No

c. Are any special tools needed to add fluid to the transmission? ☐ Yes ☐ No

d. Will you be able to check the fluid level on the transmission? ☐ Yes ☐ No

e. Shift the transmission into each gear range. Does the gear range indicator accurately indicate the gear range? ☐ Yes ☐ No, needs service

f. If the gear range indicator is not in the correct position, refer to the service information to learn how to adjust the transmission linkage.

3. Before completing your paperwork, clean your work area, clean and return tools to their proper places, and wash your hands.

4. Record your recommendations for needed service or additional repairs and complete the repair order.

____

____

STOP

ASE Education Foundation Worksheet #2-13

INSPECT SHOCK ABSORBERS/STRUTS

Name__________________________ Class__________________

Score: ☐ Excellent ☐ Good ☐ Needs Improvement **Instructor OK** ☐

Vehicle year __________ **Make** __________________ **Model** ____________________

Objective: Upon completion of this assignment, you will be able to inspect a vehicle's shock absorbers.

ASE Education Foundation Correlation

This worksheet addresses the following **MLR** tasks:

IV.B.20 Inspect, remove, and replace shock absorbers; inspect mounts and bushings. **(P-1)**

IV.B.16 Inspect, remove, and/or replace strut cartridge or assembly, strut coil spring, insulators (silencers), and upper strut bearing mount. **(P-2)**

This worksheet addresses the following **AST/MAST** tasks:

IV.D.1 Inspect, remove, and replace shock absorbers; inspect mounts and bushings. **(P-1)**

IV.C.10 Inspect, remove, and/or replace strut cartridge or assembly, strut coil spring, insulators (silencers), and upper strut bearing mount. **(P-3)**

We Support ASE | Education Foundation

Directions: Before beginning this lab assignment, review the worksheet completely. Fill in the information in the spaces provided as you complete each task.

Tools and Equipment Required: Safety glasses, shop towel

Procedure:

1. Perform a bounce test.

   a. Push down hard two or three times on the bumper at each corner of the car. If the shock absorbers are operating properly, the spring should oscillate about 1.5 cycles and then stop. Resistance should be equal from side to side.

   Note: Do not compare front operation to rear operation.

   Results: Shock absorber resists spring oscillation ☐ Yes ☐ No

   Right to left resistance equal ☐ Yes ☐ No

   b. Is there unusual noise during the bounce test? ☐ Yes ☐ No

   Note: The sound of fluid being forced through the valves in the shock absorber is normal.

TURN

2. Inspect tires for "cupping" wear.

   Cupping wear: ☐ Yes ☐ No

   Note: Cupping indicates that the tire has been hopping. Tire imbalance could also be a contributing factor.

3. Perform a visual inspection of the shocks or strut assemblies.

   a. Condition of the shock absorber mounts and rubber cushions:

      ☐ Good ☐ Bad ☐ N/A

   b. Has any fluid leaked out of a shock absorber or strut assembly?

      ☐ Dry ☐ Leaking

   Note: It is normal for a slight amount of moisture to be on the seal of a shock absorber.

4. Is the outside of the shock or strut body damaged? ☐ Yes ☐ No

   Note: Shock absorbers and struts are always replaced in pairs (front and/or rear).

5. Before completing your paperwork, clean your work area, clean and return tools to their proper places, and wash your hands.

6. Record your recommendations for needed service or additional repairs and complete the repair order.

   ____

   ____

ASE Education Foundation Worksheet #2-14

INSTRUMENT PANEL WARNING INDICATORS

Name_________________________ Class__________________

Score: ☐ Excellent ☐ Good ☐ Needs Improvement **Instructor OK** ☐

Vehicle year _________ **Make** _________________ **Model** __________________

Objective: Upon completion of this assignment, you will be able to verify the operation of the instrument panel warning indicators.

ASE Education Foundation Correlation

This worksheet addresses the following **MLR** tasks:

I.A.2 Verify operation of the instrument panel engine warning indicators. **(P-1)**

VI.E.7 Verify operation of instrument panel gauges and warning/indicator lights; reset maintenance indicators. **(P-1)**

This worksheet addresses the following **AST/MAST** tasks:

I.A.3 Verify operation of the instrument panel engine warning indicators. **(P-1)**

VI.F.2 Diagnose (troubleshoot) the causes of incorrect operation of warning devices and other driver information systems; determine needed action. **(P-1)**

We Support ASE Education Foundation

Directions: Before beginning this lab assignment, review the worksheet completely. Fill in the information in the spaces provided as you complete each task.

Tools and equipment: None required

Procedure:

1. Use the service information or owner's manual to identify the location and meaning of the various instrument panel warning indicators (lights and gauges).

   Which service information was used? ______________________________

   Note: Not all warning indicators are lights. Some are gauges, while others appear on the instrument panel message display.

2. The lamps are of different colors. In the chart below indicate the color each lamp should be: red (R), amber (A), green (G), or blue (B).

   ______ Oil pressure
   ______ Coolant temperature
   ______ Charging indicator
   ______ Low oil level
   ______ Antilock brakes
   ______ Traction control
   ______ Air bag (SRS)
   ______ Malfunction Indicator Light
   ______ High beam indicator
   ______ Turn signal/hazard indicator
   ______ Cruise control
   ______ Low fuel level
   ______ Brake warning indicator
   ______ Door open warning
   ______ Seat belt
   ______ Other (list)
   ____
   ____

TURN

3. What should a driver do if a red indicator lamp illuminates while driving?
   - ☐ Stop and get immediate service.
   - ☐ Drive to the nearest service center for help.
   - ☐ Do not worry. The lamp may go out on its own.
   - ☐ Proceed with caution and get the problem checked out soon.
4. What should a driver do if an amber indicator lamp illuminates while driving?
   - ☐ Stop and get immediate service.
   - ☐ Drive to the nearest service center for help.
   - ☐ Do not worry. The lamp may go out on its own.
   - ☐ Proceed with caution and get the problem checked out soon.
5. Place a check next to the warning indicators used on the vehicle being inspected.
   - ☐ Oil pressure
   - ☐ Coolant temperature
   - ☐ Charging indicator
   - ☐ Low oil level
   - ☐ Antilock brakes
   - ☐ Traction control
   - ☐ Air bag (SRS)
   - ☐ Malfunction Indicator Light
   - ☐ High beam indicator
   - ☐ Turn signal/hazard indicator
   - ☐ Cruise control
   - ☐ Low fuel level
   - ☐ Brake warning indicator
   - ☐ Door open warning
   - ☐ Seat belt
   - ☐ Other (list)

   ____

   ____
6. Turn the key to the on position. With the key on and engine off (KOEO), all the warning indicator lights should illuminate. This is how they are tested. Some of them might go out after a few seconds.
7. With the key on and the engine off (KOEO), did all the required warning indicator lights illuminate? Yes ☐ No ☐
8. Indicate below any of the required lamps that failed to illuminate.
   - ☐ Oil pressure
   - ☐ Coolant temperature
   - ☐ Charging indicator
   - ☐ Low oil level
   - ☐ Antilock brakes
   - ☐ Traction control
   - ☐ Air bag (SRS)
   - ☐ Malfunction Indicator Light
   - ☐ High beam indicator
   - ☐ Turn signal/hazard indicator
   - ☐ Cruise control
   - ☐ Low fuel level
   - ☐ Brake warning indicator
   - ☐ Door open warning
   - ☐ Seat belt
   - ☐ Other (list)

   ____

   ____
9. If there were one or more lamps that did not illuminate, list the indicated problem.

   ____

   ____

   ____

STOP

Part II

Lab Worksheets for ASE Maintenance and Light Repair

Service Area 3

Under-Vehicle Service

ASE Education Foundation Worksheet #3-1

TIRE INSPECTION

Name ______________________ Class ________________

Score: ☐ Excellent ☐ Good ☐ Needs Improvement **Instructor OK** ☐

Vehicle year ________ **Make** ______________ **Model** ________________

Objective: Upon completion of this assignment, you will be able to inspect the condition of a vehicle's tires.

ASE Education Foundation Correlation

This worksheet addresses the following **MLR** task:

IV.D.1 Inspect tire condition; identify tire wear patterns; check for correct tire size, application (load and speed ratings), and air pressure as listed on the tire information placard/label. **(P-1)**

This worksheet addresses the following **AST/MAST** task:

IV.F.1 Inspect tire condition; identify tire wear patterns; check for correct tire size, application (load and speed ratings), and air pressure as listed on the tire information placard/label. **(P-1)**

We Support
ASE Education Foundation

Directions: Before beginning this lab assignment, review the worksheet completely. Fill in the information in the spaces provided as you complete each task.

Tools and Equipment Required: Safety glasses, jack stands or vehicle lift, shop towel

Procedure:

Tire size ____________ Tire manufacturer ____________

1. Raising the vehicle onto jack stands or a lift allows the tires to be more easily inspected.
2. Visually inspect the tires. Check any of the following that apply.

   ☐ Nails ☐ Glass
   ☐ Screws ☐ Cracks
   ☐ Tears ☐ Bulges
   ☐ None of the above ☐ Other

3. Is there any sign that the tread is separating from the tire casing?

   ☐ Yes ☐ No

Photo by Tim Gilles

Photo by Tim Gilles

TURN

4. Tire sidewall condition:

   ☐ Good condition?

   ☐ Sun cracks?

   ☐ Whitewall scuffed?

   ☐ Bubble or separation?

   ☐ Cut?

5. Spin the wheels. Do the tires appear to spin "true"?

   Left front: ☐ Yes ☐ No Right front: ☐ Yes ☐ No

   Left rear: ☐ Yes ☐ No Right rear: ☐ Yes ☐ No

6. Does the tire show signs of underinflation or overinflation wear?

   ☐ Yes ☐ No

7. Measure the depth of the tread.

   Front: Left ______/32" Rear: Left ______/32"

   Right ______/32" Right ______/32"

8. Are the tread wear bars even with the surface of the tread (less than 1/16" of tread remaining)?

   ☐ Yes ☐ No

9. Are there signs of unusual tread wear? ☐ Yes ☐ No

10. Are any of the tires worn more than the others?

    ☐ Yes ☐ No

    If yes, list which tire(s) here. ______________

11. Should a tire rotation be recommended?

    ☐ Yes ☐ No

    If you are recommending a tire rotation, explain why.

    ____

12. Are all of the lug nuts on each wheel?

    ☐ Yes ☐ No

    If any of the lug nuts are missing, list which tire here.

    ____

13. Are any of the wheel covers missing?

    ☐ Yes ☐ No

    If any of the covers are missing, list which tire here.

    ____

• Overinflated • Underinflated

Tread depth gauge

New tread Worn tread

Tread wear indicator location marks

14. Before completing your paperwork, clean your work area, clean and return tools to their proper places, and wash your hands.

    ____

    ____

STOP

ASE Education Foundation Worksheet #3-2

ADJUST TIRE AIR PRESSURES

Name ______________________ Class ________________

Score: ☐ Excellent ☐ Good ☐ Needs Improvement **Instructor OK** ☐

Vehicle year ________ **Make** ______________ **Model** ________________

Objective: Upon completion of this assignment, you will be able to inspect the vehicle's tire condition and check and adjust tire air pressure.

ASE Education Foundation Correlation

This worksheet addresses the following **MLR** tasks:

IV.D.1 Inspect tire condition; identify tire wear patterns; check for correct tire size, application (load and speed ratings), and air pressure as listed on the tire information placard/label. **(P-1)**

IV.D.5 Inspect tire and wheel assembly for air loss; determine necessary action. **(P-1)**

This worksheet addresses the following **AST/MAST** tasks:

IV.F.1 Inspect tire condition; identify tire wear patterns; check for correct tire size, application (load and speed ratings), and air pressure as listed on the tire information placard/label. **(P-1)**

IV.F.8 Inspect tire and wheel assembly for air loss; perform needed action. **(P-1)**

We Support ASE Education Foundation

Directions: Before beginning this lab assignment, review the worksheet completely. Fill in the information in the spaces provided as you complete each task.

Tools and Equipment Required: Safety glasses, tire pressure gauge, air chuck, shop towel

Procedure:

1. Locate the tire pressure specifications.

   Front ________ psi Rear ________ psi

2. Tire pressures should be checked when the tires are:

   ☐ Hot ☐ Cold

3. Where was the tire pressure specification located?

   ☐ Driver's door or door post ☐ Owner's manual

   ☐ Glove compartment label ☐ Service manual

   If service information or the owner's manual was used to find the specifications, which one was used and in what page did you find the specification?

   Source of specification ____________ Page ____________ N/A ____________

4. Remove the valve stem caps.

TURN

5. Check the condition of rubber valve stems by bending each of them back and forth.

   Do any of the valve stems need to be replaced?

   ☐ Yes ☐ No

   Which, if any, of the valve stems need to be replaced?

   ☐ Right front ☐ Right rear

   ☐ Left front ☐ Left rear

Bend

6. Use a tire pressure gauge to check tire pressures.
7. Record tire pressures below.

	LF	RF	LR	RR
Before adjusting air pressure	_____	_____	_____	_____
After adjusting air pressure	_____	_____	_____	_____

8. Reinstall valve stem caps.
9. Check the general condition of the tires as you check the air pressure.

	LF	RF	LR	RR
Good	☐	☐	☐	☐
Fair	☐	☐	☐	☐
Unsafe	☐	☐	☐	☐

   Describe the problem with any unsafe tires.

   ____

   ____

   ____

This side is used for inner duals

Photo by Tim Gilles

10. Check the pressure and the condition of the spare tire.

    Type of spare tire: ☐ Standard tire

    ☐ Compact spare ☐ No spare

    Does it have the proper air pressure?

    ☐ Yes ☐ No

    Spare tire condition: ☐ Good ☐ Fair ☐ Needs to be replaced

11. Before completing your paperwork, clean your work area, clean and return tools to their proper places, and wash your hands.
12. Record your recommendations for needed service or additional repairs.

    ____

    ____

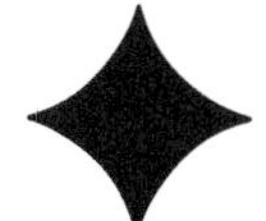

ASE Education Foundation Worksheet #3-3

TIRE WEAR DIAGNOSIS

Name______________________________ Class______________________

Score: ☐ Excellent ☐ Good ☐ Needs Improvement **Instructor OK** ☐

Vehicle year __________ **Make** ____________________ **Model** ______________________

Objective: Upon completion of this assignment, you will be able to inspect the condition of a vehicle's tires for serviceability and abnormal wear.

ASE Education Foundation Correlation

This worksheet addresses the following **MLR** task:

IV.D.1 Inspect tire condition; identify tire wear patterns; check for correct tire size, application (load and speed ratings), and air pressure as listed on the tire information placard/label. **(P-1)**

This worksheet addresses the following **AST/MAST** task:

IV.F.1 Inspect tire condition; identify tire wear patterns; check for correct tire size, application (load and speed ratings), and air pressure as listed on the tire information placard/label. **(P-1)**

We Support ASE | Education Foundation

Directions: Before beginning this lab assignment, review the worksheet completely. Fill in the information in the spaces provided as you complete each task.

Tools and Equipment Required: Safety glasses, tire depth gauge

Procedure:

Inspect the vehicle's tires and record the results below. Indicate a problem tire using the following abbreviations: right front (RF), left front (LF), right rear (RR), left rear (LR), and spare (S).

1. Do all of the tires on the vehicle have the same tread design? ☐ Yes ☐ No

   If no, which ones are different.

   ☐ RF ☐ LF ☐ RR ☐ LR ☐ S

2. Are any of the tires worn to the wear bars?

   ☐ Yes ☐ No

3. Measure the tread depth for each of the tires and list below.

   ______ RF ______ LF ______ RR ______ LR ______ S

Tread wear indicators

Photo by Tim Gilles

4. Inspect each of the tires for unusual wear and indicate problem tire(s).

   Underinflation wear ☐ RF ☐ LF ☐ RR ☐ LR ☐ S

   Overinflation wear ☐ RF ☐ LF ☐ RR ☐ LR ☐ S

   Camber wear (one side of the tread)?

   ☐ RF ☐ LF ☐ RR ☐ LR ☐ S

   Toe wear (feathered edge across the tread)?

   ☐ RF ☐ LF ☐ RR ☐ LR ☐ S

   Cupped wear?

   ☐ RF ☐ LF ☐ RR ☐ LR ☐ S ☐ None

   Cuts in tire sidewall?

   ☐ RF ☐ LF ☐ RR ☐ LR ☐ S ☐ None

5. List any tires that should be replaced.

   ☐ RF ☐ LF ☐ RR ☐ LR ☐ S ☐ All tires are good

6. Abnormal tire wear can be due to other problems with the vehicle, driver habits, or road conditions. Which of the following could cause abnormal tire wear? Check all that apply.

   a. High-speed driving ☐ b. Bad shocks ☐

   c. Bad transmission ☐ d. Mountain driving ☐

   e. Loud stereo ☐

7. Before completing your paperwork, clean your work area, clean and return tools to their proper places, and wash your hands.

8. Record your recommendations for needed service or additional repairs and complete the repair order.

   ____

   ____

STOP

ASE Education Foundation Worksheet #3-4

EXHAUST SYSTEM INSPECTION

Name________________________________ Class______________________

Score: ☐ Excellent ☐ Good ☐ Needs Improvement **Instructor OK** ☐

Vehicle year __________ **Make** ____________________ **Model** ______________________

Objective: Upon completion of this assignment, you will be able to inspect a vehicle's exhaust system for leaks and damage.

ASE Education Foundation Correlation

This worksheet addresses the following **MLR** tasks:

VIII.C.3 Inspect integrity of the exhaust manifold, exhaust pipes, muffler(s), catalytic converter(s), resonator(s), tail pipe(s), and heat shields; determine necessary action. **(P-1)**

VIII.C.4 Inspect condition of exhaust system hangers, brackets, clamps, and heat shields; determine necessary action. **(P-1)**

This worksheet addresses the following **AST** tasks:

VIII.D.8 Inspect integrity of the exhaust manifold, exhaust pipes, muffler(s), catalytic converter(s), resonator(s), tail pipe(s), and heat shields; determine needed action. **(P-1)**

VIII.D.9 Inspect condition of exhaust system hangers, brackets, clamps, and heat shields; determine needed action. **(P-1)**

This worksheet addresses the following **MAST** tasks:

VIII.D.9 Inspect integrity of the exhaust manifold, exhaust pipes, muffler(s), catalytic converter(s), resonator(s), tail pipe(s), and heat shields; perform needed action. **(P-1)**

VIII.D.10 Inspect condition of exhaust system hangers, brackets, clamps, and heat shields; determine needed action. **(P-1)**

We Support ASE | Education Foundation

Directions: Before beginning this lab assignment, review the worksheet completely. Fill in the information in the spaces provided as you complete each task.

Tools and Equipment Required: Safety glasses, jack stands or vehicle lift, shop towel

Procedure:

1. Raise the vehicle to gain easy access to the exhaust system.
2. Identify the components of the exhaust system on the vehicle that you are inspecting.

   Check all that apply.

   ☐ Single exhaust

   ☐ Dual exhaust

Upstream O_2 sensor
Converter
Resonator
Muffler
Exhaust pipe
Downstream O_2 sensor
Tailpipe

TURN

☐ Catalytic converter

☐ Muffler

☐ Resonator

3. How is the vehicle you are inspecting equipped?

   Check all that apply.

   Stock exhaust system? ☐ Yes ☐ No

   Modified exhaust system? ☐ Yes ☐ No

   Tailpipe after rear axle? ☐ Yes ☐ No

   Pipe exiting in front of rear wheels?

   ☐ Yes ☐ No

Weep hole or drain

Photo by Tim Gilles

4. Visually inspect the components of the exhaust system.

   a. Condition of muffler hangers/supports. ☐ Yes ☐ No
   b. Holes in the muffler? ☐ Yes ☐ No
   c. Is there a weep hole or drain? ☐ Yes ☐ No
   d. Holes in any of the pipes? ☐ Yes ☐ No
   e. Do the clamps appear to be tight? ☐ Yes ☐ No
   f. Catalytic converter damaged? ☐ Yes ☐ No
   g. Exhaust system shields in place? ☐ Yes ☐ No

5. Start the engine and check for leaks in each of these components:

	OK	Leaks	Parts needed	N/A
Exhaust manifold	☐	☐	☐	☐
Exhaust manifold gasket	☐	☐	☐	☐
Header pipe	☐	☐	☐	☐
Catalytic converter	☐	☐	☐	☐
Muffler	☐	☐	☐	☐
Exhaust pipe	☐	☐	☐	☐
Resonator	☐	☐	☐	☐
Tailpipe	☐	☐	☐	☐

Other ______________________________

6. Check the condition of the exhaust system clamps and hangers. ☐ OK ☐ Parts needed

7. Before completing your paperwork, clean your work area, clean and return tools to their proper places, and wash your hands.

8. Record your recommendations for needed exhaust service or additional repairs.

____

____

STOP

ASE Education Foundation Worksheet #3-5

INSPECT REAR SUSPENSION COMPONENTS

Name______________________ Class________________

Score: ☐ Excellent ☐ Good ☐ Needs Improvement **Instructor OK** ☐

Vehicle year ________ **Make** ______________ **Model** ______________

Objective: Upon completion of this assignment, you will be able to inspect a rear suspension system.

ASE Education Foundation Correlation

This worksheet addresses the following **MLR** tasks:

IV.A.3 Identify suspension and steering system components and configurations. **(P-1)**

IV.B.18 Inspect rear suspension system lateral links/arms (track bars), control (trailing) arms. **(P-1)**

IV.B.19 Inspect rear suspension system leaf spring(s), spring insulators (silencers), shackles, brackets, bushings, center pins/bolts, and mounts. **(P-1)**

This worksheet addresses the following **AST/MAST** tasks:

IV.C.11 Inspect, remove, and/or replace track bar, strut rods/radius arms, and related mounts and bushings. **(P-3)**

IV.C.12 Inspect rear suspension system leaf spring(s), spring insulators (silencers), shackles, brackets, bushings, center pins/bolts, and mounts. **(P-1)**

We Support ASE Education Foundation

Directions: Before beginning this lab assignment, review the worksheet completely. Fill in the information in the spaces provided as you complete each task.

Tools and equipment: Safety glasses, fender covers, hand tools, vehicle lift, shop towels

Procedure:

1. Use the service information to identify the parts and procedures for inspecting the rear suspension components.

   Which service information was used? ______________________________

2. Follow the recommended procedures for inspecting the rear suspension components.
3. Raise the vehicle.
4. What is the driveline type of this vehicle? FWD ☐ RWD ☐ 4WD ☐ AWD ☐

TURN

5. Inspect the suspension bushings, visually and with a prybar, as needed. Record your results below.

	Good	Need repair	N/A
Control arm bushings	☐	☐	☐
Trailing arm bushings	☐	☐	☐
Sway bar bushings	☐	☐	☐

6. Check the wheel bearing adjustment (if equipped with tapered roller bearings).

☐ OK ☐ Loose ☐ N/A

7. If equipped with coil springs, inspect the following.

Coil springs	☐ OK	☐ Damaged	☐ N/A
Spring seats	☐ OK	☐ Damaged	☐ N/A

8. If equipped with leaf springs, inspect the following:

☐ Insulators	☐ OK	☐ Damaged	☐ N/A
☐ Shackles	☐ OK	☐ Damaged	☐ N/A
☐ Center pin	☐ OK	☐ Damaged	☐ N/A
☐ U-bolts	☐ OK	☐ Damaged	☐ N/A
☐ Other	☐ OK	☐ Damaged	☐ N/A

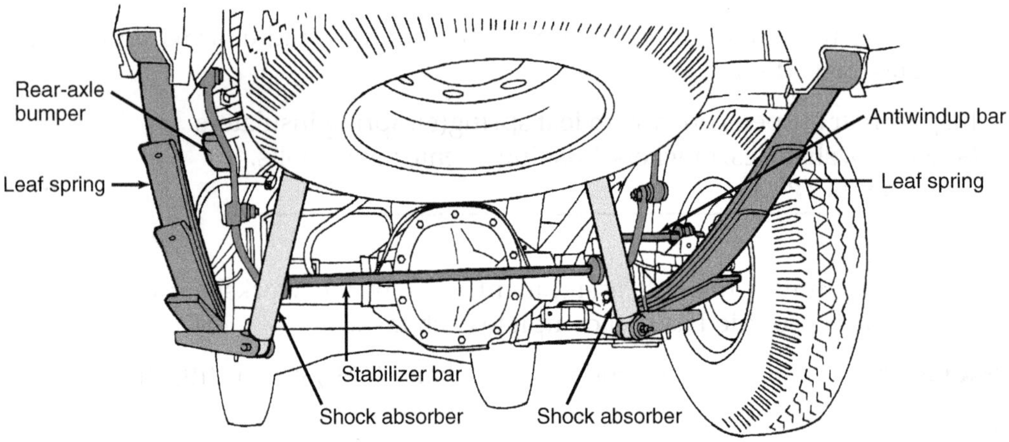

9. The rear suspension on this vehicle has:

☐ Shock absorbers ☐ MacPherson struts

10. Inspect the shock absorbers or struts.

☐ OK ☐ Damaged ☐ Leaking

11. Before completing the paperwork, clean your work area, put the tools in their proper places, and wash your hands.

12. List your recommendations for future service and/or additional repairs and complete the repair order.

____

____

STOP

ASE Education Foundation Worksheet #3-6

INSPECT FRONT SUSPENSION AND STEERING LINKAGE

Name________________________ Class________________

Score: ☐ Excellent ☐ Good ☐ Needs Improvement **Instructor OK** ☐

Vehicle year _________ **Make** ________________ **Model** ________________

Objective: Upon completion of this assignment, you will be able to inspect a vehicle's front suspension and steering linkage.

ASE Education Foundation Correlation

This worksheet addresses the following **MLR** tasks:

IV.B.1	Inspect rack and pinion steering gear inner tie rod ends (sockets) and bellows boots. **(P-1)**
IV.B.2	Inspect fluid level and condition. **(P-1)**
IV.B.5	Remove, inspect, replace, and/or adjust power steering pump drive belt. **(P-1)**
IV.B.6	Inspect and replace power steering hoses and fittings. **(P-1)**
IV.B.7	Inspect pitman arm, relay (centerlink/intermediate) rod, idler arm, mountings, and steering linkage damper. **(P-1)**
IV.B.8	Inspect tie rod ends (sockets), tie rod sleeves, and clamps. **(P-1)**
IV.B.9	Inspect upper and lower control arms, bushings, and shafts. **(P-1)**
IV.B.10	Inspect and replace rebound and jounce bumpers. **(P-1)**
IV.B.11	Inspect track bar, strut rods/radius arms, and related mounts and bushings. **(P-1)**
IV.B.12	Inspect upper and lower ball joints (with or without wear indicators). **(P-1)**

This worksheet addresses the following **AST/MAST** tasks:

IV.B.9	Determine proper power steering fluid type; inspect fluid level and condition. **(P-1)**
IV.B.11	Inspect for power steering fluid leakage; determine needed action. **(P-1)**
IV.C.3	Inspect, remove, and/or replace upper and lower control arms, bushings, shafts, and rebound bumpers. **(P-3)**
IV.C.4	Inspect, remove, and/or replace strut rods and bushings. **(P-3)**
IV.C.7	Inspect, remove and/or replace short and long arm suspension system coil springs and spring insulators. **(P-3)**
IV.C.9	Inspect, remove, and/or replace front/rear stabilizer bar (sway bar) bushings, brackets, and links. **(P-3)**

We Support ASE Education Foundation

Directions: Before beginning this lab assignment, review the worksheet completely. Fill in the information in the spaces provided as you complete each task.

Tools and Equipment Required: Safety glasses, jack stands or vehicle lift, shop towel

Procedure:

Visual Inspection

1. Check the steering system for looseness (dry park check).

   With the wheels on the ground, have an assistant turn the steering wheel back and forth a short distance while you look for looseness in steering linkage. A power steering car must have the engine running for this test.

   ☐ Loose ☐ Tight

TURN

2. Raise the vehicle.
3. Inspect suspension bushings visually and with a prybar. Record your results below.

	Good	Need repair	N/A
Upper control arm bushings	☐	☐	☐
Lower control arm bushings	☐	☐	☐
Sway bar bushings	☐	☐	☐
Strut rod bushing	☐	☐	☐
Rebound and jounce bumpers	☐	☐	☐

Courtesy of Federal-Mogul Corporation

4. Inspect steering linkage pivot connections. Firmly grasp the part and rock it to check for looseness.

	Good	Need repair	N/A
Tie-rod ends	☐	☐	☐
Idler arm	☐	☐	☐
Pitman arm	☐	☐	☐
Center link	☐	☐	☐
Rack bushings	☐	☐	☐
Rack Inner tie rod sockets	☐	☐	☐

5. Check wheel bearing adjustment. Refer to Worksheet #9-15, Adjust a Tapered Roller Wheel Bearing. ☐ OK ☐ Loose
6. Check ball joints for looseness. Refer to Worksheet #9-29, Check Ball Joint Wear. ☐ OK ☐ Loose
7. Inspect rubber grease boots on tie-rods, ball joints, and steering rack.

	Good	Damaged
Tie-rod end boots	☐	☐
Rack bushings	☐	☐
Rack boots	☐	☐

8. Inspect shock absorbers or struts. Refer to Worksheet #2-13, Inspect Shock Absorbers/Struts.

   ☐ OK ☐ Dented ☐ Leaking
9. Inspect the steering gear. ☐ OK ☐ Leaking ☐ Loose

   What type of steering gear assembly is used on this vehicle?

   ☐ Rack and pinion ☐ Recirculating ball and nut ☐ Other
10. Lower the vehicle, open the hood, and place fender covers on the fenders and front body parts.
11. Inspect the power steering system.

	Good	Needs attention	N/A
Fluid level	☐	☐	☐
Leakage	☐	☐	☐
Drive belt condition	☐	☐	☐
Drive belt tension	☐	☐	☐
Hose condition	☐	☐	☐

12. Before completing your paperwork, clean your work area, clean and return tools to their proper places, and wash your hands.
13. Record your recommendations for needed service or additional repairs and complete the repair order.

____

____

STOP

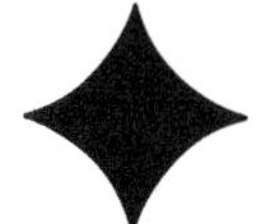

ASE Education Foundation Worksheet #3-7

CHASSIS LUBRICATION

Name________________________ Class__________________

Score: ☐ Excellent ☐ Good ☐ Needs Improvement **Instructor OK** ☐

Vehicle year ________ **Make** ______________ **Model** ______________

Objective: Upon completion of this assignment, you will be able to lubricate a vehicle's suspension and steering components.

ASE Education Foundation Correlation

This worksheet addresses the following **AST/MAST** tasks:

IV.C.1 Diagnose short- and long-arm suspension system noises, body sway, and uneven ride height concerns; determine needed action. **(P-1)**

IV.C.2 Diagnose strut suspension system noises, body sway, and uneven ride height concerns; determine needed action. **(P-1)**

We Support ASE | Education Foundation

Directions: Before beginning this lab assignment, review the worksheet completely. Fill in the information in the spaces provided as you complete each task.

Tools and Equipment Required: Safety glasses, fender covers, jack stands or vehicle lift, grease gun, shop towel, service manual

Parts and Supplies: Grease, hinge lubricant, door latch lubricant

Procedure:

1. Refer to a service manual for the number and location of the lubrication points.

   Number of fittings ______ ☐ N/A

   Name of service manual used ______ Page # ______

2. Open the hood and place fender covers on the fenders and front body parts.
3. Raise the vehicle on a hoist or with a jack. Be sure to support it with jack stands.
4. Locate all of the fittings to be greased and wipe them clean with a shop towel.

   Number of fittings located ______

   Cleaned? ☐ Yes ☐ No

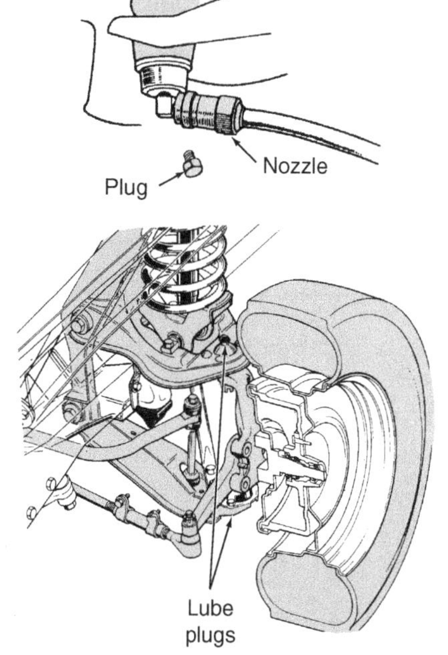

The vehicle is equipped with:

☐ Plugs ☐ Zerk fittings ☐ N/A

Note: Some vehicles have plugs installed in place of the grease fittings. These will need to be removed and a fitting temporarily installed. Future grease jobs are easier if the plugs are replaced with standard zerk grease fittings.

Note: Pumping too much grease into the fitting can damage the seals. There are two types of joint seals.

❑ *Sealed boot.* Apply only enough grease to slightly bulge the boot.

❑ *Nonsealed.* Apply grease until the old grease has been flushed from the joint.

5. Push the grease gun straight onto the fitting. Hold it firmly in place and pump the grease into the joint slowly to prevent damage to the grease seals.

   Type of grease gun used:

   ☐ Hand pump ☐ Air operated

6. Wipe excess grease from the fittings.
7. Lower the vehicle. Lubricate the moving parts of the hinges with a small amount of oil or lubricant.

   ☐ Door hinges

   ☐ Trunk hinges

   ☐ Hood hinges

Photo by Tim Gilles

8. Lubricate the hood and door latches.

   Note: The door latches are located where people rub against them when entering and exiting the vehicle. To prevent damage to clothing, use a nonstaining lubricant that does not attract dirt.

   ☐ Door latches ☐ Trunk latch ☐ Hood latch

9. Before completing your paperwork, clean your work area, clean and return tools to their proper places, and wash your hands.
10. Record your recommendations for needed service or additional repairs and complete the repair order.

____

____

STOP

ASE Education Foundation Worksheet #3-8

MANUAL TRANSMISSION/TRANSAXLE SERVICE

Name________________________ Class________________

Score: ☐ Excellent ☐ Good ☐ Needs Improvement **Instructor OK** ☐

Vehicle year ________ **Make** ______________ **Model** ________________

Objective: Upon completion of this assignment, you will be able to inspect a manual transmission for leaks, check the fluid level, and replace the fluid.

ASE Education Foundation Correlation

This worksheet addresses the following **MLR** tasks:

III.A.2 Drain and refill manual transmission/transaxle and final drive unit; use proper fluid type per manufacturer specification. **(P-1)**

III.A.3 Check fluid condition; check for leaks. **(P-2)**

This worksheet addresses the following **AST/MAST** tasks:

III.A.3 Check fluid condition; check for leaks; determine needed action. **(P-1)**

III.A.4 Drain and refill manual transmission/transaxle and final drive unit; use proper fluid type per manufacturer specification. **(P-1)**

We Support ASE Education Foundation

Directions: Before beginning this lab assignment, review the worksheet completely. Fill in the information in the spaces provided as you complete each task.

Procedure:

1. The vehicle is equipped with an:

   ☐ RWD manual transmission ☐ FWD manual transaxle

2. What type of fluid is required? ____________________
3. How much fluid is needed to refill this transmission/transaxle when empty? ____________
4. Raise the vehicle on a lift.
5. Locate the fill plug on the side of the transmission or transaxle.
6. Remove the plug and check the fluid level. The fluid should be level with the bottom of the fill plug hole.

   ☐ OK ☐ Low ☐ N/A

   Note: If the fluid is to be drained, remove the drain plug and drain the fluid into a drain pan. When the fluid has drained completely, replace the drain plug.

   Fluid drained? ☐ Yes ☐ No

TURN

7. Add the correct fluid as required.
8. How much fluid was added? __________ ☐ N/A
9. Install the fill plug.

CAUTION Do not overtighten the fill plug. The thread is usually tapered (NPT). An overtightened fill plug can cause the transmission case to crack.

10. Lower the vehicle.
11. Start the engine and shift through the gears.
12. Shut the engine off, raise the vehicle, and inspect the transmission/transaxle for leaks.
    ☐ No leaks ☐ Leaks; identify location: ____________________
13. Lower the vehicle.
14. Before completing your paperwork, clean your work area, clean and return tools to their proper places, and wash your hands.
15. Record your recommendations for needed service or additional repairs and complete the repair order.

    ____

    ____

STOP

ASE Education Foundation Worksheet #3-9

DIFFERENTIAL FLUID SERVICE

Name________________________ Class________________

Score: ☐ Excellent ☐ Good ☐ Needs Improvement **Instructor OK** ☐

Vehicle year ________ **Make** ______________ **Model** ______________

Objective: Upon completion of this assignment, you will be able to check the fluid level in a differential.

ASE Education Foundation Correlation

This worksheet addresses the following **MLR** tasks:

III.E.2 Check and adjust differential housing fluid level. **(P-1)**

III.E.3 Drain and refill differential housing. **(P-1)**

This worksheet addresses the following **AST/MAST** tasks:

III.E.1.1 Clean and inspect differential case; check for leaks; inspect housing vent. **(P-1)**

III.E.1.2 Check and adjust differential case fluid level; use proper fluid type per manufacturer specification. **(P-1)**

III.E.1.3 Drain and refill differential case; use proper fluid type per manufacturer specification. **(P-1)**

We Support ASE Education Foundation

Directions: Before beginning this lab assignment, review the worksheet completely. Fill in the information in the spaces provided as you complete each task.

Procedure:

1. What type of fluid is required? ______________
2. How much fluid is required for a refill? __________
3. Raise the vehicle on a lift.
4. Locate the fill plug on the differential (final drive) housing.
5. Remove the plug and check the fluid level. The fluid should be level with the bottom of the fill plug hole.

   ☐ OK ☐ Low ☐ N/A

   Note: If the fluid is to be drained, remove the drain plug and drain the fluid into a drain pan. When the fluid has drained completely, replace the drain plug.

   Was the fluid drained? ☐ Yes ☐ No

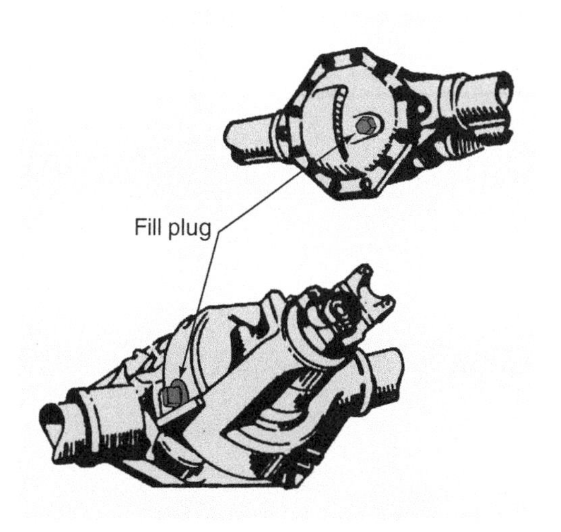

6. Add the correct fluid as required.

   How much fluid was added? ________ ☐ N/A

7. Install the fill plug.

CAUTION Do not overtighten the fill plug. The thread is usually tapered (NPT). An overtightened fill plug can damage the differential housing.

8. Lower the vehicle.
9. After the vehicle has been driven, inspect the differential area for leaks.
10. Before completing your paperwork, clean your work area, clean and return tools to their proper places, and wash your hands.
11. Record your recommendations for needed service or additional repairs and complete the repair order.

    ____

    ____

ASE Education Foundation Worksheet #3-10

COMPLETE MAINTENANCE AND INSPECTION SERVICE

Name_______________________________ Class_______________________

Score: ☐ Excellent ☐ Good ☐ Needs Improvement **Instructor OK** ☐

Vehicle year __________ **Make** ____________________ **Model** ______________________

Objective: Upon completion of this assignment, you should be able to do a complete lubrication service and comprehensive safety inspection on a vehicle.

ASE Education Foundation Correlation

This worksheet addresses the following **Required Supplemental Task (RST):**

Preparing Vehicle for Customer, Task 1: Ensure vehicle is prepared to return to customer per school/company policy (floor mats, steering wheel cover, etc.).

We Support ASE Education Foundation

Directions: Before beginning this lab assignment, review the worksheet completely. Fill in the information in the spaces provided as you complete each task.

Tools and Equipment Required: Safety glasses, fender covers, jack and jack stands or vehicle lift, drain pan, filter wrench, shop towels

Parts and Supplies: Oil, oil filter

Related Information: Lubrication/safety service is very important, both to the customer and to the service technician. Potential component failures and safety problems can be identified during the inspection. Getting the problems repaired before there is a breakdown can save the customer from the inconvenience of being without his or her vehicle, not to mention the unexpected expense. A considerable amount of service and repair work can be identified from a properly performed inspection.

During a vehicle inspection, items in need of repairs are located and documented. A lubrication/safety service includes underhood and underbody inspections. The underhood and body inspection is done while the car is on the ground. After completing inspection and maintenance under the hood, the vehicle is raised in the air to perform an undercar inspection. The oil and filter are usually changed while the vehicle is in the air. Position the vehicle correctly on the lift using the specified lift contact points.

When performing undercar service, practice developing an efficient routine. After raising the car on the lift, start at the front on the passenger side and work around the car, finishing up at the front again. Some technicians prefer to start with undercar service, including draining the oil and changing the filter. The technician would then complete underhood services while refilling the crankcase.

Procedure:

1. Fill in the Car Care Service form at the end of this assignment as you complete the lubrication/safety service.
2. Practice developing an efficient routine.

   Where did you start the job?

   ☐ Underhood ☐ Undercar
3. Before working under the hood, place fender covers on the fenders and front body parts.
4. Before starting the undercar service, raise the vehicle on a lift or place it on jack stands.
5. Before completing your paperwork, clean your work area, clean and return tools to their proper places, and wash your hands.
6. Record your recommendations for needed service or additional repairs on the back of the Car Service Checklist that follows this worksheet and complete the repair order.
7. There are two Car Care Service checklists. Complete the second checklist as you perform a Car Care Service on a different vehicle.

Certified Car Care Service #1

Customer Name ______________________________

Address ____________________ City ____________ Zip Code ____________ Phone ____________

Date ____________________ Time ____________

Vehicle ____________ Year ________ Model ____________ License Number ____________ Odometer Reading ____________

ELECTRICAL SYSTEM CHECKS
- ☐ Wiring Visual Inspection

Battery
- ☐ Top Off Water Level

Posts and Cables
- ☐ Clean ☐ Corroded
- ☐ Damaged

Battery Condition
- ☐ Good ☐ Replace
- ☐ Recharge

LIGHTS
- ☐ Back up ☐ License
- ☐ Park ☐ Brake
- ☐ Signal ☐ Emergency
- ☐ Dash Lights ☐ Back up

Headlight Operation
- ☐ High Left
- ☐ Low Left
- ☐ High Right
- ☐ Low Right
- ☐ Horn Operation

FULL SYSTEM CHECKS
- ☐ Condition of Hoses
- ☐ Gas Cap Condition
- ☐ Air Cleaner
- ☐ Crankcase Vent Filter
- ☐ Fuel Filter

COOLING SYSTEM CHECKS
- ☐ Level
- ☐ Strength of Coolant (Protection to________°)
- ☐ No Leaks

Condition of Hoses
- ☐ Radiator
- ☐ Heater
- ☐ Thermostat By-pass Hose (if so equipped)

Pressure Test
- ☐ Radiator
- ☐ Cap
- ☐ Condition of Coolant Pump Belt

BRAKE INSPECTION
- ☐ Pedal Travel
- ☐ Emergency Brake
- ☐ Brake Hoses and Lines
- ☐ Master Brake Cylinder-Fluid Level and Condition

ON-GROUND STEERING, SUSPENSION, DRIVELINE CHECKS
- ☐ Steering Wheel Freeplay
- ☐ Power Steering Fluid Level
- ☐ Shock Absorber Bounce Test

	Good	Unsafe
Front	☐	☐
Rear	☐	☐

- ☐ No Squeaks
- ☐ Ride Height Check
- ☐ Check ATF Level
- ☐ Clutch Master Cylinder Level

VISIBILITY
- ☐ Mirrors
- ☐ Wiper Blades

Wiper Operation
- ☐ Fast ☐ Slow
- ☐ Washer Fluid and Pump
- ☐ Clean and Inspect all Glass

DIESEL
- ☐ Exhaust Fluid Level

ENGINE LEAKS
- ☐ Fuel
- ☐ Oil
- ☐ Coolant
- ☐ Other

OIL SERVICE
- ☐ Drain Crankcase (if ordered)
- ☐ Remove and Replace Oil Filter
- ☐ Replace Crankcase Oil
- ☐ Inspect Undercar for Fluid Leaks
- ☐ Check Crankcase Oil Level
- ☐ Check Oil Filter for Leaks

INFLATE AND CHECK TIRES
Inflate to_____ psi

Tire Condition:

	Good	Fair	Unsafe
RF	☐	☐	☐
LF	☐	☐	☐
RR	☐	☐	☐
LR	☐	☐	☐

Inflate and Check Spare

Good ☐ Fair ☐ Unsafe ☐

SUSPENSION AND STEERING
- ☐ Inspect Steering Linkage
- ☐ Inspect Shock Absorbers
- ☐ Inspect Suspension Bushings
- ☐ Clean Lubrication Fittings
- ☐ Lubricate Fittings
- ☐ Ball Joints
- ☐ Inspect Ball Joint Seals
- ☐ Ball Joint Wear
- ☐ Inspect Ride Height
- ☐ Inspect Suspension Bumpers
- ☐ Inspect Spring seats
- ☐ Inspect Struts

UNDERCAR FUEL SYSTEM CHECKS
- ☐ Condition of Fuel Hoses
- ☐ Condition of Fuel Tank

DRIVELINE CHECKS
- ☐ Check Universal or CV Joints
- ☐ Check Clutch Linkage
- ☐ Inspect Gear Cases
- ☐ Transmission
- ☐ Differential
- ☐ Replace Drain Plugs
- ☐ Inspect Motor Mounts
- ☐ Inspect Transmission Mounts

EXHAUST SYSTEM CHECKS
- ☐ Mufflers and Pipes
- ☐ Pipe Hangers
- ☐ Exhaust Leaks
- ☐ Heat Riser

FINAL VEHICLE PREPARATION
- ☐ Clean Windows, Vacuum Interior
- ☐ Fill Out and Affix Door Jamb Record to Door Post
- ☐ Complete a Repair Order
- ☐ Lubricate Door, Hood Hinge and Latches

NOTES (specs, procedures, additional service, or repair information):

ADDITIONAL RECOMMENDATIONS FOR SERVICE OR REPAIRS:

STOP

CERTIFIED CAR CARE SERVICE #1

Customer Name ____________________

Address ____________ City ____________ Zip Code ____________ Phone ____________

Date ____________ Time ____________

Vehicle ____________ Year ____________ Model ____________ License Number ____________ Odometer Reading ____________

ELECTRICAL SYSTEM CHECKS

- ☐ Wiring Visual Inspection

Battery

- ☐ Top Off Water Level

Posts and Cables

- ☐ Clean ☐ Corroded
- ☐ Damaged

Battery Condition

- ☐ Good ☐ Replace
- ☐ Recharge

LIGHTS

- ☐ Back up ☐ License
- ☐ Park ☐ Brake
- ☐ Signal ☐ Emergency
- ☐ Dash Lights ☐ Back up

Headlight Operation

- ☐ High Left
- ☐ Low Left
- ☐ High Right
- ☐ Low Right
- ☐ Horn Operation

FULL SYSTEM CHECKS

- ☐ Condition of Hoses
- ☐ Gas Cap Condition
- ☐ Air Cleaner
- ☐ Crankcase Vent Filter
- ☐ Fuel Filter

COOLING SYSTEM CHECKS

- ☐ Level
- ☐ Strength of Coolant (Protection to________°)
- ☐ No Leaks

Condition of Hoses

- ☐ Radiator
- ☐ Heater
- ☐ Thermostat By-pass Hose (if so equipped)

Pressure Test

- ☐ Radiator
- ☐ Cap
- ☐ Condition of Coolant Pump Belt

BRAKE INSPECTION

- ☐ Pedal Travel
- ☐ Emergency Brake
- ☐ Brake Hoses and Lines
- ☐ Master Brake Cylinder-Fluid Level and Condition

ON-GROUND STEERING, SUSPENSION, DRIVELINE CHECKS

- ☐ Steering Wheel Freeplay
- ☐ Power Steering Fluid Level
- ☐ Shock Absorber Bounce Test

	Good	Unsafe
Front	☐	☐
Rear	☐	☐

- ☐ No Squeaks
- ☐ Ride Height Check
- ☐ Check ATF Level
- ☐ Clutch Master Cylinder Level

VISIBILITY

- ☐ Mirrors
- ☐ Wiper Blades

Wiper Operation

- ☐ Fast ☐ Slow
- ☐ Washer Fluid and Pump
- ☐ Clean and Inspect all Glass

DIESEL

- ☐ Exhaust Fluid Level

ENGINE LEAKS

- ☐ Fuel
- ☐ Oil
- ☐ Coolant
- ☐ Other

OIL SERVICE

- ☐ Drain Crankcase (if ordered)
- ☐ Remove and Replace Oil Filter
- ☐ Replace Crankcase Oil
- ☐ Inspect Undercar for Fluid Leaks
- ☐ Check Crankcase Oil Level
- ☐ Check Oil Filter for Leaks

INFLATE AND CHECK TIRES

Inflate to______ psi

Tire Condition:

	Good	Fair	Unsafe
RF	☐	☐	☐
LF	☐	☐	☐
RR	☐	☐	☐
LR	☐	☐	☐

Inflate and Check Spare

Good ☐ Fair ☐ Unsafe ☐

SUSPENSION AND STEERING

- ☐ Inspect Steering Linkage
- ☐ Inspect Shock Absorbers
- ☐ Inspect Suspension Bushings
- ☐ Clean Lubrication Fittings
- ☐ Lubricate Fittings
- ☐ Ball Joints
- ☐ Inspect Ball Joint Seals
- ☐ Ball Joint Wear
- ☐ Inspect Ride Height
- ☐ Inspect Suspension Bumpers
- ☐ Inspect Spring seats
- ☐ Inspect Struts

UNDERCAR FUEL SYSTEM CHECKS

- ☐ Condition of Fuel Hoses
- ☐ Condition of Fuel Tank

DRIVELINE CHECKS

- ☐ Check Universal or CV Joints
- ☐ Check Clutch Linkage
- ☐ Inspect Gear Cases
- ☐ Transmission
- ☐ Differential
- ☐ Replace Drain Plugs
- ☐ Inspect Motor Mounts
- ☐ Inspect Transmission Mounts

EXHAUST SYSTEM CHECKS

- ☐ Mufflers and Pipes
- ☐ Pipe Hangers
- ☐ Exhaust Leaks
- ☐ Heat Riser

FINAL VEHICLE PREPARATION

- ☐ Clean Windows, Vacuum Interior
- ☐ Fill Out and Affix Door Jamb Record to Door Post
- ☐ Complete a Repair Order
- ☐ Lubricate Door, Hood Hinge and Latches

NOTES (specs, procedures, additional service, or repair information):

ADDITIONAL RECOMMENDATIONS FOR SERVICE OR REPAIRS:

STOP

Part II

Lab Worksheets for ASE Maintenance and Light Repair

Service Area 4

Tire and Wheel Service

ASE Education Foundation Worksheet #4-1

TIRE IDENTIFICATION

Name______________________ Class______________

Score: ☐ Excellent ☐ Good ☐ Needs Improvement **Instructor OK** ☐

Vehicle year ________ **Make** ______________ **Model** ______________

Objective: Upon completion of this assignment, you will be able to read and understand the tire side-wall markings.

ASE Education Foundation Correlation

This worksheet addresses the following **MLR** task:

IV.D.1 Inspect tire condition; identify tire wear patterns; check for correct size and application (load and speed ratings) and adjust air pressure as listed on the tire information placard/label. **(P-1)**

This worksheet addresses the following **AST/MAST** task:

IV.F.1 Inspect tire condition; identify tire wear patterns; check for correct size and application (load and speed ratings) and adjust air pressure as listed on the tire information placard/label. **(P-1)**

We Support ASE Education Foundation

Directions: Before beginning this lab assignment, review the worksheet completely. Fill in the information in the spaces provided as you complete each task.

Tools and Equipment Required: Safety glasses

Note: Select a passenger car for this assignment. Truck tires do not have UTQG or speed ratings.

Procedure: Inspect the right front tire for the assigned vehicle and record the following information:

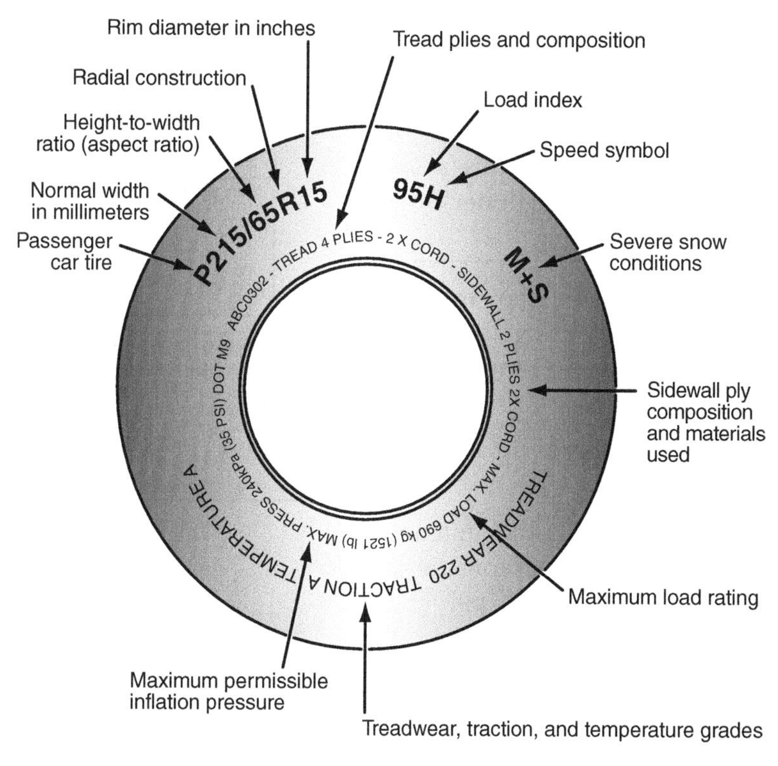

1. Identify the type of tire construction.

   ☐ Radial ☐ Belted Bias

   ☐ Blackwall ☐ Whitewall

2. Number of sidewall plies:

   ☐ 1 ☐ 2 ☐ 3 ☐ Other

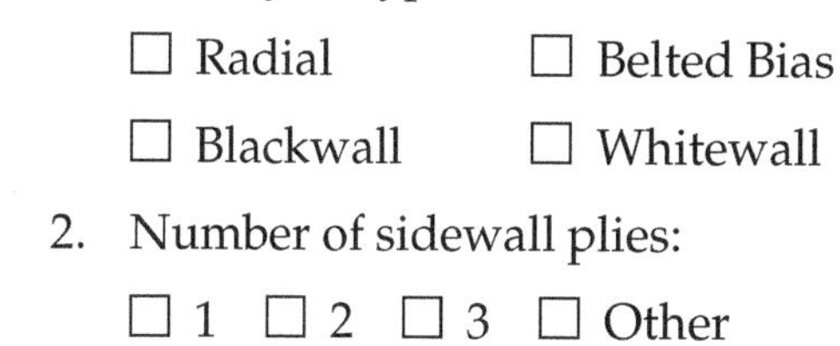

TURN

3. Number of tread plies:

☐ 2 ☐ 3 ☐ 4 ☐ Other

4. If the tire has belts, how many are there? _______
5. What material(s) are they made of (e.g., steel, rayon, nylon)?

____

6. What is the DOT Manufacturer's Code Number?

____

7. What brand name is on the tire?

____

8. What size is the tire?

____

9. What is the wheel rim diameter?

☐ 13" ☐ 14" ☐ 15" ☐ 16" ☐ 17" ☐ 18" ☐ Other

10. What is the aspect ratio of the tire? ___________________________
11. What is the maximum air pressure for the tire?___ psi
12. List the UTQG rating from the tire sidewall.

Treadwear _______ Traction _______ Temperature ______

13. Is this an appropriate tire for the vehicle being inspected?

☐ Yes ☐ No

14. Is the tire listed as M&S?

☐ Yes ☐ No

15. What does M&S mean?

____

16. What is the tire's speed rating code (letter)?

____

17. What is the maximum speed rating for the tire? _________ mph
18. What is the load code (number) for the tire?

____

19. What is the maximum load rating for the tire? ____ lb @ ____ psi

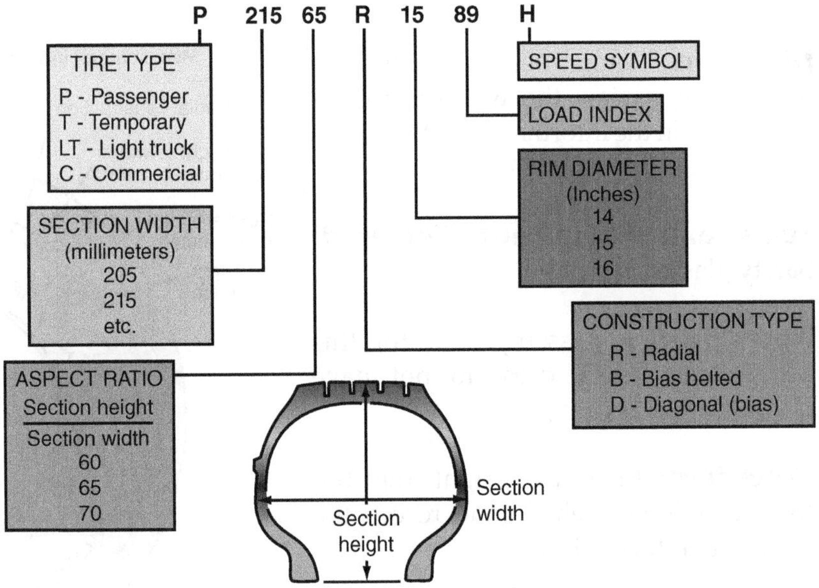

P 215 65 R 15 89 H

TIRE TYPE
P - Passenger
T - Temporary
LT - Light truck
C - Commercial

SECTION WIDTH
(millimeters)
205
215
etc.

ASPECT RATIO
Section height / Section width
60
65
70

Section height

Section width

CONSTRUCTION TYPE
R - Radial
B - Bias belted
D - Diagonal (bias)

RIM DIAMETER
(Inches)
14
15
16

LOAD INDEX

SPEED SYMBOL

20. Before completing your paperwork, clean your work area, clean and return tools to their proper places, and wash your hands.

STOP

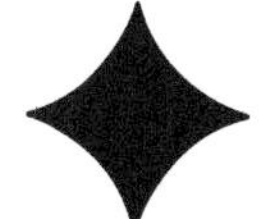

ASE Education Foundation Worksheet #4-2

TIRE MAINTENANCE/WHEEL TORQUE

Name________________________ Class________________

Score: ☐ Excellent ☐ Good ☐ Needs Improvement **Instructor OK** ☐

Vehicle year ________ **Make** ______________ **Model** ______________

Objective: Upon completion of this assignment, you will be able to rotate a vehicle's tires.

ASE Education Foundation Correlation

This worksheet addresses the following **MLR** tasks:

IV.D.2 Rotate tires according to manufacturer's recommendations including vehicles equipped with tire pressure monitoring systems (TPMS). **(P-1)**

V.A.3 Install wheel and torque lug nuts. **(P-1)**

This worksheet addresses the following **AST/MAST** tasks:

IV.F.3 Rotate tires according to manufacturer's recommendations including vehicles equipped with tire pressure monitoring systems (TPMS). **(P-1)**

V.A.4 Install wheel and torque lug nuts. **(P-1)**

We Support ASE Education Foundation

Directions: Before beginning this lab assignment, review the worksheet completely. Fill in the information in the spaces provided as you complete each task.

Tools and Equipment Required: Safety glasses, jack stands or vehicle lift, ratchet and sockets, torque wrench, shop towel, air impact wrench, impact sockets, tire gauge, air chuck

Procedure:

Tire size ____________________ Tire manufacturer ____________________

1. Check and adjust tire pressures. (Review Worksheet #3-2, Adjust Tire Air Pressures.)

   When checking the air pressure, tires should be: ☐ Hot ☐ Cold

Record Pressures:	Before	After	
Left front	_____ psi	_____ psi	
Right front	_____ psi	_____ psi	
Left rear	_____ psi	_____ psi	
Right rear	_____ psi	_____ psi	
Spare tire	_____ psi	_____ psi	_____ N/A

2. Determine proper rotation pattern.
   a. In the box labeled "Prior Positions," draw arrows to show the positions where the tires will be rotated.
   b. In the box labeled "Rotated Positions," use letters and numbers to indicate the rotated positions of the tires.

Prior Positions

RF=1 RR=3

Spare=5

LF=2 LR=4

Rotated Positions
(enter the new positions in the boxes below)

<--Front

3. How many tires will be rotated: ☐ 2 ☐ 4 ☐ 5 (spare)
4. What is the torque specification for the lug nuts? ________ ft.-lb
5. Where did you locate the lug nut torque specification? ____________________

Rotate the Tires

6. Remove the wheels.
   a. Loosen each lug nut about ¼ turn before raising the wheels off the ground.

   Note: With the air impact wrench, you will probably *not* need to loosen the lug nuts prior to lifting the vehicle off the ground.

   b. Which direction did you turn the lug nuts to loosen them?

   ☐ Clockwise ☐ Counterclockwise

   c. When did you loosen the lugs?

   ☐ Before raising the vehicle ☐ After raising the vehicle

7. Inspect the threads on the lug studs and lug nuts. ☐ Good ☐ Damaged

Reinstalling Wheels

8. Check the condition of all tubeless valve stems.

   Do any need to be replaced? ☐ Yes ☐ No

   If yes, which one(s)? ☐ LF ☐ RF ☐ RR ☐ LR ☐ Spare

9. Rotate wheels to desired positions and lift the wheel assembly onto the lug bolts.

CAUTION When lifting heavy objects, remember to lift with your legs, not with your back.

10. Start the lug nuts onto the threads and turn them by hand for at least three turns.

    Note: Several manufactures require that lug nuts be installed on clean, dry threads.

11. Which way does the tapered side of the lug nut face?

    ☐ Toward the rim ☐ Away from the rim

12. Tighten the lug nuts only until they are "snug," using the correct pattern.
13. Sketch the tightening pattern in the box to the right.
14. Lower the vehicle to the ground and use a torque wrench to tighten each lug nut to specifications. ________ ft.-lb
15. Reinstall the wheel covers.
16. Before completing your paperwork, clean your work area, clean and return tools to their proper places, and wash your hands.
17. Record your recommendations for needed service or additional repairs and complete the repair order.

    ____

    ____

STOP

ASE Education Foundation Worksheet #4-3

REPLACE A RUBBER VALVE STEM (PRE-TPMS)

Name______________________________ Class______________________

Score: ☐ Excellent ☐ Good ☐ Needs Improvement **Instructor OK** ☐

Vehicle year __________ **Make** ____________________ **Model** ______________________

Objective: Upon completion of this assignment, you will be able to replace a rubber valve stem on a pre-TPMS wheel.

ASE Education Foundation Correlation

This worksheet addresses the following **MLR** tasks:

IV.D.3 Dismount, inspect, and remount tire on wheel; balance wheel-and-tire assembly. **(P-1)**

IV.D.6 Repair tire following vehicle manufacturer approved procedure. **(P-1)**

This worksheet addresses the following **AST/MAST** tasks:

IV.F.6 Dismount, inspect, and remount tire on wheel; balance wheel-and-tire assembly. **(P-1)**

IV.F.9 Repair tire following vehicle manufacturer approved procedure. **(P-1)**

We Support
ASE | Education Foundation

Directions: Before beginning this lab assignment, review the worksheet completely. Fill in the information in the spaces provided as you complete each task.

Tools and Equipment Required: Safety glasses, tire changer, valve stem installation tool, air chuck, tire gauge, shop towel

Procedure:

Tire size _______________ Tire manufacturer _______________

1. Locate a replacement valve stem of the proper length and diameter.
2. The wheel is used with a:

   ☐ Wheel cover (long stem)

   ☐ Hub cap (short stem)

Short Long Large diameter

3. Install the wheel on the tire machine with the valve stem facing the bead breaker.

   Note: Before using the tire changer, check with your instructor.
4. Break down the outer bead.
5. Put rubber lube on the part of the valve stem that is inside the rim. Use the installation tool to remove the stem.

6. If the valve stem will not come out, use diagonal cutters or another suitable tool to cut the valve stem from outside the wheel. Hold onto the bottom end of the valve stem while cutting so it does not drop into the tire.
7. Lube the replacement valve stem with rubber lube.
8. Use the installation tool to pull the new stem into place.
9. Reinflate the tire to the proper pressure.
10. What pressure did you inflate the tire to? _____ psi
11. Before completing your paperwork, clean your work area, clean and return tools to their proper places, and wash your hands.
12. Record your recommendations for needed service or additional repairs and complete the repair order.

____

____

STOP

ASE Education Foundation Worksheet #4-4

DISMOUNT AND MOUNT TIRES WITH A TIRE CHANGER

Name______________________________ Class______________________

Score: ☐ Excellent ☐ Good ☐ Needs Improvement **Instructor OK** ☐

Vehicle year __________ **Make** ____________________ **Model** ______________________

Objective: Upon completion of this assignment, you will be able to dismount and mount tires.

ASE Education Foundation Correlation

This worksheet addresses the following **MLR** tasks:

IV.D.3 Dismount, inspect, and remount tire on wheel; balance wheel-and-tire assembly. **(P-1)**

IV.D.4 Dismount, inspect, and remount tire on wheel equipped with tire pressure monitoring system sensor. **(P-1)**

This worksheet addresses the following **AST/MAST** tasks:

IV.F.6 Dismount, inspect, and remount tire on wheel; balance wheel-and-tire assembly. **(P-1)**

IV.F.7 Dismount, inspect, and remount tire on wheel equipped with tire pressure monitoring system sensor. **(P-1)**

We Support ASE | Education Foundation

Directions: Before beginning this lab assignment, review the worksheet completely. Fill in the information in the spaces provided as you complete each task.

Tools and Equipment Required: Safety glasses, shop towel, European-style tire changer, valve core tool

Procedure:

Tire size ______________ Tire manufacturer ______________

Note: If this is the first time you are doing this job, you must have supervision.

1. **Dismounting the Tire**

   a. Unscrew the tire's valve core to allow the air to escape.

   Note: If the tire has a valve core type tire pressure sensor, loosen and remove its retaining nut at the base of the valve stem and allow the sensor to be loose in the tire while the bead is broken. Then reach in and retrieve the sensor.

   b. Place the rim under the bead breaker attachment with the "drop center" offset toward the bead breaker. Is this the valve stem side of the rim?

   ☐ Yes ☐ No

   c. Force the tire bead away from the safety ledges on both sides of the wheel.

   d. Install the wheel-and-tire assembly on the top of the tire changer. Push on the air control to clamp the wheel to the machine.

   e. Apply rubber lube to the bead surfaces.

Courtesy of Hennessy Industries

TURN

f. Adjust the top arm of the tire changer so that it almost contacts the edge of the rim.

g. Use the tire iron to pry the bead over the edge of the rim. Hold it down as you step on the foot pedal. This turns the wheel against the tool.

h. Remove the top bead. On the opposite side of the tire iron, push the bead down into the rim's drop center for easier removal.

Do not allow your fingers to become trapped in between the tire bead and the rim. Serious injury could result!

i. Remove the lower bead in the same manner and inspect the bead seat area of the rim for rust and dirt.

2. **Remounting the Tire on the Wheel**

a. Apply rubber lube to tire beads and position the tire on the wheel. Slide the lower bead over the rim by rotating it clockwise until it cannot be installed further.

Photo by Tim Gilles

b. Step on the air control to rotate the tire clockwise, forcing the remaining part of the bead over the top edge of the rim.

Photo by Tim Gilles

c. *Keep hands out of the way!* Step on the foot pedal and follow the tool with your *right* hand, keeping the bead pushed into the drop center. The lower bead is now installed.

d. Mount the upper bead in the same manner.

e. Reinstall the tire pressure sensor.

Note: If you are installing a new tire pressure sensor, record the identification number to help you reset the sensors to the vehicle.

3. **Inflating the Tire**

a. Be sure beads are coated with rubber lube.

b. The tire changer has an "air ring" that is used to fill the lower bead air gap while inflating the tire. Apply the correct foot pedal to inflate the tire. The pedal has two positions. Pushing on it all of the way forces air into the air ring during initial tire inflation.

c. *Stand to the side when inflating.* There is a connector that attaches the inflation hose to the tire valve so you do not have to hold it while inflating the tire.

d. Was there a loud "pop" when inflating the tire? ☐ Yes ☐ No

e. If more than 30 psi is required to seat the beads completely, call your instructor. More than 30 psi required? ☐ Yes ☐ No

f. When the beads are seated, install the valve core and inflate to the manufacturer's specifications.

4. Record your recommendations for needed service or additional repairs and complete the repair order.

____

____

ASE Education Foundation Worksheet #4-5

REPAIR A TIRE PUNCTURE

Name ______________________ Class ________________

Score: ☐ Excellent ☐ Good ☐ Needs Improvement **Instructor OK** ☐

Vehicle year ________ **Make** ______________ **Model** ________________

Objective: Upon completion of this assignment, you will be able to repair a leaking tire.

ASE Education Foundation Correlation

This worksheet addresses the following **MLR** task:

IV.D.6 Repair tire following vehicle manufacturer approved procedure. **(P-1)**

This worksheet addresses the following **AST/MAST** task:

IV.F.9 Repair tire following vehicle manufacturer approved procedure. **(P-1)**

We Support ASE | Education Foundation

Directions: Before beginning this lab assignment, review the worksheet completely. Fill in the information in the spaces provided as you complete each task.

Tools and Equipment Required: Safety glasses, tire soak tank, valve stem installation tool, air chuck, tire gauge, pliers, burr tool, tire probe, vulcanizing cement, shop towel, tire spreading fixture, vacuum cleaner, patch stitcher, tire patches

Procedure:

Tire size ________________ Tire manufacturer ________________

1. Locate the leak in the tire:

   a. Inflate the tire to the maximum pressure listed on the tire's sidewall.

      What is the maximum pressure? _____ psi

   b. Submerge the tire in the soak tank.

   c. Rotate the tire, watching for bubbles as the tread area leaves the water.

      Any leaks in the tread area? ☐ Yes ☐ No

   d. Inspect the bead area for bubbles in the same manner.

      Any leaks from the bead area? ☐ Yes ☐ No

Water

Photo by Tim Gilles

Note: If the leak is at the bead area, the tire will need to be dismounted and the bead area of the tire and rim cleaned and inspected.

Photo by Tim Gilles

e. While the valve stem is under water, push on it while looking for bubbles.

   Any leaks from the valve stem? ☐ Yes ☐ No

   If the valve stem is leaking, replace the valve stem. Refer to Worksheet #4-3, Replace a Rubber Valve Stem (Pre-TPMS).

   If the valve core is leaking, tighten or replace the valve core.

f. Use a marking crayon to mark the location of any leaks.

2. Repair a punctured tire:

   a. Remove the tire from the rim. Refer to Worksheet #4-4, Dismount and Mount Tires with a Tire Changer.

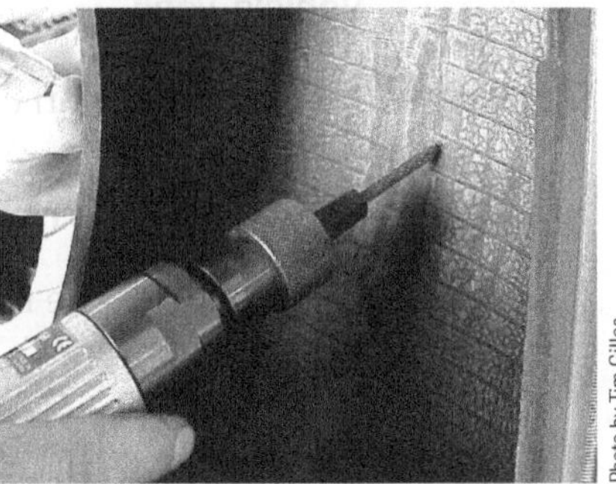
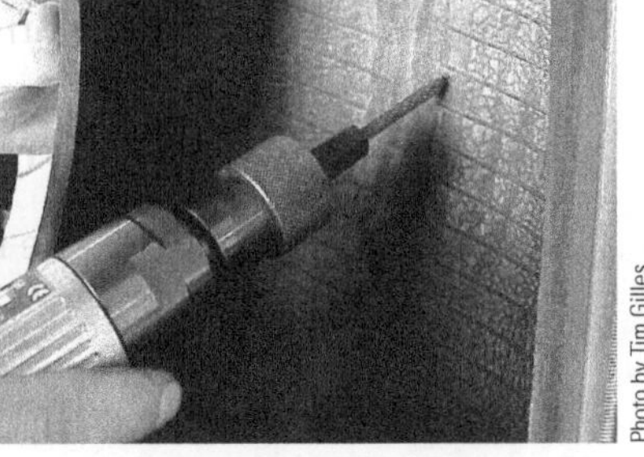
Photo by Tim Gilles

   b. Mount the tire on a tire spreading fixture.

   c. Is the item that punctured the tire still present?

      ☐ Yes ☐ No If it is, remove it.

   d. Probe the hole gently in the same direction as the nail or screw entered the tire.

   e. Clean the area to be repaired.

   f. Ream the hole with a burr tool.

   g. Install vulcanizing cement on the probe and probe the hole.

Photo by Tim Gilles

   h. Install a tire plug into the hole until it extends slightly from both the inner and the outer surfaces of the hole.

   i. Cut off the tire plug and grind it down until it is *almost* flush. Be careful not to grind into the inner surface of the tire.

   j. Select the correct type of patch.

      ☐ Radial ☐ Bias ☐ Universal

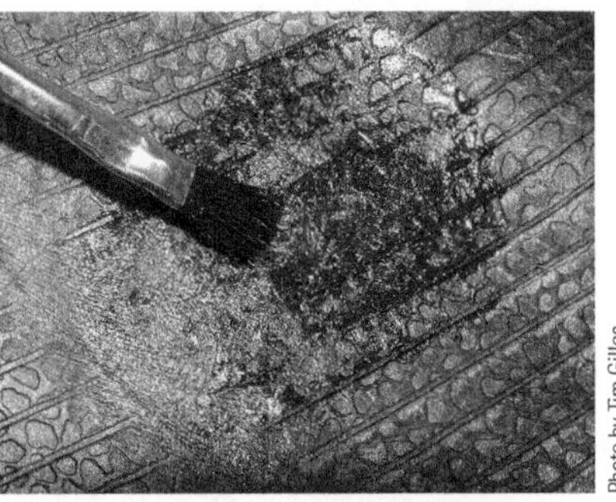
Photo by Tim Gilles

   k. Use the patch and a marking crayon to outline the repair area (slightly larger than the patch).

   l. Use liquid buffer and a scraper to clean the area to be patched.

   m. Vacuum any debris from inside the tire.

   n. Coat the repair area with vulcanizing cement.

   o. Allow the cement to dry **completely**.

Photo by Tim Gilles

   p. Apply the patch to the cemented area. Use a stitcher to seat it. Remove the plastic from the back of the patch.

   q. Remount the tire, inflate, and test for leaks.

3. Before completing your paperwork, clean your work area, clean and return tools to their proper places, and wash your hands.

4. Record your recommendations for needed service or additional repairs and complete the repair order.

____

____

STOP

ASE Education Foundation Worksheet #4-6

TIRE BALANCE

Name ______________________ Class ______________

Score: ☐ Excellent ☐ Good ☐ Needs Improvement **Instructor OK** ☐

Vehicle year ________ **Make** ______________ **Model** ______________

Objective: Upon completion of this assignment, you will be able to balance a tire.

ASE Education Foundation Correlation

This worksheet addresses the following **MLR** task:

IV.D.3 Dismount, inspect, and remount tire on wheel; balance wheel-and-tire assembly. **(P-1)**

This worksheet addresses the following **AST/MAST** task:

IV.F.6 Dismount, inspect, and remount tire on wheel; balance wheel-and-tire assembly. **(P-1)**

We Support
ASE | Education Foundation

Directions: Before beginning this lab assignment, review the worksheet completely. Fill in the information in the spaces provided as you complete each task.

Tools and Equipment Required: Safety glasses, computer wheel balancer, wheel weights, weight hammer

Hennessy Industries, Inc.

Procedure:

Tire size ______________

Tire manufacturer ______________

1. Remove all rocks and mud from the tire and rim.
2. Mount the wheel on the balancer.
3. Enter the required information into the computer as necessary.
   a. Enter wheel rim size.

      ☐ 13" ☐ 14" ☐ 15"

      ☐ 16" ☐ 17" ☐ 18"

   b. Enter the distance from the balancer to the edge of the wheel rim.

      Distance __________ " __________ mm

   c. Measure the width of wheel rim from bead to bead with the special caliper.

      Width __________ " __________ mm

Photo by Tim Gilles

4. Drop the protective cover over the wheel.
5. Depress the switch to spin the wheel-and-tire assembly.
6. When the wheel stops spinning, raise the cover and rotate the wheel until it is at the indicated position for the left side of the tire.
7. Install the specified weight onto the wheel rim in line with the weight line on the wheel balancer.

   How much weight was added to the inside of the wheel rim at this point? ______ oz ______ grams

Photo by Tim Gilles

8. Position the wheel at the indicated position for the right side and install proper weight.

   How much weight was added to the outside of the wheel rim at this point? ______ oz ______ grams
9. Spin the wheel once again to check the accuracy of the weight installation.

   ☐ OK ☐ Rebalance

   How much total weight was added to the wheel rim? ______ oz

   Note: If the tire is still not correctly balanced following retesting, do not add another weight to correct the imbalance. Move the weight halfway to the new indicated position or start the balance procedure over.
10. Before completing your paperwork, clean your work area, clean and return tools to their proper places, and wash your hands.
11. Record your recommendations for needed service or additional repairs and complete the repair order.

    ____

    ____

STOP

ASE Education Foundation Worksheet #4-7

INSPECT THE TIRE PRESSURE MONITORING SYSTEM (TPMS)

Name________________________ Class________________

Score: ☐ Excellent ☐ Good ☐ Needs Improvement **Instructor OK** ☐

Vehicle year ________ **Make** ________________ **Model** ________________

Objective: Upon completion of this assignment, you will be able to inspect the operation of the tire pressure monitoring system.

ASE Education Foundation Correlation

This worksheet addresses the following **MLR** task:

IV.D.7 Identify indirect and direct tire pressure monitoring systems (TPMS); calibrate system; verify operation of instrument panel lamps. **(P-2)**

This worksheet addresses the following **AST/MAST** tasks:

IV.F.10 Identify and test tire pressure monitoring system (indirect and direct) for operation; calibrate system; verify operation of instrument panel lamps. **(P-1)**

IV.F.11 Demonstrate knowledge of steps required to remove and replace sensors in a tire pressure monitoring system (TPMS) including relearn procedure. **(P-1)**

We Support ASE | Education Foundation

Directions: Before beginning this lab assignment, review the worksheet completely. Fill in the information in the spaces provided as you complete each task.

Tools and Equipment Required: Safety glasses, fender covers, TPMS monitor, shop rags

Procedure:

1. Use the service information to identify the type of TPMS. Then locate the testing and service procedures.

   What service information did you use?

   ____

2. What type of TPMS does this vehicle have?

   ☐ Indirect ☐ Direct ☐ N/A

3. Turn the ignition switch to the on position, but do not start the engine.

   Did the TPMS light on the instrument panel illuminate?

   ☐ Yes ☐ No

4. Start the engine. Did the TPMS light go out?

   ☐ Yes ☐ No

5. If you have a direct TPMS, use a TPMS scanner to check the system.

   Courtesy of Bartec USA, LLC

   Note: Read and follow the instructions for using the TPMS scan tool.

   Record any displayed data from the scan tool below.

Wheel	ID Hex	ID Dec	Battery State	Pressure	Temperature
Right Front					
Left Front					
Right Rear					
Left Rear					

6. Remove the TPMS scan tool from the vehicle.

7. Before completing the paperwork, clean your work area, clean and return tools to their proper places, and wash your hands.

8. Were there any recommendations for needed service or unusual conditions that you noticed while you were inspecting the TPMS?

   ☐ Yes ☐ No

9. Record your recommendations for needed service or additional repairs below and complete the repair order.

   ____

   ____

STOP

ASE Education Foundation Worksheet #4-8

REPLACE DRIVE AXLE STUDS

Name________________________ Class________________

Score: ☐ Excellent ☐ Good ☐ Needs Improvement **Instructor OK** ☐

Vehicle year _________ **Make** ________________ **Model** __________________

Objective: Upon completion of this assignment, you will be able to replace the wheel studs on a drive axle.

ASE Education Foundation Correlation

This worksheet addresses the following **MLR** tasks:

III.E.4 Inspect and replace drive axle wheel studs. **(P-1)**

V.F.6 Inspect and replace wheel studs. **(P-1)**

This worksheet addresses the following **AST** tasks:

III.E.2.1 Inspect and replace drive axle wheel studs. **(P-1)**

V.F.7 Inspect and replace wheel studs. **(P-1)**

This worksheet addresses the following **MAST** tasks:

III.E.3.1 Inspect and replace drive axle wheel studs. **(P-1)**

V.F.8 Inspect and replace wheel studs. **(P-1)**

We Support ASE | Education Foundation

Directions: Before beginning this lab assignment, review the worksheet completely. Fill in the information in the spaces provided as you complete each task.

Tools and equipment: Safety glasses, fender covers, hand tools, shop towels

Procedure:

1. Raise the vehicle on a lift or place it on jack stands.
2. Remove the wheel-and-tire assemblies.
3. Inspect the lug nuts.
   - ☐ All are OK.
   - ☐ Some need replacement. How many? ______________
4. Inspect the wheel studs.
   - ☐ All are OK.
   - ☐ Some need replacement. How many? ______________

TURN

Replace damaged wheel studs:

5. Remove the brake drum or rotor.
6. Check to see if there is clearance behind the hub for the studs to be removed from the hub.

   Is there adequate clearance for the studs? ☐ Yes ☐ No

   Note: If there is not adequate clearance, refer to the service information for stud removal procedure.
7. Use a hammer or a punch to knock the studs from the hub.

SHOP TIP If the studs do not remove easily from the hub with a hammer or a punch, a tie-rod press can sometimes be used.

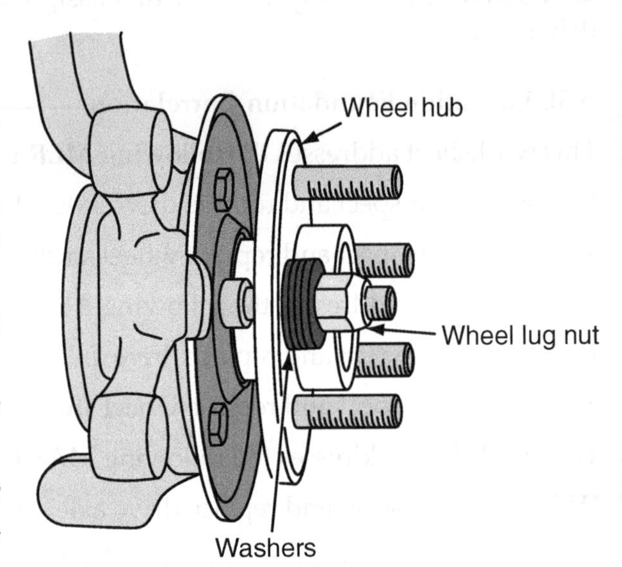

8. Insert the new stud into the hub.
9. Place several washers on the stud and install the lug nut backward on the stud.
10. Use a ratchet and socket to tighten the wheel nut onto the stud. Continue tightening the lug nut until the stud is completely installed.
11. Repeat this procedure until all the damaged lug studs have been replaced.
12. Reinstall the tire-and-wheel assembly and lower the vehicle.
13. Before completing the paperwork, clean your work area, put the tools in their proper places, and wash your hands.
14. List your recommendations for future service and/or additional repairs and complete the order.

   ____

   ____

STOP

Part II

Lab Worksheets for ASE Maintenance and Light Repair

Service Area 5

Service Information

ASE Education Foundation Worksheet #5-1

UNDERHOOD LABEL WORKSHEET

Name_______________________ Class_________________

Score: ☐ Excellent ☐ Good ☐ Needs Improvement **Instructor OK** ☐

Vehicle year _______ **Make** _______________ **Model** _________________

Objective: Completion of this assignment should prepare you to obtain important vehicle information from the underhood emission label.

ASE Education Foundation Correlation

This worksheet addresses the following **Required Supplemental Tasks (RST):**

Preparing Vehicle for Service, Task 1:	Identify information needed and the service requested on a repair order.

This worksheet addresses the following **MLR** tasks:

I.A.1; II.A.1; III.A.1; IV.A.1; V.A.1; VI.A.1; VIII.A.1	Research vehicle service information, including fluid type, vehicle service history, service precautions, and technical service bulletins. **(P-1); (P-1); (P-1); (P-1); (P-1); (P-1); (P-1)**
VII.A.1	Research vehicle service information, including refrigerant/oil type, vehicle service history, service precautions, and technical service bulletins. **(P-1)**

This worksheet addresses the following **AST/MAST** tasks:

I.A.2	Research vehicle service information including fluid type, internal engine operation, vehicle service history, service precautions, and technical service bulletins. **(P-1)**
II.A.2; III.A.2; IV.A.1; V.A.2	Research vehicle service information including fluid type, vehicle service history, service precautions, and technical service bulletins. **(P-1); (P-1); (P-1); (P-1)**
VI.A.1; VIII.A.2	Research applicable vehicle and service information, vehicle service history, service precautions, and technical service bulletins. **(P-1); (P-1)**

We Support ASE Education Foundation

Directions: Complete this worksheet on a vehicle assigned by your instructor. All of the requested information may not be found on the underhood label of the vehicle that you are inspecting. If the information is not on the label, write N/A (not available) in the answer space. Before beginning this lab task, review the worksheet completely. Fill in the information in the spaces provided as you complete this assignment.

Tools and Equipment Required: Safety glasses, fender covers

Procedure:

1. Open the vehicle's hood and put fender covers over the fenders.
2. Identify underhood labels: check below which labels are on the vehicle being inspected:

   ☐ Air conditioning ☐ Belt routing

   ☐ Vacuum hose routing ☐ Emission ☐ Other: ____________________

3. Locate the underhood **emission** label. Where is it located?

   ☐ Underside of the hood ☐ Radiator support

   ☐ Inner fender ☐ Engine ☐ Other ________________

CATALYST
DaimlerChrysler Corporation
04726351AA
VEHICLE EMISSION CONTROL INFORMATION
THIS VEHICLE CONFORMS TO U.S. EPA AND CALIFORNIA REGULATIONS APPLICABLE TO GASOLINE FUELED 2007 MODEL YEAR NEW ULEV II PASSENGER CARS PROVIDED THAT THIS VEHICLE IS ONLY INTRODUCED IN THE STATE OF CALIFORNIA OR IN STATES WHICH HAVE ADOPTED CALIFORNIA EMISSIONS VIA SECTION 177 OF THE CLEAN AIR ACT.
EGR, 2OC, 2TWC, 2HO2S(2), SFI, OBDII CERTIFIED
IF EQUIPPED WITH ALL WHEEL DRIVE, THIS VEHICLE IS EXEMPT FROM THE FOLLOWING I/M (PERFORMANCE WARRANTY SHORT TEST):
85.-2217 LOADED TEST -EPA91
85.-2219 IDLE TEST WITH LOADED PRECONDITION -EPA91
* BASIC IGNITION TIMING AND IDLE FUEL / AIR MIXTURE HAVE BEEN PRESET AT THE FACTORY. SEE THE SERVICE MANUAL FOR PROPER PROCEDURES AND OTHER ADDITIONAL INFORMATION.
* ADJUSTMENTS MADE BY OTHER THAN APPROVED SERVICE MANUAL PROCEDURES MAY VIOLATE LOCAL LAWS.
CAUTION: APPLY PARKING BRAKE WHEN SERVICING VEHICLE
3.5 LITER
7CRXV0215M80
7CRXR0150GHA
SPARK PLUGS
0.050 IN (1.27mm) GAP
ZFR5LP
NO ADJUSTMENTS NEEDED

Photo by Tim Gilles

4. Record the following information from the vehicle's underhood label.

   Note: The engine idle speed and/or the ignition timing are not adjustable on all vehicles. Also, not all underhood labels have all of the information that is requested here. If the information is not given on the underhood label or if adjustments are not necessary, indicate that by entering "N/A" for that item.

   ☐ OBD II (1996 and newer) ☐ Pre-OBD II

   Engine size ________________

   Transmission type: Automatic ________________ Manual ________________

   Engine idle speed ________________ rpm

   Emission certification year ______

   The vehicle is certified for sale by: EPA ______ California ______ Other (list) ______

   Recommended spark plug ________________

   Spark plug gap 0. ______"

   Does the engine have adjustable valve clearance? ☐ Yes ☐ No

   If the valves are adjustable, what is the lash (clearance) specification?

   Intake 0. ______" Exhaust 0. ______"

5. List the emission control devices on the underhood label.

____

____

6. List any other important information found on the underhood emission label.

____

____

7. When finished, remove the fender covers and close the hood.

ASE Education Foundation Worksheet #5-2

MITCHELL SERVICE MANUAL WORKSHEET—MAINTENANCE SPECIFICATIONS

Name_____________________________ Class______________________

Score: ☐ Excellent ☐ Good ☐ Needs Improvement **Instructor OK** ☐

Vehicle year _________ **Make** __________________ **Model** ____________________

Vintage Note: Although today's service information is typically found in electronic form, some of the service information for vehicles prior to 1990 is only available in printed service manuals. Some older vehicles are still on the road. Smog technicians in many states still need to service and repair these vehicles.

Objective: Upon completion of this assignment, you should be able to use a Mitchell's Service and Repair Manual to locate specifications required to service and repair a vehicle.

ASE Education Foundation Correlation

This worksheet addresses the following **MLR** tasks:

I.A.1; II.A.1; III.A.1; IV.A.1; V.A.1; VI.A.1; VIII.A.1	Research vehicle service information, including fluid type, vehicle service history, service precautions, and technical service bulletins. **(P-1); (P-1); (P-1); (P-1); (P-1); (P-1); (P-1)**
VII.A.1	Research vehicle service information, including refrigerant/oil type, vehicle service history, service precautions, and technical service bulletins. **(P-1)**

This worksheet addresses the following **AST/MAST** tasks:

I.A.2	Research vehicle service information including fluid type, internal engine operation, vehicle service history, service precautions, and technical service bulletins. **(P-1)**
II.A.2; III.A.2; IV.A.1; V.A.2	Research vehicle service information including fluid type, vehicle service history, service precautions, and technical service bulletins. **(P-1); (P-1); (P-1); (P-1)**
VI.A.1; VIII.A.2	Research applicable vehicle and service information, vehicle service history, service precautions, and technical service bulletins. **(P-1); (P-1)**

We Support ASE Education Foundation

Directions: Use a Mitchell's Service and Repair Manual and your own vehicle or one assigned by your instructor to locate the information requested. Most vehicles are equipped differently. If an item does not apply, write or choose N/A.

Tools and Equipment Required: Mitchell's Service and Repair Manual

Procedure:

Select a pre-1990 vehicle to complete this worksheet. Make, model, and year of vehicle selected ______________________________

Vehicle Identification Number (VIN) ______________________

Engine type: ☐ V8 ☐ V6 ☐ In-line 6 ☐ In-line 4 ☐ Other

Engine size (C.I. or liters) __________

Fuel system type: ☐ Carburetor ☐ Fuel injection

Transmission: ☐ Automatic ☐ Manual

Drive wheels:

☐ Front-wheel drive ☐ Rear-wheel drive

☐ Four-wheel drive ☐ All-wheel drive

VIN plate location

Instrument panel

Use the service manual that you have selected and answer the following questions:

1. What manual are you using? ______________________________
2. What model years does it cover? ______________
3. Are imported vehicles included in this service manual? ☐ Yes ☐ No
4. Are light-duty trucks included? ☐ Yes ☐ No

Tune-Up Specifications

5. Spark plug gap 0.____"
6. Firing order ____________________
7. Which direction does the distributor rotate? ☐ Clockwise ☐ Counterclockwise ☐ N/A
8. In the spaces below, draw a sketch of the distributor and engine that shows the firing order, cylinder numbering sequence, and the direction of distributor rotation.

9. Idle speed: Curb idle ____ Fast idle ____
10. Ignition timing: ____ degrees ____ TDC at ____ rpm ☐ N/A

11. Draw a sketch below that represents the ignition timing marks. If timing is not adjustable, mark "N/A" in the box.

Capacities

12. Cooling system capacity _____ quarts with A/C _____ quarts without A/C
13. Radiator cap pressure _____ psi _____ bar _____ Kpa
14. Fuel tank _____ gallons to fill
15. Engine oil refill _____ quarts without filter _____ quarts with filter

STOP

ASE Education Foundation Worksheet #5-3

DIGITAL SERVICE INFORMATION

Name________________________ Class__________________

Score: ☐ Excellent ☐ Good ☐ Needs Improvement **Instructor OK** ☐

Vehicle year _________ **Make** _______________ **Model** _______________

Objective: Upon completion of this assignment, you should be able to locate service specifications using a digital service information system.

ASE Education Foundation Correlation

This worksheet addresses the following **MLR** tasks:

I.A.1; II.A.1; III.A.1; IV.A.1; V.A.1; VI.A.1; VIII.A.1	Research vehicle service information, including fluid type, vehicle service history, service precautions, and technical service bulletins. **(P-1); (P-1); (P-1); (P-1); (P-1); (P-1); (P-1)**
VII.A.1	Research vehicle service information, including refrigerant/oil type, vehicle service history, service precautions, and technical service bulletins. **(P-1)**

This worksheet addresses the following **AST/MAST** tasks:

I.A.2	Research vehicle service information including fluid type, internal engine operation, vehicle service history, service precautions, and technical service bulletins. **(P-1)**
II.A.2; III.A.2; IV.A.1; V.A.2	Research vehicle service information including fluid type, vehicle service history, service precautions, and technical service bulletins. **(P-1); (P-1); (P-1); (P-1)**
VI.A.1; VIII.A.2	Research applicable vehicle and service information, vehicle service history, service precautions, and technical service bulletins. **(P-1); (P-1)**

We Support ASE Education Foundation

Directions: Before beginning this lab assignment, review the worksheet completely. Digital service information is a fee-based service that can be found on the Internet. Fill in the information in the spaces provided as you complete each task.

Tools and Equipment Required: Computer station, Internet access, subscription to one of the vehicle information services

Photo by Tim Gilles

Procedure:

1. In the shop or library, locate a computer station that has Internet access and a subscription to one of the vehicle information services.
2. Open the system to locate information for a 2010 Toyota Camry XLE that has an automatic transmission and a 2GR-FE engine.

Tune-Up Specifications

3. Spark plug gap 0.______"
4. Spark plug torque ______ ft.-lb
5. Firing order ________________
6. Cylinder arrangement. Draw a picture of the engine below that shows the cylinder arrangement.
7. Ignition timing ______ degrees BTDC

   Note: The ignition timing is not adjustable but can be checked using a scan tool.
8. What is the displacement of the engine? ______ liters

General Service Information

9. How often should each of the following be replaced during the first 100,000 miles for a vehicle that requires normal maintenance service?
   a. Air filter ________________ miles ☐ N/A
   b. Engine oil ________________ miles ☐ N/A
   c. Fuel filter ________________ miles ☐ N/A
   d. Oil filter ________________ miles ☐ N/A
   e. PCV filter ________________ miles ☐ N/A
   f. Spark plugs ________________ miles ☐ N/A
   g. Rotate tires ________________ miles ☐ N/A
   h. Replace cabin filter ________________ miles ☐ N/A
10. What is the cooling system capacity? ______ qt.
11. What type of coolant is recommended? ________________
12. What is the engine oil capacity? ______ qt.

    Does the engine oil capacity include the capacity of the oil filter? ☐ Yes ☐ No
13. What is the recommended oil viscosity ________________
14. What is the capacity of the automatic transmission? ________________

    What type of transmission fluid should be used? ________________

Tightening Specifications

a. Coolant pump bolt torque ________ ft.-lb

b. Main bearing cap torque ________ ft.-lb then turn ________ degrees

c. Connecting rod bolt torque ________ ft.-lb then turn ________ degrees

d. Cylinder head bolt torque ________ ft.-lb then turn ________ degrees

15. Draw a sketch below that shows the cylinder head bolt tightening sequence.

    Are there any technical service bulletins (TSBs) listed for this vehicle?

    ☐ Yes ☐ No. List the three most recent TSBs.

    ____

    ____

    ____

16. Exit the computer program.

STOP

ASE Education Foundation Worksheet #5-4

FLAT-RATE WORKSHEET

Name ____________________ Class ____________________

Score: ☐ Excellent ☐ Good ☐ Needs Improvement **Instructor OK** ☐

Vehicle year ________ **Make** ____________ **Model** ______________

Objective: Upon completion of this assignment, you should be able to use a computer information system or a *Parts and Time Guide* (flat-rate manual) to determine the cost of vehicle repairs.

Directions: Before beginning this lab assignment, review the worksheet completely. Fill in the information in the spaces provided as you complete each task.

Tools and Equipment Required: An online vehicle information system with estimating capability, or a flat-rate manual.

Procedure: Use an electronic information library or a *Parts and Time Guide*. Determine the estimated time for the following repairs. Then multiply the time by a shop rate of $80 per hour to determine the labor estimate for the customer.

2005 Ford F150 2-wheel drive pickup, 4-speed transmission, air conditioning, 5.0L engine.

Example:

1. List the time required to remove and replace (R&R) the engine.

Labor time	6.2 hours
Shop rate	× $80.00
Total labor estimate	$496.00
Estimated cost of parts	$2178.38
Total estimate	$2674.38

2. List the time required to remove and replace (R&R) a clutch master cylinder.

Labor time	____________ hours
Shop rate	× $80.00
Total labor estimate	____________
Estimated cost of parts	____________
Total estimate	____________

3. List the time required to evaporate and recharge an air-conditioning system.

Labor time	____________ hours
Shop rate	× $80.00
Total labor estimate	____________
Estimated cost of refrigerant @ $5 per pound	____________
Total estimate	____________

TURN

4. List the time and materials required to remove and replace (R&R) the front crankshaft seal.

Labor time	______________	hours
Shop rate	× $80.00	
Total labor estimate	______________	
Estimated cost of parts	______________	
Total estimate	______________	

5. List the time and materials required to remove and replace (R&R) the cylinder head gaskets.

Labor time	______________	hours
Shop rate	× $80.00	
Total labor estimate	______________	
Estimated cost of parts	______________	
Total estimate	______________	

6. After completing this assignment, clean your work area, put the manual in its proper place, or close the computer program.

STOP

Part II

Lab Worksheets for ASE Maintenance and Light Repair

Service Area 6

Belts, Hoses, Fuel, and Cooling System Service

ASE Education Foundation Worksheet #6-1

COOLING SYSTEM INSPECTION

Name________________________ Class________________

Score: ☐ Excellent ☐ Good ☐ Needs Improvement **Instructor OK** ☐

Vehicle year ________ **Make** ______________ **Model** ______________

Objective: Upon completion of this assignment, you should be able to do a visual inspection of the cooling system.

ASE Education Foundation Correlation

This worksheet addresses the following **MLR** tasks:

I.A.3 Inspect engine assembly for fuel, oil, coolant, and other leaks; determine necessary action. **(P-1)**

I.C.1 Perform cooling system pressure and dye tests to identify leaks; check coolant condition and level; inspect and test radiator, pressure cap, coolant recovery tank, and heater core and galley plugs; determine necessary action. **(P-1)**

This worksheet addresses the following **AST/MAST** tasks:

I.A.4 Inspect engine assembly for fuel, oil, coolant, and other leaks; determine needed action. **(P-1)**

I.D.1 Perform cooling system pressure and dye tests to identify leaks; check coolant condition and level; inspect and test radiator, pressure cap, coolant recovery tank, heater core and galley plugs; determine needed action. **(P-1)**

We Support ASE Education Foundation

Directions: Before beginning this lab assignment, review the worksheet completely. Fill in the information in the spaces provided as you complete each task.

Tools and Equipment Required: Safety glasses, fender covers, shop towel

Procedure:

1. Open the hood and place fender covers on the fenders and over the front body area.
2. Inspect the coolant pump drive belt.

   Drive belt condition ☐ OK ☐ Loose ☐ Glazed ☐ Split ☐ Damaged

   Belt alignment ☐ Correct ☐ Incorrect

   Belt size and length ☐ Correct ☐ Incorrect

CAUTION Do not remove the radiator cap if the system is hot or pressurized.

3. Inspect the radiator cap:

   Is it the correct cap? ☐ Yes ☐ No

   What is the condition of the gasket?

   ☐ OK ☐ Worn/damaged

   What is the condition of the pressure valve?

   ☐ OK ☐ Damaged

   What is the condition of the vacuum valve?

   ☐ OK ☐ Damaged

Vacuum valve

Pressure valve

Gasket

Top seal

Photo by Tim Gilles

4. Inspect radiator condition:

Coolant level	☐ OK	☐ Low		
Coolant strength	☐ OK	☐ Weak		
Leaks in the radiator	☐ Yes	☐ No		
Fan condition	☐ OK	☐ Damaged		
Radiator cap seat	☐ OK	☐ Damaged		
Drain valve	☐ OK	☐ Needs service		
Overflow tube	☐ OK	☐ Needs service		
Overflow tank	☐ Clean	☐ Needs service		
Radiator core	☐ Clean	☐ Leaves	☐ Bugs	☐ Dirt

   Automatic transmission cooler line condition? ☐ OK ☐ Damaged

5. Inspect coolant pump condition:

   At the weep hole ☐ OK ☐ Leaks ☐ N/A

   Using a mirror and flashlight, show the weep hole to your instructor.

   Instructor OK ___

   Leakage around gasket ☐ OK ☐ Leaks

   Pump bearing ☐ OK ☐ Loose

6. Inspect cooling fan condition:

Fan blades	☐ OK	☐ Damaged	
Fan clutch	☐ OK	☐ Defective	☐ N/A
Electric fan	☐ OK	☐ Defective	☐ N/A
Fan shroud	☐ OK	☐ Missing/damaged	

7. Inspect heater core condition:

   ☐ OK ☐ Leaks ☐ Debris in core housing

8. Inspect the gaskets and core plugs:

   Thermostat housing ☐ OK ☐ Leaking
   External head gasket leak ☐ Yes ☐ No
   Core plugs ☐ OK ☐ Leaking
   Intake manifold ☐ OK ☐ Leaking

9. Inspect the temperature gauge: ☐ OK ☐ Inoperative ☐ N/A

   Warning light: ☐ OK ☐ Inoperative ☐ N/A

10. Measure the coolant temperature with a mercury thermometer or digital laser thermometer. Compare the temperature of the coolant to the reading on the vehicle's temperature gauge.

    Test results: ☐ Low ☐ High ☐ OK

11. Inspect the hoses: ☐ OK ☐ Leaks ☐ Swollen ☐ Pinched ☐ Collapsed ☐ Incorrect fit

    Hose clamps: ☐ OK ☐ Need replacing

12. Before completing your paperwork, clean your work area, clean and return tools to their proper places, and wash your hands.

13. Record your recommendations for needed service or additional repairs and complete the repair order.

    ____

    ____

STOP

ASE Education Foundation Worksheet #6-2

PRESSURE TEST A RADIATOR CAP

Name_______________________ Class_________________

Score: ☐ Excellent ☐ Good ☐ Needs Improvement **Instructor OK** ☐

Vehicle year _________ **Make** _______________ **Model** _________________

Objective: Upon completion of this assignment, you should be able to pressure test a radiator cap.

ASE Education Foundation Correlation

This worksheet addresses the following **MLR** task:

I.C.1 Perform cooling system pressure and dye tests to identify leaks; check coolant condition and level; inspect and test radiator, pressure cap, coolant recovery tank, and heater core and galley plugs; determine necessary action. **(P-1)**

This worksheet addresses the following **AST/MAST** task:

I.D.1 Perform cooling system pressure and dye tests to identify leaks; check coolant condition and level; inspect and test radiator, pressure cap, coolant recovery tank, heater core and galley plugs; determine needed action. **(P-1)**

We Support ASE | Education Foundation

Directions: Before beginning this lab assignment, review the worksheet completely. Fill in the information in the spaces provided as you complete each task.

Tools and Equipment Required: Safety glasses, fender covers, radiator pressure tester, shop towel

Procedure:

1. Open the hood and install fender covers over the fenders and the front body parts.
2. Check the vehicle's temperature gauge. ☐ Cold ☐ Normal ☐ Hot ☐ N/A

CAUTION When removing the radiator cap, the engine must be off and the pressure released. The cooling system is under pressure when the system is at normal operating temperature. As the radiator cap is removed, the coolant may boil. Squeeze the radiator hose to check if the system is under pressure. If the hose is hard, do not remove the radiator cap. See your instructor.

3. Squeeze the top hose. ☐ Hard ☐ Soft
4. Fold a shop rag to ¼ size. Use it to turn the radiator cap ¼ turn until its first stop. This will allow any remaining pressure to escape.
5. Squeeze the top radiator hose to check for system pressure. ☐ Hard ☐ Soft
6. If the hose is soft, press down on the cap and remove it.

TURN

7. Visually inspect the radiator cap.

   Radiator cap pressure seal:

   ☐ OK ☐ Worn ☐ Missing

   Radiator cap pressure spring:

   ☐ OK ☐ Rusted ☐ Broken

   Radiator cap vacuum valve:

   ☐ OK ☐ Stuck ☐ Missing

   Hint: Check for signs of corrosion under the seal.

8. Wet the rubber pressure seal and install the radiator cap on the adapter.

Imprinted seal

Filler neck sealing surface

Photo by Tim Gilles

9. Attach the tester to the adapter.
10. Pump up the tester until the gauge reaches its highest point and holds.
11. What is the highest pressure that the cap will hold?

    ☐ 15 lb ☐ 13–14 lb ☐ 7 lb ☐ 0

12. Does the cap maintain pressure after the high point is reached?

    ☐ Yes ☐ No

13. Remove pressure from the tester by pushing the tester hose sideways at the cap.

    Did it release air? ☐ Yes ☐ No

14. Remove the radiator cap from the tester and return the tester to its container.
15. Put the radiator cap back on the radiator.
16. Does the radiator cap need to be replaced?

    ☐ Yes ☐ No

Photo by Tim Gilles

17. What happens to the boiling point of the coolant if a faulty radiator cap is not replaced?
18. Before completing your paperwork, clean your work area, clean and return tools to their proper places, and wash your hands.
19. Record your recommendations for needed service or additional repairs and complete the repair order.

    ____

    ____

ASE Education Foundation Worksheet #6-3

PRESSURE TEST A COOLING SYSTEM

Name________________________________ Class____________________

Score: ☐ Excellent ☐ Good ☐ Needs Improvement **Instructor OK** ☐

Vehicle year __________ **Make** ____________________ **Model** ______________________

Objective: Upon completion of this assignment, you should be able to pressure test a cooling system.

ASE Education Foundation Correlation

This worksheet addresses the following **MLR** task:

I.C.1 Perform cooling system pressure and dye tests to identify leaks; check coolant condition and level; inspect and test radiator, pressure cap, coolant recovery tank, and heater core and galley plugs; determine necessary action. **(P-1)**

This worksheet addresses the following **AST/MAST** task:

I.D.1 Perform cooling system pressure and dye tests to identify leaks; check coolant condition and level; inspect and test radiator, pressure cap, coolant recovery tank, heater core and galley plugs; determine needed action. **(P-1)**

We Support ASE Education Foundation

Directions: Before beginning this lab assignment, review the worksheet completely. Fill in the information in the spaces provided as you complete each task.

Tools and Equipment Required: Safety glasses, fender covers, radiator pressure tester, shop towel

Procedure:

1. Open the hood and install fender covers on the fenders and front body parts.
2. Before starting to work, set the parking brake firmly and put the transmission in park or neutral (manual transmissions).
3. Remove the radiator cap. Refer to Worksheet #6-2, Pressure Test a Radiator Cap, for the safe procedure.
4. Pressurize the cold system.
   a. Wet the radiator pressure tester gasket and install the tester on the radiator.
   b. Pressurize the system to the pressure marked on the cap.

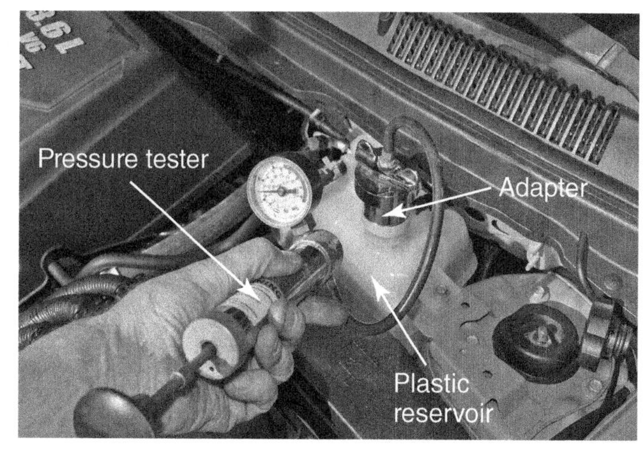

Photo by Tim Gilles

c. Wait 5 minutes. Does the pressure on the gauge remain steady? ☐ Yes ☐ No

d. Are there leaks at any of the following?

☐ Heater core ☐ Heater hoses ☐ Radiator hoses

☐ Thermostat housing ☐ Core plugs ☐ Radiator

☐ Water pump ☐ Carburetor heater ☐ No leaks found

Note: If the pressure drops but no leaks are found, do the combustion leak test (Worksheet #6-4, Perform a Cooling System Combustion Leak Test) after completing this worksheet.

e. Release the pressure from the cooling system.

5. Start the engine and let it run until it reaches operating temperature, then shut it off.

   Note: It is easier to pressure test a cold engine, but some leaks will only show up when the engine is warm.

6. Pressurize the system to the pressure rating marked on the cap.

   Radiator cap pressure _______ lb _______ bar

7. Wait 5 minutes. Does the pressure on the gauge remain steady? ☐ Yes ☐ No

   Note: A very small leak may not be evident in this short a time.

8. If the pressure reading has dropped, look for signs of external leakage.

   Note: Some pressure drop may occur as the cooling system temperature drops.

9. Are there leaks at any of the following?

   ☐ Heater core ☐ Heater hoses ☐ Radiator hoses

   ☐ Thermostat housing ☐ Core plugs ☐ Radiator

   ☐ Coolant pump ☐ No leaks found

10. Push the tester hose to the side where it connects to the radiator adapter. This will release the pressure from the system. Remove the tester.

Photo by Tim Gilles

11. Before completing your paperwork, clean your work area, clean and return tools to their proper places, and wash your hands.

12. Record your recommendations for needed service or additional repairs and complete the repair order.

____

____

STOP

ASE Education Foundation Worksheet #6-4

PERFORM A COOLING SYSTEM COMBUSTION LEAK TEST

Name ______________________ Class ______________

Score: ☐ Excellent ☐ Good ☐ Needs Improvement **Instructor OK** ☐

Vehicle year _______ **Make** _____________ **Model** _______________

Objective: Upon completion of this assignment, you should be able to test an engine for internal cooling system combustion leaks.

ASE Education Foundation Correlation

This worksheet addresses the following **MLR** task:

I.C.1 Perform cooling system pressure and dye tests to identify leaks; check coolant condition and level; inspect and test radiator, pressure cap, coolant recovery tank, and heater core and galley plugs; determine necessary action. **(P-1)**

This worksheet addresses the following **AST/MAST** task:

I.D.1 Perform cooling system pressure and dye tests to identify leaks; check coolant condition and level; inspect and test radiator, pressure cap, coolant recovery tank, heater core and galley plugs; determine needed action. **(P-1)**

We Support ASE | Education Foundation

Directions: Before beginning this lab assignment, review the worksheet completely. Fill in the information in the spaces provided as you complete each task.

Tools and Equipment Required: Safety glasses, fender covers, drain pan, combustion leak tester, shop towel

Procedure:

1. Install fender covers on the fender and over the front body parts.
2. Remove the radiator cap. Refer to Worksheet #6-2, Pressure Test a Radiator Cap, for the safe procedure.
3. Open the radiator drain valve and release some coolant into a drain pan to lower the water level in the radiator about 2". Close the drain valve.
4. Start the engine and run it until the engine is warm.

   Did the thermostat open? ☐ Yes ☐ No

   Is the upper radiator hose hot? ☐ Yes ☐ No

Photo by Tim Gilles

TURN

5. Pour the testing liquid into the tester until it reaches the "fill" line.
6. Place the tester on the radiator filler neck and pump the bulb several times to suck *air* from above the coolant.

   Note: As the coolant gets hotter, its level will rise. Allowing coolant into the test fluid will ruin the fluid and void the test.
7. Did the liquid change color? ☐ Yes ☐ No

   If the liquid did change color, what color is it?

   ☐ Blue ☐ Green ☐ Yellow

   What does each color indicate?

   Blue ____________________

   Green ____________________

   Yellow ____________________
8. Refill the radiator and put the radiator cap in place.
9. Drain the used test fluid from the tester and dry the tester.
10. If the test fluid had changed color, what repair would most likely correct the problem? ____________________
11. Before completing your paperwork, clean your work area, clean and return tools to their proper places, and wash your hands.
12. Record your recommendations for needed service or additional repairs and complete the repair order.

    ____

    ____

STOP

ASE Education Foundation Worksheet #6-5

CHECK COOLANT CONDITION

Name_______________________ Class_________________

Score: ☐ Excellent ☐ Good ☐ Needs Improvement **Instructor OK** ☐

Vehicle year ________ **Make** ________________ **Model** __________________

Objective: Upon completion of this assignment, you should be able to test the coolant for the proper strength.

ASE Education Foundation Correlation

This worksheet addresses the following **MLR** tasks:

I.C.1 Perform cooling system pressure and dye tests to identify leaks; check coolant condition and level; inspect and test radiator, pressure cap, coolant recovery tank, and heater core and galley plugs; determine necessary action. **(P-1)**

I.C.4 Inspect and test coolant; drain and recover coolant; flush and refill cooling system; use proper fluid type per manufacturer specification; bleed air as required. **(P-1)**

This worksheet addresses the following **AST/MAST** tasks:

I.D.1 Perform cooling system pressure and dye tests to identify leaks; check coolant condition and level; inspect and test radiator, pressure cap, coolant recovery tank, heater core and galley plugs; determine needed action. **(P-1)**

I.D.4 Inspect and test coolant; drain and recover coolant; flush and refill cooling system; use proper fluid type per manufacturer specification; bleed air as required. **(P-1)**

We Support ASE Education Foundation

Directions: Before beginning this lab assignment, review the worksheet completely. Fill in the information in the spaces provided as you complete each task.

Tools and Equipment Required: Safety glasses, fender covers, coolant tester, shop towel, refractometer

Procedure:

1. Open the hood and install fender covers over the fenders and front body parts.
2. Remove the radiator cap. Refer to Worksheet #6-2, Pressure Test a Radiator Cap, for the safe procedure.
3. What is the general appearance of the coolant?

   ☐ Clean ☐ Dirty
4. What color is the coolant? ______________________________

To check coolant with a hydrometer:

5. Draw some coolant into the hydrometer read the hydrometer to check coolant strength.

TURN

6. Coolant strength:

   ☐ Good ☐ Weak

   Freezing point _______ °F

   Boiling point _______ °F

   Recommendations:

   a. Coolant that is too concentrated may be diluted with water.
   b. A slightly weak concentration can be strengthened by draining off a quart of coolant and adding straight coolant. After running the engine, the strength may be checked again.
   c. If coolant strength is very weak, a coolant flush and change is recommended.

Coolant hydrometer

Photo by Tim Gilles

To check coolant with a refractometer:

7. Check instructions on the tester. Does coolant have to be at operating temperature for an accurate reading?

   ☐ Yes ☐ No

8. Place a drop of coolant on the refractometer screen.
9. Look through the refractometer eye pièce and read the coolant condition.

Photo by Tim Gilles

10. Coolant strength: ☐ Good ☐ Weak

    Freezing point _________°F ______ Boiling point °F

11. Top off the coolant level and reinstall the radiator cap.
12. Clean the coolant hydrometer and/or the refractometer by flushing it with water.
13. Before completing your paperwork, clean your work area, clean and return tools to their proper places, and wash your hands.
14. Record your recommendations for needed service or additional repairs below.

    ____

    ____

STOP

ASE Education Foundation Worksheet #6-6

TESTING COOLANT USING TEST STRIPS

Name________________________ Class__________________

Score: ☐ Excellent ☐ Good ☐ Needs Improvement **Instructor OK** ☐

Vehicle year ________ **Make** ______________ **Model** ________________

Objective: Upon completion of this assignment, you should be able to test the coolant for the proper strength using test strips.

ASE Education Foundation Correlation

This worksheet addresses the following **MLR** task:

I.C.4 Inspect and test coolant; drain and recover coolant; flush and refill cooling system; use proper fluid type per manufacturer specification; bleed air as required. **(P-1)**

This worksheet addresses the following **AST/MAST** tasks:

I.D.4 Inspect and test coolant; drain and recover coolant; flush and refill cooling system; use proper fluid type per manufacturer specification; bleed air as required. **(P-1)**

We Support ASE | Education Foundation

Directions: Before beginning this lab assignment, review the worksheet completely. Fill in the information in the spaces provided as you complete each task.

Tools and Equipment Required: Safety glasses, fender covers, shop towel, coolant test strips

Procedure:

1. Remove the radiator cap. Refer to Worksheet #6-2, Pressure Test a Radiator Cap, for the safe procedure.
2. Test strips can be used to test the condition of a coolant. Different strips are used for conventional coolants and organic acid (OAT) coolants. What kind of coolant is used in this vehicle?

   ☐ Conventional coolant

   ☐ OAT coolant

Photo by Tim Gilles

3. Some single test strips can check pH, cavitation additive protection, and coolant concentration. There are also test strips that can tell if different types of coolants have been mixed. List the capabilities of your test strips below:

   ____

   ____

TURN

4. Dip the test strip into the coolant.

   What is the pH of the coolant? ____________

   Note: Conventional coolant has a higher pH than extended-life coolant. The additives in the coolant give it a pH level of about 10.5 when new. As coolant ages, acids form. The coolant must contain a sufficient amount of corrosion inhibitor to neutralize these acids. This neutralizing ability is called reserve alkalinity. Preserving an engine's cooling system depends on changing the coolant before its reserve alkalinity is depleted.

   When the additives become depleted, the acid level rises (pH level drops) and corrosion begins. Used conventional coolant should test at a pH level of at least 9.0. Extended-life coolant, which is more acidic due to its organic acid package, should test at a pH level of at least 7.5.

   ☐ Coolant needs to be replaced.

   ☐ Coolant is acceptable for continued service.

5. Replace the radiator cap.
6. Before completing your paperwork, clean your work area, clean and return tools to their proper places, and wash your hands.
7. Record your recommendations for needed service or additional repairs and complete the repair order.

   ____

   ____

STOP

ASE Education Foundation Worksheet #6-7

CHECK COOLANT STRENGTH—VOLTMETER

Name_______________________________ Class_______________________

Score: ☐ Excellent ☐ Good ☐ Needs Improvement **Instructor OK** ☐

Vehicle year __________ **Make** ____________________ **Model** ______________________

Objective: Upon completion of this assignment, you should be able to test the coolant for the proper strength with a voltmeter.

ASE Education Foundation Correlation

This worksheet addresses the following **MLR** tasks:

I.C.4 Inspect and test coolant; drain and recover coolant; flush and refill cooling system; use proper fluid type per manufacturer specification; bleed air as required. **(P-1)**

This worksheet addresses the following **AST/MAST** tasks:

I.D.4 Inspect and test coolant; drain and recover coolant; flush and refill cooling system; use proper fluid type per manufacturer specification; bleed air as required. **(P-1)**

We Support ASE | Education Foundation

Directions: Before beginning this lab assignment, review the worksheet completely. Fill in the information in the spaces provided as you complete each task.

Tools and Equipment Required: Safety glasses, fender covers, coolant tester, shop towel

Procedure:

1. Open the hood and install fender covers over the fenders and front body parts.
2. Remove the radiator cap. Refer to Worksheet #6-2, Pressure Test a Radiator Cap, for the safe procedure.
3. What is the general appearance of the coolant?

   ☐ Clean ☐ Dirty
4. What color is the coolant? _________________________
5. Set the voltmeter to read DC voltage. Which scale did you select? ☐ AC ☐ DC
6. Connect the ground (black) lead of a voltmeter to the negative (–) terminal of the battery.
7. Insert the positive (red) voltmeter lead into the coolant.

   What is the voltage reading? ________

   What is the maximum voltage reading allowable? ________

8. What do you think the results of this test indicate? ___________________________

____

9. Before completing your paperwork, clean your work area, clean and return tools to their proper places, and wash your hands.

10. Record your recommendations for needed service or additional repairs and complete the repair order.

____

____

STOP

ASE Education Foundation Worksheet #6-8

TEST THE COOLING SYSTEM FOR COMBUSTION GASES

Name_______________________________ Class______________________

Score: ☐ Excellent ☐ Good ☐ Needs Improvement **Instructor OK** ☐

Vehicle year __________ **Make** __________________ **Model** ____________________

Objective: Upon completion of this assignment, you will be able to test a cooling system for the presence of combustion gases.

ASE Education Foundation Correlation

This worksheet addresses the following **MLR** task:

I.C.1 Perform cooling system pressure and dye tests to identify leaks; check coolant condition and level; inspect and test radiator, pressure cap, coolant recovery tank, and heater core and galley plugs; determine necessary action. **(P-1)**

This worksheet addresses the following **AST/MAST** tasks:

I.D.1 Perform cooling system pressure and dye tests to identify leaks; check coolant condition and level; inspect and test radiator, pressure cap, coolant recovery tank, heater core and galley plugs; determine needed action. **(P-1)**

I.D.2 Identify causes of engine overheating. **(P-1)**

We Support ASE Education Foundation

Directions: Before beginning this lab assignment, review the worksheet completely. Fill in the information in the spaces provided as you complete each task.

Tools and Equipment Required: Safety glasses, fender covers, cooling system combustion tester, shop towels

Procedure:

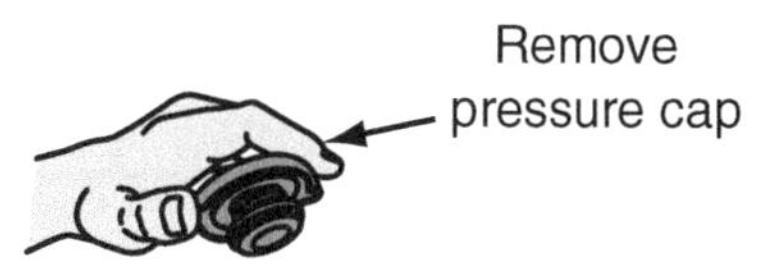

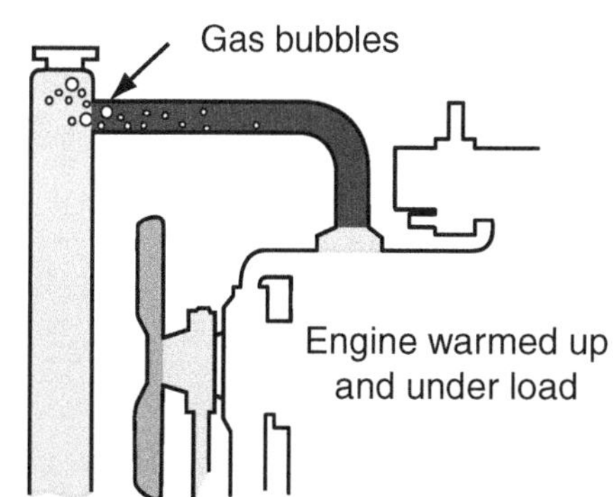

1. Review the instructions for the cooling system combustion leak tester.
2. While the engine is cool, remove the radiator cap.

   How can you determine that the engine is cool enough to remove the radiator cap? __________

   ____

3. Start the engine and bring it to normal operating temperature.
4. Make sure that the level of the coolant is approximately 2" from the filler neck. Drain some coolant if necessary.

5. Add the correct amount of the special blue fluid to the tester.
6. With the engine idling, place the tester in the filler radiator neck.
7. Pump the tester bulb several times to suck air from above the coolant.

   Note: Do not let coolant enter into the tester. It will disallow your test.

8. Did the fluid in the tester change color?

   ☐ Yes ☐ No

9. Are combustion gases entering the cooling system?

   ☐ Yes ☐ No

Photo by Tim Gilles

   Which of the following are possible causes of combustion gases in the cooling system? Mark any that apply.

   ☐ Cracked cylinder head

   ☐ Cracked block

   ☐ Faulty head gasket

10. Before completing the paperwork, clean your work area, clean and return tools to their proper places, and wash your hands.
11. Were there any recommendations for needed service or unusual conditions that you noticed while you were testing for combustion leaks?

    ☐ Yes ☐ No

12. Record your recommendations for needed service or additional repairs and complete the repair order.

    ____

    ____

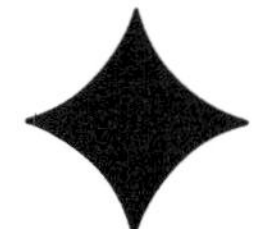

ASE Education Foundation Worksheet #6-9

REPLACE A RADIATOR HOSE

Name ______________________________ Class ______________________

Score: ☐ Excellent ☐ Good ☐ Needs Improvement **Instructor OK** ☐

Vehicle year __________ **Make** ____________________ **Model** ______________________

Objective: Upon completion of this assignment, you should be able to replace a radiator hose.

ASE Education Foundation Correlation

This worksheet addresses the following **MLR** tasks:

I.C.4 Inspect and test coolant; drain and recover coolant; flush and refill cooling system; use proper fluid type per manufacturer specification; bleed air as required. **(P-1)**

VII.C.1 Inspect engine cooling and heater systems hoses and pipes; determine necessary action. **(P-1)**

This worksheet addresses the following **AST/MAST** tasks:

I.D.4 Inspect and test coolant; drain and recover coolant; flush and refill cooling system; use proper fluid type per manufacturer specification; bleed air as required. **(P-1)**

VII.C.1 Inspect engine cooling and heater systems hoses and pipes; perform needed action. **(P-1)**

We Support ASE Education Foundation

Directions: Before beginning this lab assignment, review the worksheet completely. Fill in the information in the spaces provided as you complete each task.

Tools and Equipment Required: Safety glasses, fender covers, drain pan, slot screwdriver, razor knife, shop towel

Parts and Supplies: Radiator hose, hose clamps

Procedure:

1. Obtain the new replacement hose and hose clamps.
2. Open the hood and install fender covers over the fenders and front body parts.
3. Before starting to remove the radiator hose, compare the new hose with the one to be replaced. Do they appear to be the same size and shape?

   ☐ Yes ☐ No
4. Remove the radiator cap. Refer to Worksheet #6-2, Pressure Test a Radiator Cap, for the safe procedure.
5. Open the radiator drain valve located in the lower radiator tank and drain some coolant into a clean container. Drain the coolant until its level is below the level of the radiator hose.

6. Loosen the hose clamps on the radiator hose to be removed.

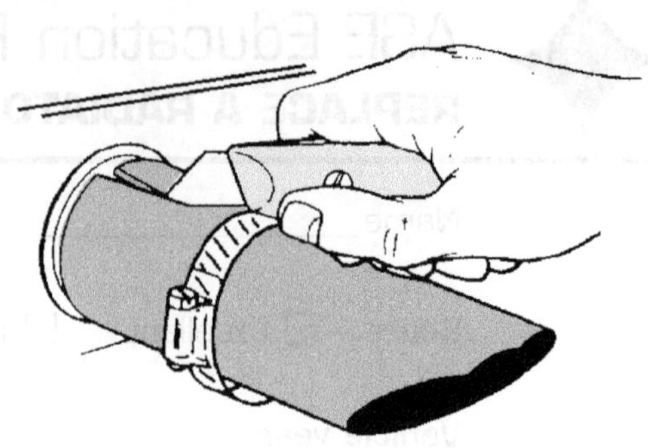

7. Twist the hose to loosen it. It may be necessary to cut the hose with a sharp knife if it is not easily removed.

   Note: Use caution when twisting the hose. The hose fitting on the radiator is easily damaged.

   Was the hose easily removed? ☐ Yes ☐ No

   Was it necessary to cut the hose? ☐ Yes ☐ No

   Was the radiator hose fitting damaged?
   ☐ Yes ☐ No

8. Clean the radiator hose fittings.

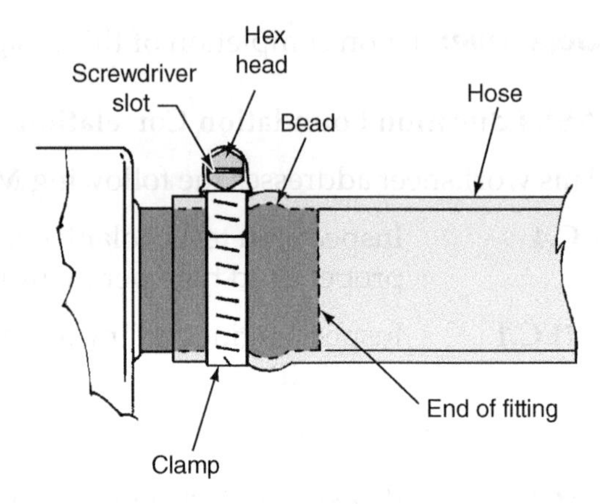

9. Install the new hose. If the hose is difficult to install, apply some rubber lube to the connection. Be sure any bends are properly located. Check to see that the hose will not be damaged by movement of the engine or its accessories and that the clamps do not interfere with the fan belts, fuel lines, fuel pump, or fan.

10. Replace rusted or damaged hose clamps with new clamps and install them on the hose. Position the screw side of the clamp for easy access.

    The hose clamps were: ☐ Reused ☐ Replaced

    What size are the hose clamps?

    ____

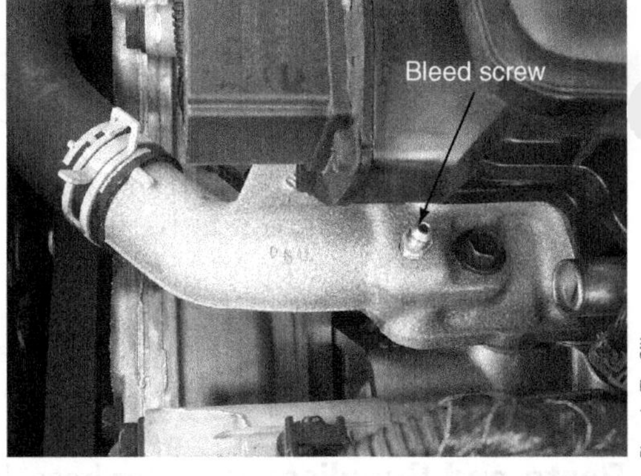

Photo by Tim Gilles

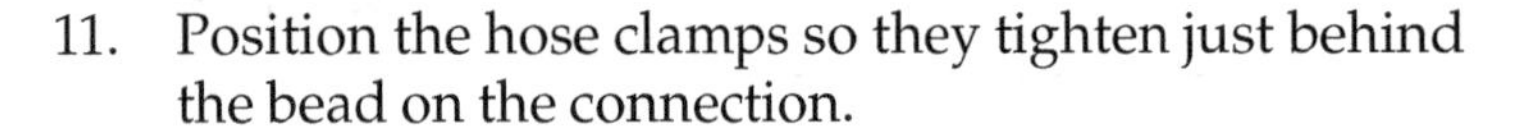

11. Position the hose clamps so they tighten just behind the bead on the connection.

    Clamps properly installed? ☐ Yes ☐ No

12. Position and tighten the hose clamps.

13. Refill the system. Be certain that no air remains in the cooling system. Some vehicles have a bleed screw like the one shown in the photo. If there is no bleed screw, sometimes you can remove a heater hose that is higher than the engine and bleed air from there.

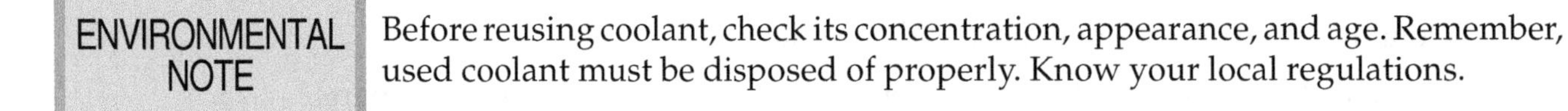

ENVIRONMENTAL NOTE Before reusing coolant, check its concentration, appearance, and age. Remember, used coolant must be disposed of properly. Know your local regulations.

14. Pressure test the cooling system before starting the engine to check for leaks. **Instructor OK** _____

15. Replace the radiator cap and run the engine until it is warm. Feel the upper radiator hose. When it is hot, the thermostat has opened. Bleed air from the system when necessary. Then check the coolant level.

    ☐ OK ☐ Low

16. Refill the system and check for leaks.
17. Replace the radiator cap.

    Note: After the cooling system has fully warmed up and then cooled again, the hose clamps should be retightened. Hoses may shrink after their first use.

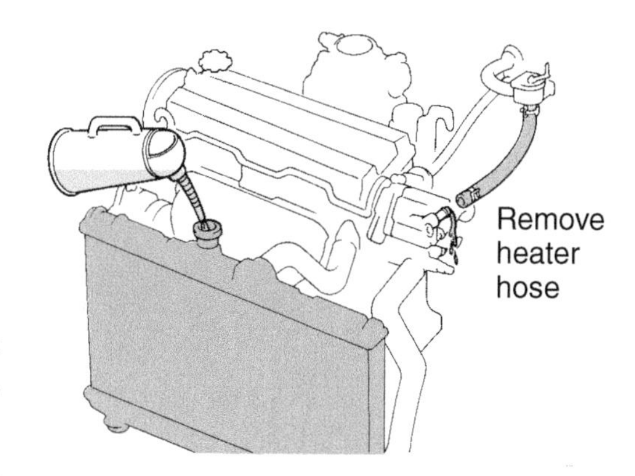

18. Retighten the hose clamps.
19. Before completing your paperwork, clean your work area, clean and return tools to their proper places, and wash your hands.
20. Record your recommendations for needed service or additional repairs and complete the repair order.

____

____

STOP

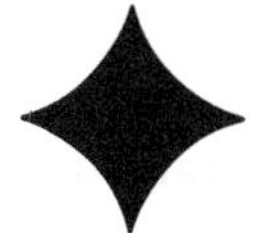

ASE Education Foundation Worksheet #6-10

INSPECT AND REPLACE AN ADJUSTABLE V-RIBBED BELT

Name_______________________________ Class_______________________

Score: ☐ Excellent ☐ Good ☐ Needs Improvement **Instructor OK** ☐

Vehicle year __________ **Make** ____________________ **Model** ______________________

Objective: Upon completion of this assignment, you should be able to inspect and replace a V-ribbed belt and adjust it to the proper tension.

ASE Education Foundation Correlation

This worksheet addresses the following **MLR** tasks:

I.C.2 Inspect, replace, and adjust drive belts, tensioners, and pulleys; check pulley and belt alignment. **(P-1)**

VII.B.1 Inspect and replace A/C compressor drive belts, pulleys, and tensioners; visually inspect A/C components for signs of leaks; determine necessary action. **(P-2)**

This worksheet addresses the following **AST** tasks:

I.D.3 Inspect, replace, and adjust drive belts, tensioners, and pulleys; check pulley and belt alignment. **(P-1)**

VII.B.1 Inspect, remove, and/or replace A/C compressor drive belts, pulleys, and tensioners; visually inspect A/C components for signs of leaks; determine needed action. **(P-2)**

This worksheet addresses the following **MAST** tasks:

I.D.3 Inspect, remove, and/or replace A/C compressor drive belts, pulleys, tensioners and visually inspect A/C components for signs of leaks; determine needed action. **(P-2)**

VII.B.1 Inspect, remove, and/or replace A/C compressor drive belts, pulleys, tensioners and visually inspect A/C components for signs of leaks; determine needed action. **(P-2)**

We Support ASE Education Foundation

Directions: Before beginning this lab assignment, review the worksheet completely. Fill in the information in the spaces provided as you complete each task.

Tools and Equipment Required: Safety glasses, fender covers, shop towel, belt tension gauge, hand tools

Parts and Supplies: V-ribbed accessory drive belt

Procedure:

1. Before starting to work, check the service manual for belt tension and adjustment procedures. There are several methods for adjustment. Always check the service manual or computer program for the proper procedure.

   Service manual or computer program used ____________________

2. Obtain the new belt(s) before starting to work.
3. Open the vehicle's hood and place fender covers on the fenders and front body parts.
4. Which accessories are driven by the belt that is being replaced?

   ☐ Alternator ☐ Air-conditioning compressor

   ☐ Water pump ☐ Power steering pump

   ☐ Air pump

5. Remove the belt by loosening the adjuster and pivot bolts. Push the component inward and remove the belt. It may be necessary to weave the belt around the fan assembly.

   Was the belt easily removed? ☐ Yes ☐ No

6. Compare the new belt to the old belt. The width, as well as the length, must be the same. Was the correct belt obtained?

   ☐ Yes ☐ No

   Note: If the old belt is not available, use a piece of string in the pulley groove to estimate the replacement belt size. The parts supplier should be able to determine the correct width for the application.

7. Install the belt by weaving it over the fan assembly and onto the pulleys.
8. Check the belt pulley alignment. The pulleys must be in alignment within $^1/_{16}$" per foot.

   Are the pulleys in alignment? ☐ Yes ☐ No

9. Inspect the new belt to be sure that it does not rub on a radiator hose, fuel hose, or another belt.

   ☐ Rubs ☐ OK

   Note: If the belt is rubbing, correct the problem.

10. Adjust the belt using the correct method.
11. Tighten the mounting bolts/nuts after the proper belt tension is reached.
12. Before completing your paperwork, clean your work area, clean and return tools to their proper places, and wash your hands.
13. Record your recommendations for needed service or additional repairs and complete the repair order.

    ____

    ____

STOP

ASE Education Foundation Worksheet #6-11

REPLACE A SERPENTINE BELT

Name________________________ Class____________________

Score: ☐ Excellent ☐ Good ☐ Needs Improvement **Instructor OK** ☐

Vehicle year __________ **Make** ________________ **Model** ________________

Objective: Upon completion of this assignment, you should be able to inspect and replace a serpentine belt.

ASE Education Foundation Correlation

This worksheet addresses the following **MLR** tasks:

I.C.2 Inspect, replace, and adjust drive belts, tensioners, and pulleys; check pulley and belt alignment. **(P-1)**

VII.B.1 Inspect and replace A/C compressor drive belts, pulleys, and tensioners; visually inspect A/C components for signs of leaks; determine necessary action. **(P-2)**

This worksheet addresses the following **AST** tasks:

I.D.3 Inspect, replace, and adjust drive belts, tensioners, and pulleys; check pulley and belt alignment. **(P-1)**

VII.B.1 Inspect, remove, and/or replace A/C compressor drive belts, pulleys, and tensioners; visually inspect A/C components for signs of leaks; determine needed action. **(P-2)**

This worksheet addresses the following **MAST** tasks:

I.D.3 Inspect, remove, and/or replace A/C compressor drive belts, pulleys, tensioners and visually inspect A/C components for signs of leaks; determine needed action. **(P-2)**

VII.B.1 Inspect, remove, and/or replace A/C compressor drive belts, pulleys, tensioners and visually inspect A/C components for signs of leaks; determine needed action. **(P-2)**

We Support ASE | Education Foundation

Directions: Before beginning this lab assignment, review the worksheet completely. Fill in the information in the spaces provided as you complete each task.

Tools and Equipment Required: Safety glasses, fender covers, shop towel, belt tension gauge, hand tools

Parts and Supplies: Serpentine belt

Procedure:

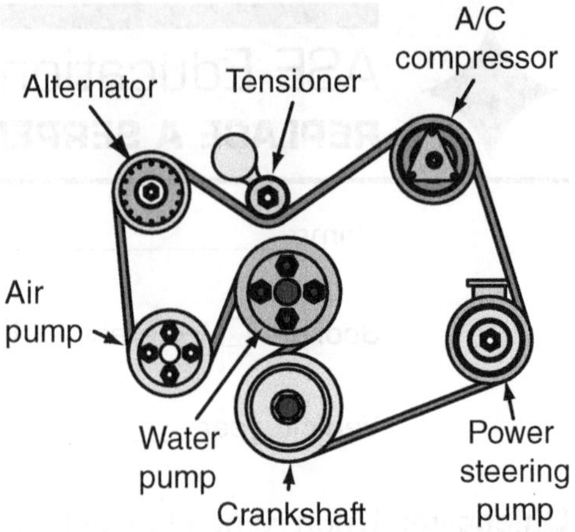

1. Before starting to work, check the service information for belt tension and adjustment procedures. There are several methods for adjustment. Always check the service manual or computer program for the proper procedure.

   Service information or computer program used _______

   ____

2. Obtain the new belt before starting to work.

   Brand ______________ Part # ______________

3. Open the hood and place fender covers on the fenders and front body parts.

4. Locate the belt routing diagram in the engine compartment. Did you find a diagram?

   ☐ Yes ☐ No

5. In the box, draw a sketch of the belt routing before removing the belt.

6. Pry against the tensioner to loosen it and remove the belt.

7. Compare the new belt to the old belt. The width, as well as the length, must be the same.

   Correct belt obtained? ☐ Yes ☐ No

   Note: If the old belt is not available, use a piece of string in the pulley grooves to estimate the replacement belt size. The parts supplier should be able to determine the correct width for the application.

8. Inspect the feel of the tensioner bearing.

   ☐ Smooth ☐ Rough

   Note: When replacing a long-life EPDM belt, the tensioner is typically replaced.

9. Install the belt by weaving it around the pulleys as pictured in the belt diagram. Pry against the tensioner and allow it to spring back against the belt.

10. Check the belt pulley alignment. The pulleys must be in alignment within $\frac{1}{16}$" per foot.

    Are the pulleys in alignment? ☐ Yes ☐ No

11. Check the new belt to be sure that it does not rub on a radiator hose, fuel hose, or another belt.

    ☐ Rubs ☐ OK

    If the belt is rubbing, correct the problem.

12. Before completing your paperwork, clean your work area, clean and return tools to their proper places, and wash your hands.

13. Record your recommendations for needed service or additional repairs and complete the repair order.

    ____

    ____

STOP

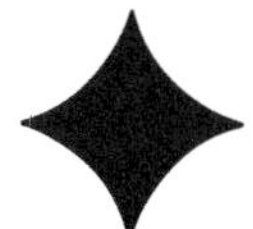

ASE Education Foundation Worksheet #6-12

FLUSH A COOLING SYSTEM AND INSTALL COOLANT

Name________________________ Class__________________

Score: ☐ Excellent ☐ Good ☐ Needs Improvement **Instructor OK** ☐

Vehicle year ________ **Make** ________________ **Model** ________________

Objective: Upon completion of this assignment, you should be able to flush a cooling system and replace the coolant.

ASE Education Foundation Correlation

This worksheet addresses the following **MLR** task:

I.C.4 Inspect and test coolant; drain and recover coolant; flush and refill cooling system; use proper fluid type per manufacturer specification; bleed air as required. **(P-1)**

This worksheet addresses the following **AST** task:

I.D.4 Inspect and test coolant; drain and recover coolant; flush and refill cooling system; use proper fluid type per manufacturer specification; bleed air as required. **(P-1)**

We Support ASE Education Foundation

Directions: Before beginning this lab assignment, review the worksheet completely. Fill in the information in the spaces provided as you complete each task.

Tools and Equipment Required: Safety glasses, service publications, fender covers, drain pan, shop towel, jack stands or vehicle lift

Parts and supplies: Cooling system flush, distilled water, coolant (50% of cooling system capacity or 100% of capacity for premixed coolant), flushing-T

Procedure:

1. Look up the cooling system capacity in a service manual or computer program.

   Cooling system capacity: ______ qt.

   Which manual or computer program was used? ______________________

2. Open the hood and install fender covers over the fenders and front body parts.

   Note: Do not remove the radiator cap if the system is hot or pressurized.

3. Squeeze the upper radiator hose to check for system pressure.

   ☐ Hard ☐ Soft

4. Open the radiator cap. Loosen the radiator drain valve and drain the cooling system.

   Note: Sometimes jacking up the rear of the vehicle will allow for a more thorough drain of the block.

TURN

Flushing the System with a Chemical

1. If the cooling system is dirty or rusty, flush it with a commercial cleaner.
   ☐ Dirty ☐ Clean (skip to "Refilling the System")
2. Lower the vehicle, close the drain valve, and refill the cooling system with water until it is about 2" to 3" below the filler neck.
3. Read the directions on the cleaning chemical.
4. Add the cleaning chemical to the radiator as directed.

Always wear eye protection (goggles) when working with chemicals.

Photo by Tim Gilles

5. Turn the heater control on. On vehicles with heater control valves, this will allow coolant to circulate through the heater.
6. With the emergency brake on and the transmission in park or neutral, start the engine and run it until operating temperature is reached.
7. Double-check to be sure the coolant level is correct.
8. Run the engine for the specified period.
9. Turn off the engine and drain the cooling system.
10. Add a "flushing-T" to the heater hose coming from the heater.
11. If the cleaning chemical used requires a neutralizer, add it to the radiator.
12. After neutralizing, flush the system again thoroughly using the flushing "T."

Refilling the System

1. Inspect the hose from the radiator fill neck to the overflow tank.
   ☐ OK ☐ Needs replacement
2. Drain and flush the recovery tank with water.
3. Add ethylene glycol coolant to the radiator. If you are not using premixed 50/50 coolant, add an amount of full-strength coolant equal to 50% of the total system capacity.
4. Top off the radiator with 50/50 coolant and fill the recovery tank to about ½ full.
5. Run the engine with the radiator cap off until it is at normal operating temperature. Double-check the coolant level.
   ☐ OK ☐ Low

   Note: Sometimes the coolant level will drop after the thermostat opens. Many new cars require bleeding air out of the system after a refill and flush.
6. Replace the radiator cap.
7. Clean your work area, clean and return tools to their proper places, and wash your hands.
8. Record your recommendations for needed service or additional repairs and complete the repair order.

   ____

   ____

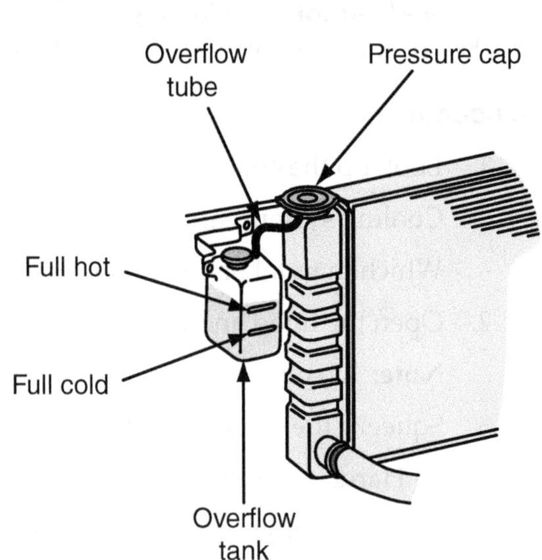

ASE Education Foundation Worksheet #6-13

REMOVE AND REPLACE A RADIATOR

Name_______________________ Class_________________

Score: ☐ Excellent ☐ Good ☐ Needs Improvement **Instructor OK** ☐

Vehicle year ________ **Make** ________________ **Model** ________________

Objective: Upon completion of this assignment, you will be able to remove and replace a radiator.

ASE Education Foundation Correlation

This worksheet addresses the following **AST/MAST** task:

I.D.6 Remove and replace radiator. **(P-2)**

We Support | Education Foundation

Directions: Before beginning this lab assignment, review the worksheet completely. Fill in the information in the spaces provided as you complete each task.

Tools and Equipment Required: Safety glasses, fender covers, shop towel, drain pan, coolant, radiator, radiator hoses, hose clamps

Procedure:

Note: Worksheet #2-4, Check and Correct Coolant Level, should be completed before attempting this worksheet.

1. Raise the hood and install fender covers on the vehicle.
2. Before removing the radiator cap, check the cooling system pressure by squeezing the upper radiator hose. Is it hard or soft when you squeeze it?

   ☐ Hose is hard.

   ☐ Hose is soft.
3. If the top radiator hose is soft, fold a shop towel and place it over the cap.
4. Hold down firmly on the cap and turn it counterclockwise ¼ turn until the cap is opened to the safety catch. Rotate the cap ¼ turn farther and remove it.

   Note: Some radiator caps are on the coolant reservoir and do not have a safety catch.

   ☐ Safety catch

   ☐ No safety catch

Let up slowly on the pressure you are exerting on the cap. If coolant escapes, press the cap back down. (On most cars, cap pressure will be no more than 17 psi.) If pressure is allowed to escape from a hot system, the coolant boiling point will be lowered and it may boil.

Radiator Removal

1. Place a *clean* drain pan under the radiator drain plug. Loosen the drain plug and drain the coolant into the pan. If there is no drain plug, remove the lower hose and aim it into the drain pan while trying to spill as little coolant as possible.

   ☐ Coolant spilled

   ☐ No coolant spilled

2. Remove both radiator hoses.

3. If the vehicle is equipped with an automatic transmission, disconnect the transmission cooler lines from the radiator. Be sure to use a flare-nut wrench on tubing fittings.

   ☐ Automatic transmission

   ☐ Manual transmission

Flare-nut wrench

Photo by Tim Gilles

4. Remove the fan shroud if necessary.

5. If applicable, disconnect the fan's electrical connector.

   ☐ Electric fan

   ☐ Belt-driven fan

6. Unbolt the radiator and remove it from the vehicle.

7. Transfer all the necessary parts from the old radiator to the new one.

Radiator Installation

1. Install the radiator in the vehicle and tighten the bolts.

2. Install the fan shroud and reconnect electrical connectors as necessary.

3. If the vehicle is equipped with an automatic transmission, connect the transmission cooler lines. Use a flare-nut wrench on tubing fittings.

4. Install the radiator hoses.

5. Refill the radiator with the correct mixture of coolant and water.

6. Install a pressure tester and check for leaks. Remove the tester and reinstall the radiator cap.

   ☐ No leaks

   ☐ Leaks

7. Start the engine and run it until it reaches normal operating temperature.

8. Recheck the coolant level in the reservoir and fill as necessary.

9. Before completing your paperwork, clean your work area, clean and return tools to their proper places, and wash your hands.

10. Record your recommendations for needed service or additional repairs and complete the repair order.

    ____

    ____

STOP

ASE Education Foundation Worksheet #6-14

REPLACE A THERMOSTAT

Name______________________ Class______________

Score: ☐ Excellent ☐ Good ☐ Needs Improvement **Instructor OK** ☐

Vehicle year ________ **Make** ______________ **Model** ______________

Objective: Upon completion of this assignment, you should be able to remove and replace a thermostat.

ASE Education Foundation Correlation

This worksheet addresses the following **MLR** task:

I.C.3 Remove, inspect, and replace thermostat and gasket/seal. **(P-1)**

This worksheet addresses the following **AST/MAST** task:

I.D.7 Remove, inspect, and replace thermostat and gasket/seal. **(P-1)**

We Support ASE Education Foundation

Directions: Before beginning this lab assignment, review the worksheet completely. Fill in the information in the spaces provided as you complete each task.

Tools and Equipment Required: Safety glasses, fender covers, drain pan, ratchet, sockets, wrenches, shop towel

Parts and Supplies: Thermostat, gasket, gasket sealer

Procedure:

1. Obtain the replacement thermostat and gasket before starting to work.

   Temperature rating: ☐ 180 ☐ 195 ☐ 200 ☐ Other (list): ______________

2. Open the hood and install fender covers on the fenders and front body parts.

CAUTION When removing the radiator cap, the engine must be off and the pressure released. The cooling system is under pressure when the system is at normal operating temperature. As the radiator cap is removed, the coolant may boil. Squeeze the radiator hose to check if the system is under pressure. If the hose is hard, do not remove the radiator cap. See your instructor.

3. Check to see if the cooling system is pressurized by squeezing the upper radiator hose.

   Is it under pressure? ☐ Yes ☐ No

4. Fold a shop rag to ¼ size. Use the rag to turn the radiator cap ¼ turn until its first stop. This will allow any remaining pressure to escape.
5. Check the cap to see if it is loose, indicating that all the pressure has escaped.
6. If the cap is loose, press down against the spring pressure and turn it to remove it.
7. Open the drain valve located in the lower radiator tank to allow some coolant to drain into a clean drain pan. Drain the coolant until its level is below the thermostat housing.

TURN

8. Disconnect the upper radiator hose from the water outlet housing.
9. Remove the bolts holding the water outlet housing.
10. Remove the water outlet housing and thermostat.
11. Thoroughly clean the gasket surfaces and the hose fitting.
12. Carefully inspect the thermostat housing for damage.

    ☐ OK ☐ Damaged
13. Compare the new thermostat to the old one.

    Is it the same temperature rating? ☐ Yes ☐ No
14. Install the thermostat. Which direction does the temperature sensing bulb face? Toward or away from the engine?

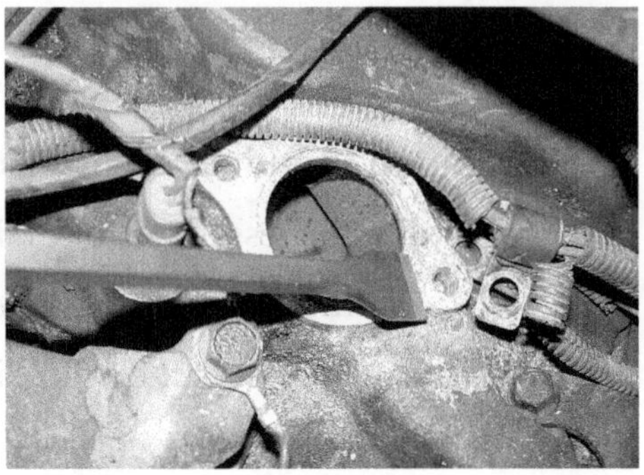

    ☐ Toward ☐ Away
15. Fit the thermostat into the recessed groove.

    The recess is in the:

    ☐ Water outlet housing ☐ Engine/manifold
16. Coat a new paper gasket with gasket sealer and install it.
17. Install the water outlet. Before tightening, attempt to rock the water outlet back and forth to be sure it is flush.

    ☐ Is the outlet flush? ☐ Yes ☐ No
18. Tighten the bolts evenly and carefully.

    ☐ Did the outlet crack? ☐ Yes ☐ No

    Note: The thermostat outlet housing can be broken during this step if the thermostat does not fit in the recess or if the bolts are not tightened evenly.
19. Connect the upper hose and refill the system.
20. Pressurize the system and check for leaks. Refer to Worksheet #6-3, Pressure Test a Cooling System.

    Are there any leaks? ☐ Yes ☐ No
21. Run the engine until it is warm to be sure that the thermostat opens and the system is operating properly.

    Did the thermostat open? ☐ Yes ☐ No

    Top off the coolant in the radiator and fill the coolant recovery tank as required.
22. Before completing your paperwork, clean your work area, clean and return tools to their proper places, and wash your hands.
23. Record your recommendations for needed service or additional repairs and complete the repair order

    ____

    ____

Photo by Tim Gilles

Photo by Tim Gilles

Photo by Tim Gilles

Photo by Tim Gilles

STOP

ASE Education Foundation Worksheet #6-15

TEST A THERMOSTAT

Name_______________________ Class_______________

Score: ☐ Excellent ☐ Good ☐ Needs Improvement **Instructor OK** ☐

Vehicle year ________ **Make** ______________ **Model** ________________

Objective: Upon completion of this assignment, you should be able to test a thermostat.

ASE Education Foundation Correlation

This worksheet addresses the following **MLR** task:

I.C.3 Remove, inspect, and replace thermostat and gasket/seal. **(P-1)**

This worksheet addresses the following **AST/MAST** task:

I.D.7 Remove, inspect, and replace thermostat and gasket/seal. **(P-1)**

We Support ASE Education Foundation

Directions: Before beginning this lab assignment, review the worksheet completely. Fill in the information in the spaces provided as you complete each task.

Tools and Equipment Required: Safety glasses, heat source, container, thermometer, feeler gauge, water

Procedure:

1. Temperature rating of the thermostat being tested: _______
2. Temperature rating of the thermostat required for the vehicle: _______
3. Locate the following items:
   - ☐ Heat source
   - ☐ Container
   - ☐ Thermometer
   - ☐ Feeler gauge
   - ☐ Water
4. Slightly open the thermostat. Slip the feeler gauge in the thermostat and let the thermostat close, holding the feeler gauge in place.

Check temperature when thermostat opens

Heat

Ford Motor Company

5. Hang the thermostat in the container of water by the feeler gauge and start heating the water.
6. Place a thermometer in the water.

7. Watch the thermostat when it starts to open. The thermostat will fall from the feeler gauge when it starts to open.
8. Note the temperature on the thermometer when the thermostat begins to open.

    ________________ Fahrenheit/Celsius.
9. Did the thermostat open at the specified temperature? ☐ Yes ☐ No
10. Did you have any problems completing this worksheet? ☐ Yes ☐ No

    If so, list them below.

    ____

    ____
11. Before completing your paperwork, clean your work area, clean and return tools to their proper places, and wash your hands.
12. Record your recommendations for needed service or additional repairs and complete the repair order.

    ____

    ____

STOP

ASE Education Foundation Worksheet #6-16

TEST A RADIATOR ELECTRIC FAN

Name ______________________ Class ________________

Score: ☐ Excellent ☐ Good ☐ Needs Improvement **Instructor OK** ☐

Vehicle year ________ **Make** ______________ **Model** ________________

Objective: Upon completion of this assignment, you should be able to test a radiator electric fan for proper operation.

ASE Education Foundation Correlation

This worksheet addresses the following **AST/MAST** task:

I.D.8 Inspect and test fan(s) (electrical or mechanical), fan clutch, fan shroud, and air dams. **(P-1)**

We Support ASE | Education Foundation

Directions: Before beginning this lab assignment, review the worksheet completely. Fill in the information in the spaces provided as you complete each task.

Tools and Equipment Required: Safety glasses

Procedure:

Check the operation of a radiator electric fan.

CAUTION Be careful when working around an electric radiator fan. It may come on at any time, whether the engine is running or not.

1. How many electric fans are there on the radiator? ______
2. Check for obvious problems:

   Electrical connectors: ☐ OK ☐ Problem

   Fan fuse in the fuse panel: ☐ OK ☐ Problem
3. If the engine is cold, use an ohmmeter to check the coolant temperature switch.

   Results: ☐ Continuity ☐ Infinite resistance

Engine
A/C condenser fan
Cooling fan

4. Run the engine until it reaches normal operating temperature. Turn the engine off and recheck the coolant temperature switch.

   Results: ☐ Continuity ☐ Infinite resistance
5. Coolant temperature switch condition: ☐ Good ☐ Bad

6. Before completing your paperwork, clean your work area, clean and return tools to their proper places, and wash your hands.
7. Record your recommendations for needed service or additional repairs and complete the repair order.

____

____

ASE Education Foundation Worksheet #6-17

REPLACE A HEATER HOSE

Name________________________________ Class________________________

Score: ☐ Excellent ☐ Good ☐ Needs Improvement **Instructor OK** ☐

Vehicle year __________ **Make** ____________________ **Model** ______________________

Objective: Upon completion of this assignment, you should be able to replace a heater hose.

ASE Education Foundation Correlation

This worksheet addresses the following **MLR** tasks:

I.C.4 Inspect and test coolant; drain and recover coolant; flush and refill cooling system; use proper fluid type per manufacturer specification; bleed air as required. **(P-1)**

VII.C.1 Inspect engine cooling and heater systems hoses and pipes; determine necessary action. **(P-1)**

This worksheet addresses the following **AST/MAST** tasks:

I.D.4 Inspect and test coolant; drain and recover coolant; flush and refill cooling system; use proper fluid type per manufacturer specification; bleed air as required. **(P-1)**

VII.C.1 Inspect engine cooling and heater systems hoses and pipes; perform needed action. **(P-1)**

We Support ASE Education Foundation

Directions: Before beginning this lab assignment, review the worksheet completely. Fill in the information in the spaces provided as you complete each task.

Tools and Equipment Required: Safety glasses, fender covers, drain pan, shop towel

Parts and Supplies: Heater hose, hose clamps

Procedure:

1. Open the hood and install fender covers over the fenders and front body parts.

CAUTION When removing the radiator cap, the engine must be off and the pressure released. The cooling system is under pressure when the system is at normal operating temperature. As the radiator cap is removed, the coolant may boil. Squeeze the radiator hose to check if the system is under pressure. If the hose is hard, do not remove the radiator cap. See your instructor.

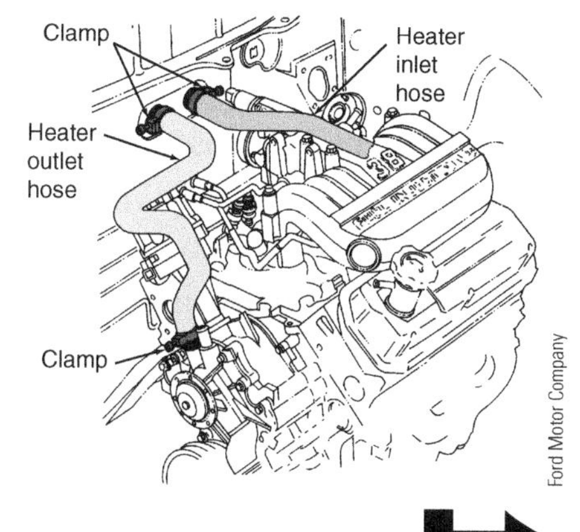

2. Squeeze the top hose. Results: ☐ Hard ☐ Soft

3. Fold a shop towel to ¼ size. Use it to turn the radiator cap ¼ turn until its first stop. This will allow any remaining pressure to escape.

TURN

4. Check the cap to see that it is loose, indicating that all the pressure has escaped.

   ☐ Loose ☐ Tight

5. If the cap is loose, press down against the spring pressure and turn the cap to remove it.
6. Open the radiator drain valve located in the lower radiator tank and drain some coolant into a clean container. Drain the coolant until the coolant level is below the level of the heater hose.

ENVIRONMENTAL NOTE Before reusing coolant, check its concentration, appearance, and age. Remember, used coolant must be disposed of properly.

7. Remove the hose clamps from the heater hose.

CAUTION The heater core inlet or outlet is easily damaged or deformed from rough handling of the hoses.

8. Twist the hose to loosen it. It may be necessary to cut the hose with a sharp knife if it does not come off easily.
9. The diameter of the hose is determined by its:

   ☐ Outside diameter (O.D.) ☐ Inside diameter (I.D.)

10. What is the diameter of the heater hose? _______
11. Cut a piece of replacement hose slightly longer than the old one. It can be trimmed later if necessary.

    How long is the new replacement hose? _______

    Note: Many vehicles use molded heater hoses. Match the new hose to the old one before installation.

12. Clean the hose connection prior to installing the new hose.

    Note: If the hose is difficult to install, apply some soap to the connection.

CAUTION Be certain that the hose clamps do not interfere with manifolds, belts, or spark plug wiring, and that they will not be damaged by movement of the engine or its accessories.

13. Replace rusted or damaged hose clamps with new clamps and install them on the hose. Position the screw side of the clamp for easy access.

    What size are the replacement hose clamps? _______"

14. Position and tighten the clamps.
15. Refill the system and check for leaks.

    Note: After the cooling system has fully warmed up and then cooled again, the hose clamps should be retightened. Hoses may shrink after their first use.

Screwdriver slot
Hex head
Bead
Hose
End of fitting
Clamp
Reprinted with permission from Gates Corporation

16. Before completing your paperwork, clean your work area, clean and return tools to their proper places, and wash your hands.
17. Record your recommendations for needed service or additional repairs and complete the repair order.

    ____

    ____

STOP

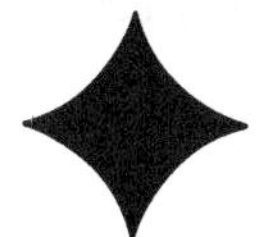

ASE Education Foundation Worksheet #6-18

CHECK AIR-CONDITIONING SYSTEM PERFORMANCE

Name________________________ Class__________________

Score: ☐ Excellent ☐ Good ☐ Needs Improvement **Instructor OK** ☐

Vehicle year ________ **Make** ________________ **Model** __________________

Objective: Upon completion of this assignment, you should be able to check the performance of the air-conditioning system.

ASE Education Foundation Correlation

This worksheet addresses the following **MLR** task:

VII.A.2 Identify heating, ventilation and air conditioning (HVAC) components and configuration. **(P-1)**

This worksheet addresses the following **AST/MAST** tasks:

VII.A.1 Identify and interpret heating and air conditioning problems; determine needed action. **(P-1)**

VII.A.3 Performance test A/C system; identify problems. **(P-1)**

We Support ASE Education Foundation

Directions: Before beginning this lab assignment, review the worksheet completely. Fill in the information in the spaces provided as you complete each task.

Tools and Equipment Required: Safety glasses, fender covers, shop towel, thermometer

Procedure:

1. Open the hood and place fender covers on the fenders and front body panels.
2. Is there an air conditioning underhood label? ☐ Yes ☐ No
3. Identify the refrigerant type.
   - ☐ R-12
   - ☐ R-134A
   - ☐ R-1234yf
   - ☐ Other (list): ______
4. How is the air conditioning compressor driven?
   - ☐ Belt driven
   - ☐ Electrically powered
   - ☐ Both electric and belt driven

CAUTION When working on or around electric air conditioning compressors be aware that they are likely to be powered by high voltage. High voltage circuits can be identified by orange wires, connectors, or labels, either on or near the compressor.

TURN

5. Verify the compressor engagement by turning the A/C on and listening for the compressor clutch engagement.

   ☐ Compressor clutch engages.

   ☐ Electric radiator fans start.

6. Check the A/C condenser for blockage and remove as needed. What did you find?

   ☐ Nothing ☐ Leaves

   ☐ Dirt ☐ Bugs

7. Check for wetness around and under the A/C lines and hoses.

   ☐ Dry

   ☐ Moist

8. Check the cowl grille for leaves and other debris.

   ☐ Clear ☐ Required cleaning

9. Check the evaporator housing drain hose.

   ☐ Attached ☐ Clear ☐ Blocked

10. Before doing the temperature tests, complete the following checklist:

    ☐ Set the temperature control to maximum.

    ☐ Adjust the blower to its highest speed.

    ☐ Allow the vehicle's inside temperature to stabilize.

    ☐ Run the engine at 1,500 rpm.

    ☐ Check that the compressor is engaged.

11. Run the A/C system for 5–10 minutes and then feel the temperature on both the high- and low-pressure sides of the system.

    High side temperature: ☐ Hot ☐ Cool ☐ Normal

    Low side temperature: ☐ Hot ☐ Cool ☐ Normal

12. Check the A/C outlet duct temperature by placing a thermometer in the center A/C outlet.

    List the outlet duct temperature below:

    Outlet duct temperature: ______ °F

    Is this temperature: ☐ High ☐ Low ☐ Normal

13. If the A/C system is electrically powered, inspect the wiring and connectors for damage.

    ☐ OK ☐ Needs attention ☐ N/A

14. Before completing your paperwork, clean your work area, clean and return tools to their proper places, and wash your hands.

15. Record your recommendations for needed service or additional repairs and complete the repair order.

    ____

    ____

STOP

ASE Education Foundation Worksheet #6-19

REPLACE A CABIN AIR FILTER

Name_______________________ Class_______________

Score: ☐ Excellent ☐ Good ☐ Needs Improvement **Instructor OK** ☐

Vehicle year ________ **Make** ______________ **Model** ______________

Objective: Upon completion of this assignment, you will be able to replace a cabin air filter.

ASE Education Foundation Correlation

This worksheet addresses the following **MLR** task:

VII.D.1 Inspect A/C-heater ducts, doors, hoses, cabin filters, and outlets; perform necessary action. **(P-1)**

This worksheet addresses the following **AST/MAST** task:

VII.D.6 Inspect A/C-heater ducts, doors, hoses, cabin filters, and outlets; perform needed action. **(P-1)**

We Support
ASE Education Foundation

Directions: Before beginning this lab assignment, review the worksheet completely. Fill in the information in the spaces provided as you complete each task.

Tools and equipment: Safety glasses, fender covers, hand tools, and shop towels.

Procedure:

Glove compartment

1. Use the service information to identify the location of the cabin air filter and instructions for its replacement.

   Which service information did you use?

   ____

2. The cabin air filter is located:

   ☐ Under the hood ☐ Behind the glove box

   ☐ Other (list): ______________________

3. Remove the cabin air filter.

4. Identify the condition of the cabin air filter below:

   ☐ Clean

   ☐ Needs cleaning

   ☐ Needs to be replaced

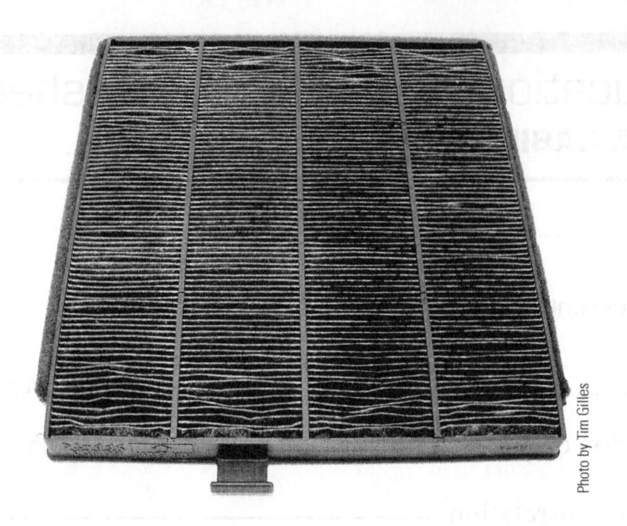

5. Replace the cabin air filter if necessary.

   ☐ Yes ☐ No

6. Reinstall the cabin air filter.
7. Inspect the air inlet for the cabin filter. It is usually located in the cowl below the windshield.

   ☐ Clean ☐ Needs cleaning

8. Before completing the paperwork, clean your work area, put the tools in their proper places, and wash your hands.
9. List your recommendations for future service and/or additional repairs and complete the repair order.

   ____

   ____

STOP

ASE Education Foundation Worksheet #6-20

METAL TUBING SERVICE

Name_______________________ Class_______________

Score: ☐ Excellent ☐ Good ☐ Needs Improvement **Instructor OK** ☐

Vehicle year ________ **Make** ______________ **Model** ______________

Objective: Upon completion of this assignment, you should be able to fabricate a replacement metal tube with double flared fittings.

ASE Education Foundation Correlation

This worksheet addresses the following **AST/MAST** tasks:

V.B.7 Replace brake lines, hoses, fittings, and supports. **(P-2)**

V.B.8 Fabricate brake lines using proper material and flaring procedures (double flare and ISO types). **(P-2)**

We Support
ASE | Education Foundation

Directions: Before beginning this lab assignment, review the worksheet completely. Fill in the information in the spaces provided as you complete each task.

Tools and Equipment Required: Safety glasses, flaring tool set, tubing cutter, tube bender, shop towel

Parts and Supplies: Six inches of copper or steel tubing

Procedure:

1. Locate the following:
   - ☐ Six inches of tubing
   - ☐ Tubing cutter
   - ☐ Flaring tool
   - ☐ Tube bender

2. What is the diameter of the tubing? _______"

   Note: The size of tubing is determined by its outside diameter. The size of hose is determined by the inside diameter.

Step one

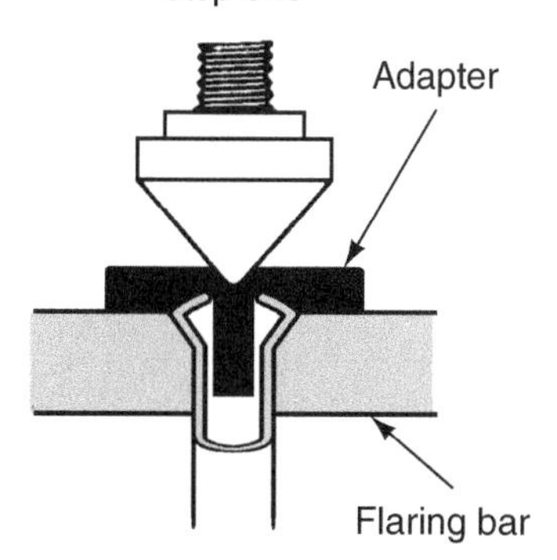

3. Use the tubing cutter to cut off 1" from the tubing and discard it.
4. Deburr the end of the tubing using the deburring tool.
5. Clamp the tubing into the flaring tool bar. It should protrude above the bar by the width of the flaring tool adapter.
6. Insert the adapter into the end of the tubing and tighten down on it with the threaded flaring tool until it bottoms out.

7. Remove the adapter and tighten the flaring tool against the tubing again to complete the flare.

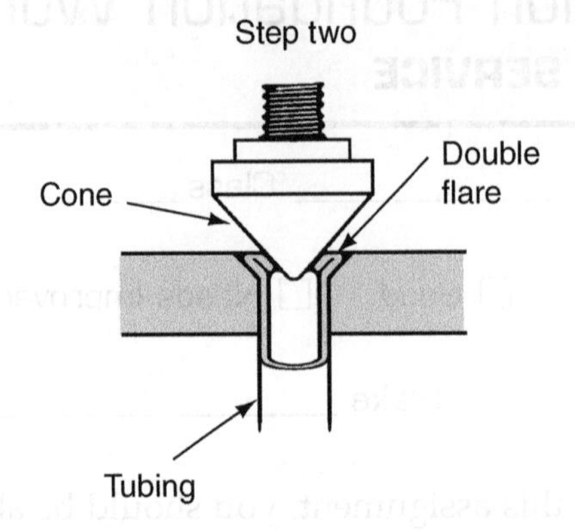

8. Use the tubing bender to bend the middle of the tubing to a 90-degree angle.

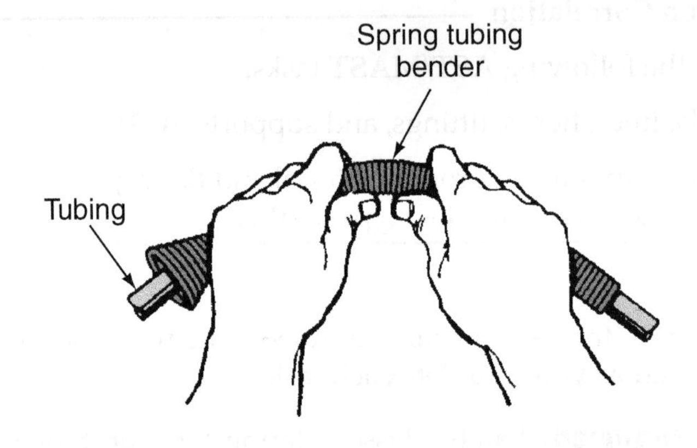

9. Did you have any problems completing this worksheet? ☐ Yes ☐ No

Explain ____

____

____

STOP

ASE Education Foundation Worksheet #6-21

FUEL FILTER SERVICE (FUEL INJECTION)

Name_______________________________ Class_______________________

Score: ☐ Excellent ☐ Good ☐ Needs Improvement **Instructor OK** ☐

Vehicle year __________ **Make** ____________________ **Model** ______________________

Objective: Upon completion of this assignment, you should be able to replace a fuel filter on a fuel-injected vehicle.

ASE Education Foundation Correlation

This worksheet addresses the following **MLR** task:

VIII.C.1 Replace fuel filter(s) where applicable. **(P-2)**

This worksheet addresses the following **AST** task:

VIII.D.3 Replace fuel filter(s) where applicable. **(P-2)**

This worksheet addresses the following **MAST** task:

VIII.D.4 Replace fuel filter(s) where applicable. **(P-2)**

We Support ASE | Education Foundation

Directions: Before beginning this lab assignment, review the worksheet completely. Fill in the information in the spaces provided as you complete each task.

Tools and Equipment Required: Safety glasses, fender covers, shop towel, screwdriver, drain pan, tape measure, hose cutter, hand tools, spring clock clamp tool, flare-nut wrenches

Parts and Supplies: In-line fuel filter

Procedure:

1. Open the hood and place fender covers on the fenders and front body parts.
2. Locate and inspect all synthetic rubber hoses in the fuel system.

   ☐ OK ☐ Need replacement

   Note: Fuel injection systems can use one of several methods to attach the fuel lines to the fuel filter. Special hose clamps and banjo fittings are common. Check the service information for the correct procedure for removing the special clamps. Use flare-nut wrenches when removing the banjo bolts.

   Photo by Tim Gilles

   How are the fuel lines attached to the fuel filter on the vehicle that you are servicing?

   ☐ Flare fitting ☐ Banjo bolts ☐ Hose clamps ☐ Spring lock ☐ Other (list): ______

3. How many synthetic rubber fuel hoses are there on the vehicle? ______

SAFETY NOTE Any fuel hose that appears to be old or damaged should be replaced immediately after getting the customer's approval. Always start the engine and check for leaks after replacing any fuel lines.

TURN

4. Obtain the correct filter.
5. Relieve any pressure from the fuel tank by removing the fuel filler cap.

   Fuel filler cap removed? ☐ Yes ☐ No
6. Position a drain pan under the filter to catch any leaking fuel.
7. Most fuel injection systems maintain some residual pressure in the fuel system. It is recommended that the pressure be relieved before the fuel hoses are removed. There are several ways that the fuel pressure can be bled from the system. Check the service manual for the recommended procedure. Indicate which method below is recommended for the vehicle you are servicing:

   ☐ Remove the fuel pump fuse and crank the engine for a few seconds.

   ☐ Use a jumper wire to bypass the fuel pump relay.

   ☐ Some fuel injection systems have a *Schrader valve* that can be used to bleed off pressure from the system before disassembly. (A Schrader valve is the kind of valve that is found on tire valve stems.)

   ☐ Use jumper wires to energize an injector.

   ☐ Other (describe): ____

   ____
8. If there is a synthetic rubber hose on the fuel tank side of the filter, apply pinch pliers to the hose. This will minimize leakage of fuel.

   Is there a synthetic rubber hose?

   ☐ Yes ☐ No

   Were pinch pliers applied to the hose?

   ☐ Yes ☐ No
9. Slowly loosen the clamp, fitting, or banjo bolt and hose on the injector end of the filter.
10. Does the filter have an arrow that indicates the correct direction of installation?

    ☐ Arrow ☐ No arrow
11. Install the filter on the injector end of the fuel line first. 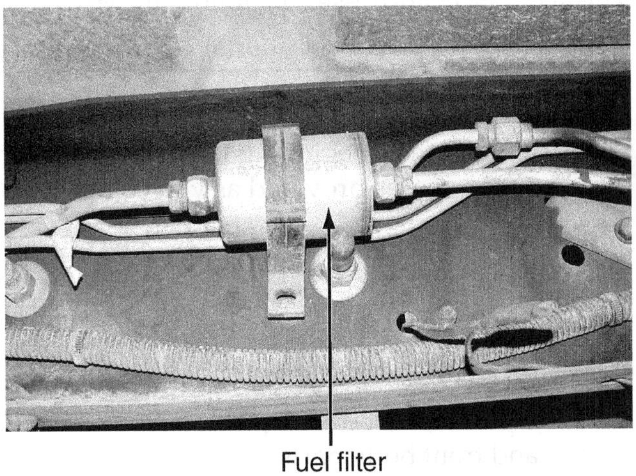
12. Loosen and remove the old fuel filter. Then *quickly* install the new filter.

Fuel filter

13. Tighten the hose clamps, fittings, or banjo bolts and clean up any fuel that leaked while you were changing the fuel filter.
14. Replace the fuel filler cap and run the engine to check for leaks. ☐ Leaks ☐ No leaks
15. Before completing your paperwork, clean your work area, clean and return tools to their proper places, and wash your hands.
16. Record your recommendations for needed service or additional repairs and complete the repair order.

    ____

    ____

STOP

ASE Education Foundation Worksheet #6-22

CHECK FUEL PRESSURE—FUEL INJECTION

Name________________________ Class________________

Score: ☐ Excellent ☐ Good ☐ Needs Improvement **Instructor OK** ☐

Vehicle year ________ **Make** ______________ **Model** ______________

Objective: Upon completion of this assignment, you should be able to check the fuel pressure in a fuel injection system.

ASE Education Foundation Correlation

This worksheet addresses the following **AST** task:

VIII.D.2 Inspect and test fuel pump(s) and pump control system for pressure, regulation, and volume; perform needed action. **(P-1)**

This worksheet addresses the following **MAST** task:

VIII.D.3 Inspect and test fuel pump(s) and pump control system for pressure, regulation, and volume; perform needed action. **(P-1)**

We Support ASE Education Foundation

Directions: Before beginning this lab assignment, review the worksheet completely. Fill in the information in the spaces provided as you complete each task.

Tools and Equipment Required: Safety glasses, fender covers, shop towel

Procedure:

1. Open the hood and place fender covers on the fenders and front body parts.
2. Locate the appropriate place to attach the fuel pressure tester. Indicate which you choose.
   - ☐ Fuel line pressure port (Schrader valve)
   - ☐ Fuel line (in series)
   - ☐ Cold start injector

Most fuel injection systems maintain some residual pressure in the fuel system. It is recommended that the pressure be relieved before the fuel system is opened to attach the fuel pressure gauge.

3. There are several ways that the fuel pressure can be bled from the system. Check the service information for the recommended procedure. Indicate which method below is recommended for the vehicle that you are servicing.

   ☐ Remove the fuel pump fuse and crank the engine for a few seconds.

   ☐ Use a jumper wire to bypass the fuel pump relay.

   ☐ Bleed the Schrader valve.

   Note: Cover the valve with a clean rag to prevent fuel from spraying on you or the engine.

   ☐ Use jumper wires to energize an injector.

   ☐ Other (describe): ______________________________

   ____

4. Connect the fuel pressure gauge to the fuel injection system.

   Note: The pressure gauge must be able to read at least 100 psi.

5. Start the engine and let it run at idle. If the engine will not run, check the service manual for the proper way to energize the fuel pump.

6. Read the pressure on the fuel gauge.

   What did it read? ______

   What is the fuel pressure specification? ______

7. Was the pressure: ☐ High ☐ Low ☐ OK

8. What could cause high fuel pressure? ______________________________

9. What could cause the fuel pressure to be too low? ______________________________

   ____

10. Remove the fuel gauge and replace the fuel line or valve cap.

11. Start the engine and check for fuel leaks. Is there any fuel leaking? ☐ Yes ☐ No

12. Before completing your paperwork, clean your work area, clean and return tools to their proper places, and wash your hands.

13. Record your recommendations for needed service or additional repairs and complete the repair order.

    ____

    ____

STOP

ASE Education Foundation Worksheet #6-23

PCV VALVE INSPECTION AND REPLACEMENT

Name_______________________ Class_______________

Score: ☐ Excellent ☐ Good ☐ Needs Improvement **Instructor OK** ☐

Vehicle year ________ **Make** ______________ **Model** ______________

Objective: Upon completion of this assignment, you should be able to inspect and service the positive crankcase ventilation (PCV) system.

ASE Education Foundation Correlation

This worksheet addresses the following **MLR** task:

VIII.D.1 Inspect, test, and service positive crankcase ventilation (PCV) filter/breather cap, valve, tubes, orifices, and hoses; perform necessary action. **(P-2)**

This worksheet addresses the following **AST/MAST** task:

VIII.E.2 Inspect, test, and service positive crankcase ventilation (PCV) filter/breather cap, valve, tubes, orifices, and hoses; perform needed action. **(P-2)**

We Support ASE Education Foundation

Directions: Before beginning this lab assignment, review the worksheet completely. Fill in the information in the spaces provided as you complete each task.

Tools and Equipment Required: Safety glasses, fender covers, shop towel

Procedure:

1. Open the hood and place fender covers on the fenders and front body parts.
2. Locate the PCV system air filter. Where is it located?

   ☐ Air filter housing ☐ Oil filler cap ☐ Other (describe): ______
3. Condition of crankcase vent filter:

   ☐ OK ☐ Needs service

   Note: Oil in the air cleaner can indicate a plugged PCV system or excessive blowby. Further testing of the PCV system or engine will be required.
4. Locate the PCV valve. Where is it located? ______
5. Which of the following engine gaskets show signs of leakage?

   ☐ Valve cover ☐ Rear crankshaft seal

   ☐ Oil pan ☐ Camshaft seal

   ☐ Timing cover ☐ Other (describe): ______

   ☐ Front crankshaft seal

TURN

6. Inspect the condition of the PCV hoses.

   ☐ Good ☐ Cracked

   ☐ Deteriorated ☐ Loose connections

7. Disconnect the PCV valve from the engine.

8. Shake the valve. Does it rattle?
   ☐ Yes ☐ No

Photo by Tim Gilles

Rpm Drop Test

9. Be certain the vehicle is in park (automatic transmission)/neutral (manual transmission) with the parking brake firmly set.

   ☐ Park ☐ Neutral ☐ Parking brake set

10. Start the engine and let it idle.

11. Cover the end of the PCV valve with your thumb. Does the engine idle change? On older engines, speed should drop.

    ☐ Yes ☐ No

12. Did you feel a strong vacuum when the valve was restricted? ☐ Yes ☐ No

13. Reinstall the PCV valve. How is it attached to the engine?

    ☐ Rubber grommet

    ☐ Threaded connection

    ☐ Other (describe): ______

Photo by Tim Gilles

Vacuum Test (Pushrod Engine only)

14. Remove the oil filler cap.

15. Position a piece of paper or a dollar bill over the oil filler opening. If the system is operating properly, engine vacuum will pull the paper against the opening.

    Note: This test does not work on overhead cam engines with the oil filler cap in the valve cover. The rotation of the camshaft creates positive pressure in the immediate area.

    If the system fails to operate properly, replace the valve.

    ☐ OK ☐ Needs valve replaced

16. Before completing your paperwork, clean your work area, clean and return tools to their proper places, and wash your hands.

17. Record your recommendations for needed service or additional repairs and complete the repair order.

    ____

    ____

STOP

ASE Education Foundation Worksheet #6-24

OXYGEN SENSOR TEST (ZIRCONIUM-TYPE SENSOR)

Name_______________________ Class_________________

Score: ☐ Excellent ☐ Good ☐ Needs Improvement **Instructor OK** ☐

Vehicle year _________ **Make** _______________ **Model** _______________

Objective: Upon completion of this assignment, you should be able to inspect and test an oxygen sensor(s).

ASE Education Foundation Correlation

This worksheet addresses the following **MAST** task:

VIII.B.7 Inspect and test computerized engine control system sensors, powertrain/engine control module (PCM/ECM), actuators, and circuits using a graphing multimeter (GMM)/digital storage oscilloscope (DSO); perform necessary action. **(P-2)**

We Support ASE Education Foundation

Directions: Before beginning this lab assignment, review the worksheet completely. Fill in the information in the spaces provided as you complete each task.

Tools and Equipment Required: Safety glasses, digital multimeter, fender covers, shop towel

Note: A *high impedance voltmeter* must be used to perform the following tests. Most *digital* meters are high impedance.

Procedure:

1. Open the hood and install fender covers over the fenders.
2. Locate the oxygen sensor(s).

   How many oxygen sensors are used on the vehicle? ______
3. How many wires does each of the sensors have?

   ☐ 1 wire ☐ 2 wires

   ☐ 3 wires ☐ 4 wires

   Note: Oxygen sensors have many variations in the number of wires they use. Single-wire systems use the wire for signal voltage. In multiwire systems, one of the wires provides power to heat the sensor, one or two provide the ground paths, and the other wire provides the computer signal.

TURN

4. Connect the voltmeter ground lead to ground and the positive lead to the signal wire coming from the sensor.

   Note: If there is more than one wire, checking with a voltmeter will determine the signal lead. The engine must be running at normal operating temperature to make this test.

5. Start the engine and run it at fast idle (2,000 rpm) for 2 minutes to heat the exhaust system.

   Hint: Feel the top radiator hose to see if the engine is getting hot. Also, vehicles with smog pumps often have switching for the pump that changes when the fuel system goes into closed loop. This can be heard or felt through the hose from the smog pump.

6. When the oxygen sensor is hot enough, it will begin to operate.

   Note the voltage readings.

   Does voltage move constantly between 0.2 and 0.8 of a volt? ☐ Yes ☐ No

   Does the voltage hold steady? ☐ Yes ☐ No

   Note: A good oxygen sensor will vary between low and high voltage at least 10 times a second at 2,000 rpm. This is too fast to measure with a voltmeter, but a scan tool or oscilloscope can measure this "crosscount" speed.

Test Results:

☐ Good oxygen sensor, operating correctly

☐ Requires more testing, the system is not operating correctly

7. Before completing your paperwork, clean your work area, clean and return tools to their proper places, and wash your hands.

8. Record your recommendations for needed service or additional repairs and complete the repair order.

   ____

   ____

ASE Education Foundation Worksheet #6-25

IDENTIFY AND INSPECT EMISSION CONTROL SYSTEMS

Name_______________________ Class_______________

Score: ☐ Excellent ☐ Good ☐ Needs Improvement **Instructor OK** ☐

Vehicle year ________ **Make** ______________ **Model** ______________

Objective: Upon completion of this assignment, you should be able to identify emission control systems and inspect for their presence on a vehicle.

ASE Education Foundation Correlation

This worksheet addresses the following **MLR** task:

VIII.D.1 Inspect, test, and service positive crankcase ventilation (PCV) filter/breather, valve, tubes, orifices, and hoses; perform necessary action. **(P-2)**

This worksheet addresses the following **AST** tasks:

VIII.E.2 Inspect, test, service and/or replace positive crankcase ventilation (PCV) filter/breather, valve, tubes, orifices, and hoses; determine needed action. **(P-2)**

VIII.E.3 Diagnose emissions and driveability concerns caused by the exhaust gas recirculation (EGR) system; inspect, test, service and/or replace electrical/electronic sensors, controls, wiring, tubing, exhaust passages, vacuum/pressure controls, filters, and hoses of exhaust gas recirculation (EGR) system; determine needed action. **(P-3)**

VIII.E.4 Inspect and test electrical/electronically-operated components and circuits of secondary air injection systems; determine needed action. **(P-3)**

VIII.E.5 Diagnose emissions and driveability concerns caused by the catalytic converter system; determine needed action. **(P-3)**

VIII.E.6 Inspect and test components and hoses of the evaporative emissions control (EVAP) system; determine needed action. **(P-1)**

This worksheet addresses the following **MAST** tasks:

VIII.E.2 Inspect, test, service, and/or replace positive crankcase ventilation (PCV) filter/breather, valve, tubes, orifices, and hoses; perform needed action. **(P-2)**

VIII.E.3 Diagnose emissions and driveability concerns caused by the exhaust gas recirculation (EGR) system; inspect, test, service and/or replace electrical/electronic sensors, controls, wiring, tubing, exhaust passages, vacuum/pressure controls, filters, and hoses of exhaust gas recirculation (EGR) systems; determine needed action. **(P-2)**

VIII.E.4 Diagnose emissions and driveability concerns caused by the secondary air injection system; inspect, test, repair, and/or replace electrical/electronically-operated components and circuits of secondary air injection systems; determine needed action. **(P-2)**

VIII.E.5 Diagnose emissions and driveablility concerns caused by the evaporative emissions control (EVAP) system; determine needed action. **(P-1)**

VIII.E.6 Diagnose emissions and driveability concerns caused by catalytic converter system; determine needed action. **(P-2)**

We Support ASE | Education Foundation

Directions: Before beginning this lab assignment, review the worksheet completely. Fill in the information in the spaces provided as you complete each task.

Tools and Equipment Required: Safety glasses, fender covers

Procedure:

1. Open the hood and place fender covers on the fenders and front body parts.
2. Locate the vehicle's underhood emission label.
3. Record the information from the underhood emission label below:

   Note: Not all vehicles require all of the adjustments requested. If requested information is not on the underhood label, write N/A (not applicable) in the answer space.

   Engine size ____________________

   Timing specifications ____________________

   Special ignition timing instructions ____

   ____

   Fast idle speed ____________________

   Curb idle speed ____________________

   Valve lash: Intake ______ Exhaust ______

   Spark plug gap ______

   Other ____________________

4. Read the underhood emission label to see which emission control systems are required on the vehicle. Check them off on the following list:

   Note: Not all emission control systems are identified on the underhood label. It may be necessary to use a service manual to identify which systems are required on the vehicle you are inspecting.

   Positive Crankcase Ventilation (PCV) ☐ Yes ☐ No

   Thermostatic Air Cleaner (TAC) ☐ Yes ☐ No

   Fuel Evaporation System (EVAP) ☐ Yes ☐ No

   Catalytic Converter (CAT) ☐ Yes ☐ No

   Exhaust Gas Recirculation System (EGR) ☐ Yes ☐ No

   Air Injection (AIR) ☐ Yes ☐ No

   Early Fuel Evaporation (EFE) ☐ Yes ☐ No

5. Does the vehicle have an underhood vacuum routing label? ☐ Yes ☐ No

6. Without removing the air cleaner, check the routing of the vacuum hoses. Do they match the routing on the underhood label?

   ☐ Yes ☐ No

7. Locate and inspect each of the emission control systems. Indicate below the condition of the system. Use the following terms to describe the condition of the systems: pass, missing, disconnected, defective, or not used.

   Positive Crankcase Ventilation (PCV) ______

   Fuel Evaporation System (EVAP) ______

   Catalytic Converter (CAT) ______

   Exhaust Gas Recirculation System (EGR) ______

   Air Injection (AIR) ______

8. Before completing your paperwork, clean your work area, clean and return tools to their proper places, and wash your hands.

9. Record your recommendations for needed service or additional repairs and complete the repair order.

   ____

   ____

STOP

ASE Education Foundation Worksheet #6-26

CHECK EXHAUST EMISSIONS

Name_______________________________ Class_______________________

Score: ☐ Excellent ☐ Good ☐ Needs Improvement **Instructor OK** ☐

Vehicle year __________ **Make** ____________________ **Model** ______________________

Objective: Upon completion of this assignment, you will be able to check exhaust emissions.

ASE Education Foundation Correlation

This worksheet addresses the following **MLR** task:

VIII.D.1 Inspect, test, and service positive crankcase ventilation (PCV) filter/breather, valve, tubes, orifices, and hoses; perform necessary action. **(P-2)**

This worksheet addresses the following **AST** tasks:

VIII.E.2 Inspect, test, service and/or replace positive crankcase ventilation (PCV) filter/breather, valve, tubes, orifices, and hoses; determine needed action. **(P-2)**

VIII.E.3 Diagnose emissions and driveability concerns caused by the exhaust gas recirculation (EGR) system; inspect, test, service and/or replace electrical/electronic sensors, controls, wiring, tubing, exhaust passages, vacuum/pressure controls, filters, and hoses of exhaust gas recirculation (EGR) system; determine needed action. **(P-3)**

VIII.E.4 Inspect and test electrical/electronically-operated components and circuits of secondary air injection systems; determine needed action. **(P-3)**

VIII.E.5 Diagnose emissions and driveability concerns caused by the catalytic converter system; determine needed action. **(P-3)**

VIII.E.6 Inspect and test components and hoses of the evaporative emissions control (EVAP) system; determine needed action. **(P-1)**

This worksheet addresses the following **MAST** tasks:

VIII.E.2 Inspect, test, service, and/or replace positive crankcase ventilation (PCV) filter/breather, valve, tubes, orifices, and hoses; perform needed action. **(P-2)**

VIII.E.3 Diagnose emissions and driveability concerns caused by the exhaust gas recirculation (EGR) system; inspect, test, service and/or replace electrical/electronic sensors, controls, wiring, tubing, exhaust passages, vacuum/pressure controls, filters, and hoses of exhaust gas recirculation (EGR) systems; determine needed action. **(P-2)**

VIII.E.4 Diagnose emissions and driveability concerns caused by the secondary air injection system; inspect, test, repair, and/or replace electrical/electronically-operated components and circuits of secondary air injection systems; determine needed action. **(P-2)**

VIII.E.5 Diagnose emissions and driveablility concerns caused by the evaporative emissions control (EVAP) system; determine needed action. **(P-1)**

VIII.E.6 Diagnose emissions and driveability concerns caused by catalytic converter system; determine needed action. **(P-2)**

We Support

Education Foundation

TURN

Directions: Before beginning this lab assignment, review the worksheet completely. Fill in the information in the spaces provided as you complete each task.

Tools and Equipment Required: Safety glasses, shop towels, 4 or 5 gas analyzer

Procedure:

1. Open the hood and place fender covers on the fenders and front body parts.
2. Inspect the emission control systems and indicate if the system passes inspection or not.

Positive Crankcase Ventilation (PCV)	☐ Yes	☐ No	☐ N/A
Thermostatic Air Cleaner (TAC)	☐ Yes	☐ No	☐ N/A
Fuel Evaporation System (EVAP)	☐ Yes	☐ No	☐ N/A
Catalytic Converter (CAT)	☐ Yes	☐ No	☐ N/A
Exhaust Gas Recirculation (EGR)	☐ Yes	☐ No	☐ N/A
Air Injection (AIR)	☐ Yes	☐ No	☐ N/A
Early Fuel Evaporation (EFE)	☐ Yes	☐ No	☐ N/A

3. Start the vehicle and run the engine until it reaches operating temperature.
4. Enter any necessary information into the analyzer.
5. Insert the analyzer probe into the vehicle's exhaust pipe.
6. Record the analyzer readings below:

   Hydrocarbons (HC) ________ parts per million (ppm)

   Carbon Monoxide (CO) ________ %

   Carbon Dioxide (CO_2) ________ %

   Oxygen (O_2) ________ %

   Oxides of Nitrogen (NO_x) ________ ppm

7. Remove the analyzer probe from the exhaust pipe.
8. What are the limits for each of the exhaust gases?

   Hydrocarbons (HC) ________ ppm

   Carbon Monoxide (CO) ________ %

   Carbon Dioxide (CO_2) ________ %

   Oxygen (O_2) ________ %

   Oxides of Nitrogen (NO_x) ________ ppm

9. Are the emission readings within the required limits? ☐ Yes ☐ No

   If the readings are outside of limits, what repair(s) might correct the problem?

   ____

10. Before completing your paperwork, clean your work area, clean and return tools to their proper places, and wash your hands.
11. Record your recommendations for needed service or additional repairs and complete the repair order.

____

____

STOP

Part II

Lab Worksheets for ASE Maintenance and Light Repair

Service Area 7

Electrical Services

ASE Education Foundation Worksheet #7-1

BLADE FUSE TESTING AND SERVICE

Name______________________________ Class______________________

Score: ☐ Excellent ☐ Good ☐ Needs Improvement **Instructor OK** ☐

Vehicle year __________ **Make** ____________________ **Model** ______________________

Objective: Upon completion of this assignment, you will be able to determine if a blade fuse is faulty and be able to replace it.

ASE Education Foundation Correlation

This worksheet addresses the following **MLR** tasks:

VI.A.6 Check operation of electrical circuits with a test light. **(P-2)**

VI.A.9 Inspect and test fusible links, circuit breakers, and fuses; determine necessary action. **(P-1)**

This worksheet addresses the following **AST/MAST** tasks:

VI.A.5 Demonstrate proper use of a test light on an electrical circuit. **(P-1)**

VI.A.9 Inspect and test fusible links, circuit breakers, and fuses; determine needed action. **(P-1)**

We Support ASE | Education Foundation

Directions: Before beginning this lab assignment, review the worksheet completely. Fill in the information in the spaces provided as you complete each task.

Tools and Equipment Required: Safety glasses, fender covers, Voltmeter shop towel

Parts and Supplies: Blade fuses

Procedure:

1. Open the hood and place fender covers on the fenders and front body parts.
2. Locate the fuse panels (usually under the hood or under the instrument panel).

   Where are they located? ____________________
3. Are there any other components in the fuse panel?

   ☐ Relays ☐ Diodes ☐ Short pins ☐ Circuit breakers ☐ Other __________

   Note: The preferred method of testing an electrical circuit is to use a voltmeter instead of a test light. Test lights can damage electronic circuits.
4. Use a voltmeter to check battery voltage by connecting the meter leads to the terminals of the battery.

   Does the meter show battery voltage? ☐ Yes ☐ No Record the battery voltage ________
5. Locate the windshield wiper fuse. Fuse identification and rating is usually labeled inside of the fuse box cover.
6. What fuse rating is required for the wiper circuit? ________ amps

TURN

7. What is the rating of the fuse in the windshield wiper socket? ________ amps

   Is it the correct fuse? ☐ Yes ☐ No

8. What color is the fuse? ____________________

9. Blade fuses are color coded for easy identification. Write the identifying color on the lines provided for each of the fuses. Choose from the following colors; Red, Blue, Green, Yellow, Clear, Yellowish Brown, or Brown.

   5 amps ____________________
   7.5 amps ____________________
   10 amps ____________________
   15 amps ____________________
   20 amps ____________________
   25 amps ____________________
   30 amps ____________________

10. Connect the voltmeter (–) to ground. Probe both sides of the fuse with the meter positive lead. Does the voltmeter show voltage when each sides of the fuse are probed?

    ☐ Yes ☐ No

    Note: If the voltmeter does not indicate voltage when either side of the fuse is probed, turn the ignition to the *ON* position and try the test again.

11. Does the meter show voltage when the ends of the fuse are probed with the key on?

    ☐ Yes ☐ No

    Note: If the voltmeter still does not show voltage, the most likely cause is a bad fusible link. Locate the fusible link. Then test before and after the link. If there is voltage before, but not after, the fusible link, the fusible link is faulty.

12. Remove the wiper fuse with a fuse removal tool.

13. Visually inspect the fuse. Does the fuse appear to be good? ☐ Yes ☐ No

14. Carefully probe both sides of the fuse socket with the voltmeter positive lead. Was there voltage when both sides of the fuse socket were probed? ☐ Yes ☐ No

15. Replace the windshield wiper fuse.

16. Turn the wipers on. Do they work? ☐ Yes ☐ No

Interpretation of Fuse Test Results

- If there is voltage present to both sides of an installed fuse, the fuse is good and the current is flowing.
- If there is voltage only at one end of the fuse, the fuse is defective. (The side that does not light is the ground side of the circuit.)
- If there is no voltage on either side of the circuit, the circuit is shut off, the circuit is faulty, or the tester does not have a good ground connection (there also could be nothing wrong).

17. Record your recommendations for needed service or additional repairs and complete the repair order.

    ____

    ____

STOP

ASE Education Foundation Worksheet #7-2

SPLICE A WIRE WITH A CRIMP CONNECTOR

Name_______________________ Class_______________

Score: ☐ Excellent ☐ Good ☐ Needs Improvement **Instructor OK** ☐

Vehicle year ________ **Make** ______________ **Model** ______________

Objective: Upon completion of this assignment, you will be able to repair a damaged wire by splicing it together with a crimp connector.

ASE Education Foundation Correlation

This worksheet addresses the following **MLR** task:

VI.A.10 Repair and/or replace connectors, terminal ends, and wiring of electrical/electronic systems (including solder repair) **(P-1)**

This worksheet addresses the following **AST/MAST** task:

VI.A.10 Inspect, test, repair, and/or replace components, connectors, terminals, harnesses, and wiring in electrical/electronic systems (including solder repairs); determine needed action. **(P-1)**

We Support ASE | Education Foundation

Directions: Before beginning this lab assignment, review the worksheet completely. Fill in the information in the spaces provided as you complete each task.

Tools and Equipment Required: Safety glasses, fender covers, crimping–stripper tool, shop towel

Parts and Supplies: Crimp-type wire terminals

Procedure:

Note: If you are doing this worksheet for practice, write N/A in the spaces for vehicle identification.

1. Locate the two wires to be spliced together.

   Are the wires part of a repair or are the wires being spliced together for practice?

   ☐ Repair ☐ Practice

2. Use a crimping tool to strip the insulation (about ¼") from the ends of the two wires that are to be spliced. If the wire is not clean and shiny, cut the wire back until the wires are clean and shiny.

Photo by Tim Gilles

3. Select the proper crimp connector.

   What type of crimp connector was chosen?

   ____

4. Locate the proper crimp area on the crimp tool that corresponds to the gauge of the wire being spliced.

Terminal

Photo by Tim Gilles

Ring terminal

Spade terminal

Hook terminal

Butt splice

3-way "Y" connector

Snap plug terminal

Quick disconnect terminal

5. Crimp the wire into the connector. The dimple from the crimp tool should be opposite the seam in the connector.
6. Insert the other wire end into the crimp.
7. Test the crimp by lightly pulling on the wires.

   Did the crimp repair hold? ☐ Yes ☐ No
8. Before completing your paperwork, clean your work area, clean and return tools to their proper places, and wash your hands.
9. Record your recommendations for needed service or additional repairs and complete the repair order.

   ____

   ____

ASE Education Foundation Worksheet #7-3

SOLDER A WIRE CONNECTION

Name______________________________ Class______________________

Score: ☐ Excellent ☐ Good ☐ Needs Improvement **Instructor OK** ☐

Vehicle year __________ **Make** ________________ **Model** ________________

Objective: Upon completion of this assignment, you will be able to solder a wire to make an electrical repair.

ASE Education Foundation Correlation

This worksheet addresses the following **MLR** task:

VI.A.10 Repair and/or replace connectors, terminal ends, and wiring of electrical/electronic systems (including solder repair) **(P-1)**

This worksheet addresses the following **AST/MAST** task:

VI.A.10 Inspect, test, repair, and/or replace components, connectors, terminals, harnesses, and wiring in electrical/electronic systems (including solder repairs); determine needed action. **(P-1)**

We Support ASE | Education Foundation

Directions: Before beginning this lab assignment, review the worksheet completely. Fill in the information in the spaces provided as you complete each task.

Tools and Equipment Required: Safety glasses, shop towel, soldering gun, wire strippers, heat gun

Parts and Supplies: Shrink tube or electrical tape, rosin core solder, 6″ piece of wire

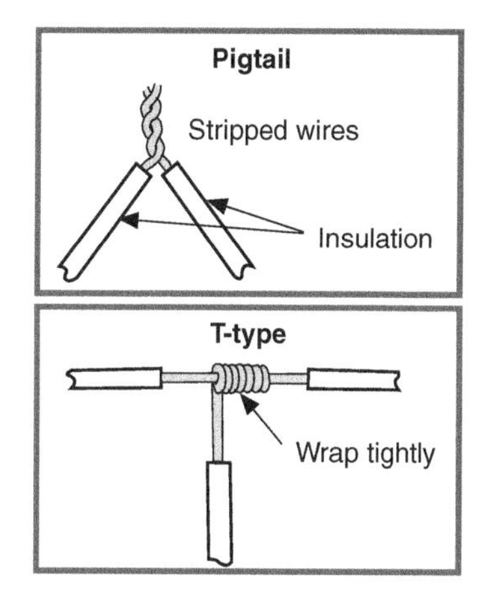

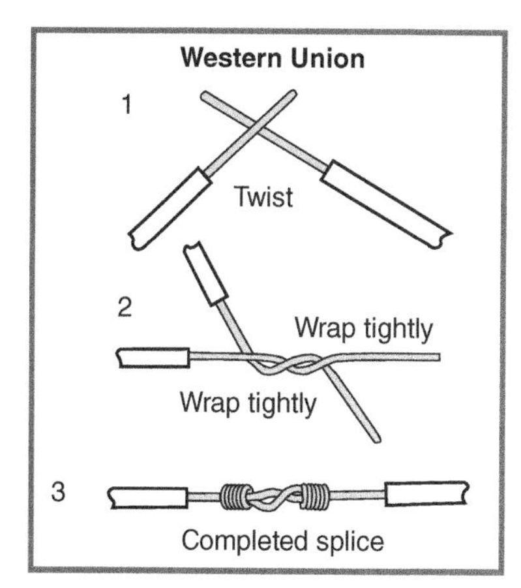

Procedure:

1. Obtain a 6″ length of wire.
2. Strip the ends of the wire using wire strippers.
3. Bring the ends of the wire together to form a loop and twist the wires together like a pigtail.

   Note: Shrink tubing must be installed on the wire before splicing and soldering the connection.

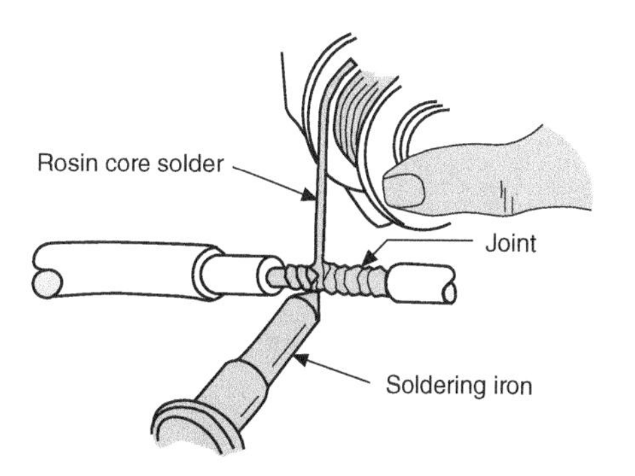

4. When the soldering gun is hot, apply solder to the soldering tip. Clean the hot tip with a damp towel. Reheat the tip and place a little solder (tinning) on the tip.

   Note: Use rosin core or solid core solder for electrical connections.

5. Hold the tip of the soldering gun against the wire. Depress the trigger to heat the wire.
6. Hold the solder against the opposite side of the connection from the soldering gun.
7. When the wires are heated, the solder will melt, saturating the connection.

   Note: Crimp connections can be soldered to wire in a similar manner.

8. After the soldered wire cools, insulate the connection with electrical tape or shrink tube.

   What insulation was used?

   ☐ Electrical tape

   ☐ Shrink tubing

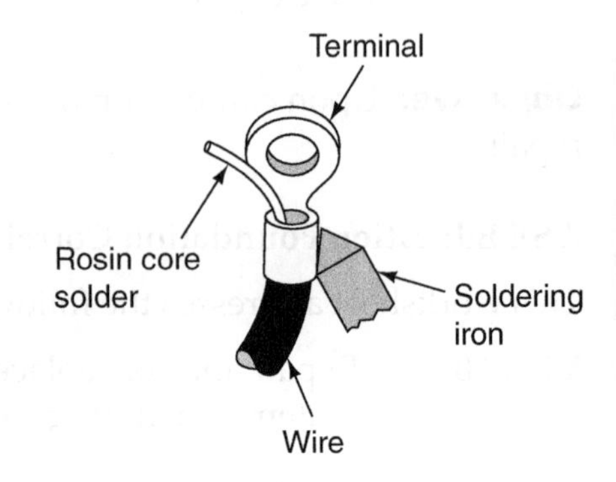

9. Before completing your paperwork, clean your work area, clean and return tools to their proper places, and wash your hands.
10. Record your recommendations for needed service or additional repairs and complete the repair order.

____

____

STOP

ASE Education Foundation Worksheet #7-4

BATTERY SERVICE

Name ______________________ Class ________________

Score: ☐ Excellent ☐ Good ☐ Needs Improvement **Instructor OK** ☐

Vehicle year ________ **Make** ______________ **Model** ______________

Objective: Upon completion of this assignment, you will be able to service a battery.

ASE Education Foundation Correlation

This worksheet addresses the following **MLR/AST/MAST** tasks:

VI.B.3 Maintain or restore electronic memory functions. **(P-1)**

VI.B.4 Inspect and clean battery; fill battery cells; check battery cables, connectors, clamps, and hold-downs. **(P-1)**

VI.B.8 Identify electronic modules, security systems, radios, and other accessories that require reinitialization or code entry after reconnecting vehicle battery. **(P-1)**

We Support ASE Education Foundation

Directions: Before beginning this lab assignment, review the worksheet completely. Fill in the information in the spaces provided as you complete each task.

Tools and Equipment Required: Safety glasses, fender covers, shop towel, computer memory retaining tool, battery pliers, battery terminal cleaner, battery terminal puller, battery carrier or strap

Parts and Supplies: Baking soda, paint

Procedure:

Note: Parking the vehicle outdoors near a water drain for battery tray washing will make cleanup easier.

1. Open the hood and place fender covers on the fenders and front body parts.
2. Check that all electrical circuits are off.
3. If there are any electronic memory circuits (radio stations, radio security, seats, computers, etc.), install a computer memory retaining tool into the cigarette lighter socket. Otherwise the memory circuits will need to be reset after the battery service.
4. Identify any electronic memory circuits that may require resetting after service.

   ☐ Radio ☐ Clock ☐ Seats ☐ Mirrors

   ☐ Other ____________

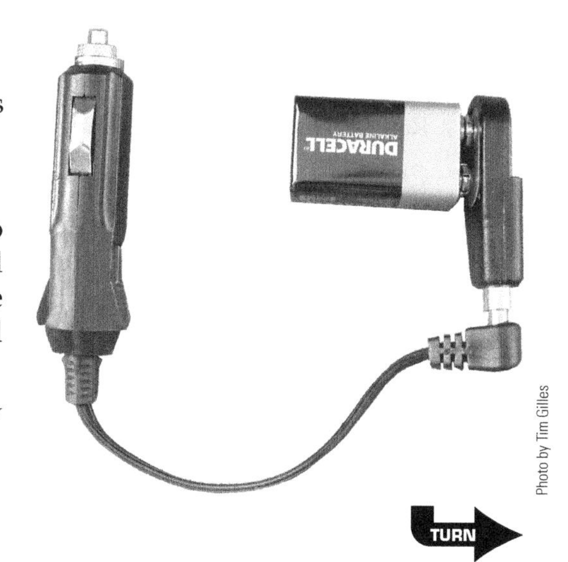

Photo by Tim Gilles

TURN

5. Loosen the ground cable first. Which is the ground cable?

   ☐ (+) ☐ (–)

6. Inspect the condition of the terminal clamp nut. If the nut is worn, use battery pliers to loosen it. Remove the terminal clamp.

   Note: If the clamp does not come off easily, use a battery terminal puller.

   Note: If the clamp is not serviceable, the cable should be replaced. Refer to Worksheet #7-5, Replace a Battery Cable.

Photo by Tim Gilles

7. Remove the remaining terminal clamp.
8. Inspect the condition of the terminal clamp and post.

   ☐ Oxidized ☐ Clean ☐ Worn

9. Use a battery terminal cleaning tool to carefully clean both posts and the insides of the terminal clamps.
10. Remove the battery holddown clamp and lift the battery from the vehicle.
11. Clean the battery, its tray, and the battery holddown with baking soda and water. Hose them off thoroughly. Do not allow baking soda solution to enter the battery cells.
12. Repaint the tray and holddown after they are completely dry.
13. Check the electrolyte level and refill the battery as needed.

    ☐ Electrolyte level OK ☐ Water added

14. Reinstall the battery and holddown.
15. Tighten the holddown fasteners until they are snug. *Do not overtighten.*

    Note: Overtightening the holddown fasteners can damage the battery.

16. Reinstall the positive (+) and ground (–) battery cables.
17. Use battery spray, treated felt washers, or silicone RTV, or apply some grease around the base of the terminal to prevent oxidation. Method used:

    ☐ Battery spray ☐ Felt washers ☐ Grease ☐ Silicone RTV ☐ Other (describe): ______

18. Remove the computer memory retaining tool from the cigarette lighter.
19. Check that the vehicle starts.
20. Before completing your paperwork, clean your work area, clean and return tools to their proper places, and wash your hands.
21. Record your recommendations for needed service or additional repairs and complete the repair order.

    ____

    ____

STOP

ASE Education Foundation Worksheet #7-5

REPLACE A BATTERY CABLE

Name______________________ Class________________

Score: ☐ Excellent ☐ Good ☐ Needs Improvement **Instructor OK** ☐

Vehicle year ________ **Make** ______________ **Model** ________________

Objective: Upon completion of this assignment, you will be able to replace a battery cable.

ASE Education Foundation Correlation

This worksheet addresses the following **MLR/AST/MAST** task:

VI.B.4 Inspect and clean battery; fill battery cells; check battery cables, connectors, clamps, and hold-downs. **(P-1)**

We Support Education Foundation

Directions: Before beginning this lab assignment, review the worksheet completely. Fill in the information in the spaces provided as you complete each task.

Tools and Equipment Required: Safety glasses, fender covers, shop towel, computer memory retaining tool, battery pliers, battery terminal cleaner, battery terminal puller

Procedure:

1. Which cable is being replaced?

   ☐ Positive ☐ Negative

2. How did you determine that the battery cable needed to be replaced?

   ____

3. Open the hood and place fender covers on the fenders and front body parts.

4. Remove the terminal clamp from the battery post. Refer to Worksheet #7-4, Battery Service.

   Note: Always remove the ground cable before working with the positive battery cable. Also, remember to protect the memory circuits.

5. Remove the other end of the cable from the:

   ☐ Starter solenoid

   ☐ Cylinder block

   ☐ Other (describe): ____________________

6. Clean the battery post with a terminal cleaner.

7. Expand the eye of the terminal using the expanding tool.

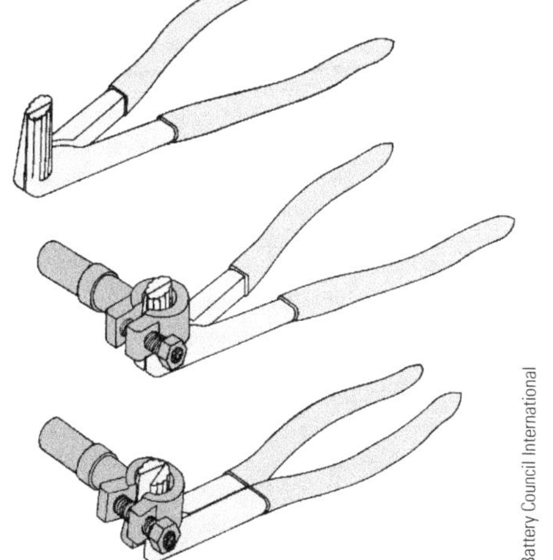

Battery Council International

TURN

8. Install the battery terminal clamp over the battery post.
9. Install the other end of the cable to the:

   ☐ Cylinder block

   ☐ Starter solenoid

   ☐ Other (describe): ______________________________

10. Which cable do you always install last?

    ☐ Positive ☐ Negative ☐ Ground

11. Check that the cables are properly routed so that they will not be damaged.

    Are the cables properly routed?

    ☐ Yes ☐ No

12. Treat the terminal connection to prevent corrosion. Which method was used?

    ☐ Grease

    ☐ Felt washers

    ☐ Spray

13. Does the vehicle start?

    ☐ Yes ☐ No

14. Before completing your paperwork, clean your work area, clean and return tools to their proper places, and wash your hands.
15. Record your recommendations for needed service or additional repairs and complete the repair order.

    ____

    ____

STOP

ASE Education Foundation Worksheet #7-6

REPLACE A BATTERY TERMINAL CLAMP

Name______________________________ Class____________________

Score: ☐ Excellent ☐ Good ☐ Needs Improvement **Instructor OK** ☐

Vehicle year __________ **Make** ____________________ **Model** ______________________

Objective: Upon completion of this assignment, you will be able to replace a battery terminal clamp.

ASE Education Foundation Correlation

This worksheet addresses the following **MLR/AST/MAST** task:

VI.B.4 Inspect and clean battery; fill battery cells; check battery cables, connectors, clamps, and hold-downs. **(P-1)**

We Support ASE Education Foundation

Directions: Before beginning this lab assignment, review the worksheet completely. Fill in the information in the spaces provided as you complete each task.

Tools and Equipment Required: Safety glasses, fender covers, shop towel, diagonal cutting pliers, sharp knife, computer memory retaining tool, battery pliers, battery terminal cleaner, battery terminal puller. A propane torch will be needed if a permanent terminal is being installed.

Parts and Supplies: Battery terminal clamp

Procedure:

1. Obtain a new battery terminal clamp.
2. Which type of terminal clamp is being used as the replacement?

   ☐ Temporary bolt-on clamp

   ☐ Solder-type clamp
3. Open the hood and place fender covers on the fenders and front body parts.

   Note: If there are any electronic memory circuits (radio stations, radio security, seats, computers, etc.), install a computer memory retaining tool into the cigarette lighter socket. Refer to Worksheet #7-4, Battery Service.
4. Remove the battery ground cable.

   Note: Always remove the ground cable first when working on the battery.
5. Remove the terminal clamp that will be replaced.
6. Use diagonal cutting pliers to cut the old clamp from the cable.

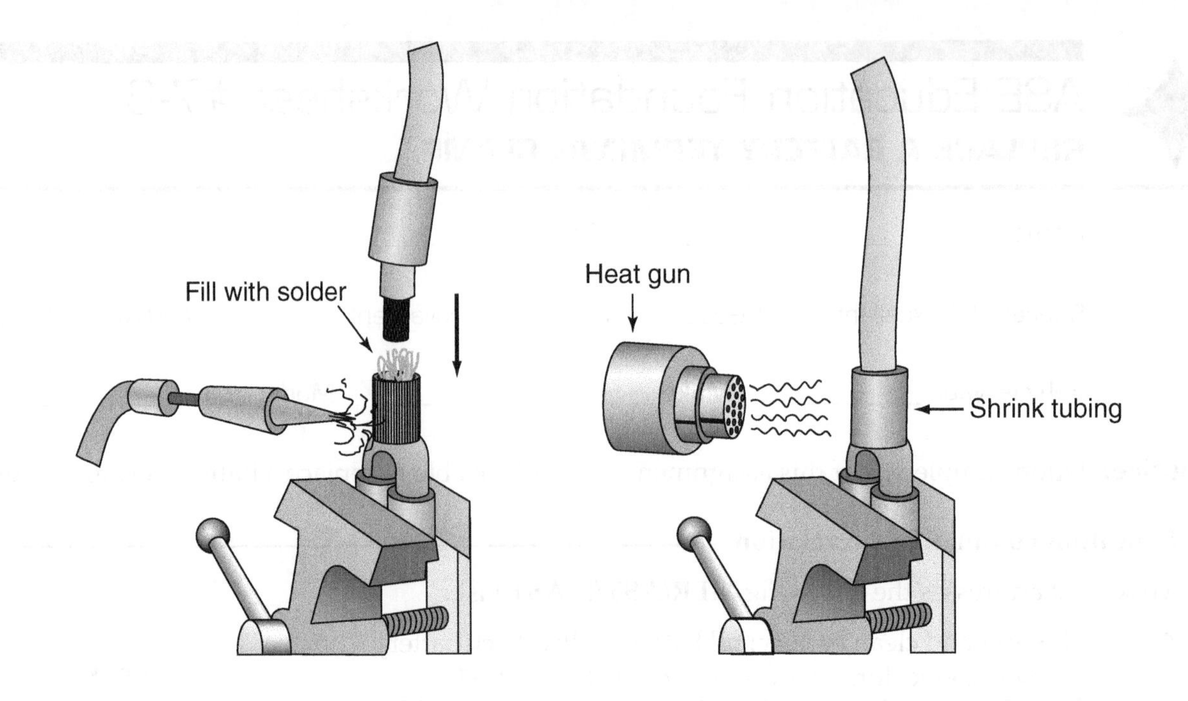

7. Use a sharp knife to strip back about ½″ of insulation from the end of the cable.

   Note: Before either type of terminal clamp is used, the end of the cable must be bright and clean.

8. Are the connecting surfaces clean and bright? ☐ Yes ☐ No

9. Install the new clamp on the cable, following the terminal clamp manufacturer's installation instructions.

10. Install the clamp onto the battery post.

11. Which cable is always connected last? ☐ Positive ☐ Negative ☐ Ground

12. After both cables have been connected to the battery, start the vehicle.

    Does it start? ☐ Yes ☐ No

13. Before completing your paperwork, clean your work area, clean and return tools to their proper places, and wash your hands.

14. Record your recommendations for needed service or additional repairs and complete the repair order.

    ____

    ____

STOP

ASE Education Foundation Worksheet #7-7

BATTERY SPECIFIC GRAVITY TEST

Name_______________________________ Class_______________________

Score: ☐ Excellent ☐ Good ☐ Needs Improvement **Instructor OK** ☐

Vehicle year __________ **Make** ____________________ **Model** ______________________

Objective: Upon completion of this assignment, you will be able to test a battery's state of charge using a hydrometer.

ASE Education Foundation Correlation

This worksheet addresses the following **MLR** task:

VI.B.1 Perform battery state-of-charge test; determine necessary action. **(P-1)**

This worksheet addresses the following **AST/MAST** task:

VI.B.1 Perform battery state-of-charge test; determine needed action. **(P-1)**

We Support
ASE | Education Foundation

Directions: Before beginning this lab assignment, review the worksheet completely. Fill in the information in the spaces provided as you complete each task.

Tools and Equipment Required: Safety glasses, fender covers, shop towel, hydrometer

Related Information: Not all batteries have vent caps. There are two ways to test a battery's state of charge. One method is to measure the specific gravity of the electrolyte. The other method is to measure the open-circuit voltage of the battery. Either method will give a good indication of the battery's state of charge.

If the battery being tested has vent caps, perform both tests. When a battery has no vent caps, do only the open-circuit voltage test.

Be sure to wear eye protection when performing this test. Electrolyte can cause serious eye injuries.

Procedure:

Hydrometer Test (specific gravity of electrolyte)

1. Install a fender cover on the fender nearest the battery.
2. Remove the vent caps and set them on top of the battery.
3. Check the level of the electrolyte.

   Note: If the electrolyte level is too low to perform this test, add water to the cells and recharge the battery. Refer to Worksheet #7-9, Battery Charging: Fast and Slow.

   Electrolyte level: ☐ OK ☐ Add water ☐ Recharge

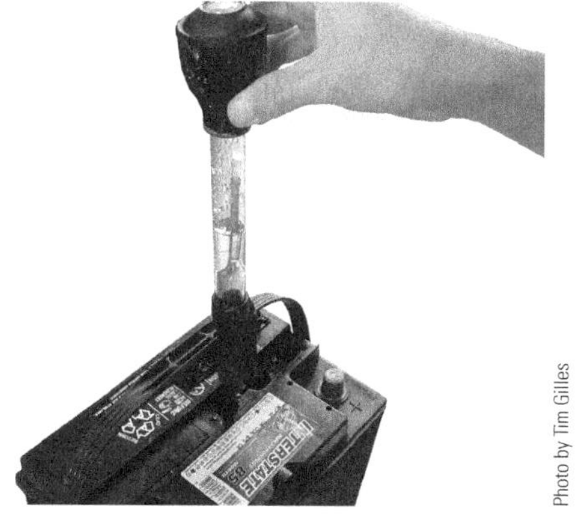

Photo by Tim Gilles

TURN

4. Draw electrolyte into the hydrometer to the line on the tester bulb (if so equipped). The gauge must float freely.
5. Hold the hydrometer vertically so the float can rise to its proper level.
6. Read the specific gravity on the float and record it below. Make a temperature correction, if necessary. The compensation factor is more important at temperature extremes.

   Note: Most hydrometers are temperature compensated. If not, a correction of +0.004 is made for each 10°F change above 80°F. Subtract 0.004 for each 10°F drop in temperature below 80°F.
7. Put the hydrometer tube back into the cell and squeeze it gently to return the electrolyte to the cell. Repeat steps 4 through 6 for each of the cells.

Photo by Tim Gilles

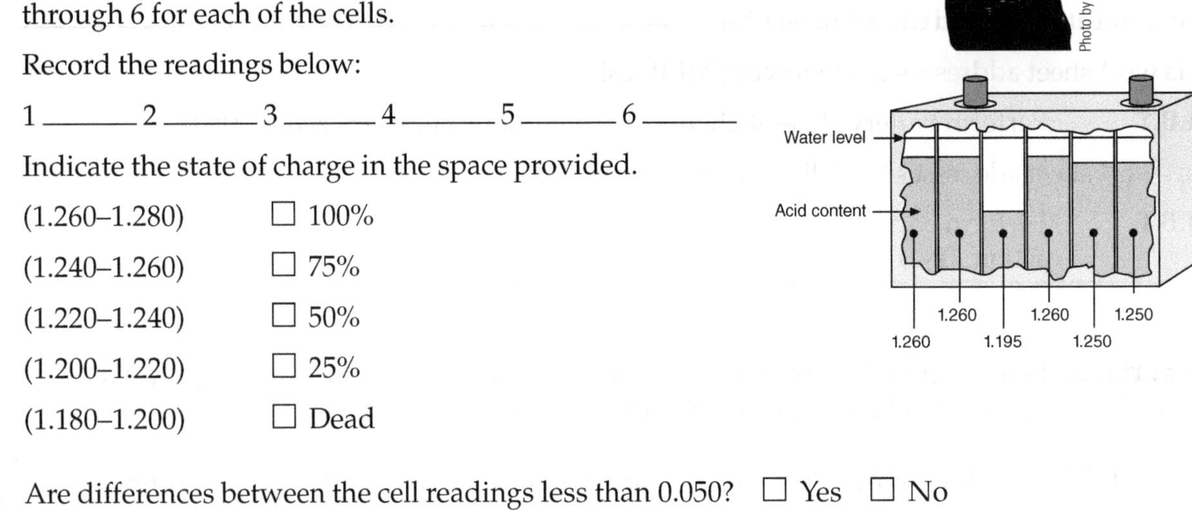

   Record the readings below:

   1 ______ 2 ______ 3 ______ 4 ______ 5 ______ 6 ______

   Indicate the state of charge in the space provided.

   (1.260–1.280) ☐ 100%

   (1.240–1.260) ☐ 75%

   (1.220–1.240) ☐ 50%

   (1.200–1.220) ☐ 25%

   (1.180–1.200) ☐ Dead

8. Are differences between the cell readings less than 0.050? ☐ Yes ☐ No

   Note: A faulty battery can be determined by the specific gravity readings. If the differences among any of the cells are greater than 0.050, a cell is damaged and the battery must be replaced.
9. What service does the battery need?

   None. ☐ The battery is fully charged.

   Recharging. ☐ The specific gravity is below 75%.

   Replacement. ☐ The battery has a damaged cell.
10. Before completing your paperwork, clean your work area, clean and return tools to their proper places, and wash your hands.
11. Record your recommendations for needed service or additional repairs and complete the repair order.

____

____

ASE Education Foundation Worksheet #7-8

BATTERY OPEN-CIRCUIT VOLTAGE TEST

Name ____________________ Class ____________

Score: ☐ Excellent ☐ Good ☐ Needs Improvement **Instructor OK** ☐

Vehicle year ________ **Make** ____________ **Model** ____________

Objective: Upon completion of this assignment, you will be able to test a battery using a digital voltmeter and determine the battery's state of charge.

ASE Education Foundation Correlation

This worksheet addresses the following **MLR** task:

VI.B.1 Perform battery state-of-charge test; determine necessary action. **(P-1)**

This worksheet addresses the following **AST/MAST** task:

VI.B.1 Perform battery state-of-charge test; determine needed action. **(P-1)**

We Support Education Foundation

Directions: Before beginning this lab assignment, review the worksheet completely. Fill in the information in the spaces provided as you complete each task.

Tools and Equipment Required: Safety glasses, fender covers, shop towel, digital voltmeter

Related Information: Not all batteries have vent caps. There are two ways to test a battery's state of charge. One method is to measure the specific gravity of the electrolyte. The other method is to measure the open-circuit voltage of the battery. Either method will give a good indication of the battery's state of charge.

If the battery being tested has vent caps, perform both tests. When a battery has no vent caps, do only the open-circuit voltage test.

Procedure:

Open-Circuit Voltage Test

Photo by Tim Gilles

1. Install a fender cover on the fender nearest the battery.
2. Turn on the headlights for 30 seconds to remove the surface charge from the battery.
3. Set the digital voltmeter to the DC voltage scale higher than the rated voltage of the battery.

   Voltage scale selected ______
4. Connect the voltmeter in parallel with the battery. This means to connect the positive meter lead to the positive battery post and the negative meter lead to the negative battery post.
5. Read and record the voltage. ______ volts

6. The battery's state of charge relates to the measured voltage as follows:

   12.6 = 100% charged

   12.4 = 75% charged

   12.2 = 50% charged

   12.0 = discharged

   What is the state of charge of the battery being tested? _______%

7. What service does the battery need?

   ☐ None. The battery is fully charged.

   ☐ Recharge. The battery state of charge is below 75%.

   Note: If the battery measured below 75% charge and the reason is not known, more complete testing will be necessary. A problem with the battery, the charging system, or the vehicle's electrical system could be the reason why the battery was not completely charged.

8. Is more testing necessary to determine the reason for the battery's low state of charge?

   ☐ Yes ☐ No

9. Before completing your paperwork, clean your work area, clean and return tools to their proper places, and wash your hands.

10. Record your recommendations for needed service or additional repairs and complete the repair order.

    ____

    ____

STOP

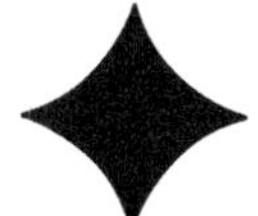

ASE Education Foundation Worksheet #7-9

BATTERY CHARGING: FAST AND SLOW

Name______________________________ Class______________________

Score: ☐ Excellent ☐ Good ☐ Needs Improvement **Instructor OK** ☐

Vehicle year __________ **Make** ____________________ **Model** ______________________

Objective: Upon completion of this assignment, you will be able to return a battery to service by fast or slow charging.

ASE Education Foundation Correlation

This worksheet addresses the following **MLR/AST/MAST** tasks:

VI.B.5 Perform slow/fast battery charge according to manufacturer's recommendations. **(P-1)**

VI.B.9 Identify hybrid vehicle auxiliary (12v) battery service, repair, and test procedures. **(P-2)**

We Support ASE Education Foundation

Directions: Before beginning this lab assignment, review the worksheet completely. Fill in the information in the spaces provided as you complete each task.

Tools and Equipment Required: Safety glasses, fender covers, battery charger, shop towel

SAFETY NOTE: A battery being charged gives off hydrogen gas. A spark can cause a dangerous explosion!

Procedure:

1. Open the hood and place fender covers on the fenders and front body parts.
2. Check the electrolyte level and fill the battery only enough to cover the plates.

   ☐ Low ☐ Plates are covered

   If the battery is the sealed type and its electrolyte level is low, do *not* attempt to recharge it.

   Is it a sealed battery? ☐ Yes ☐ No
3. Disconnect the ground cable from the battery.

   Which is the ground cable? ☐ (+) ☐ (–)
4. Connect the charger to the battery (before plugging it into wall current).
5. What color charger clamp goes to the positive battery terminal?

   ☐ Red ☐ Black

Photo by Tim Gilles

TURN

6. Plug the charger into wall current.

Slow Charge

1. Turn the charger on.
2. Set the charging current to approximately 1% of the battery's cold cranking amps (CCA). For example, a 650 CCA battery should be charged at 6.5 amps.

   What is the CCA of the battery? ______ amps

   What is the charging rate for your battery? ______ amps
3. Check the voltage at regular intervals. When the voltage does not change for 1 hour, the battery is fully charged.

   Did the voltage stabilize for 1 hour? ☐ Yes ☐ No

Fast Charge

1. Set the charger control to maximum. Turn on the timer to start charging.

   How many amps does the gauge indicate? ☐ 20–30 amps ☐ Under 20 amps
2. If the amp gauge reads over 30 amps, adjust charger output to a lower setting.

   Is the charger reading over 30 amps? ☐ Yes ☐ No

Complete Battery Charging

1. If the battery will not take a charge, double-check to see that the polarity is correct. Hold down the jump-start button for 1 minute and then release it. The battery should begin to take a charge.
2. When the desired charge is completed, shut off the charger.

   Note: Do not fast charge a battery for an extended time. Heat is generated in the battery during charging. Overheating a battery can severely shorten its useful life.
3. Unplug the charger from the wall current.
4. Disconnect the charger cables and reattach the battery ground cable.
5. Adjust the battery electrolyte level as needed. ☐ Low ☐ OK

Charging an AGM Battery

Absorbed glass matt (AGM) batteries are used as the auxiliary battery in hybrid vehicles. They are sometimes installed in conventional vehicles, as well. A common battery charger will not charge an AGM battery. These batteries are charged using a smart charger. This is one that monitors battery voltage and cycles the charging current and voltage during multiple stages that are controlled by a computer algorithm.

- First the charger checks continuity and polarity.
- Then it charges at maximum output until the battery is approximately 80% recharged.
- Next, the charger applies current, while it holds output voltage constant at about 15 V. The charge current decreases gradually during this stage, which can take a long time. Battery manufacturers specify different recommendations; eight hours of charging time is one example. The battery will be 95–100% charged following this stage.
- The next step applies a small constant current for a short time, typically 5–30 minutes.

Using a smart charger is easy and straightforward. Turn on the machine and follow the instructions.

1. Does your shop have a smart charger? ☐ Yes ☐ No
2. Before completing your paperwork, clean your work area, clean and return tools to their proper places, and wash your hands.
3. Record your recommendations for needed service or additional repairs and complete the repair order.

   ____

   ____

STOP

ASE Education Foundation Worksheet #7-10

BATTERY JUMP-STARTING (LOW-MAINTENANCE BATTERY)

Name_______________________________ Class_______________________

Score: ☐ Excellent ☐ Good ☐ Needs Improvement **Instructor OK** ☐

Vehicle year __________ **Make** ________________ **Model** ________________

Objective: Upon completion of this assignment, you will be able to jump-start a vehicle that has a low-maintenance battery.

ASE Education Foundation Correlation

This worksheet addresses the following **MLR/AST/MAST** task:

VI.B.6 Jump-start vehicle using jumper cables and a booster battery or an auxiliary power supply. **(P-1)**

We Support ASE Education Foundation

Directions: Before beginning this lab assignment, review the worksheet completely. Fill in the information in the spaces provided as you complete each task.

Tools and Equipment Required: Safety glasses, fender covers, jumper cables, shop towel

Procedure:

CAUTION Use extra care when jump-starting a vehicle with electronic components. Misconnected cables or electrical surges can damage electronic components.

1. Wear eye protection. Batteries produce explosive gases, which may accidentally explode.

   Type of eye protection worn:

   ☐ Safety glasses

   ☐ Goggles

   ☐ Face shield

2. Be sure the transmission is in neutral or park and the emergency brake is set firmly.

   Transmission: ☐ Park ☐ Neutral

   Parking brake: ☐ On ☐ Off

3. Are all electrical loads off? ☐ Yes ☐ No

4. Check the battery's electrolyte level. If the level is low, do not attempt to jump-start the car. Refill and recharge the battery as needed.

   Electrolyte level: ☐ OK ☐ Needs water

NEG − + POS

Smaller Larger

Top-terminal batteries

5. Keep the vent caps in place on the battery. The vent caps act as spark arrestors.

TURN

6. Attach one end of the jumper cable to the positive terminal on the discharged battery and the other end to the positive terminal on the booster battery.

CAUTION The vehicles should not be touching each other. This could provide an unwanted ground path and a spark could result.

7. Attach one end of the negative cable to the negative terminal on the fully charged booster battery.
8. Connect the other end of the negative cable to a ground on the engine. A metal bracket or the end of the negative battery cable that is attached to the block is a good ground point.

CAUTION A spark can occur as the negative cable is attached to the dead battery as it tries to equalize its voltage with the booster battery. Making a connection at a point away from the battery avoids the possibility of a dangerous spark near the battery.

Note: Low-maintenance batteries have higher internal resistance than conventional lead–antimony batteries. The jumper cables may need to remain in place for a minute or so before attempting to start the disabled vehicle. This will allow the dead battery to take on a charge.

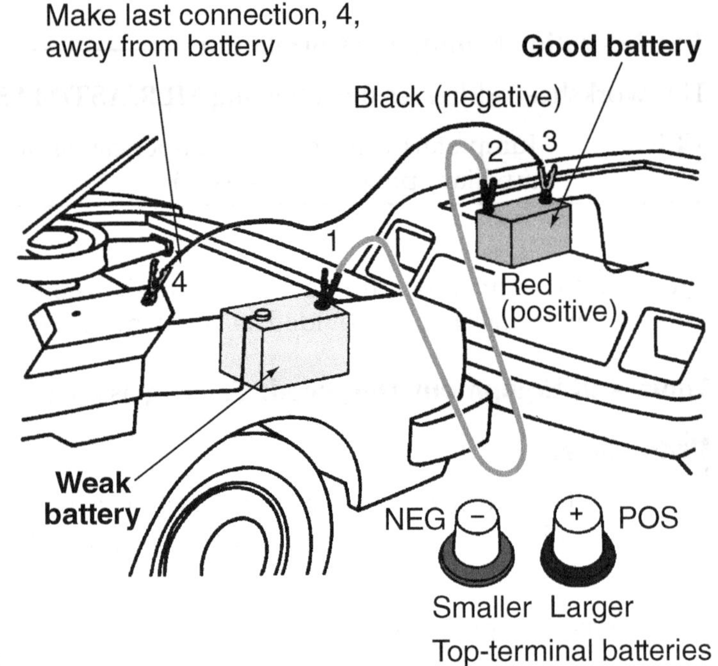

9. When the dead vehicle starts, immediately disconnect the jumper cable from the block.

   Note: If the vehicle does not start quickly, do not crank the engine for longer than 30 seconds. The starter motor can overheat, resulting in starter failure.

10. Run the host vehicle at 2,000 rpm to allow its charging system to recharge the battery.
11. Before completing your paperwork, clean your work area, clean and return tools to their proper places, and wash your hands.
12. Record your recommendations for needed service or additional repairs and complete the repair order.

____

____

STOP

ASE Education Foundation Worksheet #7-11

BATTERY CAPACITY/LOAD TESTING (WITH VAT)

Name________________________ Class__________________

Score: ☐ Excellent ☐ Good ☐ Needs Improvement **Instructor OK** ☐

Vehicle year _________ **Make** _______________ **Model** _______________

Objective: Upon completion of this assignment, you will be able to load test a battery using a volt-amp tester (VAT).

ASE Education Foundation Correlation

This worksheet addresses the following **MLR** task:

VI.B.2 Confirm proper battery capacity for vehicle application; perform battery capacity and load test; determine necessary action. **(P-1)**

This worksheet addresses the following **AST/MAST** task:

VI.B.2 Confirm proper battery capacity for vehicle application; perform battery capacity and load test; determine needed action. **(P-1)**

We Support
ASE | Education Foundation

Directions: Before beginning this lab assignment, review the worksheet completely. Fill in the information in the spaces provided as you complete each task.

Tools and Equipment Required: Safety glasses, fender covers, shop towel, volt/amp tester

Procedure:

1. Open the hood and place fender covers on the fender nearest the battery.
2. Inspect the battery for the following:

Cracks in the case	☐ OK	☐ Needs attention
Battery tray condition	☐ OK	☐ Needs attention
Battery cables	☐ OK	☐ Need attention
Posts and terminals	☐ OK	☐ Need attention
Electrolyte level	☐ OK	☐ Needs attention
Caps and/or covers	☐ In place	☐ Missing
Holddown	☐ Secure	☐ Loose
Corrosion	☐ Clean	☐ Built up

Photo by Tim Gilles

3. The battery must be at least 75% charged to continue load testing. What is the battery's state of charge? Refer to Worksheets #7-7, Battery Specific Gravity Test, and #7-8, Battery Open-Circuit Voltage Test.

   More than 75% charged ☐ Less than 75% charged

   How was the state of charge measured? ☐ Specific gravity ☐ Open-circuit voltage

Battery Capacity Test Using a Carbon Pile (VAT)

4. **Note:** The battery should be disconnected during a battery capacity test. This is to protect electronic circuits. Remove the ignition key and disconnect the battery ground cable. If the vehicle has computer controls, connect a memory retaining tool to the vehicle. Also, this test is only accurate if the battery is more than 75% charged. 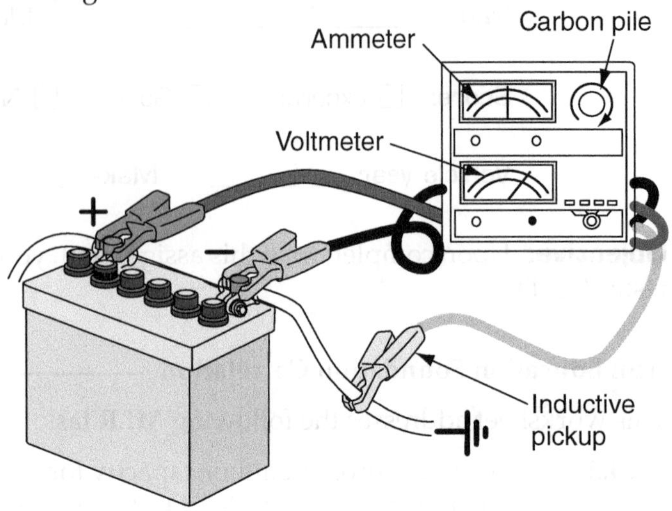
5. Check that the control knob on the carbon pile is off (counterclockwise).
6. Turn test selector switch to **battery test or starting system.**
7. Turn voltage selector switch to a voltage higher than battery voltage. What range was selected?

   ☐ Int. 18V ☐ Ext. 18V ☐ Ext. 3V
8. Connect the large red (positive) tester lead to the positive post of the battery and connect the large black (negative) tester lead to the negative post of the battery.

9. Clamp the inductive ammeter pickup around the negative tester lead.
10. Determine the load to use during the battery capacity test in each of the following ways:
    a. Amperage needed to crank the engine: ______ amps
    b. The CCA of the battery divided by 2 = ______ amps

    Note: Use the vehicle CCA specifications. Do not use the CCA listed on the battery.

    Vehicle CCA specification ______

    Note: The large knob on the tester will cause a rapid discharge of the battery when turned clockwise.
11. Turn the large control knob clockwise to load the battery. The ammeter needle should move. If not, determine the cause before proceeding. Maintain the desired amperage for 15 seconds while watching the *voltmeter.* Then shut off the amp load by turning the control knob counterclockwise.
12. What is the voltmeter reading at the end of the test (before turning off the carbon-pile)?

    ______ volts

Test Results

The battery passes the capacity test if the voltage remains above 9.6 volts during the test.

Note: For every 10° below 70°F, voltage may be 0.1 less than 9.6.

Results: ☐ Passed ☐ Failed

13. If the battery *passed* the capacity test, disconnect the tester and reconnect the battery negative cable. If the battery *failed* the capacity test, charge the battery and retest or replace the battery.
14. Record your recommendations for needed service or additional repairs and complete the repair order.

____

____

STOP

ASE Education Foundation Worksheet #7-12

BATTERY CAPACITY/LOAD TESTING (WITHOUT VAT)

Name_______________________________ Class______________________

Score: ☐ Excellent ☐ Good ☐ Needs Improvement **Instructor OK** ☐

Vehicle year __________ **Make** ____________________ **Model** ______________________

Objective: Upon completion of this assignment, you will be able to load test a battery.

ASE Education Foundation Correlation

This worksheet addresses the following **MLR** task:

VI.B.2 Confirm proper battery capacity for vehicle application; perform battery capacity and load test; determine necessary action. **(P-1)**

This worksheet addresses the following **AST/MAST** task:

VI.B.2 Confirm proper battery capacity for vehicle application; perform battery capacity and load test; determine needed action. **(P-1)**

We Support Education Foundation

Directions: Before beginning this lab assignment, review the worksheet completely. Fill in the information in the spaces provided as you complete each task.

Tools and Equipment Required: Safety glasses, fender covers, shop towel

Procedure:

Note: Before doing a load/capacity test on a battery, the battery must be in good physical condition and be more than 75% charged.

1. Open the hood and place a fender cover on the fender nearest the battery.
2. Inspect the battery for the following:

Cracks in the case	☐ OK	☐ Needs attention
Leaking electrolyte	☐ OK	☐ Needs attention
Battery tray condition	☐ OK	☐ Needs attention
Battery cables	☐ OK	☐ Need attention
Posts and terminals	☐ OK	☐ Need attention
Electrolyte level	☐ OK	☐ Needs attention
Caps and/or covers	☐ In place	☐ Missing
Holddown	☐ Secure	☐ Loose
Corrosion	☐ Clean	☐ Built up

Photo by Tim Gilles

3. If the battery has cracks or leaks, the battery must be replaced.

   ☐ Cracks ☐ Leaks ☐ OK

TURN

4. What is the battery's state of charge? Refer to Worksheets #7-7, Battery Specific Gravity Test, and #7-8, Battery Open-Circuit Voltage Test.

   ☐ 100%

   ☐ 75%

   ☐ Less than 75%

   Note: The battery must be 75% or more charged to continue load testing.

   How was the state of charge measured?

   ☐ Specific gravity ☐ Open-circuit voltage

Battery Capacity Test Without a Carbon-Pile

Note: If the battery is less than 75% charged, charge the battery before testing.

The battery requires:

☐ Recharging

☐ A 3-minute charge test

☐ Nothing

5. If the battery is more than 75% charged, continue by disabling the engine's ignition system. See an instructor for specific instructions.

   Ignition disabled? ☐ Yes ☐ No

6. Connect a voltmeter to the battery.

7. Use the starter motor to crank the engine for 15 seconds.

   Record the voltmeter reading. ______ volts

   a. If the battery voltage is 9.6 volts or higher during the test, the battery is good.

   b. If battery voltage falls below 9.6 volts, recharge the battery and repeat the test. If the result is still below 9.6 volts, the starter motor draw should be checked to see that its current draw is not excessive. If not, the starter is good, replace the battery.

Testing Battery Condition with a Conductance Tester.

8. Connect the positive and negative tester clamps to the battery.

9. Enter the battery's cold cranking amps rating into the tester.

10. Press the test button and read the results.

    Test Result: ☐ Good ☐ Recharge ☐ Replace battery

11. Before completing your paperwork, clean your work area, clean and return tools to their proper places, and wash your hands.

12. Record your recommendations for needed service or additional repairs and complete the repair order.

    ____

    ____

STOP

ASE Education Foundation Worksheet #7-13

BATTERY DRAIN TEST

Name ______________________ Class ______________

Score: ☐ Excellent ☐ Good ☐ Needs Improvement **Instructor OK** ☐

Vehicle year ________ **Make** ______________ **Model** ______________

Objective: Upon completion of this assignment, you will be able to test a vehicle's electrical system for excessive drain.

ASE Education Foundation Correlation

This worksheet addresses the following **MLR** task:

VI.A.8 Measure key-off battery drain (parasitic draw). **(P-1)**

This worksheet addresses the following **AST/MAST** task:

VI.A.8 Diagnose the cause(s) of excessive key-off battery drain (parasitic draw); determine needed action. **(P-1)**

We Support ASE | Education Foundation

Directions: Before beginning this lab assignment, review the worksheet completely. Fill in the information in the spaces provided as you complete each task.

Tools and Equipment Required: Safety glasses, fender covers, shop towel, test light, digital multimeter, amp probe

Procedure:

Using a test light:

1. Open the hood and place fender covers on the fenders and front body parts.
2. With the key off, be sure all lights and accessories are off.

   ☐ Off ☐ On
3. Check all courtesy lights. These are the ones that come on when a car door is open. They should not be on.

   ☐ Off ☐ On
4. Disconnect the battery ground cable.
5. Connect a test light in series between the battery ground cable end and the battery post.

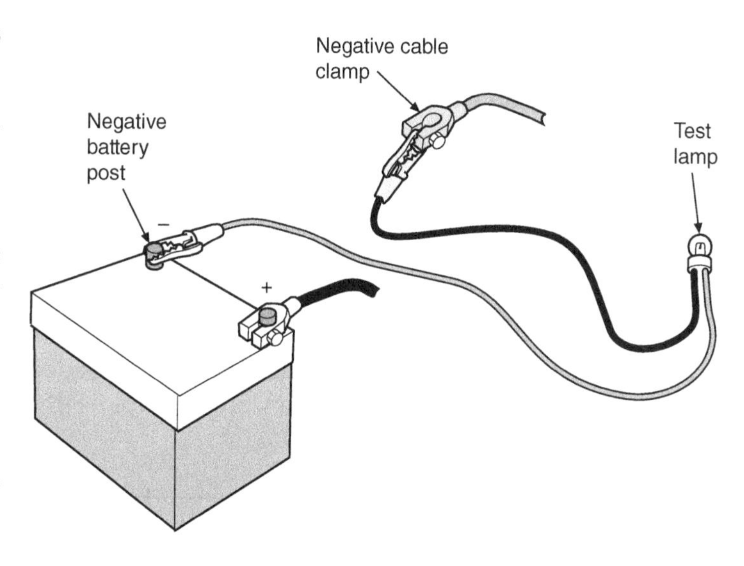

Open the door to turn on the dome light. The test light should come on.

☐ No light ☐ The light illuminates

Note: If the bulb lights, additional testing is necessary to locate the problem.

Another method to more accurately check for a drain would be to use an ammeter in place of the test light.

Using an ammeter:

6. Connect an ammeter between the negative cable and the negative post of the battery. The meter should show 0 amps.

   Ammeter reading: ______

SAFETY NOTE

Note: Excessive amps can damage the meter or blow the meter fuse.

Do not try to start the vehicle or turn on any of the accessories while the ammeter is connected.

Note: A drain of less than 50 milliamps (0.050 amp) is acceptable on **some** computer-controlled vehicles.

7. Is the drain excessive? ☐ Yes ☐ No

   Is further testing required to locate the problem?

   ☐ Yes ☐ No

   Note: An amp probe is a popular alternative to using an ammeter.

Using an Amp Probe:

8. Make sure the key is in the *Off* position and all the accessories are off.
9. Clamp the amp probe around the negative battery cable.
10. Read the meter and record the parasitic draw. ______ amps.

    Note: A drain of less than 50 milliamps (0.050 amp) is acceptable on **some** computer-controlled vehicles.

11. Is the drain excessive? ☐ Yes ☐ No

    Is further testing required to locate the problem?

    ☐ Yes ☐ No

12. Before completing your paperwork, clean your work area, clean and return tools to their proper places, and wash your hands.
13. Record your recommendations for needed service or additional repairs and complete the repair order.

____

____

STOP

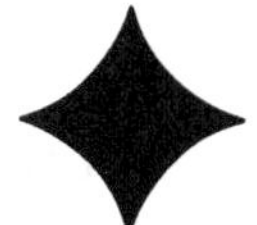

ASE Education Foundation Worksheet #7-14

STARTER CIRCUIT VOLTAGE DROP TEST

Name______________________________ Class____________________

Score: ☐ Excellent ☐ Good ☐ Needs Improvement **Instructor OK** ☐

Vehicle year __________ **Make** ____________________ **Model** ______________________

Objective: Upon completion of this assignment, you will be able to test a starter circuit for high resistance by measuring the voltage drop.

ASE Education Foundation Correlation

This worksheet addresses the following **MLR** task:

VI.C.2 Perform starter circuit voltage drop tests; determine necessary action. **(P-1)**

This worksheet addresses the following **AST/MAST** task:

VI.C.2 Perform starter circuit voltage drop tests; determine needed action. **(P-1)**

We Support ASE Education Foundation

Directions: Before beginning this lab assignment, review the worksheet completely. Fill in the information in the spaces provided as you complete each task.

Tools and Equipment Required: Safety glasses, fender covers, voltmeter, shop towel

Procedure:

Note: Refer to your textbook for more information on voltage drop testing.

1. Open the hood and place fender covers on the fenders and front body parts.

Note: The starter circuit voltage drop test is a valuable test for locating hard starting problems.

Positive Side Voltage Drop Test

Note: Do not crank the starter for periods longer than 30 seconds or the starter can overheat.

Note: With a manual transmission, have someone depress the clutch while cranking. Keep the transmission in neutral.

2. Set the parking brake. Position the transmission selector in Park or Neutral.

3. Disable the ignition system so that the engine can be cranked but will not start. See your instructor if you need help.

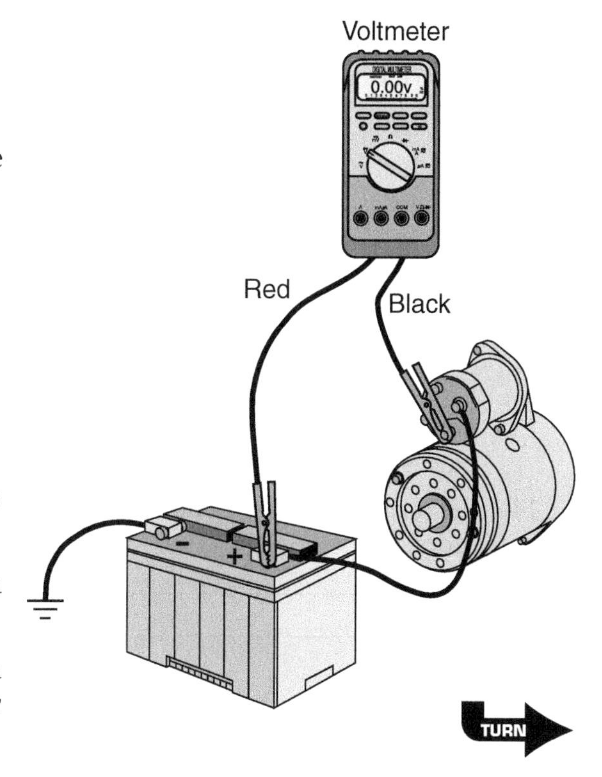

TURN

4. Connect the positive lead of the voltmeter to the positive post of the battery. Be sure it is connected to the post, not the clamp.
5. Connect the negative voltmeter lead to the point where the positive lead enters the starter. (This will be beyond the solenoid or relay.) The circuit is open at the solenoid, so the meter should read battery voltage.

   Voltmeter reading: ______ volts
6. Set the voltmeter to a low scale (2 volts).
7. Crank the engine. The voltage drop will appear on the meter while cranking. An acceptable reading is less than 0.6 volt.

   Voltmeter reading: ______ volts
8. Move the negative voltmeter lead from the starter side of the solenoid to the solenoid terminal that is connected to the battery. Crank the engine. The voltage drop will appear on the meter while cranking. An acceptable reading is less than 0.6 volt.

   Voltmeter reading: ______ volts

   Note: The difference between the voltage readings in steps #7 and #8 indicates the resistance of the solenoid switch circuit.

Negative Side Voltage Drop Test

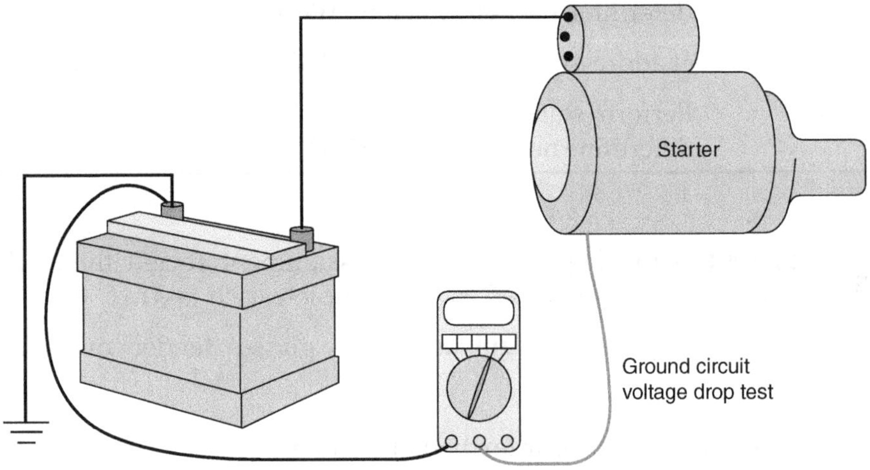

9. Connect the negative lead of the voltmeter to the negative post of the battery. Be sure it is connected to the post, not the clamp.
10. Connect the positive voltmeter lead to the case of the starter. The meter will read "0" volt because there is not any electrical potential difference between the connections.

   Voltmeter reading: ______ volts
11. Crank the engine. The voltage drop will appear on the meter while cranking. An acceptable reading is less than 0.2 volt.

   Voltmeter reading: ______ volts

   Note: Total voltage drop for the entire circuit (negative and positive) should not exceed 0.6 volts.
12. Reconnect the ignition system and start the vehicle.

   Did the engine start? ☐ Yes ☐ No
13. Before completing your paperwork, clean your work area, clean and return tools to their proper places, and wash your hands.
14. Record your recommendations for needed service or additional repairs and complete the repair order.

____

____

STOP

ASE Education Foundation Worksheet #7-15

PERFORM A STARTER DRAW TEST

Name______________________________ Class_____________________

Score: ☐ Excellent ☐ Good ☐ Needs Improvement **Instructor OK** ☐

Vehicle year __________ **Make** ____________________ **Model** ______________________

Objective: Upon completion of this assignment, you will be able to test starter draw.

ASE Education Foundation Correlation

This worksheet addresses the following **MLR** task:

VI.C.1 Perform starter current draw tests; determine necessary action. **(P-1)**

This worksheet addresses the following **AST/MAST** task:

VI.C.1 Perform starter current draw tests; determine needed action. **(P-1)**

We Support ASE Education Foundation

Directions: Before beginning this lab assignment, review the worksheet completely. Fill in the information in the spaces provided as you complete each task.

Tools and Equipment Required: Safety glasses, shop towels, voltmeter, inductive ammeter

Procedure:

1. Open the hood and place fender covers on the fenders and front body parts.
2. Visually inspect the battery, battery cables, and starter.

Battery	☐ OK	☐ Service required
Battery Cables	☐ OK	☐ Service required
Starter	☐ OK	☐ Service required

   Describe any service required: ____________________
3. Conduct a battery load/capacity test.

   ☐ Passed ☐ Failed

   Note: If the battery fails the load/capacity test, it must be recharged or replaced before testing the starter circuit.
4. Disable the engine to prevent it from starting. If you are unsure how to the safely disable the engine ask your instructor for help.
5. Connect a voltmeter to the battery and check the battery voltage.
6. Clamp an inductive ammeter around the negative battery cable.

TURN

7. Turn the ignition key to the start position while reading volts and amps.

   Voltmeter reading: _______________

   Ammeter reading: _______________

8. What are the acceptable cranking voltage and amperage readings for this test.

   Voltage _______________

   Amperage _______________

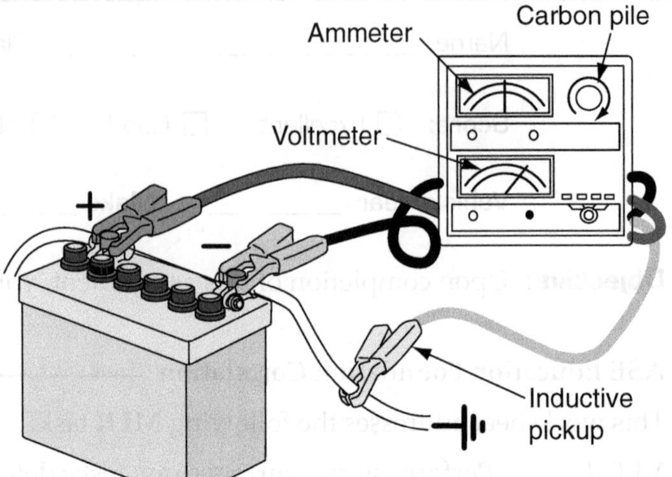

9. Are the readings within the required limits?

   ☐ Yes ☐ No

   If the readings are outside of the limits, what repair(s) might correct the problem?

   ____

10. Before completing your paperwork, clean your work area, clean and return tools to their proper places, and wash your hands.

11. Record your recommendations for needed service or additional repairs and complete the repair order.

   ____

   ____

STOP

ASE Education Foundation Worksheet #7-16

TESTING A STARTER CONTROL CIRCUIT

Name ______________________ Class ______________

Score: ☐ Excellent ☐ Good ☐ Needs Improvement **Instructor OK** ☐

Vehicle year ________ **Make** ______________ **Model** ______________

Objective: Upon completion of this assignment, you will be able to test a starter control circuit.

ASE Education Foundation Correlation

This worksheet addresses the following **MLR** tasks:

VI.C.3 Inspect and test starter relays and solenoids; determine necessary action. **(P-2)**

VI.C.5 Inspect and test switches, connectors, and wires of starter control circuits; determine necessary action. **(P-2)**

This worksheet addresses the following **AST/MAST** tasks:

VI.C.3 Inspect and test starter relays and solenoids; determine needed action. **(P-2)**

VI.C.5 Inspect and test switches, connectors, and wires of starter control circuits; determine needed action. **(P-2)**

We Support ASE Education Foundation

Directions: Before beginning this lab assignment, review the worksheet completely. Fill in the information in the spaces provided as you complete each task.

Tools and equipment: Safety glasses, fender covers, hand tools, jumper leads, voltmeter, and shop towels

Procedure:

1. Inspect the battery terminals: ☐ Good ☐ Need service

   Note: Refer to your textbook for more information on voltage drop testing.

2. Check the voltage drop between the battery post and the battery terminals.

   Voltage drop between the positive post and the positive battery terminal ______

   Voltage drop between the negative post and the negative battery terminal ______

   Voltage drop test results: ☐ Good ☐ Needs service

3. Before proceeding, service the battery terminals as needed.

   ☐ Serviced ☐ No service needed

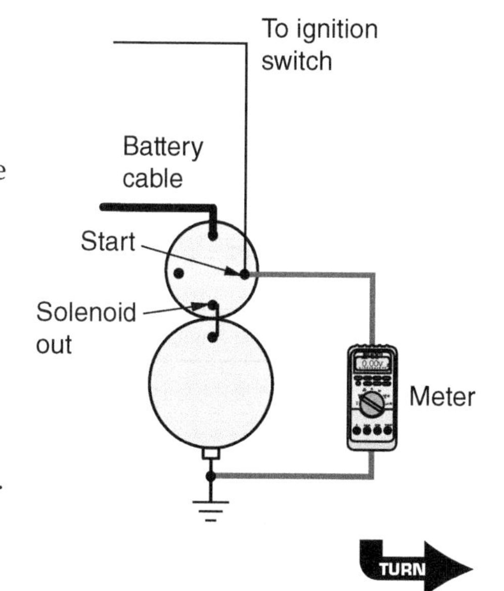

TURN

4. Measure the voltage at the solenoid ignition terminal on the starter while the ignition key is in the *Start* position.

   _______ volts

5. Measure the voltage at the solenoid out (M+) terminal on the starter while the ignition key is in the *Start* position.

   Note: The M+ terminal is the motor connection from the solenoid output.

   Test results: _______ volts

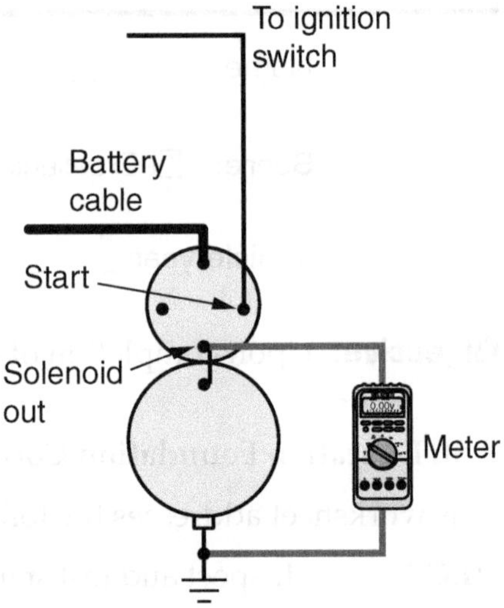

To ignition switch
Battery cable
Start
Solenoid out
Meter

6. Measure the voltage drop in the starting control circuit.

- Disable the fuel or ignition system so the engine will not start during testing.
- Connect the voltmeter positive lead to the battery positive post and the negative lead to the ignition terminal on the starter.
- Turn the ignition key to the start position and read the voltmeter.
- What was the voltage reading while the key was in the *Start* position _______________

  Test Results: ☐ Good ☐ Excessive Voltage Drop

  Note: If the voltage drop is excessive, there is a problem in the ignition switch circuit.

To locate a problem in the ignition switch circuit:

7. Visually inspect the wires and connectors in the ignition switch circuit.

   ☐ OK ☐ Loose ☐ Damaged

8. Test the voltage drop through the ignition switch.

- Connect the positive lead to positive side of the ignition switch and the negative lead to the start wire out of the switch.
- Turn the ignition switch to the start position and read the voltmeter.
- List the voltmeter reading with key in the *Start* position _______________

  Test results: ☐ Good ☐ Excessive Voltage Drop

  Note: If the voltage drop is excessive, replace the ignition switch.

9. Before completing the paperwork, clean your work area, put the tools in their proper places, and wash your hands.

10. List your recommendations for future service and/or additional repairs and complete the repair order.

    ____

    ____

STOP

ASE Education Foundation Worksheet #7-17

TEST A STARTER SOLENOID

Name______________________ Class________________

Score: ☐ Excellent ☐ Good ☐ Needs Improvement **Instructor OK** ☐

Vehicle year ________ **Make** ______________ **Model** ______________

Objective: Upon completion of this assignment, you will be able to test a starter solenoid.

ASE Education Foundation Correlation

This worksheet addresses the following **MLR** tasks:

VI.C.3 Inspect and test starter relays and solenoids; determine necessary action. **(P-2)**

VI.C.5 Inspect and test switches, connectors, and wires of starter control circuits; determine necessary action. **(P-2)**

This worksheet addresses the following **AST/MAST** tasks:

VI.C.3 Inspect and test starter relays and solenoids; determine needed action. **(P-2)**;

VI.C.5 Inspect and test switches, connectors, and wires of starter control circuits; determine needed action. **(P-2)**

We Support ASE Education Foundation

Directions: Before beginning this lab assignment, review the worksheet completely. Fill in the information in the spaces provided as you complete each task.

Tools and equipment: Safety glasses, fender covers, hand tools, jumper cables, small jumper wire and shop towels

Procedure:

In the following solenoid tests, the ground connections are made first. When (B+) is applied to the solenoid ignition terminal, the solenoid will operate. The pull-in winding will also open when B+ is applied to the solenoid ignition terminal. When the starter ground is opened (disconnected), the pinion gear will return.

1. Remove the starter from the vehicle or obtain a starter solenoid assembly from your instructor.
2. Carefully mount the starter in a vise.
3. Disconnect the wire or strap that connects the starter to the solenoid (M+).
4. Connect a small jumper wire from the solenoid starter terminal (M+) to a good ground.
5. Ground the starter by connecting the negative jumper cable from the battery negative to the starter body.

When testing the solenoid, the starter drive pinion will move out rapidly.

6. To test the solenoid:

- Connect the positive (red) jumper cable clamps to the battery positive and the ignition connecter on the solenoid.
- Remove the jumper wire from the solenoid "out" terminal (M+) and make a mental note of what happened.
- Next remove the ground wire from the starter body and make a mental note of what happened.
- Finally, remove the positive (red) jumper lead from the ignition terminal of the starter.

7. Did the solenoid move the starter pinion gear out when the positive (red) jumper clamp was attached?

   ☐ Yes ☐ No

8. When the first jumper wire was removed, what was tested?

   ☐ Pull-in winding ☐ Hold-in winding

9. What happened when the ground cable was removed from the starter? Did the starter drive pinion gear move in?

   ☐ Yes ☐ No

   This tested the: ☐ Pull-in ☐ Hold-in winding

10. Did the solenoid pass all the checks?

    ☐ Yes ☐ No

11. If the starter was removed from a vehicle, reinstall it. If not, return the shop-supplied starter to your instructor.

12. Before completing the paperwork, clean your work area, put the tools in their proper places, and wash your hands.

13. List your recommendations for future service and/or additional repairs and complete the repair order.

    ____

    ____

ASE Education Foundation Worksheet #7-18

REMOVE AND REPLACE A STARTER MOTOR

Name ______________________ Class ________________

Score: ☐ Excellent ☐ Good ☐ Needs Improvement **Instructor OK** ☐

Vehicle year ________ **Make** ______________ **Model** ______________

Objective: Upon completion of this assignment, you will be able to remove and replace a starter.

ASE Education Foundation Correlation

This worksheet addresses the following **MLR/AST/MAST** task:

VI.C.4 Remove and install starter in a vehicle. **(P-1)**

We Support ASE Education Foundation

Directions: Before beginning this lab assignment, review the worksheet completely. Fill in the information in the spaces provided as you complete each task.

Tools and Equipment Required: Safety glasses, fender covers, hand tools, shop towels

Procedure:

1. Use the service information to locate the procedures for removing and replacing the starter.
2. Which service information was used? ______________________________

   Follow the recommended procedures to remove the starter.

CAUTION Before removing the starter, disconnect the negative cable from the battery.

3. Check the condition of the battery cable at the starter.

   ☐ Good ☐ Needs to be replaced ☐ N/A
4. Replace the starter if necessary.

   ☐ New ☐ Used ☐ Rebuilt ☐ Remanufactured ☐ N/A
5. Reinstall the starter.
6. Before completing the paperwork, clean your work area, clean and return tools to their proper places, and wash your hands.
7. Record your recommendations for needed service or additional repairs and complete the repair order.

   ____

   ____

STOP

ASE Education Foundation Worksheet #7-19

MEASURE CHARGING SYSTEM VOLTAGE DROP

Name________________________ Class__________________

Score: ☐ Excellent ☐ Good ☐ Needs Improvement **Instructor OK** ☐

Vehicle year _________ **Make** ______________ **Model** ______________

Objective: Upon completion of this assignment, you will be able to measure voltage drop in the charging system.

ASE Education Foundation Correlation

This worksheet addresses the following **MLR** task:

VI.D.4 Perform charging circuit voltage drop tests; determine necessary action. **(P-2)**

This worksheet addresses the following **AST/MAST** task:

VI.D.2 Diagnose (troubleshoot) charging system for causes of undercharge, no-charge, or overcharge conditions. **(P-1)**

VI.D.5 Perform charging circuit voltage drop tests; determine needed action. **(P-1)**

We Support ASE Education Foundation

Directions: Before beginning this lab assignment, review the worksheet completely. Fill in the information in the spaces provided as you complete each task.

Tools and equipment: Safety glasses, fender covers, hand tools, voltmeter, shop towels

Procedure:

1. Measure the positive side voltage drop.
   - Set the voltmeter to measure DC voltage.
   - Connect the voltmeter positive lead to the battery.
   - Connect the negative voltmeter lead to the AC generator (alternator) voltage output terminal.
   - Start the engine. Accelerate the engine to 2,000 rpm and read the voltmeter.

   Positive side voltage drop ______________

   ☐ Good ☐ Needs service

   Note: If there is excessive voltage drop, track down the problem by measuring the voltage drop at various parts of the positive side of the charging circuit.

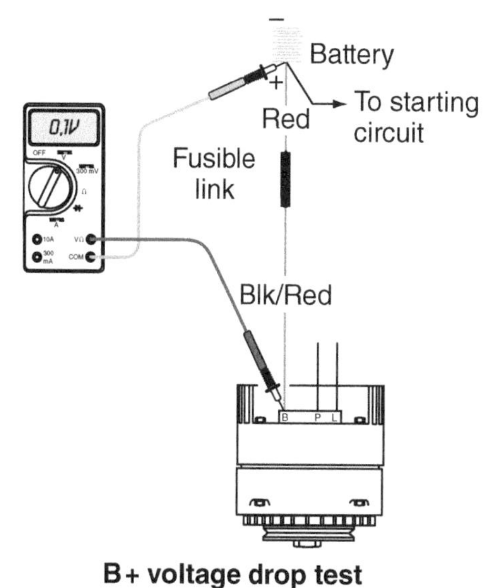

B+ voltage drop test

2. Measure the negative side voltage drop.
   - Connect the voltmeter positive lead to the AC generator housing.
   - Connect the negative voltmeter lead to the negative terminal of the battery.
   - Start the engine. Accelerate to 2,000 rpm and read the voltmeter.

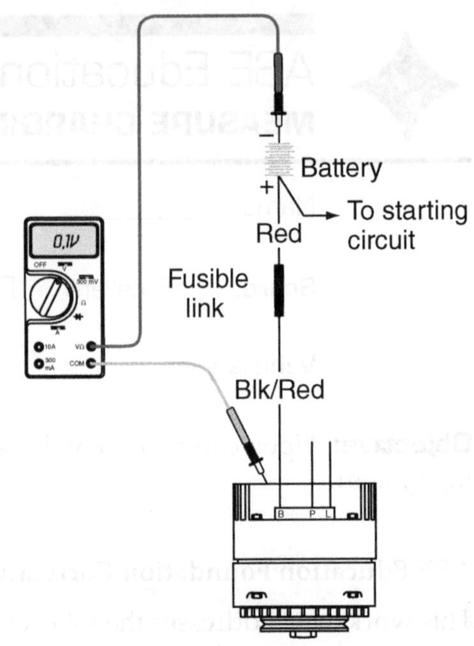

Ground side voltage drop test

Negative side voltage drop ______________

☐ Good ☐ Needs service

Note: If there is excessive voltage drop, track down the problem by measuring the voltage drop at various parts of the negative side of the charging circuit.

3. Before completing the paperwork, clean your work area, put the tools in their proper places, and wash your hands.
4. List your recommendations for future service and/or additional repairs and complete the repair order.

____

____

STOP

ASE Education Foundation Worksheet #7-20

PERFORM A CHARGING SYSTEM OUTPUT TEST

Name______________________ Class________________

Score: ☐ Excellent ☐ Good ☐ Needs Improvement **Instructor OK** ☐

Vehicle year ________ **Make** ______________ **Model** ______________

Objective: Upon completion of this assignment, you will be able to test a generator's output.

ASE Education Foundation Correlation

This worksheet addresses the following **MLR** task:

VI.D.1 Perform charging system output test; determine necessary action. **(P-1)**

This worksheet addresses the following **AST/MAST** task:

VI.D.1 Perform charging system output test; determine needed action. **(P-1)**

We Support ASE | Education Foundation

Directions: Before beginning this lab assignment, review the worksheet completely. Fill in the information in the spaces provided as you complete each task.

Tools and Equipment Required: Safety glasses, shop towels, voltmeter, inductive ammeter

Procedure:

1. Open the hood and place fender covers on the fenders and front body parts.
2. Visually inspect the battery, battery cables, generator, and drive belts.

Battery	☐ OK	☐ Needs service
Battery cables	☐ OK	☐ Needs service
Generator	☐ OK	☐ Needs service
Drive belt	☐ OK	☐ Needs service

   If any service is needed, describe the service required. ______________

   ____

3. Connect a voltmeter to the battery and check the battery voltage.

   Battery voltage: ____________

4. Connect an inductive ammeter around the negative battery cable.
5. Start the engine and maintain the idle speed at 2,000 rpm.
6. Read the voltmeter and ammeter.

   Voltmeter reading: ____________

   Ammeter reading: ____________

TURN

7. Turn on as many electrical accessories as possible (heater, lights, etc.).
8. Maintain the idle speed at 2,000 rpm.

   Read the voltmeter and ammeter.

   Voltmeter reading: ____________

   Ammeter reading: ____________
9. What are the acceptable generator voltage and amperage readings for this test.

   Voltage ____________

   Amperage ____________
10. Are the readings within the required limits?

    ☐ Yes ☐ No

    If the readings are outside of limits, what repair(s) might correct the problem?

    ____
11. Before completing your paperwork, clean your work area, clean and return tools to their proper places, and wash your hands.
12. Record your recommendations for needed service or additional repairs and complete the repair order.

    ____

    ____

STOP

ASE Education Foundation Worksheet #7-21

REMOVE AND REPLACE AN AC GENERATOR (ALTERNATOR)

Name ______________________ Class ______________________

Score: ☐ Excellent ☐ Good ☐ Needs Improvement **Instructor OK** ☐

Vehicle year __________ **Make** ______________ **Model** ______________

Objective: Upon completion of this assignment, you will be able to remove and replace an AC generator (alternator).

ASE Education Foundation Correlation

This worksheet addresses the following **MLR** task:

VI.D.3 Remove, inspect, and/or replace generator (alternator). **(P-2)**

This worksheet addresses the following **AST/MAST** task:

VI.D.4 Remove, inspect, and/or replace generator (alternator). **(P-1)**

We Support ASE Education Foundation

Directions: Before beginning this lab assignment, review the worksheet completely. Fill in the information in the spaces provided as you complete each task.

Tools and Equipment Required: Safety glasses, fender covers, hand tools, shop towels

Procedure:

1. Use the service information to identify the procedures for removing and replacing the generator (alternator).
2. Which service information was used? ______________
3. Follow the recommended procedures to remove the generator.

CAUTION Before removing the generator, disconnect the negative cable from the battery.

4. Loosen and remove the drive belt.
5. Check the condition of the generator drive belt.

   ☐ Good condition ☐ Needs to be replaced
6. Remove the wires from the generator and mark them for future reference.
7. Unbolt the generator. Carefully watch for any spacers that are used in the mounting.
8. Remove the generator from the vehicle.
9. Replace the generator if necessary.

   ☐ New ☐ Used ☐ Rebuilt ☐ Remanufactured ☐ N/A

TURN

10. Before attempting to install the generator compare it to the old one. Make sure that the terminals are the same size and thread pitch and that the mounting flanges are in the same location.

    Does everything match? ☐ Yes ☐ No

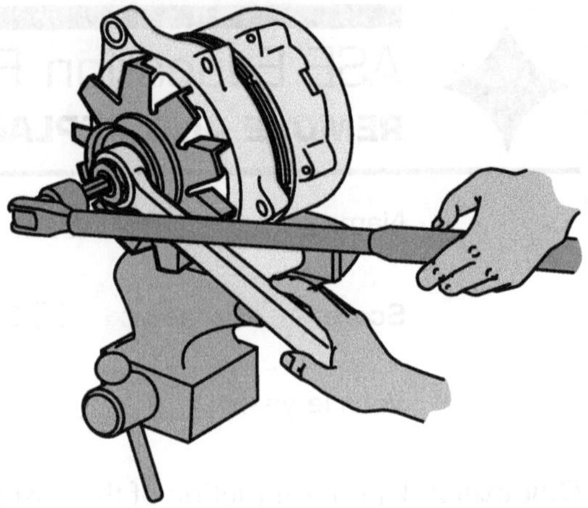

11. It may be necessary to install the drive pulley from the old generator to the new one.

    Was it necessary to switch the pulley?

    ☐ Yes ☐ No

12. Reinstall the generator.

13. Before completing the paperwork, clean your work area, clean and return tools to their proper places, and wash your hands.

14. Record your recommendations for needed service or additional repairs and complete the repair order.

    ____

    ____

STOP

ASE Education Foundation Worksheet #7-22

REPLACE A TAIL/BRAKE LIGHT BULB

Name________________________ Class__________________

Score: ☐ Excellent ☐ Good ☐ Needs Improvement **Instructor OK** ☐

Vehicle year ________ **Make** ________________ **Model** __________________

Objective: Upon completion of this assignment, you will be able to replace a signal, tail, or brake light bulb.

ASE Education Foundation Correlation

This worksheet addresses the following **MLR** task:

VI.E.1 Inspect interior and exterior lamps and sockets including headlights and auxiliary lights (fog lights/driving lights); replace as needed. **(P-1)**

This worksheet addresses the following **AST/MAST** task:

VI.E.2 Inspect interior and exterior lamps and sockets including headlights and auxiliary lights (fog lights/driving lights); replace as needed. **(P-1)**

We Support ASE | Education Foundation

Directions: Before beginning this lab assignment, review the worksheet completely. Fill in the information in the spaces provided as you complete each task.

Tools and Equipment Required: Safety glasses, shop towel, screwdriver, ohmmeter

Parts and Supplies: Light bulb(s)

Procedure:

Note: Signal lights (lights used to signal a driver's intentions) must be checked on a regular basis.

1. Check all exterior lights. Does each of the bulbs light when it is switched on?

   ☐ Yes ☐ No

   Remember that the key must be on for some of the lights to work.

2. List four of the exterior rear light systems on the vehicle on the lines below.

   a. ____________ ☐ OK ☐ Bad ____________

   b. ____________ ☐ OK ☐ Bad ____________

   c. ____________ ☐ OK ☐ Bad ____________

   d. ____________ ☐ OK ☐ Bad ____________

   Note: To replace a bulb, it may be necessary to remove the lens. Some bulbs can be replaced from the back of the light assembly, either under the vehicle or in the fender well. Many rear lights are replaced by accessing them from inside the trunk. Usually if the lens has visible screws, the bulb is replaced by removing the lens. Before removing the lens, check the back of the assembly. If there is an easily removed cover or if the bulb socket is exposed, the bulb is probably removed from the back. Check carefully before proceeding.

   How does the bulb that you are changing appear to be removed?

   ☐ From the trunk ☐ Removable lens cover ☐ Other (describe): ______

TURN

Note: Many bulbs are replaced by pushing them in and turning counterclockwise. Some smaller bulbs are removed by pulling them straight out of their sockets.

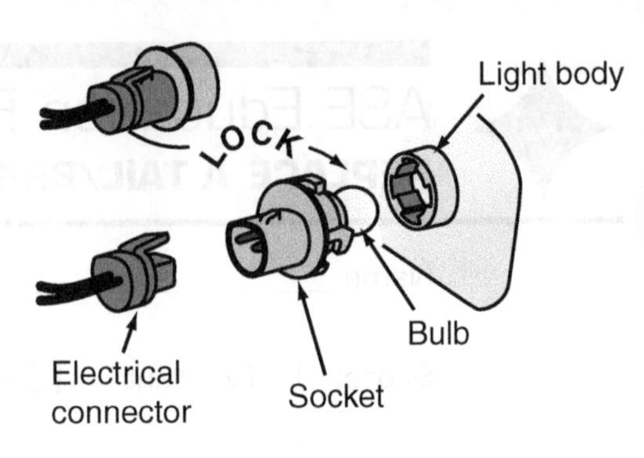

3. To replace a bulb that is removed from the back of the assembly, twist the bulb socket counterclockwise. The socket and bulb assembly will come out of the light housing. Hold the socket and push in on the bulb as you turn it counterclockwise. Reverse the procedure to install the new bulb.

4. To replace a bulb that requires removal of the lens, remove the mounting screws and carefully remove the lens. Be careful not to damage the gasket. Now, push in on the bulb and turn it counterclockwise. Reverse the procedure to install the new bulb.

   Note: When installing the lens, be careful not to overtighten the lens screws. If the lens cracks, moisture will be able to enter the light assembly, which can damage the light and socket.

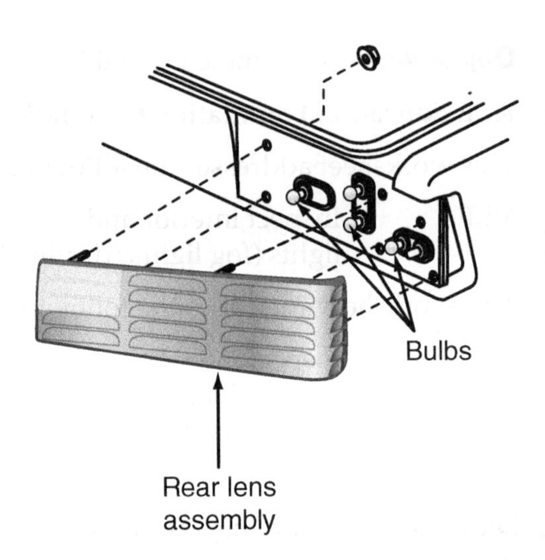

5. Some larger double filament bulbs are used for both the taillights and the brake lights. Do not put a single filament bulb in a dual filament socket.

   The bulb that was changed was a:

   ☐ Single filament ☐ Double filament

6. Check the operation of all of the lights.

7. Use an ohmmeter to check the bulb that was replaced. Set the meter to the highest ohm scale. On a single-filament bulb, connect the red lead to the center electrode of the bulb and the black lead to the side of the bulb or connect one ohmmeter lead to each of the bulb terminals.

   Is the resistance: ☐ High (infinity) ☐ Low

   Bulb is: ☐ Good ☐ Bad

   On a dual-filament bulb, compare the resistances of the two filaments and list them below.

   Brightest filament resistance: ___ ohms

   Dimmest filament resistance: ___ ohms

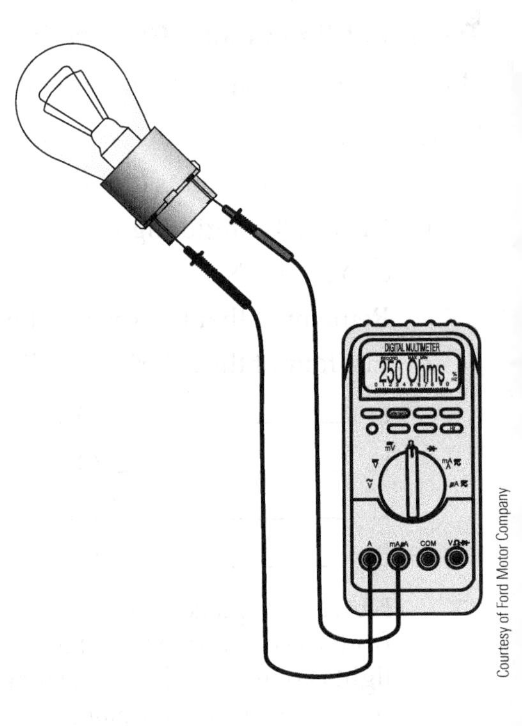

Courtesy of Ford Motor Company

8. Before completing your paperwork, clean your work area, clean and return tools to their proper places, and wash your hands.

9. Record your recommendations for needed service or additional repairs and complete the repair order.

____

____

ASE Education Foundation Worksheet #7-23

REPLACE A TURN SIGNAL FLASHER

Name________________________________ Class________________________

Score: ☐ Excellent ☐ Good ☐ Needs Improvement **Instructor OK** ☐

Vehicle year __________ **Make** ______________________ **Model** ________________________

Objective: Upon completion of this assignment, you should be able to replace a turn signal flasher.

ASE Education Foundation Correlation

This worksheet addresses the following **AST/MAST** task:

VI.E.1 Diagnose (troubleshoot) the causes of brighter-than-normal, intermittent, dim, or no light operation; determine needed action. **(P-1)**

We Support ASE Education Foundation

Directions: Before beginning this lab assignment, review the worksheet completely. Fill in the information in the spaces provided as you complete each task.

Tools and Equipment Required: Safety glasses, shop towels

Procedure:

1. Use the service information to locate the turn signal flasher and the hazard flasher.
   a. Which of the following did you use to find the flasher locations?
      ☐ Service manual
      ☐ Computer program
   b. Where is the turn signal flasher located?
      ____
   c. Where is the hazard flasher located?
      ____
2. Is a flasher located in the fuse panel?
   ☐ Yes ☐ No
3. Turn on the ignition switch and operate the turn signal.
   Can you hear the flasher?
   ☐ Yes ☐ No

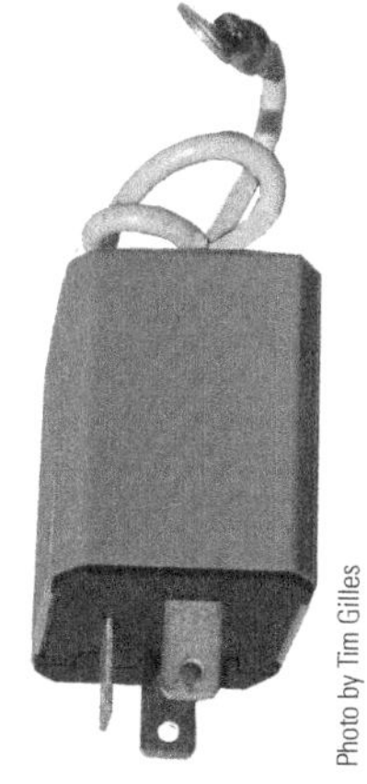

Photo by Tim Gilles

4. Remove the turn signal flasher from its mount. How many electrical connector prongs does it have?

   ☐ Two ☐ Three

5. Reinstall the turn signal flasher.
6. Check the operation of the turn signals.

Right front	☐ OK	☐ Bad
Right rear	☐ OK	☐ Bad
Left front	☐ OK	☐ Bad
Left rear	☐ OK	☐ Bad

   Note: Turn signal flashers that do not flash or that flash too rapidly can indicate a defective turn signal bulb.

7. Before completing your paperwork, clean your work area, clean and return tools to their proper places, and wash your hands.
8. Record your recommendations for needed service or additional repairs and complete the repair order.

   ____

   ____

STOP

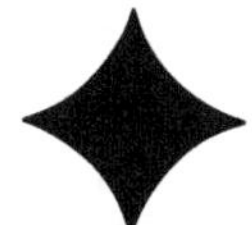

ASE Education Foundation Worksheet #7-24

REPLACE A SEALED BEAM HEADLAMP

Name_______________________________ Class_______________________

Score: ☐ Excellent ☐ Good ☐ Needs Improvement **Instructor OK** ☐

Vehicle year __________ **Make** ____________________ **Model** ______________________

Objective: Upon completion of this assignment, you should be able to replace a sealed beam headlamp.

ASE Education Foundation Correlation

This worksheet addresses the following **MLR** task:

VI.E.1 Inspect interior and exterior lamps and sockets including headlights and auxiliary lights (fog lights/driving lights); replace as needed. **(P-1)**

This worksheet addresses the following **AST/MAST** task:

VI.E.2 Inspect interior and exterior lamps and sockets including headlights and auxiliary lights (fog lights/driving lights); replace as needed. **(P-1)**

We Support ASE | Education Foundation

Directions: Before beginning this lab assignment, review the worksheet completely. Fill in the information in the spaces provided as you complete each task.

Tools and Equipment Required: Safety glasses, fender covers, shop towel, screwdrivers

Parts and Supplies: Sealed beam headlamp

Procedure:

Note: Vehicles use either two or four sealed beam headlamps. A two-headlamp system has both high and low beams in one lamp.

1. Verify that the headlights operate on low and high beams.

Right high beam	☐ OK	☐ Bad
Right low beam	☐ OK	☐ Bad
Left high beam	☐ OK	☐ Bad
Left low beam	☐ OK	☐ Bad

Note: If both high beams or both low beams fail to operate, or if one beam is bright and the other is dim, perform basic electrical testing. This is done to determine whether the bulbs are faulty or there is a problem with an electrical circuit before replacing the bulb(s).

2. Turn off the lights.

3. Open the hood and place fender covers on the fenders.

4. Are the headlight bulbs of the sealed beam design or the composite design?

   ☐ Sealed beam ☐ Composite

   Note: Use Worksheet #7-25, Replace a Composite Headlamp Bulb.

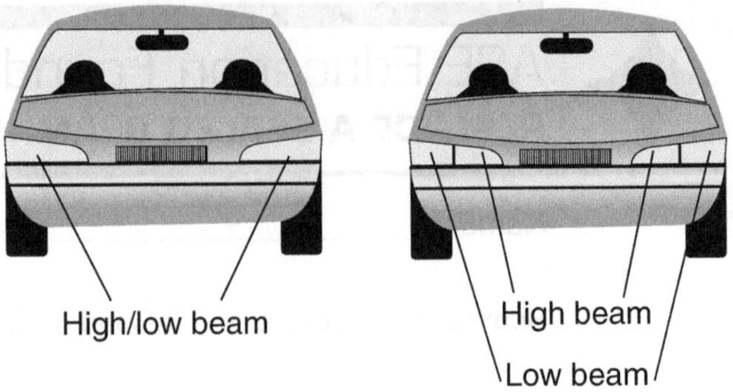

High/low beam

High beam

Low beam

5. An identification number is embossed on the lens of the sealed beam bulb. List the number on each of the bulbs.

   Left ___ Right ___

6. Obtain the replacement bulb before removal of the original. Is it the correct replacement?

   ☐ Yes ☐ No

7. Remove any trim (bezel) or grille pieces that will interfere with the removal of the headlight.

8. Locate the headlight adjusting screws. Do *not* turn these screws.

   Did you locate the adjusting screws? ☐ Yes ☐ No

   Note: Turning the adjustment screws will make it necessary to adjust the headlights after the bulb(s) are replaced.

9. Locate and remove the small screws that hold the headlight retaining ring.

10. If necessary, unhook the headlight retaining ring from its spring and remove it.

11. Remove the headlight and disconnect its electrical connection.

12. Locate the alignment tabs on the rear of the new bulb.

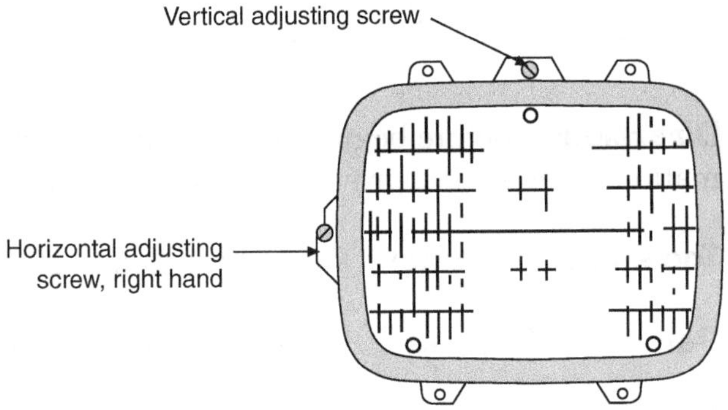

Vertical adjusting screw

Horizontal adjusting screw, right hand

13. Install the bulb in the bracket.

    Note: Align the bulb carefully. It only fits properly one way.

14. Reinstall the headlight retaining ring and reconnect the bulb to the electrical connector.

15. Reinstall any trim or grille pieces that were removed.

16. Check the operation of the headlight. ☐ Good ☐ Bad

17. Before completing your paperwork, clean your work area, clean and return tools to their proper places, and wash your hands.

18. Record your recommendations for needed service or additional repairs and complete the repair order.

    ____

    ____

STOP

ASE Education Foundation Worksheet #7-25

REPLACE A COMPOSITE HEADLAMP BULB

Name_______________________ Class________________

Score: ☐ Excellent ☐ Good ☐ Needs Improvement **Instructor OK** ☐

Vehicle year _________ **Make** ______________ **Model** ______________

Objective: Upon completion of this assignment, you should be able to replace a composite headlamp bulb.

ASE Education Foundation Correlation

This worksheet addresses the following **MLR** task:

VI.E.1 Inspect interior and exterior lamps and sockets including headlights and auxiliary lights (fog lights/driving lights); replace as needed. **(P-1)**

This worksheet addresses the following **AST/MAST** task:

VI.E.2 Inspect interior and exterior lamps and sockets including headlights and auxiliary lights (fog lights/driving lights); replace as needed. **(P-1)**

We Support ASE Education Foundation

Directions: Before beginning this lab assignment, review the worksheet completely. Fill in the information in the spaces provided as you complete each task.

Tools and Equipment Required: Safety glasses, fender covers, shop towel

Parts and Supplies: Composite headlight bulb(s)

Procedure:

1. Verify that the headlights operate on low and high beams.

Right high beam	☐ OK	☐ Bad
Right low beam	☐ OK	☐ Bad
Left high beam	☐ OK	☐ Bad
Left low beam	☐ OK	☐ Bad

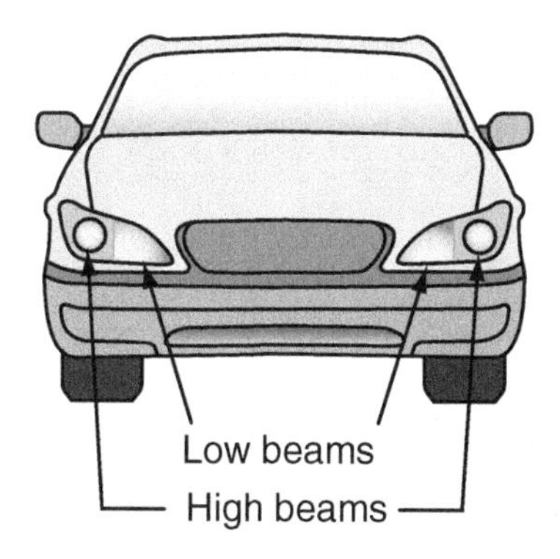

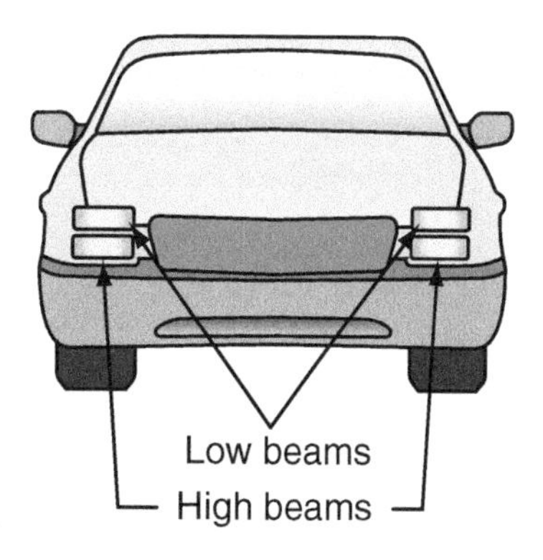

Note: If both high beams or both low beams fail to operate, or if one beam is bright and the other is dim, perform basic electrical testing. This is done to determine whether the bulbs are faulty or there is a problem with an electrical circuit before replacing the bulb(s).

2. Turn off the lights.
3. Open the hood and place fender covers on the fenders.
4. Are the headlight bulbs of the sealed beam design or the composite design?

   ☐ Sealed beam ☐ Composite

   Note: Use Worksheet #7-24, Replace a Sealed Beam Headlamp.

5. Obtain the replacement bulb(s) before removal of the originals. Check that they are the correct replacements. Are they?

   ☐ Yes ☐ No

6. Disconnect the electrical connection to the bulb.
7. Unscrew the bulb retaining ring and remove the bulb.

   Note: Handle the new bulb carefully. Do not touch the glass part of the bulb. The oil from your hands may cause it to explode as it heats up.

8. Locate the alignment tabs on the new bulb.
9. Install the bulb in the bracket.

   Note: Align the bulb carefully; it only fits properly one way.

10. Reinstall the retaining ring and reconnect the bulb to the electrical connector.
11. Check the operation of the headlight. ☐ Good ☐ Bad
12. Before completing your paperwork, clean your work area, clean and return tools to their proper places, and wash your hands.
13. Record your recommendations for needed service or additional repairs and complete the repair order.

    ____

    ____

STOP

ASE Education Foundation Worksheet #7-26

HEADLIGHT ADJUSTING WITH PORTABLE AIMERS

Name ______________________ Class ______________

Score: ☐ Excellent ☐ Good ☐ Needs Improvement **Instructor OK** ☐

Vehicle year ________ **Make** ______________ **Model** ______________

Objective: Upon completion of this assignment, you should be able to use headlight aiming tools to adjust a vehicle's headlights.

ASE Education Foundation Correlation

This worksheet addresses the following **MLR** task:

VI.E.2 Aim headlights. **(P-2)**

This worksheet addresses the following **AST/MAST** task:

VI.E.3 Aim headlights. **(P-2)**

We Support ASE | Education Foundation

Directions: Before beginning this lab assignment, review the worksheet completely. Fill in the information in the spaces provided as you complete each task.

Tools and Equipment Required: Safety glasses, fender covers, headlight aimer, Phillips screwdriver, shop towel

Procedure:

Number of headlights: ☐ Two ☐ Four

Headlight shape: ☐ Round ☐ Rectangular ☐ Curved

Bulb type: ☐ Sealed beam ☐ Sealed halogen ☐ Halogen

Photo by Tim Gilles

1. Select a level work area and park the vehicle there.
2. Check that there are no unusual loads in the vehicle. Is the trunk empty? ☐ Yes ☐ No
3. Fuel level: ☐ Full ☐ ¾ full ☐ ½ full ☐ ¼ full ☐ Empty

   Note: Ideally the vehicle should have a half of a tank of fuel.
4. Jounce the vehicle to settle the suspension.
5. Gently rock the headlight to see if there is any slack between the adjusting screw and the headlight mounting bracket.

   Note: If the headlight is loose, the problem must be corrected before adjusting it.
6. Select the correct adapter for the size and shape of the headlight. Mount it on the aimer.

TURN

7. Clean the headlamp lenses.

   Hold the suction cup against the headlight while pushing forward on the handle on the bottom of the aimer. This forces the suction cup against the headlight lens.

   Pull the handle back until it locks in place and rock the aimer back and forth gently to check that it is attached to the headlight securely.

   Repeat the procedure for the other side.

8. If the aimers have not already been calibrated to the slope of the floor, follow the manufacturer's instructions to calibrate the aimers.

   Are they calibrated? ☐ Yes ☐ No

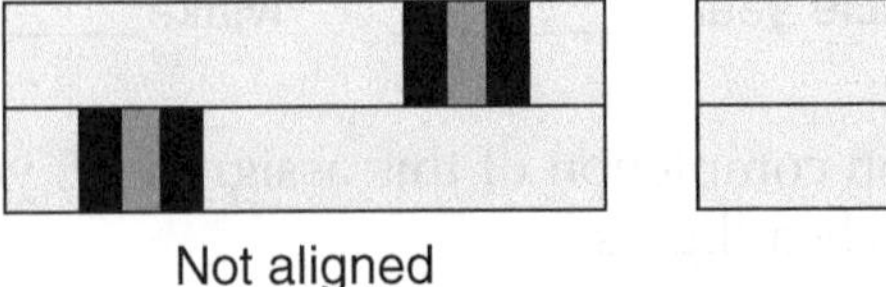

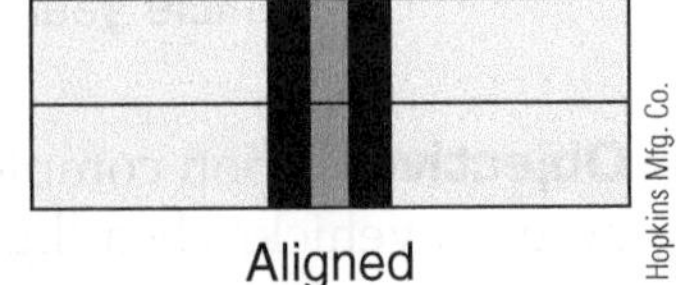

9. Horizontal adjustment:

   Observe the split image lines through the viewing port on the top of the aimer.

   Note: If the lines are not showing, the bulb may be improperly installed or the headlights are far out of adjustment.

   Use a screwdriver to turn the horizontal adjustment screw until the split image aligns. Repeat the adjustment for the other headlight.

   Note: When making adjustments, the final adjustment should be made by turning the adjusting screw clockwise.

10. Vertical adjustment:

    Turn the vertical adjusting screw until the vertical level bubble is centered. Repeat the procedure on the other headlight.

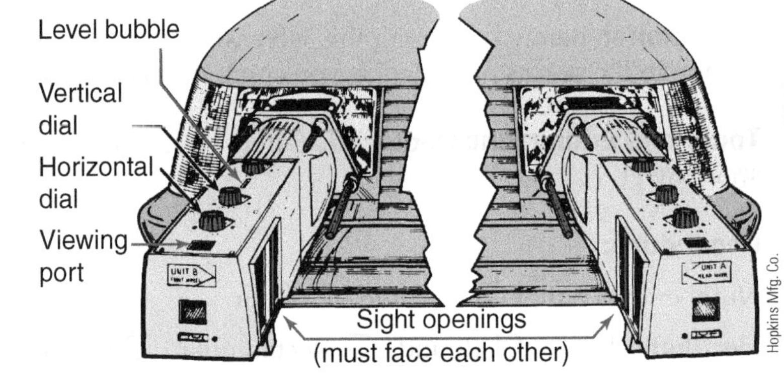

11. Double-check to see that the split images are still aligned. If they are not aligned, readjust the horizontal adjustment.

    Image aligned? ☐ Yes ☐ No

    Additional adjustment required? ☐ Yes ☐ No

12. When a vehicle has four headlights, repeat the adjusting procedure for the remaining two headlights.

13. To remove the aimers, press the vacuum handle.

14. Carefully put the aimers in their box.

15. Before completing your paperwork, clean your work area, clean and return tools to their proper places, and wash your hands.

16. Record your recommendations for needed service or additional repairs and complete the repair order.

    ____

    ____

ASE Education Foundation Worksheet #7-27

HEADLIGHT ADJUSTING WITH AN ELECTRONIC AIMER

Name ______________________ Class ______________

Score: ☐ Excellent ☐ Good ☐ Needs Improvement **Instructor OK** ☐

Vehicle year ________ **Make** ______________ **Model** ______________

Objective: Upon completion of this assignment, you should be able to use an electronic headlight aimer to adjust a vehicle's headlights.

ASE Education Foundation Correlation

This worksheet addresses the following **MLR** task:

VI.E.2 Aim headlights. **(P-2)**

This worksheet addresses the following **AST/MAST** task:

VI.E.3 Aim headlights. **(P-2)**

We Support ASE Education Foundation

Directions: Before beginning this lab assignment, review the worksheet completely. Fill in the information in the spaces provided as you complete each task.

Tools and Equipment Required: Safety glasses, fender covers, electronic headlight aimer, Phillips screwdriver, shop towel

Procedure:

Bulb type: ☐ Sealed beam ☐ Sealed halogen ☐ Halogen ☐ High intensity discharge

1. Park the vehicle in a level work area.
2. Check that there are no unusual loads in the vehicle.

   Is the trunk empty? ☐ Yes ☐ No
3. Fuel level: ☐ Full ☐ ¾ full ☐ ½ full ☐ ¼ full ☐ Empty

   Note: Ideally the vehicle should have a half of a tank of fuel.
4. Check the tire pressure.

   A. What is the recommended tire pressure?

   Front right ______ Left ______

   Rear right ______ Left ______

   B. What is the measured pressure?

   Front right ______ Left ______

   Rear right ______ Left ______
5. Jounce the vehicle to settle the suspension.

TURN

6. Gently rock the headlight to see if there is any slack between the adjusting screw and the headlight mounting bracket.

   Is the head light loose? ☐ Yes ☐ No

   Note: If the headlight is loose, the problem must be corrected before it can be properly adjusted.

7. Clean the headlamp lenses and check lens condition.

   Right: ☐ Clear ☐ Foggy ☐ Cracked

   Left: ☐ Clear ☐ Foggy ☐ Cracked

   Note: Foggy headlight lenses can be polished to restore their clarity.

Align the Aimer to the Vehicle

8. Place the aimer on its track in front of the vehicle.

   Note: The front of the aimer should be within 12″ to 24″ of the headlight.

9. Turn the aimer on and press the button that says "align to vehicle." Then move the aimer to the center of the vehicle and align the laser dot with the center of the grille.
10. Rotate the laser upward to align the dot with the center of the rear view mirror mount on the windshield.
11. Next pivot the laser back to the grille.
12. Continue doing this and moving the aimer head until an imaginary straight line can be drawn between the center of the grille and the rear view mirror. At this point, the aimer is aligned to the vehicle.

Aligning the Aimer to the Headlight

13. Turn the headlights on (low beam) and press the button that says "Align to Lamp."
14. Position the aimer in front of the right headlamp. When both Xs are displayed the center of the lamp has been located. Lock the handle on the aimer head.

Aim the Headlight

15. Open the front cover and press the button that says "Aim Lamp."
16. An X will appear in both the horizontal and vertical positions if the lamps are aimed properly.

    Did the horizontal X appear? ☐ Yes ☐ No

    Did the vertical X appear? ☐ Yes ☐ No

    If either answer is no, the lamp aim will need to be adjusted.

17. Then aim the lamp, rotate the horizontal and vertical adjusting screws until the X appears for both directions.

    Note: When making adjustments, the final adjustment should be made by turning the adjusting screw clockwise.

18. After aiming the right headlamp, move the aimer to the left head lamp and repeat the adjustment procedure.
19. Carefully replace the aimer to its storage container and put it in it's proper place.
20. Record your recommendations for needed service or additional repairs and complete the repair order.

    ____

    ____

STOP

ASE Education Foundation Worksheet #7-28

HEADLAMP ADJUSTING WITHOUT AIMING TOOLS

Name________________________ Class__________________

Score: ☐ Excellent ☐ Good ☐ Needs Improvement **Instructor OK** ☐

Vehicle year ________ **Make** ______________ **Model** ______________

Objective: Upon completion of this assignment, you should be able to adjust a vehicle's headlights without the use of aiming tools.

ASE Education Foundation Correlation

This worksheet addresses the following **MLR** task:

VI.E.2 Aim headlights. **(P-2)**

This worksheet addresses the following **AST/MAST** task:

VI.E.3 Aim headlights. **(P-2)**

We Support ASE Education Foundation

Directions: Before beginning this lab assignment, review the worksheet completely. Fill in the information in the spaces provided as you complete each task.

Tools and Equipment Required: Safety glasses, fender covers, Phillips screwdriver, shop towel, chalk or masking tape

Procedure:

Number of headlights: ☐ Two ☐ Four

Headlight shape: ☐ Round ☐ Rectangular ☐ Curved

Bulb type: ☐ Sealed beam ☐ Sealed halogen ☐ Halogen

1. Select a level work area in front of a wall.
2. Position the vehicle about 3 feet from the wall.
3. Mark the wall with large crosses (+) directly opposite the center of each headlight bulb. There should be one cross for each headlight.
4. Move the vehicle back 25 feet from the wall.
5. Clean the headlight lenses.
6. Gently rock the headlight to see if there is any slack between the adjusting screw and the headlight mounting bracket.

   Is there any slack or looseness? ☐ Yes ☐ No

   Note: If the headlight is loose, the problem must be corrected before adjusting the headlights.

TURN

7. Turn on the headlights to low beam. The low beam lights should be shining within 6 inches below and to the right of the crosses on the wall.

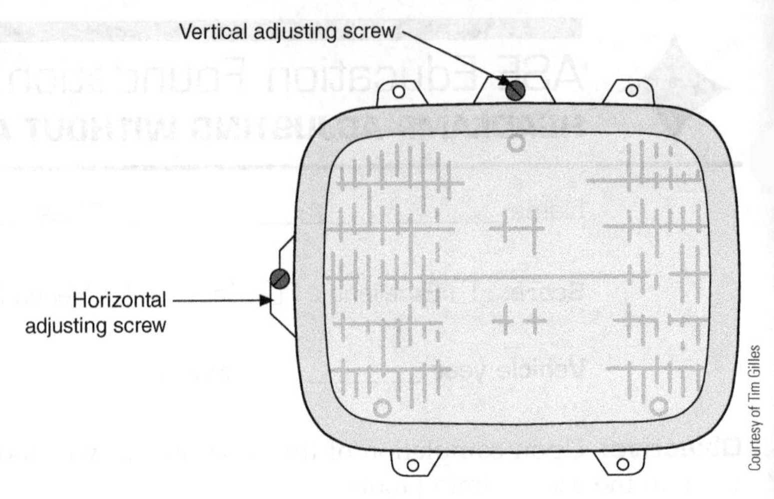

8. To adjust the low beams, start with the light on the right side of the vehicle.
9. Turn the horizontal adjusting screw in or out to move the beam on the wall.

   Note: When making adjustments, the final adjustment should always be made by turning the screw clockwise.
10. Next, turn the vertical adjusting screw in or out to move the beam on the wall.
11. Repeat the procedure on the left side headlight.
12. Check the high beam adjustment. The centers of the light from the high beams should be located near the center of the crosses.

    Note: On two-bulb systems, the high beam will be adjusted when the low beam is adjusted.
13. On four-bulb systems, the high beams will need to be adjusted. Repeat the adjustment process for each of the high beams.

    Note: Since the low beams will be on while the high beams are being adjusted, it may be necessary to cover the low-beam lights. The high beams will be easier to see on the wall with the low beams covered.
14. Before completing your paperwork, clean your work area, clean and return tools to their proper places, and wash your hands.
15. Record your recommendations for needed service or additional repairs and complete the repair order.

    ____

    ____

ASE Education Foundation Worksheet #7-29

TESTING ELECTRICAL CIRCUITS AND COMPONENTS

Name______________________ Class________________

Score: ☐ Excellent ☐ Good ☐ Needs Improvement **Instructor OK** ☐

Vehicle year ________ **Make** ______________ **Model** ______________

Objective: Upon completion of this assignment, you should be able to test electrical circuits and components.

ASE Education Foundation Correlation

This worksheet addresses the following **MLR** tasks:

VI.A.4 Demonstrate proper use of a digital multimeter (DMM) when measuring source voltage, voltage drop (including grounds), current flow, and resistance. **(P-1)**

VI.A.7 Use fused jumper wires to check operation of electrical circuits. **(P-2)**

VI.A.9 Inspect and test fusible links, circuit breakers, and fuses; determine necessary action. **(P-1)**

This worksheet addresses the following **AST/MAST** tasks:

VI.A.3 Demonstrate proper use of a digital multimeter (DMM) when measuring source voltage, voltage drop (including grounds), current flow, and resistance. **(P-1)**

VI.A.6 Use fused jumper wires to check operation of electrical circuits. **(P-1)**

VI.A.9 Inspect and test fusible links, circuit breakers, and fuses; determine needed action. **(P-1)**

VI.A.10 Inspect, test, repair, and/or replace components, connectors, terminals, harnesses, and wiring in electrical/electronic systems (including solder repairs); determine needed action. **(P-1)**

We Support ASE Education Foundation

Directions: Before beginning this lab assignment, review the worksheet completely. Fill in the information in the spaces provided as you complete each task.

Tools and Equipment Required: Safety glasses, fuse, circuit breaker, switch, relay, jumper wires, multimeter

Procedure:

Testing Fuses, Circuit Breakers, and Switches

1. Measure the resistances of the following components obtained from your instructor.

	Good (continuity)	Open (infinite resistance)
Fusible link	☐	☐
Circuit breaker	☐	☐
Fuse	☐	☐
Brake switch	☐	☐

TURN

2. Test a relay (ISO).

   Note: ISO relays have standard code numbers for the relay terminals.

   Code numbers:

   86 = Power to coil

   85 = Ground for the coil

   87 = Normally open contact

   87a = Normally closed contact

   30 = Power for the switch contact

   Note: When voltage is applied to terminal 86, the movable wiper (contact) arm moves to terminal 87. When voltage is removed, it moves back to terminal 87a.

   With no power applied to the relay, use an ohmmeter to:

   Measure the resistance of the coil by connecting an ohmmeter to terminals 86 and 85.

	Open	Closed
Check the continuity between terminals 87a and 30.	☐	☐
Check the continuity between terminals 30 and 87.	☐	☐

   Resistance:
   - ☐ Infinite
   - ☐ 50 to 150 ohms
   - ☐ 150 to 500 ohms
   - ☐ 0 to 50 ohms

   Connect a fused jumper wire from a battery + (power source) to terminal 86. Connect another jumper wire from terminal 85 to ground. Then measure the continuity between:

	Open	Closed
Terminals 87a and 30	☐	☐
Terminals 30 and 87	☐	☐

3. Before completing your paperwork, clean your work area, clean and return tools to their proper places, and wash your hands.

4. Record your recommendations for needed service or additional repairs and complete the repair order.

   ____

   ____

ASE Education Foundation Worksheet #7-30

REMOVE AND INSTALL AN INTERIOR DOOR PANEL

Name ______________________ Class ________________

Score: ☐ Excellent ☐ Good ☐ Needs Improvement **Instructor OK** ☐

Vehicle year ________ **Make** ______________ **Model** ______________

Objective: Upon completion of this assignment, you will be able to inspect, remove, and install a door panel.

ASE Education Foundation Correlation

This worksheet addresses the following **MLR** task:

VI.E.5 Remove and reinstall door panel. **(P-1)**

We Support ASE Education Foundation

Directions: Before beginning this lab assignment, review the worksheet completely. Fill in the information in the spaces provided as you complete each task.

Tools and Equipment Required: Safety glasses, fuse, circuit breaker, switch, relay, jumper wires, multimeter

Procedure:

1. Use the service information to identify service procedures for removing and installing a door panel.

   Which service information was used? ______________________________

Remove the Door Panel

2. Locate and remove any screws or fasteners.
3. Remove door handles, armrest, door lock button, and other trim as needed.
4. Use a door panel tool to carefully pry the door panel away from the door while dislodging the trim clips.
5. Lift the door panel upward to clear the top of the door.

   Note: As the door panel is removed be careful not to damage the plastic moisture barrier located behind the door panel.

6. Carefully disconnect any electrical connectors and set the door panel aside where it will not be damaged.
7. Was the door panel damaged during removal? ☐ Yes ☐ No
8. Was the moisture barrier damaged during removal? ☐ Yes ☐ No

TURN

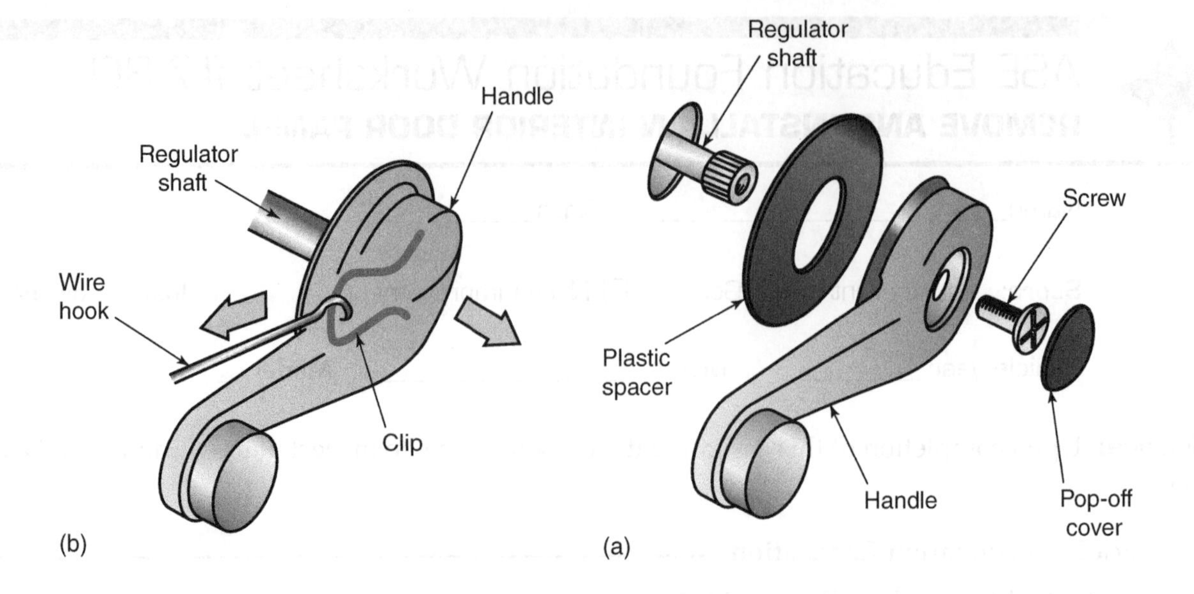

Install a Door Panel

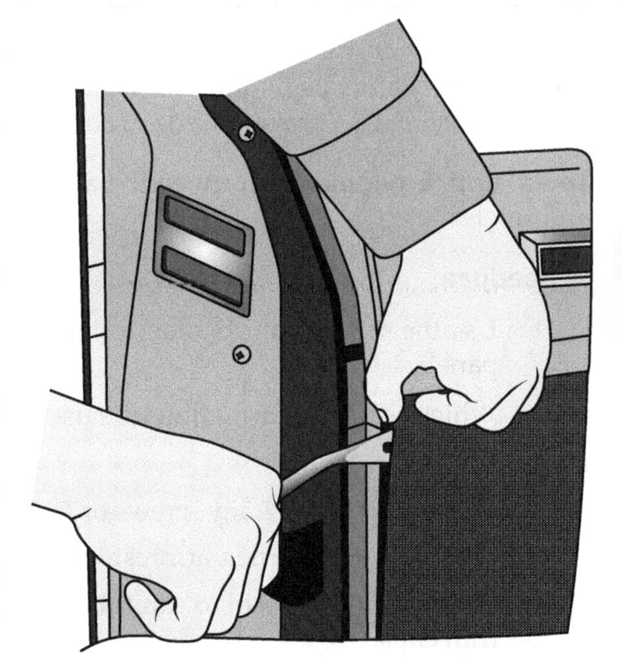

9. Inspect the trim clips and replace as needed.
10. Connect any electrical connectors for power windows or door locks.
11. Slide the panel into the top door edge.
12. Carefully align and tap the trim clips into the holes in the door.
13. Tap on the door panel to set the trim clips.
14. Reinstall door handles, armrest, door lock button, and any other trim as needed.
15. Does the panel fit securely and appear to be properly installed? ☐ Yes ☐ No
16. Before completing the paperwork, clean your work area, put the tools in their proper places, and wash your hands.
17. List your recommendations for future service and/or additional repairs and complete the repair order.

____

____

Part II

Lab Worksheets for ASE Maintenance and Light Repair

Service Area 8

Engine Performance and Maintenance Service

ASE Education Foundation Worksheet #8-1

INSPECT SPARK PLUG CABLES

Name_______________________________ Class____________________

Score: ☐ Excellent ☐ Good ☐ Needs Improvement **Instructor OK** ☐

Vehicle year __________ **Make** ____________________ **Model** ______________________

Objective: Upon completion of this assignment, you should be able to test and replace spark plug cables.

ASE Education Foundation Correlation

This worksheet addresses the following **MLR** tasks:

VI.A.4 Demonstrate proper use of a digital multimeter (DMM) when measuring source voltage, voltage drop (including grounds), current flow, and resistance. **(P-1)**

VIII.A.7 Remove and replace spark plugs; inspect secondary ignition components for wear and damage. **(P-1)**

This worksheet addresses the following **AST/MAST** tasks:

VI.A.3 Demonstrate proper use of a digital multimeter (DMM) when measuring source voltage, voltage drop (including grounds), current flow, and resistance. **(P-1)**

VIII.C.4 Remove and replace spark plugs; inspect secondary ignition components for wear and damage. **(P-1)**

We Support ASE | Education Foundation

Directions: Before beginning this lab assignment, review the worksheet completely. Fill in the information in the spaces provided as you complete each task.

Tools and Equipment Required: Safety glasses, digital multimeter, shop towels

Procedure:

Engine size __________________ # of cylinders __________________

Before starting to test the spark plug cables, identify the type of ignition system used on the vehicle. There are many different types of ignition systems. The basic test procedures are the same for most of them, but knowledge of the systems and the location of their components is important for successful completion of the task. Most of today's vehicles do not have a distributor. On those that do, it may be easier to test the distributor end of the wires from the inside of the distributor cap. Open the hood and place fender covers on the fenders and front body parts.

Correct
Wrong

1. Identify the type of ignition system on the vehicle.

   ☐ Breaker points ☐ Electronic ignition

   ☐ Distributorless ignition (DIS) ☐ Coil-on-plug (COP) ☐ Coil-near-plug

TURN

2. Remove and test the spark plug wires, one cylinder at a time. This prevents them from being accidentally installed on the wrong spark plugs.

CAUTION When removing a spark plug cable from a spark plug, remember to hold onto the spark plug boot and give it a twist before trying to pull the wire off the spark plug. Some spark plug cables are permanently attached to the distributor cap. Do not try to remove them.

3. Visually inspect the condition of each of the spark plug cables. List any defects.

☐ Dirty ☐ Insulation burned ☐ Boot damaged

☐ Terminals corroded ☐ Other

Note: Boots are often cracked. This can result in a misfire under hard acceleration or heavy load.

4. Measure the resistances of the spark plug cables. If your meter is not auto-ranging, select the ohms × 1,000 scale.

5. Connect the ohmmeter leads to both ends of the cable.

Note: It is sometimes necessary to insert a paper clip or jumper wire into the plug boot end of the cable to complete the connection.

6. If there is no specification, the resistance should measure approximately 1,000 ohms per inch of cable for each of the cables. The maximum total resistance must be 20,000 ohms or less.

Measure and record resistances below:

Coil Wire

Length ______" Reading ______ Ω Results ☐ Good ☐ Bad ☐ N/A

Spark Plug Cables

#1 Length ______" Reading ______ Ω Results ☐ Good ☐ Bad ☐ N/A

#2 Length ______" Reading ______ Ω Results ☐ Good ☐ Bad ☐ N/A

#3 Length ______" Reading ______ Ω Results ☐ Good ☐ Bad ☐ N/A

#4 Length ______" Reading ______ Ω Results ☐ Good ☐ Bad ☐ N/A

#5 Length ______" Reading ______ Ω Results ☐ Good ☐ Bad ☐ N/A

#6 Length ______" Reading ______ Ω Results ☐ Good ☐ Bad ☐ N/A

#7 Length ______" Reading ______ Ω Results ☐ Good ☐ Bad ☐ N/A

#8 Length ______" Reading ______ Ω Results ☐ Good ☐ Bad ☐ N/A

7. Condition of the cables: ☐ Good ☐ Replacement required

Number of cables to be replaced ______

Note: If more than one cable is "open" (infinite resistance), replacement of all the cables is recommended.

8. Record your recommendations for needed service or additional repairs and complete the repair order.

____

____

STOP

ASE Education Foundation Worksheet #8-2

REPLACE SPARK PLUGS

Name ______________________ Class ________________

Score: ☐ Excellent ☐ Good ☐ Needs Improvement **Instructor OK** ☐

Vehicle year ________ **Make** ______________ **Model** ________________

Objective: Upon completion of this assignment, you should be able to replace an engine's spark plugs.

ASE Education Foundation Correlation

This worksheet addresses the following **MLR** task:

VIII.A.7 Remove and replace spark plugs; inspect secondary ignition components for wear and damage. **(P-1)**

This worksheet addresses the following **AST/MAST** task:

VIII.C.4 Remove and replace spark plugs; inspect secondary ignition components for wear and damage. **(P-1)**

We Support ASE | Education Foundation

Directions: Before beginning this lab assignment, review the worksheet completely. Fill in the information in the spaces provided as you complete each task.

Tools and Equipment Required: Safety glasses, fender covers, spark plug wrench, ratchet, extension, vacuum hose, antiseize, shop towels

Parts and Supplies: Spark plugs

Procedure:

Engine size ______________ # of cylinders ______________

1. Open the hood and place fender covers on the fenders and front body parts.
2. List the part number and brand name of one of the spark plugs installed in the engine.

   Number ______________ Brand ______________
3. Purchase the specified spark plugs for the vehicle.
4. Be certain that the plugs are the correct heat range.

   Note: Changing heat ranges can result in serious engine damage or poor cold engine operation.
5. Open each of the packaged spark plugs. Check the gap to verify it is set at the manufacturer's specification. Adjust as needed.

   Gap specification 0. ______"

R45TS

R = resistor
4 = 14 mm thread
5 = heat range
T = taper seat
S = extended tip

TURN

NOTE: With many newer spark plugs, the gaps are not adjustable. Be sure to check the service information. Attempting to change the gap on these spark plugs can result in a ruined part.

Note: Removing and replacing spark plugs is best done on a cold engine.

6. Blow compressed air around the spark plugs. This prevents dirt from entering the cylinders when the plugs are removed and is especially important on transverse mounted engines.
7. Grasp the rubber boot on the spark plug wire and twist it loose before pulling it off the spark plug.
8. Use the correct size spark plug socket to remove the plug.
9. Inspect each spark plug and note its condition.

Cylinder	#1	#2	#3	#4	#5	#6	#7	#8
Normal	☐	☐	☐	☐	☐	☐	☐	☐
Blistered	☐	☐	☐	☐	☐	☐	☐	☐
Gray/white	☐	☐	☐	☐	☐	☐	☐	☐
Carbon fouled	☐	☐	☐	☐	☐	☐	☐	☐
Oil fouled	☐	☐	☐	☐	☐	☐	☐	☐
Damaged	☐	☐	☐	☐	☐	☐	☐	☐

10. Install a new plug. Do *not* use a ratchet until the plug has been hand-tightened.

SHOP TIP

A short piece of vacuum hose installed on the top of the plug makes a handy installation tool. Install the spark plug carefully. Cross-threading a spark plug can result in a very expensive repair.

Photo by Tim Gilles

11. Is the cylinder head made of aluminum or cast iron?

☐ Aluminum ☐ Cast iron

Note: If in doubt about whether or not the head is aluminum, check it with a magnet.

12. Tighten the spark plug to the recommended torque. Torque specification __________

Tightening late model spark plugs to the correct torque is very important. If there is no torque specification, the following recommendation can be used:

- Tighten gasketed plugs ¼ turn (90 degrees or 0–3 o'clock) past finger-tight.
- Tighten tapered seat plugs $^{1}/_{16}$ turn (23 degrees or 12–12:45 o'clock) past finger-tight.

The spark plugs are: ☐ Gasketed ☐ Tapered seat

13. Install the spark plug wire, making sure that the metal clip inside the boot firmly grasps the metal end of the spark plug.

Note: Some manufacturers recommend the use of a dielectric compound in spark plug boots.

14. Repeat the spark plug replacement procedure for the remaining plugs.
15. Before completing your paperwork, clean your work area, clean and return tools to their proper places, and wash your hands.
16. Record your recommendations for needed service or additional repairs and complete the repair order.

____

____

STOP

ASE Education Foundation Worksheet #8-3

REPLACE SPARK PLUG CABLES

Name________________________ Class________________

Score: ☐ Excellent ☐ Good ☐ Needs Improvement **Instructor OK** ☐

Vehicle year ________ **Make** ______________ **Model** ______________

Objective: Upon completion of this assignment, you should be able to replace an engine's spark plug cables.

ASE Education Foundation Correlation

This worksheet addresses the following **MLR** task:

VIII.A.7 Remove and replace spark plugs; inspect secondary ignition components for wear and damage. **(P-1)**

This worksheet addresses the following **AST/MAST** task:

VIII.C.4 Remove and replace spark plugs; inspect secondary ignition components for wear and damage. **(P-1)**

We Support ASE Education Foundation

Directions: Before beginning this lab assignment, review the worksheet completely. Fill in the information in the spaces provided as you complete each task.

Tools and Equipment Required: Safety glasses, shop towel

Procedure:

Engine size ______________ # of cylinders ______________

Note: Spark plug cables should be replaced one at a time. This will prevent mixing them up or misrouting them. If you choose to remove all of the cables at once, this worksheet will help you properly reinstall the cables.

1. Check the service information for the correct firing order, cylinder numbering, and direction of distributor rotation.

   Firing order ______________________________

   Distributor rotation: ☐ Clockwise ☐ Counterclockwise ☐ N/A

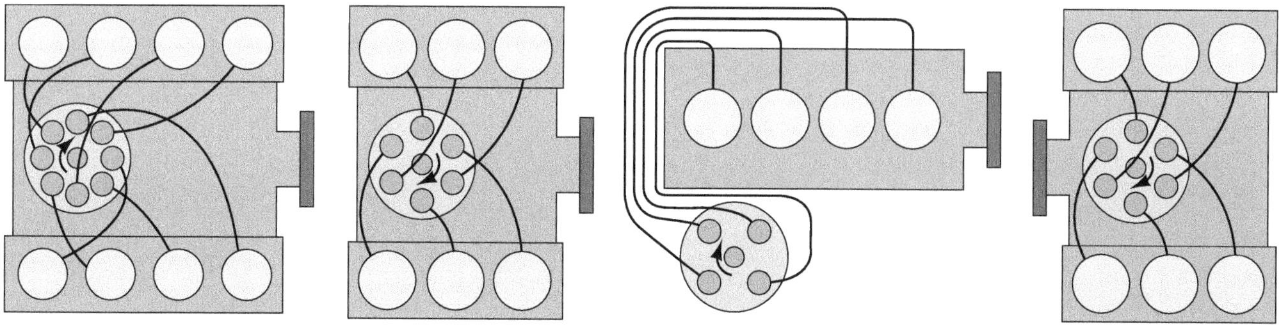

TURN

Cylinder numbering: Draw a sketch showing cylinder numbering.

2. Crank the engine until the #1 cylinder is at top dead center on the compression stroke.
3. The distributor rotor should now be pointing at the distributor cap terminal for the #1 cylinder's cable.
4. Install the #1 spark plug cable in the distributor cap and carefully route it to the #1 spark plug. Install the next cable in the firing order in the distributor cap. Make sure it follows the direction of distributor rotation. Connect the cable to the second spark plug in the firing order. Repeat this procedure until all of the cables are installed.

Clockwise rotation distributor

Note: Route the cables so that they are clear of the drive belt(s) and exhaust manifold(s). Cylinders that fire one after the other should not be positioned next to each other. Keep the plug cables away from the vehicle's electrical wiring.

5. Start the vehicle. Is the idle smooth? ☐ Yes ☐ No
6. Accelerate the engine. Does it accelerate smoothly? ☐ Yes ☐ No
7. Before completing your paperwork, clean your work area, clean and return tools to their proper places, and wash your hands.
8. Record your recommendations for needed service or additional repairs and complete the repair order.

____

____

STOP

ASE Education Foundation Worksheet #8-4

MEASURE ENGINE VACUUM

Name_______________________ Class_______________

Score: ☐ Excellent ☐ Good ☐ Needs Improvement **Instructor OK** ☐

Vehicle year ________ **Make** ______________ **Model** ______________

Objective: Upon completion of this assignment, you should be able to measure engine vacuum and analyze the results.

ASE Education Foundation Correlation

This worksheet addresses the following **MLR** task:

VIII.A.2 Perform engine absolute manifold pressure tests (vacuum/boost); document results **(P-2)**

This worksheet addresses the following **AST/MAST** tasks:

VIII.A.5 Perform engine absolute manifold pressure tests (vacuum/boost); determine needed action. **(P-1)**

VIII.A.9 Diagnose engine mechanical, electrical, electronic, fuel, and ignition concerns; determine needed action. **(P-2)**

This task addresses the following **AST** task:

VIII.D.10 Perform exhaust system back-pressure test; determine needed action. **(P-2)**

This task addresses the following **MAST** task:

VIII.D.11 Perform exhaust system back-pressure test; determine needed action. **(P-2)**

We Support ASE Education Foundation

Directions: Before beginning this lab assignment, review the worksheet completely. Fill in the information in the spaces provided as you complete each task.

Tools and Equipment Required: Safety glasses, shop towel, vacuum gauge

Procedure:

1. Open the hood and place fender covers on the fenders.
2. Locate a vacuum source at the intake manifold.
3. Contact a vacuum gauge to the intake manifold.
4. Start the engine and read the gauge. What is the vacuum reading? ______

   If the reading is 0, open the throttle. Did the reading increase?

   ☐ Yes ☐ No

If the reading increased when the engine was accelerated, connect the gauge to another vacuum source.

Does the vacuum gauge now have a good reading at idle?

☐ Yes ☐ No

5. What is the vacuum reading on the gauge? ________
6. Is it within specifications? ☐ Yes ☐ No

   Is the reading steady or jumping around?

   ☐ Steady ☐ Jumping

   If it is not steady and within specifications, what is indicated? ________
7. To check cranking vacuum, turn off the engine, plug the breather hose into the air cleaner, and disable the ignition system.
8. Crank the engine and read the gauge. What is the reading? ________

   Was the reading steady or jumping? ☐ Steady ☐ Jumping

   What would be indicated if the reading were jumping around? ________
9. Check for a restricted exhaust by holding the engine at 2,000 rpm and watching the gauge. The reading should be approximately the same as the reading at idle and hold steady. If the reading starts to fall, the exhaust is restricted.
10. Did the gauge maintain its reading or did it begin to fall? ☐ Maintained ☐ Dropped
11. Before completing your paperwork, clean your work area, clean and return tools to their proper places, and wash your hands.
12. Record your recommendations for needed service or additional repairs and complete the repair order.

____

____

Photo by Tim Gilles

STOP

ASE Education Foundation Worksheet #8-5

POWER BALANCE TESTING

Name________________________________ Class_______________________

Score: ☐ Excellent ☐ Good ☐ Needs Improvement **Instructor OK** ☐

Vehicle year __________ **Make** ___________________ **Model** ____________________

Objective: Upon completion of this assignment, you should be able to test the power balance between cylinders in an engine.

ASE Education Foundation Correlation

This worksheet addresses the following **MLR** task:

VIII.A.3 Perform cylinder power balance test; document results. **(P-2)**

This worksheet addresses the following **AST/MAST** tasks:

VIII.A.6 Perform cylinder power balance test; determine needed action. **(P-2)**

VIII.A.9 Diagnose engine mechanical, electrical, electronic, fuel, and ignition concerns; determine needed action. **(P-2)**

We Support ASE Education Foundation

Directions: Before beginning this lab assignment, review the worksheet completely. Fill in the information in the spaces provided as you complete each task.

Tools and Equipment Required: Safety glasses, fender covers, shop towel, scan tool, oscilloscope or power balance tester

Procedure: Power balance testing, also called load testing, is a quick way to compare the output of an engine's cylinders. The test is best done using a scan tool, a lab scope, or a portable electronic cylinder balance tester. These testers will avoid damage to the ignition module or the catalytic converter. Engine management systems have power balance testing capability that is accessible with a scan tool. Scan tools perform power balance tests by shutting off fuel injectors one at a time. This allows cylinders to be disabled without adding raw fuel to the exhaust stream.

1. Open the hood and place fender covers on the fenders and front body parts.
2. Connect a scan tool, a lab scope, or a power balance tester to the vehicle.
3. Power balance testing automatically shorts each cylinder to ground for a specified period before shorting the next one. The drop in rpm is recorded as each cylinder misfires.
4. Record the rpm drop for each cylinder below.

Cylinder	#1	#2	#3	#4	#5	#6	#7	#8
	______	______	______	______	______	______	______	______

5. If a cylinder did not drop in rpm or had an increase in rpm while it was misfiring, a problem is indicated.

   Which cylinders, if any, failed the test?

   List the cylinders by number that failed ________________

   ☐ All cylinders tested good

6. If the test indicates that there is a problem, there are several follow-up tests that may be done, including checking valve lash, compression testing, vacuum testing, cylinder leakage testing, and oscilloscope analysis.

7. If a problem was noticed during the test, check with your instructor before proceeding with additional tests. What did your instructor recommend? ________________

   ☐ N/A (There was no problem.)

8. Before completing your paperwork, clean your work area, clean and return tools to their proper places, and wash your hands.

9. Record your recommendations for needed service or additional repairs and complete the repair order.

   ____

   ____

STOP

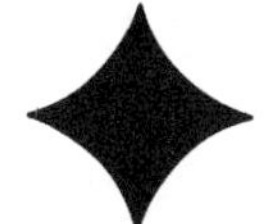

ASE Education Foundation Worksheet #8-6

COMPRESSION TEST

Name________________________ Class________________

Score: ☐ Excellent ☐ Good ☐ Needs Improvement **Instructor OK** ☐

Vehicle year ________ **Make** ______________ **Model** ______________

Objective: Upon completion of this assignment, you should be able to check an engine's compression pressure.

ASE Education Foundation Correlation

This worksheet addresses the following **MLR** tasks:

VIII.A.4 Perform cylinder cranking and running compression tests; document results. **(P-2)**

VIII.A.6 Verify engine operating temperature. **(P-1)**

This worksheet addresses the following **AST/MAST** tasks:

VIII.A.7 Perform cylinder cranking and running compression tests; determine needed action. **(P-1)**

VIII.A.9 Diagnose engine mechanical, electrical, electronic, fuel, and ignition concerns; determine needed action. **(P-2)**

VIII.A.10 Verify engine operating temperature; determine needed action. **(P-1)**

We Support ASE Education Foundation

Directions: Before beginning this lab assignment, review the worksheet completely. Fill in the information in the spaces provided as you complete each task.

Tools and Equipment Required: Safety glasses, fender covers, spark plug socket, ratchet, compression gauge, shop towel, oil

Procedure: Before attempting this worksheet, complete Worksheet #8-2, Replace Spark Plugs.

1. Open the hood and place fender covers on the fenders and front body parts.
2. Is the engine at normal operating temperature? ☐ Yes ☐ No
3. Use a pedal depressor to block the throttle wide open.
4. Remove the spark plugs one at a time for this test.

CAUTION Failure to disable the ignition system could result in a damaged ignition module during engine cranking.

5. Disable the engine using one of the following methods.

   Check the method used:

   ☐ Remove the ignition fuse. This method works well with newer distributorless ignition systems.

   ☐ On fuel-injected vehicles, disconnect the fuel pump fuse or fuel pump relay

   ☐ Distributor ignition coil: Use a jumper wire to ground the coil high-tension wire at the distributor cap end.

   ☐ Coil wire not accessible: Disconnect the battery power wire to the ignition system.

6. Select the correct compression gauge adapter and thread it into the cylinder to be tested. Tighten it just enough to compress the O-ring seal. Next install the compression gauge onto the adapter.

Photo by Tim Gilles

7. Crank the engine through at least four compression strokes.

   Note: The needle will advance with each compression stroke in response to pressure buildup in the cylinder. Engines with very good compression rings will sometimes reach the highest reading in as little as two compression strokes. With poor rings, six compression strokes might be necessary.

   Record the compression reading for each of the cylinders:

   #1 ______ #2 ______ #3 ______ #4 ______
   #5 ______ #6 ______ #7 ______ #8 ______

   Note: Variations between cylinders should be no more than 20%.

   Are any of the compression readings below acceptable levels? ☐ Yes ☐ No

8. Is a wet test required? ☐ Yes ☐ No

Wet Compression Testing

9. If a wet compression test is required, remove the compression gauge and squirt approximately 1 tablespoon of oil into the cylinder to be tested.

10. Reinstall the compression gauge and crank the engine through at least four compression strokes.

    Record the gauge reading and compare it to the dry test reading.

    Dry reading ______ Wet reading ______

    What might happen if too much oil is squirted into the cylinder?

    ____

    Note: Wet testing is done only on cylinders that may have a problem.

    Why would you not wet test all of the cylinders?

    ____

11. Which of the following diagnoses can be made from the compression test just completed?

    ❏ Worn piston rings or cylinders
    ❏ Leaking valves
    ❏ Blown head gasket
    ❏ Excessive compression

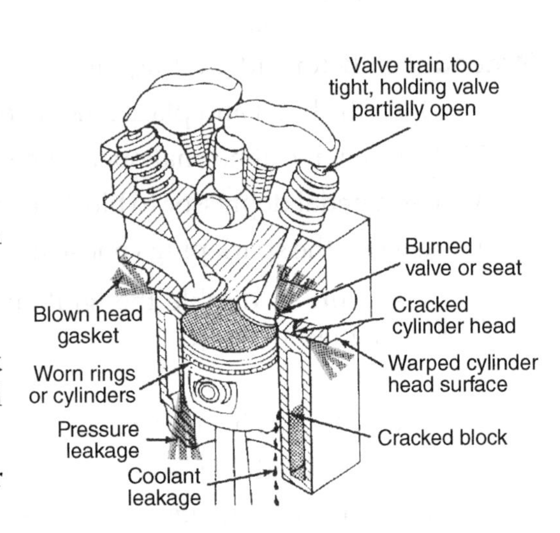

12. Install the spark plugs and enable the engine to run.

13. Are the engine's compression test results within normal limits?

    ☐ Yes ☐ No

14. Before completing your paperwork, clean your work area, clean and return tools to their proper places, and wash your hands.

15. Record your recommendations for needed service or additional repairs and complete the repair order.

    ____

    ____

STOP

ASE Education Foundation Worksheet #8-7

RUNNING COMPRESSION TEST

Name_______________________________ Class_______________________

Score: ☐ Excellent ☐ Good ☐ Needs Improvement **Instructor OK** ☐

Vehicle year __________ **Make** ___________________ **Model** _____________________

Objective: Upon completion of this assignment, you should be able to check an engine's running compression pressure.

ASE Education Foundation Correlation

This worksheet addresses the following **MLR** tasks:

VIII.A.4 Perform cylinder cranking and running compression tests; document results. **(P-2)**

VIII.A.6 Verify engine operating temperature. **(P-1)**

This worksheet addresses the following **AST/MAST** tasks:

VIII.A.7 Perform cylinder cranking and running compression tests; determine needed action. **(P-1)**

VIII.A.9 Diagnose engine mechanical, electrical, electronic, fuel, and ignition concerns; determine needed action. **(P-2)**

VIII.A.10 Verify engine operating temperature; determine needed action. **(P-1)**

We Support ASE Education Foundation

Directions: Before beginning this lab assignment, review the worksheet completely. Fill in the information in the spaces provided as you complete each task.

Tools and Equipment Required: Safety glasses, fender covers, spark plug socket, ratchet, compression gauge, shop towel, oil

Procedure: Before attempting this worksheet, complete Worksheet #8-2, Replace Spark Plugs.

1. Open the hood and place fender covers on the fenders and front body parts.
2. Is the engine at normal operating temperature? ☐ Yes ☐ No
3. Remove the spark plugs one at a time for this test.
4. Select the correct compression gauge adapter and thread it into the cylinder to be tested. Tighten it just enough to compress the O-ring seal. Next install the compression gauge onto the adapter.
5. Start the engine and let it come to an idle. Bleed pressure off through the pressure relief valve and allow the reading to stabilize.

   Note: Some technicians remove the Schrader valve from the end of the compression tester when performing this test. If you choose to do that, do not forget to reinstall it following the test. Remember that the reading must be taken while the engine runs. With no Schrader valve, the tester will not retain the reading.

TURN

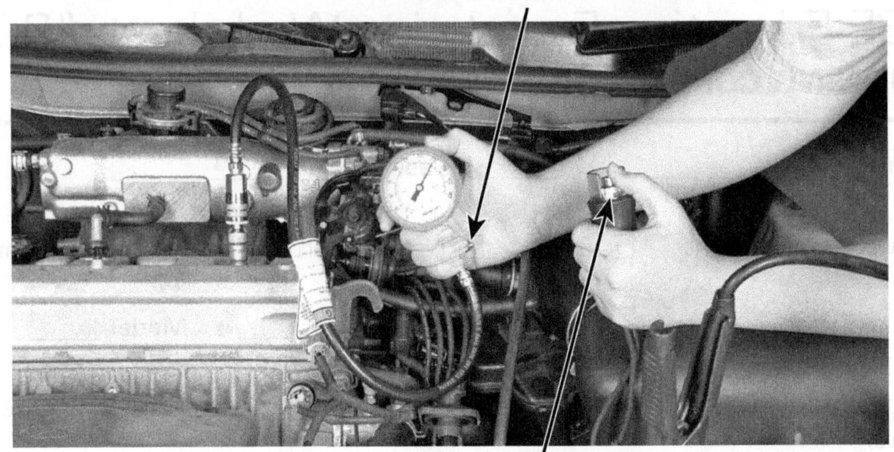

6. Read the compression gauge and record the pressure in the chart below.
7. Snap the throttle to WOT and let the engine return to idle. Record the pressure in the chart below.
8. Turn the engine off, remove the compression gauge and reinstall the spark plug.
9. Continue testing the running compression of each of the remaining cylinders and record the compression readings in the chart below.

Running Compression Test Results

	#1	#2	#3	#4	#5	#6	#7	#8
Idle								
WOT								

10. Which of the following diagnoses can be made from the running compression test just completed?

    ☐ All cylinders tested good

    ☐ Intake restriction

    ☐ Exhaust restriction

    ☐ Carbon buildup, worn cam, or broken spring

11. Before completing your paperwork, clean your work area, clean and return tools to their proper places, and wash your hands.
12. Record your recommendations for needed service or additional repairs and complete the repair order.

____

____

STOP

ASE Education Foundation Worksheet #8-8

PERFORM A CYLINDER LEAKAGE TEST

Name____________________________ Class____________________

Score: ☐ Excellent ☐ Good ☐ Needs Improvement **Instructor OK** ☐

Vehicle year __________ **Make** ___________________ **Model** ____________________

Objective: Upon completion of this assignment, you should be able to perform a cylinder leakage test.

ASE Education Foundation Correlation

This worksheet addresses the following **MLR** task:

VIII.A.5 Perform cylinder leakage test; document results. **(P-2)**

This worksheet addresses the following **AST/MAST** tasks:

VIII.A.8 Perform cylinder leakage test; determine needed action. **(P-1)**

We Support
ASE | Education Foundation

Directions: Before beginning this lab assignment, review the worksheet completely. Fill in the information in the spaces provided as you complete each task.

Tools and Equipment Required: Safety glasses, fender covers, shop towels, cylinder leakage tester, spark plug socket, ratchet

Procedure:

1. Open the hood and place fender covers on the fenders and front body parts. The cylinder leakage test is more accurate if the engine is at normal operating temperature.

   Engine temperature:

   ☐ Normal operating temperature

   ☐ Warm

   ☐ Cold

Photo by Tim Gilles

2. Carefully remove the spark plug cables or coils (if coil-on-plug). Twist the cable boots on the spark plugs to break any seal before removing the cables from the spark plugs.

   Ignition type:

   ☐ Coil-on-plug

   ☐ Distributor ignition

   ☐ Distributorless ignition

   ☐ Coil-near-plug

3. Be sure you are wearing eye protection, and carefully blow compressed air around each of the spark plugs.
4. Remove the spark plugs, checking the condition of each one as it is removed so you relate any abnormally looking spark plug to the cylinder from which it was removed.

   Spark plug condition:

   ☐ Normal

   ☐ Abnormal: List cylinder number(s) ______
5. Disable the ignition system by removing the ignition system fuse.
6. Rotate the crankshaft until the cylinder to be tested is at top dead center (TDC) on the power stroke.
7. Connect the cylinder leakage tester to the shop air hose and calibrate the tester to zero its gauge pointer.
8. Install the cylinder leakage test adapter into the spark plug hole of the cylinder to be tested.
9. Connect the cylinder leakage tester to the cylinder adapter.
10. Read and record the tester reading. ______%

    Is this an acceptable result?

    ☐ Acceptable.

    ☐ Engine requires repair.
11. If the leakage is not acceptable, note the location where you hear air leaking:

    ☐ Crankcase (oil filler cap)

    ☐ Engine air intake

    ☐ Exhaust pipe

    ☐ Radiator

    Repeat the leakage test for any other cylinders in need of testing and record any abnormalities below:

    ☐ Good condition.

    ☐ Further repair needed. List cylinder number(s): ______________________________
12. Before completing your paperwork, clean your work area, clean and return tools to their proper places, and wash your hands.
13. Record your recommendations for needed service or additional repairs and complete the repair order.

    ____

    ____

STOP

ASE Education Foundation Worksheet #8-9

RETRIEVE OBD II DIAGNOSTIC TROUBLE CODES USING A SCAN TOOL

Name______________________________ Class______________________

Score: ☐ Excellent ☐ Good ☐ Needs Improvement **Instructor OK** ☐

Vehicle year __________ **Make** ____________________ **Model** ______________________

Objective: Upon completion of this assignment, you should be able to retrieve diagnostic trouble codes (DTCs) with an OBD II scan tool.

ASE Education Foundation Correlation

This worksheet addresses the following **MLR/AST/MAST** task:

VIII.B.1 Retrieve and record diagnostic trouble codes (DTC), OBD monitor status, and freeze frame data; clear codes when applicable. **(P-1)**

We Support ASE Education Foundation

Directions: Before beginning this lab assignment, review the worksheet completely. Fill in the information in the spaces provided as you complete each task.

Tools and Equipment Required: OBD II scan tool

Procedure:

1. What is the brand name and model of the scan tool you are using?

   ____

2. Select the correct cartridge for the vehicle.

   ☐ Cartridge required ☐ No cartridge required

3. Turn the ignition to the "on" position and look at the instrument panel. Is the malfunction indicator lamp (MIL) on?

   ☐ MIL is on.
   ☐ MIL is off.

Photo by Tim Gilles

4. Start the engine. Does the MIL go out after the engine starts?

   ☐ Yes ☐ No

5. Turn the engine off and locate the data link connector. Where is it located?

   ____

6. Connect the scan tool to the data link connector. The engine should not be running when making electrical connections. Turn on the scan tool and select the OBD II page.

TURN

7. Turn on the ignition. This is commonly called KOEO (key on, engine off).
8. Check for DTCs. If there are none, cause one to set by disconnecting a primary sensor while the engine is idling. Consult your instructor before doing this.
9. Read all DTCs and record them below:

DTC number	DTC description
____________	______________________________
____________	______________________________
____________	______________________________
____________	______________________________

10. Before completing your paperwork, clean your work area, clean and return tools to their proper places, and wash your hands.
11. Record your recommendations for needed service or additional repairs and complete the repair order.

____

____

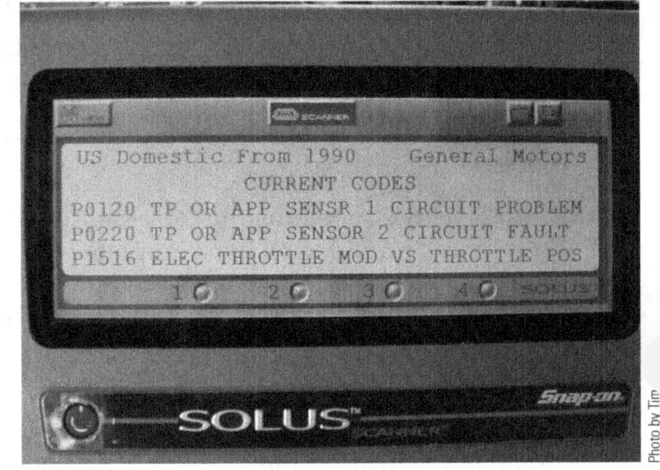

Photo by Tim

STOP

ASE Education Foundation Worksheet #8-10

INTERPRET OBD II SCAN TOOL DATA

Name______________________________ Class______________________

Score: ☐ Excellent ☐ Good ☐ Needs Improvement **Instructor OK** ☐

Vehicle year __________ **Make** ____________________ **Model** ______________________

Objective: Upon completion of this assignment, you should be able to interpret data from an OBD II scan tool.

ASE Education Foundation Correlation

This worksheet addresses the following **MLR** task:

VIII.B.2 Describe the use of OBD monitors for repair verification. **(P-1)**

This worksheet addresses the following **AST/MAST** tasks:

VIII.B.4 Describe the use of OBD monitors for repair verification. **(P-1)**

VIII.B.3 Perform active tests of actuators using a scan tool; determine needed action. **(P-2)**

VIII.E.7 Interpret diagnostic trouble codes (DTCs) and scan tool data related to the emissions control systems; determine needed action. **(P-2)**

This worksheet addresses the following **MLR/AST/MAST** task:

VIII.B.1 Retrieve and record diagnostic trouble codes (DTC), OBD monitor status, and freeze frame data; clear codes when applicable. **(P-1)**

This worksheet addresses the following **MAST** task:

VIII.B.5 Diagnose the causes of emissions or driveability concerns with stored or active diagnostic trouble codes (DTC); obtain, graph, and interpret scan tool data. **(P-1)**;

We Support ASE | Education Foundation

Directions: Before beginning this lab assignment, review the worksheet completely. Fill in the information in the spaces provided as you complete each task.

Tools and Equipment Required: OBD II scan tool and OBD II-equipped vehicle

Photo by Tim Gilles

Procedure:

1. Start the engine. Does the MIL remain on after the engine is started?

   ☐ MIL is on. ☐ MIL is off.

   ☐ MIL is flashing.

2. Connect the scan tool to the data link connector. Turn it on and select the OBD II page.
3. Move the key switch to the "on" position (KOEO).

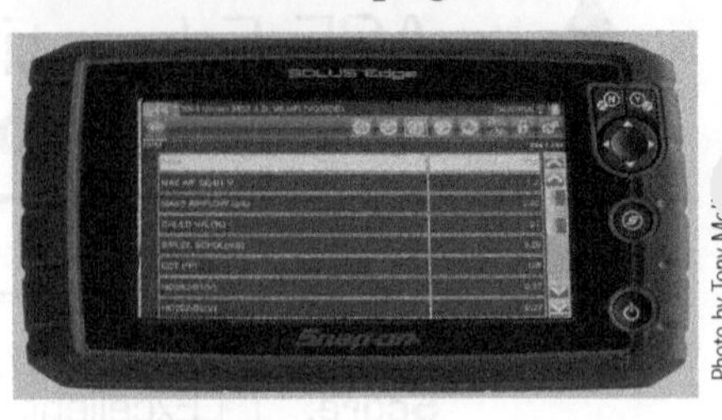

4. Check for diagnostic trouble codes (DTCs). If there are no codes, cause one to set by disconnecting a primary sensor while the engine is idling. Consult with your instructor before doing this.
5. Read all DTCs and record them below:

   DTC number __________ DTC description ______________________
   DTC number __________ DTC description ______________________
   DTC number __________ DTC description ______________________

6. What does each of the following DTCs indicate?

   PO304 ____
   PO440 ____
   PO133 ____
   PO119 ____

7. Was freeze frame data provided? ☐ Yes ☐ No

   If yes, which DTC had the freeze frame data? __________

8. Check for pending DTCs.

   ☐ There are no pending codes.

   ☐ There are pending codes. If so, list below:

DTC number	DTC description
______________	______________
______________	______________
______________	______________
______________	______________

9. Locate the I/M readiness menu and list the monitors that have completed below:

   ____
   ____
   ____

10. Were there any monitors that had not yet completed? If so, list them here:

   ____
   ____
   ____

11. Describe why it is important to verify that the monitors have run after completing a repair.

   ____

12. Perform an output control test on the EGR valve if available.

    ☐ Engine idle is rough. ☐ Engine idle remains steady. ☐ N/A.

13. Locate the oxygen sensor tests on the scan tool and record the following values below:

    Rich to lean threshold volts _______ Rich to lean switch time _______

    What is your interpretation of these results?

    ____

14. Before completing your paperwork, clean your work area, clean and return tools to their proper places, and wash your hands.

15. Record your recommendations for needed service or additional repairs and complete the repair order.

    ____

    ____

STOP

ASE Education Foundation Worksheet #8-11

REPLACE A TIMING BELT

Name______________________________ Class______________________

Score: ☐ Excellent ☐ Good ☐ Needs Improvement **Instructor OK** ☐

Vehicle year __________ **Make** ____________________ **Model** ______________________

Objective: Upon completion of this assignment, you should be able to replace a timing belt.

ASE Education Foundation Correlation

This worksheet addresses the following **AST/MAST** task:

I.B.5 Inspect and replace camshaft and drive belt/chain; includes checking drive gear wear and backlash, end play, sprocket and chain wear, overhead cam drive sprocket(s), drive belt(s), belt tension, tensioners, camshaft reluctor ring/tone-wheel, and valve timing components; verify correct camshaft timing. **(P-1)**

VIII.A.11 Verify correct camshaft timing including engines equipped with variable valve timing systems (VVT). **(P-1)**

We Support ASE Education Foundation

Directions: Before beginning this lab assignment, review the worksheet completely. Fill in the information in the spaces provided as you complete each task.

Tools and Equipment Required: Safety glasses, fender covers, shop towel, hand tools, damper puller, torque wrench

Parts and Supplies: Timing belt, tensioner, water pump, idler pulley bearing (some vehicles will not require all of these parts)

Procedure:

1. Locate the service information for the vehicle. Find the instructions for the replacement of the timing belt.

   List the estimated labor time for this job: _______ hours
2. Locate the drawing that shows the valve timing marks for the engine. Draw a sketch in the box to the right.
3. Open the hood and place fender covers on the fenders and front body parts.
4. Remove the battery ground cable and any accessory drive belts.
5. Remove the spark plugs.
6. Turn the engine by hand until the #1 cylinder is at TDC on its compression stroke and the timing marks are aligned.

   Note: Always turn the engine in its normal direction of rotation.
7. Remove the crankshaft pulley bolt. When an impact wrench is used, the crankshaft does not have to be restrained from turning while loosening the bolt.
8. Remove the crankshaft damper (pulley) using the correct type of puller.

   Note: A three-piece vibration damper can be damaged if a puller is attached to its outer ring rather than to the screw holes in its inner hub.

9. Remove the timing cover.
10. Loosen the belt tension adjustment and remove the belt. What kind of adjustment does the belt tensioner use?

    ☐ Jackscrew ☐ Hydraulic pressure

    ☐ Spring-loaded lock center ☐ Coolant pump pulley

PILOT

PULLER

11. Check the bearing in the tensioner or idler pulley for roughness. A rough or loose bearing must be replaced.

    Bearing condition? ☐ OK ☐ Bad
12. Some engines have the coolant pump located inside the timing cover. If this is the case, now is the time to check the coolant pump for leakage, bearing wear, or damage, and replace it if necessary.

    Water pump condition:

    ☐ OK ☐ Leaks ☐ Bad bearing ☐ N/A
13. Install the belt and adjust its belt tension.
14. Check that the sprocket timing marks are correctly located.
15. Reinstall any guides or spacers that were removed and reinstall the timing cover.
16. Install the crankshaft pulley and torque the crankshaft bolt.

    Torque specification: ______________________________
17. Rotate the engine through two complete revolutions. Do the timing marks line up as they did previously?

    ☐ Yes ☐ No

    Instructor OK before proceeding ______________________

    Note: If the timing marks do not line up after rotating the engine two revolutions, there is a problem. Stop, remove the timing belt, and start over at step #13.

Check the valve timing.

18. Install accessory drive belts and adjust their tension.
19. Install the spark plugs and the negative battery cable, and start the engine.
20. Put a sticker on the valve cover, listing the mileage when the belt was replaced.

    Mileage listed ______________________________

    Note: Keeping a record of timing belt replacement is important because the replacement of the timing belt is a required maintenance item for most vehicles.
21. Before completing your paperwork, clean your work area, clean and return tools to their proper places, and wash your hands.
22. Record your recommendations for needed service or additional repairs and complete the repair order.

    ____

    ____

STOP

ASE Education Foundation Worksheet #8-12

CHECK ENGINE OIL PRESSURE

Name________________________________ Class______________________

Score: ☐ Excellent ☐ Good ☐ Needs Improvement **Instructor OK** ☐

Vehicle year __________ **Make** ____________________ **Model** ______________________

Objective: Upon completion of this assignment, you should be able to connect an oil pressure gauge to measure an engine's oil pressure.

ASE Education Foundation Correlation

This worksheet addresses the following **AST/MAST** task:

I.D.9 Perform oil pressure tests; determine needed action. **(P-1)**

We Support ASE Education Foundation

Directions: Before beginning this lab assignment, review the worksheet completely. Fill in the information in the spaces provided as you complete each task.

Tools and Equipment Required: Safety glasses, fender covers, hand tools, flare-nut wrenches, oil pressure gauge, shop towel

Procedure:

1. Locate the oil pressure specification in the service information and record it below.

   Minimum: ______ psi at ______ rpm

   Maximum: ______ psi at ______ rpm

   Which type of service information was used? ____________________________________

2. Open the hood and place fender covers on the fenders and front body parts.

3. What is the engine temperature?

   Check one:

   ☐ Cold ☐ Warm

4. The vehicle is equipped with an:

   ☐ Oil pressure gauge ☐ Indicator light

5. Check the operation of the oil pressure light. Turn on the ignition and leave the engine off. Does the indicator light glow?

   ☐ Yes ☐ No ☐ Not equipped

6. Locate the sending unit.

   On what part of the engine is it located?

   ____

Oil sending unit removed

Oil pressure gauge

TURN

Photo by Tim Gilles

Note: If there is a malfunction in the gauge electrical circuit, try disconnecting the wire to the oil pressure sending unit. When it is connected to ground with the key on, the gauge or light function should change.

7. Remove the sending unit.

   Note: Some sending units require a special socket. In some cases, a 12-point socket will work.

8. Locate the proper fitting in the oil pressure gauge kit and install the gauge.

   Note: The fitting is national pipe thread (NPT), which is tapered. Tighten it only until it is snug. Do not overtighten it.

9. Start the engine and note the oil pressure while the engine temperature is cold:

   Idle ______ psi
   Fast idle ______ psi

10. Run the engine until it reaches operating temperature. Then record the oil pressure gauge reading.

    Idle ______ psi
    Fast idle ______ psi

    Note: Low oil pressure at idle that goes up with increased engine rpm usually indicates excessive bearing oil clearance.

11. Remove the oil pressure gauge and reinstall the sending unit.

12. Before completing your paperwork, clean your work area, clean and return tools to their proper places, and wash your hands.

13. Record your recommendations for needed service or additional repairs and complete the repair order.

    ____

    ____

STOP

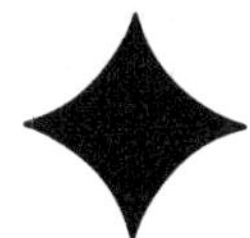

ASE Education Foundation Worksheet #8-13

REPLACE A VALVE COVER GASKET

Name______________________ Class______________

Score: ☐ Excellent ☐ Good ☐ Needs Improvement **Instructor OK** ☐

Vehicle year ________ **Make** ______________ **Model** ______________

Objective: Upon completion of this assignment, you should be able to replace a valve cover gasket.

ASE Education Foundation Correlation

This worksheet addresses the following **MLR** task:

I.A.4 Install engine covers using gaskets, seals, and sealers as required. **(P-1)**

This worksheet addresses the following **AST/MAST** task:

I.A.5 Install engine covers using gaskets, seals, and sealers as required. **(P-1)**

We Support ASE Education Foundation

Directions: Before beginning this lab assignment, review the worksheet completely. Fill in the information in the spaces provided as you complete each task.

Tools and Equipment Required: Safety glasses, fender covers, hand tools, shop towel

Parts and Supplies: Valve cover gasket

Procedure:

1. Open the hood and place fender covers on the fenders and front body parts.
2. Remove and label any spark plug wires, vacuum hoses, or electrical wires that interfere with valve cover removal.
3. Remove the screws holding the valve cover in place.
4. Rap the valve cover on its corner with a rubber mallet to loosen it. If it does not come loose easily, see your instructor.
5. Use a gasket scraper to remove the old gasket from the valve cover. In the solvent tank, clean the dirt and oil from the valve cover. Blow it dry with compressed air.

Photo by Tim Gilles

Note: Remember to blow down and away from yourself when using compressed air. Blow the solvent *into* the solvent tank, not onto the floor.

The solvent was blown: ☐ Into the tank ☐ On the floor

6. If the valve cover is made from sheet metal, flatten the area around the bolt holes using a hammer on a flat surface.

TURN

7. Position the new gasket on the valve cover. If there are no locating lugs to hold the gasket in place, use a gasket adhesive or pieces of string to fasten the gasket to the valve cover in four places.
8. Clean any old valve cover gasket residue from the cylinder head.

   Note: Do not allow any pieces of the old gasket to fall into the engine.
9. Reinstall the valve cover. Finger-tighten all the bolts.

   Note: Always start the threads on *all* of the bolts that fasten a part before tightening *any* of them.
10. Tighten all of the screws to 5–6 ft.-lb (60–72 in.-lb) using a torque wrench.

    Which torque wrench was used? ☐ Inch-pound ☐ Foot-pound

    Note: Most foot-pound torque wrenches are not accurate below 15–20 ft.-lb.
11. Reinstall any wires or hoses and start the engine. Check for smooth engine operation and look for oil leaks around the valve cover gasket.

    Does the engine run smoothly? ☐ Yes ☐ No

    Are there any oil leaks? ☐ Yes ☐ No
12. Before completing your paperwork, clean your work area, clean and return tools to their proper places, and wash your hands.
13. Record your recommendations for needed service or additional repairs and complete the repair order.

    ____

    ____

STOP

ASE Education Foundation Worksheet #8-14

READ A STANDARD MICROMETER

Name____________________ Class____________

Score: ☐ Excellent ☐ Good ☐ Needs Improvement **Instructor OK** ☐

Vehicle year ________ **Make** ____________ **Model** ____________

Objective: Upon completion of this assignment, you should be able to use and read a standard micrometer.

ASE Education Foundation Correlation

This worksheet addresses the following **Required Supplemental Task (RST)**:

Tools and Equipment

Task 5: Demonstrate proper use of precision measuring tools (i.e. micrometer, dial-indicator, dial-caliper).

We Support ASE Education Foundation

Directions: Before beginning this lab assignment, review the worksheet completely.

Tools and Equipment Required: Micrometer

Parts and Supplies: Six parts to measure (provided by your instructor)

Procedure:

1. Familiarize yourself with the micrometer by identifying its parts. Place the correct letter next to the name of the micrometer part.

   Thimble ______
   Anvil ______
   Sleeve ______
   Spindle ______
   Frame ______
   Ratchet ______
   Lock ______

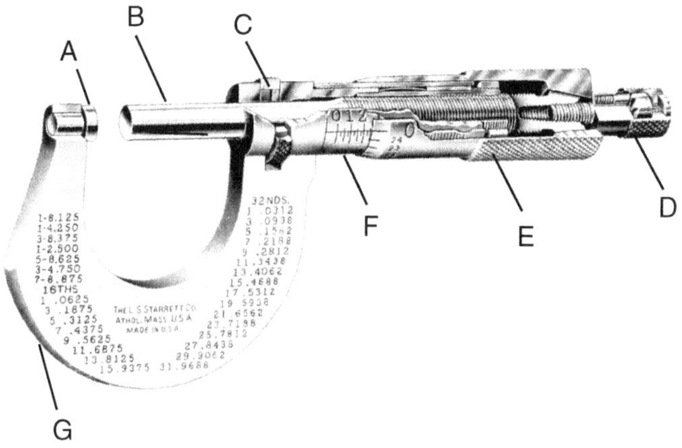

2. Locate a micrometer.

   What size is it? (Example: 0–1")

   ____

3. Find the zero (0) line on the micrometer sleeve. Turn the thimble in until the zero on the thimble is aligned with the index line (at the zero line on the sleeve). The distance from the anvil to the end of the spindle is now exactly 0 (or 1", 2", 3", 4", etc.), depending on the size of the micrometer.

   a. What is the distance from the anvil to the spindle end?

   ☐ 0 ☐ 1" ☐ 2" ☐ 3" ☐ 4" Other ____________

TURN

Turn the thimble exactly 40 turns counterclockwise.

Are any numbers on the sleeve uncovered by the thimble? ____________

4. Look at the numbers on the *sleeve*. Each of these numbers is divided into four parts.

   How many thousandths of an inch are represented by each of the numbered lines?

   0. ________"

   Each of the graduated lines on the sleeve is equal to: 0. ________"

5. Each turn of the thimble equals what fraction of an inch? 0. ________"

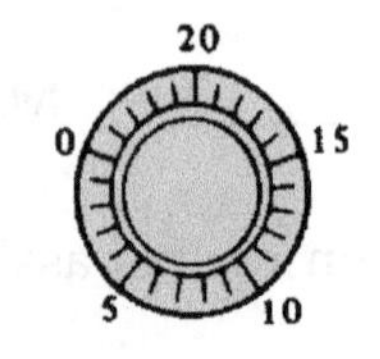

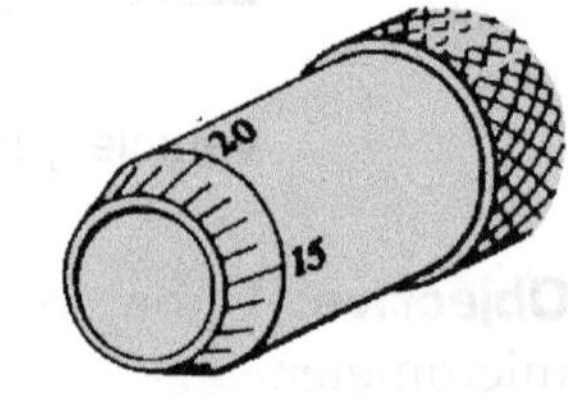

6. Next look at the *thimble*. The thimble has numbers and lines around its end.

   How many graduations are around the thimble? ____________

   Each of the graduations on the thimble represents: 0. ________"

Reading the Micrometer

7. Read the micrometer in the picture. Start by reading the last number on the sleeve that is visible (A).

   Record your reading: 0. ________"

   Next read the number of lines that can be seen from the last number visible on the sleeve to the edge of the thimble (B).

   Record your reading: 0. ________"

   Now see which graduation line on the thimble is aligned with the index line (C).

   Record your reading: 0. ________"

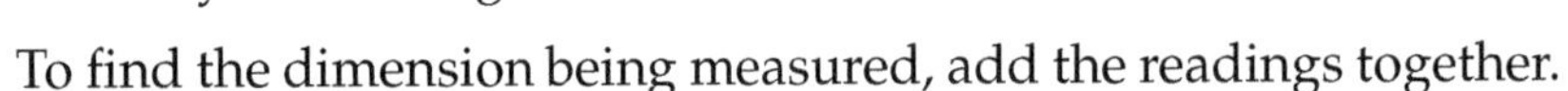

   To find the dimension being measured, add the readings together.

0. ________"
0. ________"
0. ________"

Your total should be 0.283"

If your total was two hundred and eighty-three thousandths, then you are ready to read the micrometers on Worksheet #8-15, Micrometer Practice. Remember to show your work.

STOP

ASE Education Foundation Worksheet #8-15

MICROMETER PRACTICE

Name____________________________ Class____________________

Score: ☐ Excellent ☐ Good ☐ Needs Improvement **Instructor OK** ☐

Vehicle year __________ **Make** ___________________ **Model** _____________________

Objective: Upon completion of this assignment, you should be able to use and accurately read an inch system micrometer.

ASE Education Foundation Correlation

This worksheet addresses the following **Required Supplemental Task (RST)**:

Tools and Equipment

Task 5: Demonstrate proper use of precision measuring tools (i.e. micrometer, dial-indicator, dial-caliper).

We Support ASE Education Foundation

Directions: Record the readings in the spaces for micrometers.

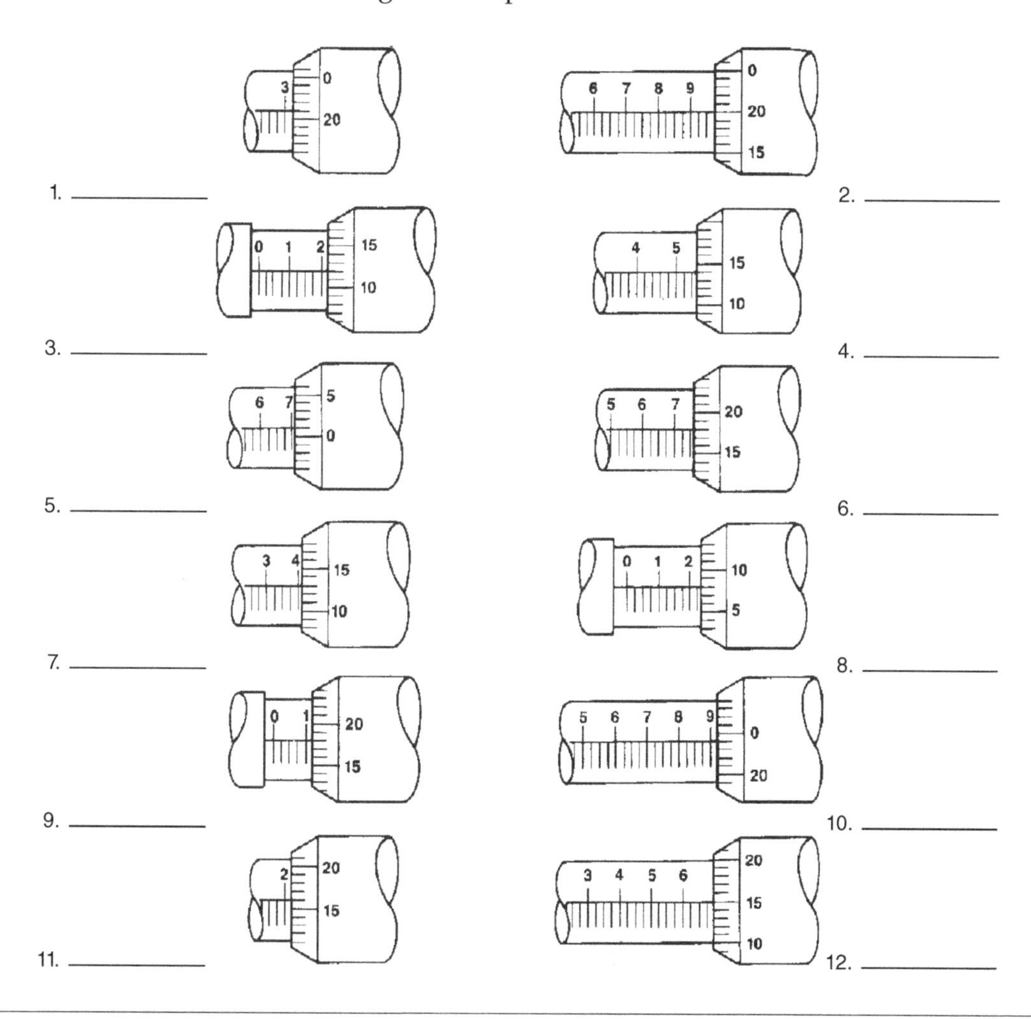

TURN

Directions: Measure six parts provided by your instructor with a micrometer and record the sizes of the parts below.

Example:

Describe the part being measured.	**Piston**
Describe where the part is being measured.	**Piston skirt**
Record the measurement of the part.	**3.875"**

1. Part #1

   Describe the part being measured. ____________________
   Describe where the part is being measured. ____________________
   Record the measurement of the part. ____________________

2. Part #2

   Describe the part being measured. ____________________
   Describe where the part is being measured. ____________________
   Record the measurement of the part. ____________________

3. Part #3

   Describe the part being measured. ____________________
   Describe where the part is being measured. ____________________
   Record the measurement of the part. ____________________

4. Part #4

   Describe the part being measured. ____________________
   Describe where the part is being measured. ____________________
   Record the measurement of the part. ____________________

5. Part #5

   Describe the part being measured. ____________________
   Describe where the part is being measured. ____________________
   Record the measurement of the part. ____________________

6. Part #6

   Describe the part being measured. ____________________
   Describe where the part is being measured. ____________________
   Record the measurement of the part. ____________________

STOP

ASE Education Foundation Worksheet #8-16

READ A DIAL CALIPER

Name ______________________ Class ________________

Score: ☐ Excellent ☐ Good ☐ Needs Improvement **Instructor OK** ☐

Vehicle year ________ **Make** ______________ **Model** ________________

Objective: Upon completion of this assignment, you should be able to use and read a dial caliper.

ASE Education Foundation Correlation

This worksheet addresses the following **Required Supplemental Task (RST)**:

Tools and Equipment

Task 5: Demonstrate proper use of precision measuring tools (i.e. micrometer, dial-indicator, dial-caliper).

We Support Education Foundation

Directions: Before beginning this lab assignment, review the worksheet completely.

Tools and Equipment Required: Dial Caliper

Parts and Supplies: Six parts to measure (provided by your instructor)

Procedure:

1. Familiarize yourself with the dial by identifying its parts. Place the correct letter next to the name of the dial caliper part.

   Dial ______

   Outside ______

   Inside ______

   Depth ______

2. Locate a dial caliper.

   What size is it? (Example: 6", 8", 12") ______

3. Completely close the caliper. It should now read "0" on the dial. If it does not read zero, rotate the outside of the dial to zero the caliper.

   Did you zero the caliper? ☐ Yes ☐ No

4. Look at the lines and numbers on the slide bar.

   How many thousandths of an inch are represented by each of the numbered lines?

   0. ______ "

TURN

5. Look at the lines and numbers on the dial.

   How many thousandths of an inch are represented by each of the lines on the dial?

   0. _______ "

   One revolution of the caliper dial represents: 0. _______ "

6. Set the dial caliper to read 0.650". Show it to another student. Was your setting correct?

   ☐ Yes ☐ No

7. Set the dial caliper to read 0.425". Show it to another student. Was your setting correct?

   ☐ Yes ☐ No

8. Set the dial caliper to read 0.324". Show it to another student. Was your setting correct?

   ☐ Yes ☐ No

STOP

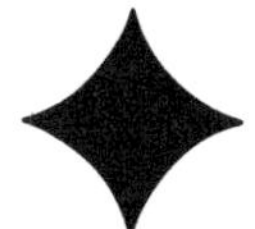

ASE Education Foundation Worksheet #8-17

DIAL CALIPER PRACTICE

Name ______________________ Class ________________

Score: ☐ Excellent ☐ Good ☐ Needs Improvement **Instructor OK** ☐

Vehicle year ________ **Make** ______________ **Model** ________________

Objective: Upon completion of this assignment, you should be able to use and read a dial caliper.

ASE Education Foundation Correlation

This worksheet addresses the following **Required Supplemental Task (RST)**:

Tools and Equipment

Task 5: Demonstrate proper use of precision measuring tools (i.e. micrometer, dial-indicator, dial-caliper).

We Support ASE | Education Foundation

Directions: Record the readings in the spaces using a dial caliper to make the measurements.

Directions: Measure six parts provided by your instructor with a dial caliper and record the sizes of the parts below.

Example:

Describe the part being measured. **Piston**
Describe where the part is being measured. **Piston skirt**
Record the measurement of the part. **3.875"**

1. Part #1
   Describe the part being measured. ________________
   Describe where the part is being measured. ________________
   Record the measurement of the part. ________________
2. Part #2
   Describe the part being measured. ________________
   Describe where the part is being measured. ________________
   Record the measurement of the part. ________________
3. Part #3
   Describe the part being measured. ________________
   Describe where the part is being measured. ________________
   Record the measurement of the part. ________________
4. Part #4
   Describe the part being measured. ________________
   Describe where the part is being measured. ________________
   Record the measurement of the part. ________________

TURN

5. Part #5

   Describe the part being measured. ____________________

   Describe where the part is being measured. ____________________

   Record the measurement of the part. ____________________

6. Part #6

   Describe the part being measured. ____________________

   Describe where the part is being measured. ____________________

   Record the measurement of the part. ____________________

ASE Education Foundation Worksheet #8-18

VALVE LASH MEASUREMENT

Name ________________________ Class ________________

Score: ☐ Excellent ☐ Good ☐ Needs Improvement **Instructor OK** ☐

Vehicle year _________ **Make** ________________ **Model** ________________

Objective: Upon completion of this assignment, you should be able to check valve clearance on an overhead cam engine that has shim-type adjustment.

ASE Education Foundation Correlation

This worksheet addresses the following **MLR** task:

I.B.1 Adjust valves (mechanical or hydraulic lifters). **(P-3)**

This worksheet addresses the following **AST/MAST** task:

I.B.4 Adjust valves (mechanical or hydraulic lifters). **(P-1)**

We Support ASE Education Foundation

Directions: Before beginning this lab assignment, review the worksheet completely. Fill in the information in the spaces provided as you complete each task.

Tools and Equipment Required: Safety glasses, fender covers, hand tools, feeler gauges, special adjustment tools, shop towel

Procedure:

Engine size ________________ # of cylinders ________________

1. Open the hood and place fender covers on the fenders and front body parts.
2. Locate the valve lash specifications in the repair manual and record them below.

	Hot		Cold
Intake	0. ______"	Intake	0. ______"
Exhaust	0. ______"	Exhaust	0. ______"

3. Record the engine's firing order below.

______ ______ ______ ______

______ ______ ______ ______

4. List the companion cylinders below by writing the first half of the firing order above the second half.

______ ______ ______ ______

______ ______ ______ ______

Photo by Tim Gilles

5. Remove the valve cover(s).
6. Inspect the runners on the intake and exhaust manifolds to determine which valves are intake and which are exhaust.

Show your instructor before proceeding. **Instructor OK** ________________

7. How many valves does this engine have per cylinder? ________________

8. Locate the large bolt on the front of the crankshaft pulley.

9. Use a large (½″-drive) ratchet to turn the bolt (and crankshaft) in its normal direction of rotation.

   Note: Always turn an engine in its normal direction of rotation.

   Which way does the crankshaft on this engine normally turn?

   ☐ Clockwise ☐ Counterclockwise

10. As the crankshaft is turned, watch the number 1 cylinder's exhaust valve. As it begins to close, continue turning the crankshaft until the number 1 intake valve just starts to open. This is the beginning of the cylinder's next four-stroke cycle. Show your instructor.

    Instructor OK ________________

    Note: At this point, the valve clearance for the *companion* cylinder to number 1 can be checked or adjusted.

    Which cylinder is the companion to cylinder #1? ______

11. Measure the clearance on the companion cylinder with a feeler gauge. Record your measurements in the chart.

12. Turn the crankshaft until the next cylinder in the firing order has its valves rocking. Then measure the clearance for its companion.

13. Following the firing order, repeat the process for the remaining cylinders.

    Record your measurements on the following chart.

Cam lobe in valve closed position

Feeler gauge

Cam follower (bucket)

Lash pad adjuster

Cylinder	#1	#2	#3	#4	#5	#6	#7	#8
Intake	______	______	______	______	______	______	______	______
Exhaust	______	______	______	______	______	______	______	______

14. In the chart above, circle the valves that need to be adjusted.

15. Before completing your paperwork, clean your work area, clean and return tools to their proper places, and wash your hands.

16. Record your recommendations for needed service or additional repairs and complete the repair order.

    ____

    ____

STOP

ASE Education Foundation Worksheet #8-19

VALVE LASH ADJUSTMENT

Name_______________________ Class_______________

Score: ☐ Excellent ☐ Good ☐ Needs Improvement **Instructor OK** ☐

Vehicle year ________ **Make** ______________ **Model** ______________

Objective: Upon completion of this assignment, you should be able to adjust the valve clearance on an overhead cam engine that has shim-type adjustment.

ASE Education Foundation Correlation

This worksheet addresses the following **MLR** task:

I.B.1 Adjust valves (mechanical or hydraulic lifters). **(P-3)**

This worksheet addresses the following **AST/MAST** task:

I.B.4 Adjust valves (mechanical or hydraulic lifters). **(P-1)**

We Support ASE Education Foundation

Directions: Before beginning this lab assignment, review the worksheet completely. Fill in the information in the spaces provided as you complete each task.

Tools and Equipment Required: Safety glasses, fender covers, hand tools, feeler gauges, special adjustment tools, shop towel

Procedure: Before attempting this worksheet, complete Worksheet #8-18, Valve Lash Measurement.

Note: The method for adjusting valve clearance (lash) varies among manufacturers. Check the service information for the proper adjustment method for this engine. The method presented here is typical of many manufacturers.

OHC Engine with Shim Adjustment

1. Turn the crankshaft to position the cam lobe for the valve to be adjusted so that the cam lobe faces away from the valve.
2. Using a special tool, press down on the valve and remove the shim. Remove the shim by lifting it with a small screwdriver. Then use a magnet to remove the shim.

   Note: Compressed air from a rubber-tipped blow gun carefully directed under the shim will help to break the seal formed by the engine oil.

Photo by Tim Gilles

3. Determine the thickness of the replacement shim by measuring the thickness of the removed shim. Then add the measured clearance, minus the clearance specification.

   $R + [C - S] = N$

   R = Thickness of the removed shim

   C = Measured valve clearance

   S = Valve lash specification

   N = New shim thickness

4. Select the new shim and install it by reversing the removal procedure.

   Replacement shim thickness: 0. _______"

5. Recheck the valve clearance. Is it now correct?

   ☐ Yes ☐ No

6. If the clearance is within the specified tolerance, continue adjusting the valves as necessary.

7. Record the shims replaced below. If you need to change more than four shims, see your instructor.

Cylinder#	Valve (int/ex?)	Specification	Measured Clearance	Old Shim Thickness	New Shim Thickness
________	________	________	________	________	________
________	________	________	________	________	________
________	________	________	________	________	________
________	________	________	________	________	________

8. Install the valve cover and any other components that were removed. Start the engine.

Rocker Arm-Type Valve Adjustment

1. Before making a valve clearance measurement, position the companion cylinder at TDC at the beginning of the power stroke.

2. Check the valve clearance with a feeler gauge.

3. Clearance is adjusted by loosening a locknut and turning an adjusting screw until the feeler gauge has a slight drag.

SHOP TIP You can use a feeler gauge as a "go-no-go" gauge. After adjusting the clearance, try to insert a feeler gauge that is 0.001" thicker. If it fits, your adjustment is too loose.

Photo by Tim Gilles

4. Before completing your paperwork, clean your work area, clean and return tools to their proper places, and wash your hands.

5. Record your recommendations for needed service or additional repairs and complete the repair order.

____

____

STOP

ASE Education Foundation Worksheet #8-20

RESTORE A BROKEN SCREW THREAD

Name______________________ Class______________________

Score: ☐ Excellent ☐ Good ☐ Needs Improvement **Instructor OK** ☐

Vehicle year __________ **Make** ______________ **Model** ______________

Objective: Upon completion of this assignment, you should be able to remove a broken fastener and restore damaged threads.

ASE Education Foundation Correlation

This worksheet addresses the following **MLR** task:

I.A.6 Perform common fastener and thread repair, to include: remove broken bolt, restore internal and external threads, and repair internal threads with thread insert. **(P-1)**

This worksheet addresses the following **AST/MAST** task:

I.A.7 Perform common fastener and thread repair, to include: remove broken bolt, restore internal and external threads, and repair internal threads with thread insert. **(P-1)**

We Support ASE Education Foundation

Directions: Before beginning this lab assignment, review the worksheet completely. Fill in the information in the spaces provided as you complete each task.

Tools and Equipment Required: Safety glasses, shop towel, drill motor, drills, taps

Preliminary Questions:

1. What caused the bolt to break?

____

2. What size is the broken bolt? ______________

Procedure:

1. Carefully file the broken bolt flat.
2. Centerpunch the top of the broken bolt.

   Note: Be absolutely certain that the centerpunch mark is exactly on center. If not, pound the centerpunch mark deeper until it is on the center.

3. Use a sharp, small drill to drill a pilot hole exactly in the center of the bolt.

   Note: If the pilot hole is off-center, the restored hole will be off-center.

Photo by Tim Gilles

TURN

4. What size drill bit and drill motor chuck are being used?

   Drill bit size ______

   Drill motor chuck size ______

   Remember: Bolts are made of steel, so cutting oil is *required.*

Photo by Tim Gilles

5. After starting to drill the hole, double-check to see that the hole is being drilled *exactly* in the center of the bolt.

   If not, use a file or a die grinder and burr to remove the small amount of hole already drilled. Then centerpunch again and drill exactly in the center of the bolt.

6. Was the hole drilled on-center? ☐ Yes ☐ No

7. If possible, drill the pilot hole all the way through the broken bolt.

   Note: This will relieve some of the internal tension in the bolt. Many times after the bolt has been drilled through, it is easily removed.

Photo by Tim Gilles

8. Finish drilling the hole with the largest size drill bit that can be used without damaging the original threads.

9. Run a tap through the hole to clean out the threads. Tap size ______

10. If the threads are not in good condition after being chased with the tap, install a thread insert.

    Thread condition: ☐ Good ☐ Bad

    Is a thread insert required? ☐ Yes ☐ No

11. Before completing your paperwork, clean your work area, clean and return tools to their proper places, and wash your hands.

12. Record your recommendations for needed service or additional repairs and complete the repair order.

    ____

    ____

STOP

ASE Education Foundation Worksheet #8-21

INSTALL A HELI-COIL® THREAD INSERT

Name________________________ Class__________________

Score: ☐ Excellent ☐ Good ☐ Needs Improvement **Instructor OK** ☐

Vehicle year ________ **Make** ______________ **Model** ______________

Objective: Upon completion of this assignment, you should be able to repair a damaged thread by installing a Heli-Coil insert.

ASE Education Foundation Correlation

This worksheet addresses the following **MLR** task:

I.A.6 Perform common fastener and thread repair, to include: remove broken bolt, restore internal and external threads, and repair internal threads with thread insert. **(P-1)**

This worksheet addresses the following **AST/MAST** task:

I.A.7 Perform common fastener and thread repair, to include: remove broken bolt, restore internal and external threads, and repair internal threads with thread insert. **(P-1)**

We Support ASE | Education Foundation

Directions: Before beginning this lab assignment, review the worksheet completely. Fill in the information in the spaces provided as you complete each task.

Tools and Equipment Required: Safety glasses, shop towel, Heli-Coil tool set

Parts and Supplies: Heli-Coil insert

Emhart Fastening Technologies

Preliminary Questions:

1. What part is being repaired? _______
2. What is the size of the threads being repaired? _______

Procedure:

1. Drill the damaged hole with a drill bit of the specified size.

   Drill size: _______

   Note: The success of the job depends on the drill being held straight (perpendicular to the hole).

2. Tap the hole using a tap of the correct size.

   Tap size: _______

   Note: Turn the tap counterclockwise after each revolution to break off the cuttings.

3. Install the thread insert on the mandrel.

TURN

4. Put a *small* amount of thread locking adhesive on the thread.

   Note: Thread sealer is sometimes provided with the insert kit. Use only a small amount. This is an *anaerobic* sealer (which hardens only without the presence of air). It will not work properly if too much is used.

5. Thread the insert into the hole until it is just below the surface of the part.

6. The bottom of the insert has a tang to assist in turning the insert into the hole. Break the tang off with a punch to complete the job.

7. Thread a new bolt into the repair hole. Does the fastener turn in easily at least three full turns?

   ☐ Yes ☐ No

   Note: If the bolt did not turn in easily, there may be a problem. Carefully check your work.

8. Before completing your paperwork, clean your work area, clean and return tools to their proper places, and wash your hands.

9. Record your recommendations for needed service or additional repairs and complete the repair order.

   ____

   ____

ASE Education Foundation Worksheet #8-22

DRILL AND TAP A HOLE

Name________________________________ Class________________________

Score: ☐ Excellent ☐ Good ☐ Needs Improvement **Instructor OK** ☐

Vehicle year __________ **Make** ____________________ **Model** ______________________

Objective: Upon completion of this assignment, you should be able to drill a hole and tap it to accept a fastener.

ASE Education Foundation Correlation

This worksheet addresses the following **MLR** task:

I.A.6 Perform common fastener and thread repair, to include: remove broken bolt, restore internal and external threads, and repair internal threads with thread insert. **(P-1)**

This worksheet addresses the following **AST/MAST** task:

I.A.7 Perform common fastener and thread repair, to include: remove broken bolt, restore internal and external threads, and repair internal threads with thread insert. **(P-1)**

We Support ASE Education Foundation

Directions: Before beginning this lab assignment, review the worksheet completely. Fill in the information in the spaces provided as you complete each task.

Tools and Equipment Required: Safety glasses, fender covers, shop towels, drill index, tap and die set, drill motor, lubricant, centerpunch, countersink.

Preliminary Questions:

1. What part is being repaired? ____
2. Determine the size of the hole to be threaded. This is done by determining the size of the mating fastener.

   What is the size of its screw thread? _________

Procedure:

1. If the part to be repaired is on a vehicle, open the hood and place fender covers on the fenders and front body parts.
2. Select the correct drill from the tap/drill size chart.

   What is the correct drill size? _________

   What is its decimal equivalent? _________

   What is the closest fractional drill size? _________

   Note: It is important that the correct drill be used. A hole that is drilled too large will not leave enough material to provide sufficient threads. A hole that is too small could bind the tap and possibly break it.
3. Use a centerpunch to put a mark where the hole is to be drilled.

   Is the mark exactly on center? ☐ Yes ☐ No

TURN

4. Before drilling the finished hole, use a small drill to make a pilot hole. This will make it easier to drill the hole.

   Size of pilot drill used: ________

5. Drill the hole to its finished size.

   Note: If the hole is being drilled in cast iron, no lubricant is needed. Remember, the hole must be drilled perpendicular to the surface of the part. If the hole is not drilled all the way through the part, be sure to leave space at the bottom of the hole for bolt clearance.

6. Use a countersink or burr to chamfer the top of the hole.

7. Select the proper size and type of tap.

   Type of tap selected:

   ☐ Bottom tap ☐ Plug tap ☐ Taper tap ☐ Pipe tap

   Tap size? ________

8. Begin tapping by carefully turning the tap clockwise while gently pushing down. Be sure to keep the tap perpendicular to the part.

   Did the tap start straight? ☐ Yes ☐ No

9. After the tap has started correctly, continue tapping using both hands to turn the tap. This helps to ensure that the tap will continue perpendicular into the hole.

   Note: Advance the tap ½ turn clockwise and then turn it back ¼ turn. Do this until the thread is completely cut. If the tap binds, stop immediately and check with your instructor.

Size	Threads per inch			Outside Diameter Inches	Tap Drill Approx. 75% Full Thread	Decimal Equivalent of Tap Drill
	NC	NF	NS			
0	...	80	...	.0600	3/64	.0469
1	...	...	56	.0730	54	.0550
1	64	...	...	.0730	53	.0595
1	...	72	...	.0730	53	.0595
2	56	...	...	.0860	50	.0700
2	...	64	...	.0860	50	.0700
3	48	...	...	.0990	47	.0785
3	...	56	...	.0990	45	.0820
4	...	...	32	.1120	45	.0820
4	...	...	36	.1120	44	.0860
4	40	...	...	.1120	43	.0890
4	...	48	...	.1120	42	.0935
5	...	...	36	.1250	40	.0980
5	40	...	...	.1250	38	.1015
5	...	44	...	.1250	37	.1040
6	32	...	...	.1380	36	.1065
6	...	...	36	.1380	34	.1110
6	...	40	...	.1380	33	.1130
8	...	...	30	.1640	30	.1285
8	32	...	...	.1640	29	.1360
8	...	36	...	.1640	29	.1360
8	...	...	40	.1640	28	.1405
10	24	...	...	.1900	25	.1495
10	...	...	28	.1900	23	.1540
10	...	...	30	.1900	22	.1570
10	...	32	...	.1900	21	.1590
12	24	...	...	.2160	16	.1770
12	...	28	...	.2160	14	.1820
12	...	...	32	.2160	13	.1850
1/4	20	...	...	.2500	7	.2010
1/4	...	28	...	.2500	3	.2130
5/16	18	...	...	.3125	F	.2570
5/16	...	24	...	.3125	I	.2720
3/8	16	...	...	.3750	5/16	.3125
3/8	...	24	...	.3750	Q	.3320
7/16	14	...	...	.4375	U	.3680
7/16	...	20	...	.4375	25/64	.3906
1/2	13	...	...	.5000	27/64	.4219
1/2	...	20	...	.5000	29/64	.4531
9/16	12	...	...	.5625	31/64	.4844
9/16	...	18	...	.5625	33/64	.5156
5/8	11	...	...	.6250	17/32	.5312
5/8	...	18	...	.6250	37/64	.5781
3/4	10	...	...	.7500	21/32	.6562
3/4	...	16	...	.7500	11/16	.6875
7/8	9	...	...	.8750	49/64	.7656

The L. S. St... ...mpany

10. Clean the completed thread with compressed air. Test the thread by turning a new bolt into the hole. It should turn easily into the hole a minimum of three complete turns without the use of tools.

    Does the bolt turn into the new threads easily?
    ☐ Yes ☐ No

    Note: If the bolt does not turn in easily, there may be a problem. Carefully check the new threads and the bolt size. Repair the problem as necessary.

Taper

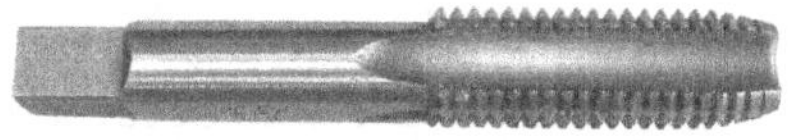

Plug

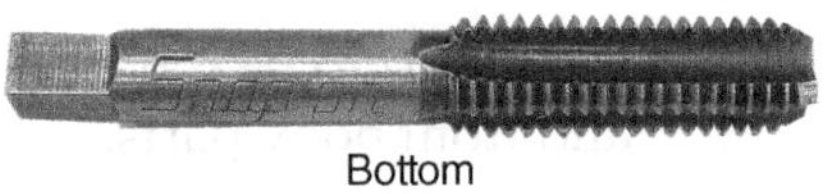

Bottom

Photo by Tim Gilles

11. Before completing your paperwork, clean your work area, clean and return tools to their proper places, and wash your hands.

12. Record your recommendations for needed service or additional repairs and complete the repair order.

    ____

    ____

STOP

Part II

Lab Worksheets for ASE Maintenance and Light Repair

Service Area 9

Chassis Service

ASE Education Foundation Worksheet #9-1

MANUALLY BLEED BRAKES AND FLUSH THE SYSTEM

Name_______________________ Class_______________

Score: ☐ Excellent ☐ Good ☐ Needs Improvement **Instructor OK** ☐

Vehicle year __________ **Make** ________________ **Model** ________________

Objective: Upon completion of this assignment, you will be able to manually bleed brakes and flush the system.

ASE Education Foundation Correlation

This worksheet addresses the following **MLR** tasks:

V.B.4 Select, handle, store, and fill brake fluids to proper level; use proper fluid type per manufacturer specification. **(P-1)**

V.B.6 Bleed and/or flush brake system. **(P-1)**

This worksheet addresses the following **AST/MAST** tasks:

V.B.9 Select, handle, store, and fill brake fluids to proper level; use proper fluid type per manufacturer specification. **(P-1)**

V.B.12 Bleed and/or flush brake system. **(P-1)**

We Support ASE Education Foundation

Directions: Before beginning this lab assignment, review the worksheet completely. Fill in the information in the spaces provided as you complete each task.

Tools and Equipment Required: Safety glasses, fender cover, jack stands or vehicle lift, shop towels, bleeder wrench, hose, container

Parts and Supplies: Brake fluid

CAUTION This worksheet is not intended for use with vehicles that have antilock brake systems. Consult the service information for the proper procedure before bleeding antilock brake systems.

Procedure:

1. Use the service information to locate the proper bleeding sequence for the vehicle being serviced. The proper bleeding sequence is important. Vehicle manufacturers publish brake bleeding sequence charts.

   Which service information did you use? ________________________________

2. On the sketch below, place numbers next to the wheels in the order that they are to be bled.

   Note: Traditional systems are bled beginning with the farthest cylinder from the master cylinder. Many front-wheel-drive vehicles use a diagonally split brake system. These systems are bled by bleeding the right rear first, followed by the left front, left rear, and right front.

3. Open the hood and place fender covers on the fenders and front body parts.

CAUTION Brake fluid will damage vehicle paint. *Always* use fender covers and clean spills immediately. Water will clean up brake fluid spills.

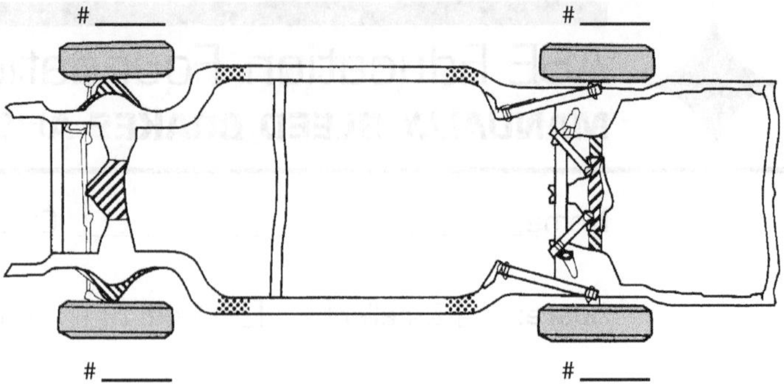

4. Open the master cylinder and use a suction tool to remove the old fluid.
5. Refill the master cylinder with the correct brake fluid.

   Fluid Type: ☐ DOT 3 ☐ DOT 4 ☐ DOT 5 ☐ DOT 5.1

#_____ #_____

#_____ #_____

6. Raise and support the vehicle and inspect the bleed screws.

   ☐ Free from dirt ☐ Turn freely ☐ Frozen ☐ OK

7. Place a small hose over the bleed screw and direct it into a container to catch the old brake fluid.
8. Find an assistant to push down on the brake pedal while the bleed screw is loose. Ask your assistant to let up on the pedal only when the bleed screw is closed. Good communication is important to successful completion of this procedure.

CAUTION Do not depress the brake pedal beyond one-half travel. **Do not push the pedal to the floor!**

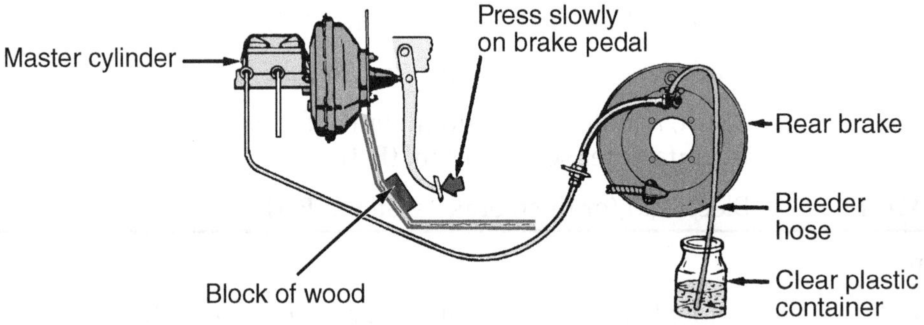

Master cylinder

Press slowly on brake pedal

Rear brake

Bleeder hose

Clear plastic container

Block of wood

9. Restrict the pedal from full travel to the floorboard by placing a short length of 2 × 4 beneath the pedal.
10. Bleed each wheel cylinder in the correct order until the fluid is clean, clear, and contains no sign of air.

    Was all of the air removed? ☐ Yes ☐ No

11. Check and refill the master cylinder after each wheel cylinder is bled. Be certain that all the bleed screws are tight.

    Note: Some master cylinders have smaller reservoirs. Check the fluid level more often when servicing these systems.

12. Apply the brake pedal and check pedal feel and height.

    ☐ OK ☐ Spongy ☐ Firm ☐ Low

13. Recheck the master cylinder fluid level.
14. Before completing your paperwork, clean your work area, clean and return tools to their proper places, and wash your hands.
15. Record your recommendations for needed service or additional repairs and complete the repair order.

____

____

STOP

ASE Education Foundation Worksheet #9-2

REMOVE A BRAKE DRUM

Name ______________________ Class ________________

Score: ☐ Excellent ☐ Good ☐ Needs Improvement **Instructor OK** ☐

Vehicle year ________ **Make** ______________ **Model** ________________

Objective: Upon completion of this assignment, you will be able to remove and replace a brake drum.

ASE Education Foundation Correlation

This worksheet addresses the following **MLR** task:

V.C.1 Remove, clean, and inspect brake drum; measure brake drum diameter; determine serviceability. **(P-1)**

This worksheet addresses the following **AST/MAST** task:

V.C.2 Remove, clean, and inspect brake drum; measure brake drum diameter; determine serviceability. **(P-1)**

We Support ASE Education Foundation

Directions: Before beginning this lab assignment, review the worksheet completely. Fill in the information in the spaces provided as you complete each task.

Tools and Equipment Required: Safety glasses, jack and safety stands or vehicle lift, shop towel, ½" impact wrench, impact sockets, hammer, torque wrench, rubber mallet

Procedure:

1. Raise the rear of the vehicle and place it on safety stands or raise it on a lift.

   How is the vehicle supported? ☐ Safety stands ☐ Lift

2. Remove the wheel cover or hubcap.

   ☐ Wheel cover ☐ Hubcap ☐ N/A

   Note: Some vehicles have lug nuts that extend through holes in the wheel cover or hubcap. On these vehicles it is not necessary to remove the hubcap.

3. Select the correct size *impact* socket. Socket size ______

4. Thread loosening direction:

   ☐ Clockwise (left-hand thread)

   ☐ Counterclockwise (right-hand thread)

   Note: Check to see that the impact wrench turns the proper direction before loosening the lug nuts.

TURN

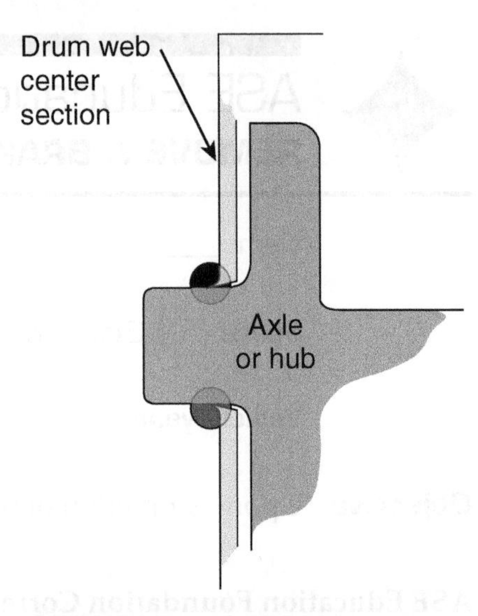

5. Loosen the lug nuts and remove the wheel.
6. Some vehicles have special fasteners that retain the brake drum.

   Were any fasteners retaining the brake drum?

   ☐ Yes ☐ No
7. Verify that the parking brake is released.
8. Use a large hammer to rap sharply on the brake drum *between* the lug bolts. The drum should pop free. If not, seek instructor assistance.

   Note: The dust around the brake sometimes contains asbestos residue. It is recommended that a respirator be worn when working around brakes for protection from breathing asbestos.

Photo by Tim Gilles

9. Reinstall the drum and wheel.
10. Tighten all of the lug nuts by hand.
11. Tighten the lug nuts in the proper order with a torque wrench.

    What is the wheel torque specification for the vehicle being serviced?

    ____

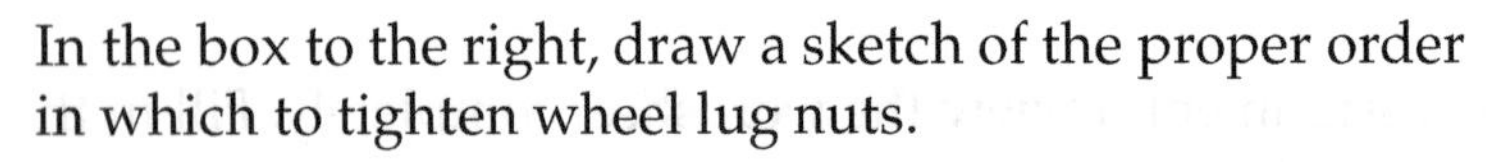

    In the box to the right, draw a sketch of the proper order in which to tighten wheel lug nuts.

12. Install the wheel cover or hubcap using a rubber mallet.
13. Lower the vehicle.
14. Before completing your paperwork, clean your work area, clean and return tools to their proper places, and wash your hands.
15. Record your recommendations for needed service or additional repairs and complete the repair order.

    ____

    ____

ASE Education Foundation Worksheet #9-3

INSPECT DRUM BRAKES

Name ______________________ Class ________________

Score: ☐ Excellent ☐ Good ☐ Needs Improvement **Instructor OK** ☐

Vehicle year ________ **Make** ______________ **Model** ______________

Objective: Upon completion of this assignment, you will be able to inspect drum brakes for needed service.

ASE Education Foundation Correlation

This worksheet addresses the following **MLR** tasks:

V.C.1 Remove, clean, and inspect brake drum; measure brake drum diameter; determine serviceability. **(P-1)**

V.C.3 Remove, clean, and inspect brake shoes, springs, pins, clips, levers, adjusters/ self-adjusters, other related brake hardware, and backing support plates; lubricate and reassemble. **(P-1)**

V.C.4 Inspect wheel cylinders for leaks and proper operation; remove and replace as needed. **(P-2)**

This worksheet addresses the following **AST/MAST** tasks:

V.C.2 Remove, clean, and inspect brake drum; measure brake drum diameter; determine serviceability. **(P-1)**

V.C.4 Remove, clean, and inspect brake shoes, springs, pins, clips, levers, adjusters/ self-adjusters, other related brake hardware, and backing support plates; lubricate and reassemble. **(P-1)**

V.C.5 Inspect wheel cylinders for leaks and proper operation; remove and replace as needed. **(P-2)**

We Support ASE Education Foundation

Directions: Before beginning this lab assignment, review the worksheet completely. Fill in the information in the spaces provided as you complete each task.

Tools and Equipment Required: Safety glasses, jack and safety stands or vehicle lift, shop towel, ½" impact wrench, impact sockets, hammer, drum gauge, torque wrench

Procedure:

1. Sit in the vehicle with the engine off and apply the brake pedal.

   Does the pedal feel firm? ☐ Yes ☐ No

   A spongy pedal indicates:

   ☐ A normal condition ☐ Brakes that need bleeding ☐ Brakes in need of adjustment

   Pump the pedal twice in rapid succession. Does the pedal height change? ☐ Yes ☐ No

TURN

If the pedal height is higher on the second application, this indicates:

☐ A normal condition ☐ Brakes that need bleeding ☐ Brakes in need of adjustment

2. Check the brake master cylinder reservoir fluid level. ☐ Full ☐ Low

   Note: Low fluid level in the disc brake reservoir is an indication that disc pads may be worn. Use Worksheet #9-7, Inspect Front Disc Brakes, to check the disc brakes.

3. Raise the vehicle and remove the wheel and drum as described in Worksheet #9-2, Remove a Brake Drum.

4. Inspect the inner surface of the brake drum.

   ☐ Smooth ☐ Scored ☐ Other

5. Measure the drum size with a drum gauge.

   Standard drum specification: ______ . ______ "

   What size is the drum? ______ . ______ "

   Is the drum oversize? ☐ Yes ☐ No

   Drum service required:

   ☐ None ☐ Machine oversize ☐ Replace

Photo by Tim Gilles

6. Use a brake parts washer to clean the brake assembly. Carefully pull back the rubber dust boots on the wheel cylinder to check for excessive leakage.

   ☐ Slightly moist (OK)

   ☐ Wet or soaked (needs service)

   Note: If a wheel cylinder is leaking, replace it.

Photo by Tim Gilles

7. Inspect the rubber brake hose leading from the axle housing to the metal tube on the chassis.

   ☐ Cracked/weathered ☐ OK

8. Inspect the linings.

   Lining construction: ☐ Bonded ☐ Riveted

   Grease or brake fluid contamination: ☐ Yes ☐ No

   Lining thickness: ☐ Thicker than the metal shoe ☐ Thinner than the metal shoe

9. Have an assistant operate the parking brake while you check its operation.

   ☐ Works properly ☐ Needs service

10. Reinstall the brake drum and wheel and lower the vehicle.

11. Before completing your paperwork, clean your work area, clean and return tools to their proper places, and wash your hands.

12. Record your recommendations for needed service or additional repairs and complete the repair order.

    ____

    ____

STOP

ASE Education Foundation Worksheet #9-4

SERVICE DRUM BRAKES—LEADING/TRAILING (FRONT-WHEEL DRIVE)

Name____________________________ Class____________________

Score: ☐ Excellent ☐ Good ☐ Needs Improvement **Instructor OK** ☐

Vehicle year __________ **Make** ___________________ **Model** ____________________

Objective: Upon completion of this assignment, you will be able to service leading/trailing-type drum brakes.

ASE Education Foundation Correlation

This worksheet addresses the following **MLR** tasks:

V.C.1 Remove, clean, and inspect brake drum; measure brake drum diameter; determine serviceability. **(P-1)**

V.C.3 Remove, clean, and inspect brake shoes, springs, pins, clips, levers, adjusters/self-adjusters, other related brake hardware, and backing support plates; lubricate and reassemble. **(P-1)**

V.C.4 Inspect wheel cylinders for leaks and proper operation; remove and replace as needed. **(P-2)**

V.C.5 Pre-adjust brake shoes and parking brake; install brake drums or drum/hub assemblies and wheel bearings; make final checks and adjustments. **(P-1)**

This worksheet addresses the following **AST/MAST** tasks:

V.C.2 Remove, clean, and inspect brake drum; measure brake drum diameter; determine serviceability. **(P-1)**

V.C.4 Remove, clean, and inspect brake shoes, springs, pins, clips, levers, adjusters/self-adjusters, other related brake hardware, and backing support plates; lubricate and reassemble. **(P-1)**

V.C.5 Inspect wheel cylinders for leaks and proper operation; remove and replace as needed. **(P-2)**

V.C.6 Pre-adjust brake shoes and parking brake; install brake drums or drum/hub assemblies and wheel bearings; make final checks and adjustments. **(P-1)**

We Support
ASE | Education Foundation

Directions: Before beginning this lab assignment, review the worksheet completely. Fill in the information in the spaces provided as you complete each task.

Tools and Equipment Required: Safety glasses, shop towels, vehicle lift, drain pan, hand tools, return spring tool, shoe retaining spring tool

Procedure: There are several types of drum brake systems. Refer to appropriate service information for the correct method of removing and installing brake shoes for the vehicle you are servicing.

Note: Worksheet #9-3, Inspect Drum Brakes, should be completed before attempting this worksheet.

Before repairing brakes on a vehicle with ABS, consult the service information for precautions and procedures. Damaged components and expensive repairs can result from failing to follow procedures.

Check one of the following: ☐ Conventional brakes/No ABS ☐ ABS

1. Raise the vehicle on a lift or place it on jack stands.
2. Remove the tire and wheel assembly.
3. What retains the brake drum to the axle?

   ☐ The tire and wheel assembly ☐ Screws ☐ Special fasteners
4. Remove any drum retaining fasteners and remove the drum.
5. Remove the brake drum and inspect it for damage. Measure the diameter of the drum, compare to specification, and determine serviceability.

   Maximum drum diameter specification _______________ Measured size _______________

   Is the drum serviceable? ☐ Yes ☐ No
6. Clean the brake components. Which of the following did you use?

   ☐ Brake parts washer ☐ Drain pan and brush
7. Remove the brake return spring.

Disassemble only one brake at a time. This will allow you to use the other side as a template for correct reassembly of the parts.

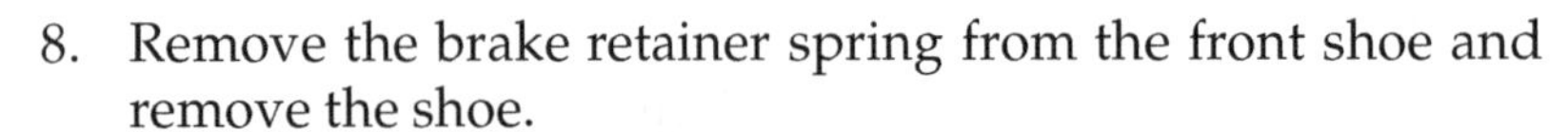

8. Remove the brake retainer spring from the front shoe and remove the shoe.
9. Remove the brake retainer spring from the rear shoe.
10. Detach the rear shoe from the self-adjuster mechanism and parking brake cable or lever.
11. Remove the parking brake lever from the rear shoe, if necessary.
12. Inspect the wheel cylinder dust boot for damage or excessive brake fluid accumulation. Replace a cylinder that is leaking.
13. Clean the backing plate and lubricate the area where the brake shoes ride.

Compare the new shoes to the old shoes. They should be exactly alike, except that the new shoes *may* have additional lining material.

14. Install the new shoes in the reverse order of removal.
15. Disassemble, clean, and lube the self-adjuster. Screw the self-adjuster all the way in and then back it off one turn.

16. Install the self-adjuster by pulling the front shoe forward and inserting the self-adjuster between the two shoes.
17. Use a brake gauge to adjust the brake shoes.
18. Install the drum. If the drum is too loose or too tight, additional adjustment may be needed.
19. Install the wheel assembly and torque the lug nuts to specification.

    Lug nut torque specification: _______ ft.-lb
20. Lower the vehicle almost to the floor and press the brake pedal several times to center the shoes.

CAUTION Before moving the vehicle after a brake repair, always pump the pedal several times. Failure to do so may cause an accident with damage to vehicles or the facility. Injuries to persons can also result.

21. Check the brake fluid level and add fluid as necessary.
22. Perform a brake test to ensure that the brakes will stop and hold the vehicle. Do this test before moving the vehicle from the service bay.

CAUTION If you noticed a problem, have your instructor check the brakes before driving the vehicle.

23. Record your recommendations for needed service or additional repairs and complete the repair order.

    ____

    ____

STOP

ASE Education Foundation Worksheet #9-5

SERVICE DRUM BRAKES—SELF-ENERGIZING (BENDIX) TYPE (REAR-WHEEL DRIVE)

Name_______________________________ Class______________________

Score: ☐ Excellent ☐ Good ☐ Needs Improvement **Instructor OK** ☐

Vehicle year __________ **Make** ____________________ **Model** ______________________

Objective: Upon completion of this assignment, you will be able to service self-energizing (Bendix) type drum brakes.

ASE Education Foundation Correlation

This worksheet addresses the following **MLR** tasks:

V.A.3 Install wheel and torque lug nuts. **(P-1)**

V.C.1 Remove, clean, and inspect brake drum; measure brake drum diameter; determine serviceability. **(P-1)**

V.C.3 Remove, clean, and inspect brake shoes, springs, pins, clips, levers, adjusters/self-adjusters, other related brake hardware, and backing support plates; lubricate and reassemble. **(P-1)**

V.C.4 Inspect wheel cylinders for leaks and proper operation; remove and replace as needed. **(P-2)**

V.C.5 Pre-adjust brake shoes and parking brake; install brake drums or drum/hub assemblies and wheel bearings; make final checks and adjustments. **(P-1)**

This worksheet addresses the following **AST/MAST** tasks:

V.A.4 Install wheel and torque lug nuts. **(P-1)**

V.C.2 Remove, clean, and inspect brake drum; measure brake drum diameter; determine serviceability. **(P-1)**

V.C.4 Remove, clean, and inspect brake shoes, springs, pins, clips, levers, adjusters/self-adjusters, other related brake hardware, and backing support plates; lubricate and reassemble. **(P-1)**

V.C.5 Inspect wheel cylinders for leaks and proper operation; remove and replace as needed. **(P-2)**

V.C.6 Pre-adjust brake shoes and parking brake; install brake drums or drum/hub assemblies and wheel bearings; make final checks and adjustments. **(P-1)**

We Support ASE Education Foundation

Directions: Before beginning this lab assignment, review the worksheet completely. Fill in the information in the spaces provided as you complete each task.

Tools and Equipment Required: Safety glasses, shop towels, vehicle lift, drain pan, hand tools, return spring tool, shoe retaining spring tool

Procedure: There are several types of drum brake systems. Refer to appropriate service information for the correct method of removing and installing brake shoes for the vehicle you are servicing.

Note: Worksheet #9-3, Inspect Drum Brakes, should be completed before attempting this worksheet.

CAUTION Before repairing brakes on a vehicle with ABS, consult the service information for precautions and procedures. Damaged components and expensive repairs can result from failure to follow procedures to protect ABS components during routine brake work.

Check one of the following: ☐ Conventional brakes/No ABS ☐ ABS

1. Raise the vehicle on a lift or place it on safety stands.
2. Remove the tire and wheel assembly.
3. Remove the brake drum and inspect it for damage. Measure the diameter of the drum, compare to specification, and determine serviceability.

   Maximum diameter specification ______ " ______mm Measured size ______ " ______mm

   Is the drum serviceable? ☐ Yes ☐ No
4. Clean the brake components. Which of the following did you use?

   ☐ Brake parts washer ☐ Drain pan and brush
5. Inspect the wheel cylinder dust boot for damage or brake fluid.
6. Remove the brake shoes.

SHOP TIP Disassemble only one brake at a time. This will allow you to use the other side as a template for correct reassembly of the assembly parts.

Compare the new shoes to the old shoes. They should be exactly alike, except that the new shoes *may* have additional lining material.

7. Clean the backing plate and lubricate the area where the brake shoes ride.
8. Disassemble, clean, and lube the self-adjuster. Screw the self-adjuster all the way in and then back it off one turn.
9. Install the brake shoes, self-adjuster, and springs.
10. Use a brake gauge to adjust the brake shoes.
11. Install the drum. If the drum is too loose or too tight, additional adjustment may be needed.
12. Install the wheel assembly and torque lug nuts to specification.

    Lug nut torque specification: ______ ft.-lb
13. Lower the vehicle almost to the floor and press the brake pedal several times to center the shoes.

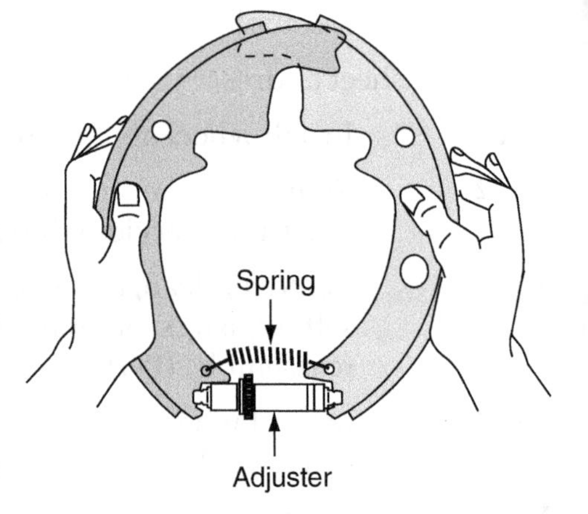

CAUTION Before moving the vehicle after a brake repair, pump the pedal several times to test the brake. Failure to do so may cause an accident with damage to vehicles or the facility or personal injury.

14. Check the brake fluid level and add fluid as necessary.

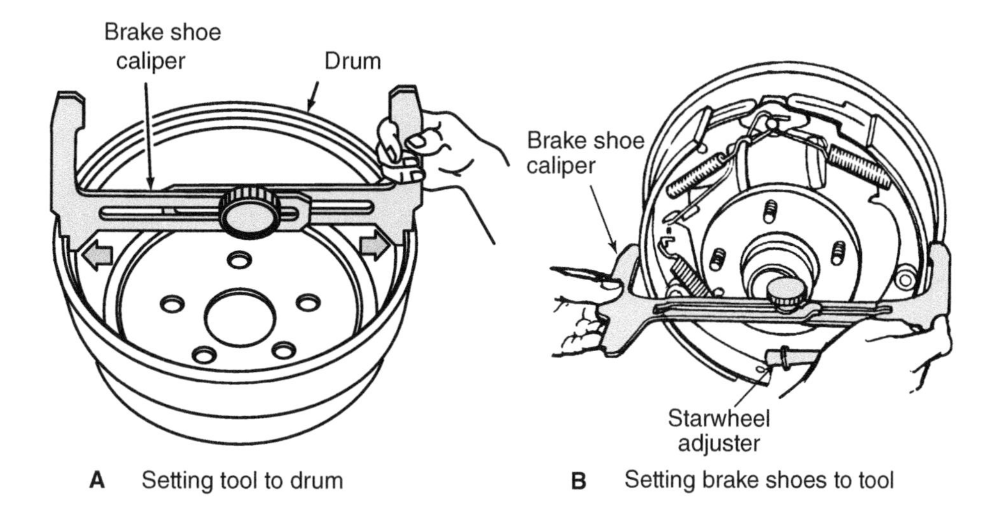

A Setting tool to drum **B** Setting brake shoes to tool

15. Perform a brake test to ensure that the brakes will stop and hold the vehicle. Do this test before moving the vehicle from the bay.

If you noticed a problem, have the instructor check the brakes before driving the vehicle.

16. Record your recommendations for needed service or additional repairs and complete the repair order.

____

____

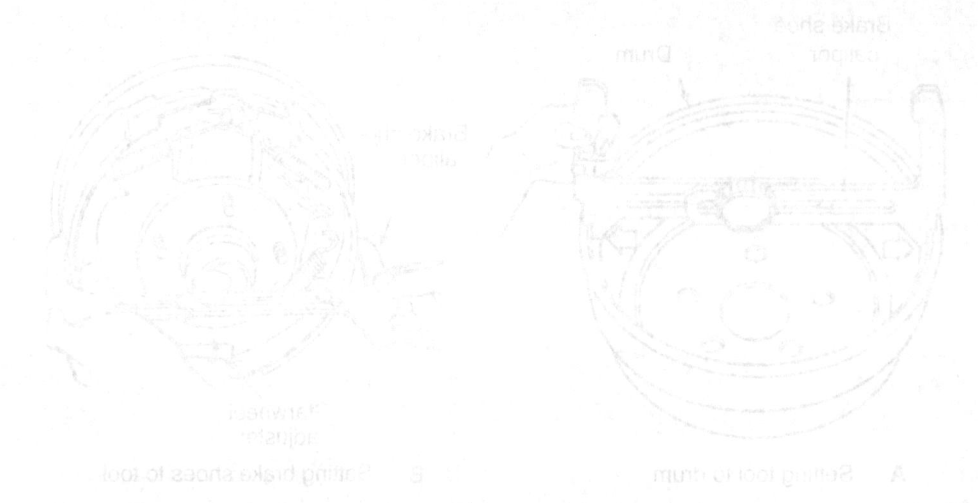

ASE Education Foundation Worksheet #9-6

MACHINE A BRAKE DRUM

Name ______________________ Class ______________

Score: ☐ Excellent ☐ Good ☐ Needs Improvement **Instructor OK** ☐

Vehicle year __________ **Make** ______________ **Model** ______________

Objective: Upon completion of this assignment, you will be able to machine a brake drum.

ASE Education Foundation Correlation

This worksheet addresses the following **MLR** task:

V.C.2 Refinish brake drum and measure final drum diameter; compare with specifications. **(P-1)**

This worksheet addresses the following **AST/MAST** task:

V.C.3 Refinish brake drum and measure final drum diameter; compare with specifications. **(P-1)**

We Support ASE Education Foundation

Directions: Before beginning this lab assignment, review the worksheet completely. Fill in the information in the spaces provided as you complete each task.

Tools and Equipment Required: Safety glasses, shop towels, vehicle lift, brake lathe, brake drum micrometer, hand tools

Procedure:

1. Before beginning to work, locate and record the following information:

   Drum discard dimension ____________ Drum maximum machining dimension ____________

   Are there any special precautions to be followed for this drum? ☐ Yes ☐ No
2. Remove the wheel assembly and remove the brake drum.
3. Inspect the drum. Does it appear to be machinable? ☐ Yes ☐ No

   If no, explain. ______________________________
4. Measure the inside diameter of the drum (measure at least two points equally spaced around the drum):

   A ____________ B ____________

   Note: The following steps are generally based on a typical bench brake lathe. Refer to the lathe operating instructions for the lathe you will be using for specific operating procedures.

TURN

5. Select a centering cone (arbor) that fits about halfway into the center hole of the drum.
6. Select two adapters that fit the drum without interfering with the cutting head of the lathe.
7. Slide one adapter onto the lathe shaft, open end out.
8. Slide a spring onto the shaft, followed by the centering cone (arbor).
9. Slide the drum onto the shaft, followed by the outer adapter, bushing, spacer (if needed), and the nut. Tighten the nut, but do not overtighten.
10. Install the damping strap with its edge overlapping the outer edge of the drum.

Note: There are two hand wheels. One moves the spindle to position the drum and the other moves the cross-feed to position the cutting tip.

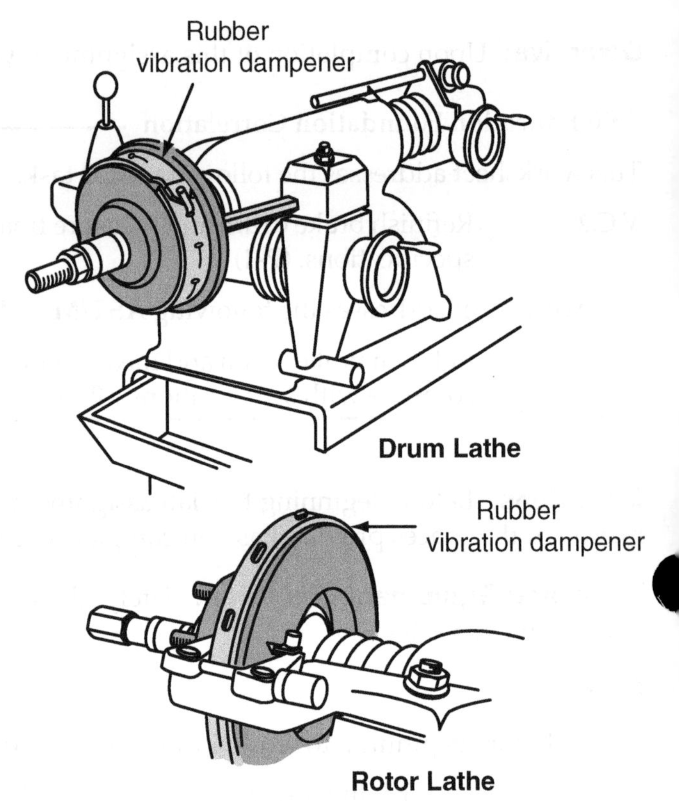

11. Turn the cutter head cross-feed hand wheel to move the cutter head assembly inward toward the lathe body until it stops. Then reverse direction for two turns.
12. Continue to adjust the cutting head so it will reach the inner edge of the machined surface of the drum.
13. Turn the spindle-feed hand wheel to move the drum out (away from the lathe) until the cutting tip is about halfway through the machined area of the drum.
14. Adjust the cutting tip toward the machined surface of the drum until it meets the drum. Then reverse direction for about half a turn of the hand wheel.
15. Verify that the area around the lathe is clear and that the lathe's drive mechanism is in neutral. Then switch on the motor to begin rotating the drum.
16. Slowly turn the cutter head hand wheel until the cutting tip comes into contact with the rotating drum. Hold the hand wheel steady while you set the sliding scale to zero. Then turn the hand wheel the opposite direction to move the cutting bit away from the drum.

Note: Sometimes a ridge appears due to wear on the inside of the drum surface. The ridge is removed prior to machining the drum wear surface.

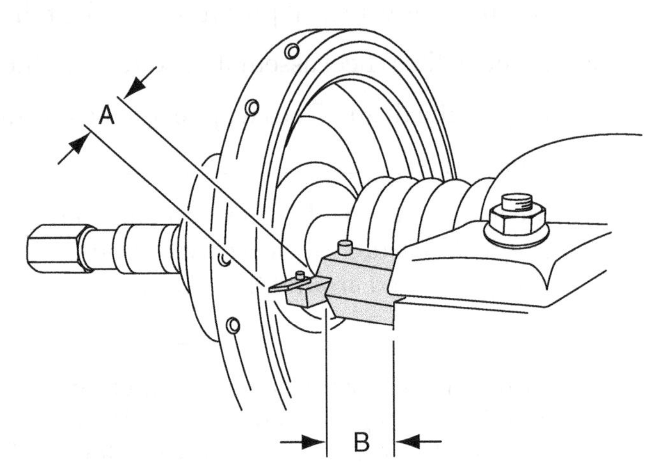

Keep distances A and B as short as possible.

17. If there is a lip on the outside of the drum, use the other hand wheel to move the drum until the cutting tip is aligned with the inside edge of the lip. Slowly remove the lip by manually adjusting the cutting depth and moving the drum past the cutting tip.

18. Use the spindle-feed hand wheel to move the cutting tip to the rear of the drum in preparation for machining the drum. Adjust the cutting tip until it contacts the drum.
19. Continue adjusting the tip until the scale is set at 0.002 inch.
20. Engage fast speed on the lathe.
21. Observe the drum as it is being machined. Are there dark (uncut) areas? ☐ Yes ☐ No
22. If the answer to step #21 is no, skip to step #26. If the answer to step #21 is yes, proceed to step #23.
23. When the cutting tip clears the outside of the machined surface of the drum, disengage the automatic feed drive and move the cutting tip back to the starting point.
24. Adjust the cutting tip to cut 0.002 inch deeper and engage fast cut.
25. Engage the drive mechanism in slow speed.
26. When the cutting bit clears the drum, disengage the drive and stop the motor.
27. When the drum stops turning, remove it from the lathe.
28. Wash the drum in hot, running soapy water using a bristle brush if possible to clean the machined surfaces.
29. Rinse with clean water and blow dry with an OSHA-approved blowgun.
30. Install the drum on the vehicle.
31. Before completing your paperwork, clean your work area, clean and return tools to their proper places, and wash your hands.
32. Record your recommendations for needed service or additional repairs and complete the repair order.

____

____

STOP

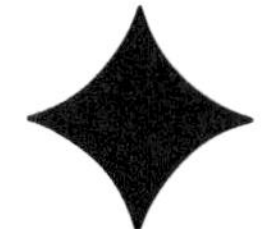

ASE Education Foundation Worksheet #9-7

INSPECT FRONT DISC BRAKES

Name_______________________ Class_______________

Score: ☐ Excellent ☐ Good ☐ Needs Improvement **Instructor OK** ☐

Vehicle year ________ **Make** ______________ **Model** ________________

Objective: Upon completion of this assignment, you will be able to inspect the front disc brakes for needed repairs.

ASE Education Foundation Correlation

This worksheet addresses the following **MLR** tasks:

V.D.3 Remove, inspect, and replace pads and retaining hardware; determine necessary action. **(P-1)**

V.D.5 Clean and inspect rotor, measure rotor thickness, thickness variation, and lateral runout; determine necessary action. **(P-1)**

V.D.10 Check brake pad wear indicator; determine necessary action. **(P-1)**

This worksheet addresses the following **AST/MAST** tasks:

V.D.1 Diagnose poor stopping, noise, vibration, pulling, grabbing, dragging, or pulsation concerns; determine needed action. **(P-1)**

V.D.4 Remove, inspect, and replace pads and retaining hardware; determine needed action. **(P-1)**

V.D.6 Clean and inspect rotor, measure rotor thickness, thickness variation, and lateral runout; determine needed action. **(P-1)**

V.D.11 Check brake pad wear indicator; determine needed action. **(P-1)**

We Support
ASE | Education Foundation

Directions: Before beginning this lab assignment, review the worksheet completely. Fill in the information in the spaces provided as you complete each task.

Tools and Equipment Required: Safety glasses, jack and jack stands or vehicle lift, shop towel, ½" impact wrench, impact sockets, flashlight, torque wrench

- ❑ If a vehicle is found to have unsafe brakes, it should not be driven from the shop. It should be towed either to a repair facility or to the owner's home.
- ❑ When brake work is performed, wheels should not be reinstalled on the vehicle until all brake work has been satisfactorily completed.
- ❑ For liability reasons, an owner may only remove an unsafe vehicle if the state police or highway patrol has been notified and issues an equipment repair citation.

Procedure:

1. Check the fluid level in the brake master cylinder disc brake reservoir.

   ☐ Full ☐ Low

   Note: Low fluid level in the disc brake reservoir indicates that the disc pads may be worn.

2. Raise the vehicle and rotate the tire and wheel assembly. Can a scraping noise be heard while the wheel is rotating?

   ☐ Yes ☐ No

3. Remove the wheel.

4. Use a flashlight to inspect both the leading and trailing edges of both linings. Linings sometimes wear unevenly.

   The lining thickness at its thinnest point is:

   ☐ Thicker than the metal shoe

   ☐ Worn evenly

   ☐ Thinner than the metal shoe

   ☐ Worn unevenly

   Do the linings need to be replaced?

   ☐ Yes ☐ No

Lining

Metal back Brake caliper

Photo by Tim Gilles

5. Inspect the pad wear indicator. Is it scraping on the rotor?

   ☐ Yes ☐ No ☐ N/A

6. Inspect the condition of the rotor.

   ☐ Smooth ☐ Wavy ☐ Scored

   Is there a lip at the rotor's outer edge? ☐ Yes ☐ No

   Rotate the rotor. Is any warpage (runout) evident?

   ☐ Yes ☐ No

7. Use a micrometer to check the rotor thickness variation. Did you measure any thickness variation?

   ☐ Yes ☐ No If not, how much is the variation? 0. ____________"

8. Use a dial indicator to check the rotor runout. Is there any runout?

   ☐ Yes How much? 0. _____" ☐ No

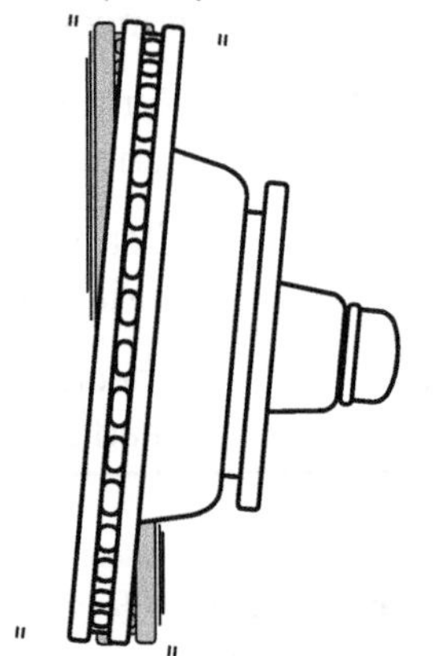

Disc brake "rotor runout".

STOP

9. Inspect the condition of the rubber brake hose leading to each brake caliper.

   ☐ OK

   ☐ Cracked/weathered

   ☐ Twisted

10. Reinstall the wheel(s) and lower the vehicle to the floor.

11. Before completing your paperwork, clean your work area, clean and return tools to their proper places, and wash your hands.

12. Record your recommendations for needed service or additional repairs and complete the repair order.

Tighten lug nuts

Photo by Tim Gilles

____

____

STOP

ASE Education Foundation Worksheet #9-8

REPLACE FRONT DISC BRAKE PADS

Name_______________________________ Class_______________________

Score: ☐ Excellent ☐ Good ☐ Needs Improvement **Instructor OK** ☐

Vehicle year __________ **Make** ____________________ **Model** ______________________

Objective: Upon completion of this assignment, you will be able to replace disc pads.

ASE Education Foundation Correlation

This worksheet addresses the following **MLR** tasks:

V.D.1 Remove and clean caliper assembly; inspect for leaks and damage/wear; determine necessary action. **(P-1)**

V.D.2 Inspect caliper mounting and slides/pins for proper operation, wear, and damage; determine necessary action. **(P-1)**

V.D.3 Remove, inspect, and/or replace brake pads and retaining hardware; determine necessary action. **(P-1)**

V.D.4 Lubricate and reinstall caliper, brake pads, and related hardware; seat brake pads and inspect for leaks. **(P-1)**

V.D.5 Clean and inspect rotor and mounting surface, measure rotor thickness, thickness variation, and lateral runout; determine necessary action. **(P-1)**

V.D.11 Describe importance of operating vehicle to burnish/break-in replacement brake pads according to manufacturer's recommendation. **(P-1)**

This worksheet addresses the following **AST/MAST** tasks:

V.D.2 Remove and clean caliper assembly; inspect for leaks, damage, and wear; determine needed action. **(P-1)**

V.D.3 Inspect caliper mounting and slides/pins for proper operation, wear, and damage; determine needed action. **(P-1)**

V.D.4 Remove, inspect, and/or replace brake pads and retaining hardware; determine needed action. **(P-1)**

V.D.5 Lubricate and reinstall caliper, brake pads, and related hardware; seat brake pads; inspect for leaks. **(P-1)**

V.D.6 Clean and inspect rotor and mounting surface; measure rotor thickness, thickness variation, and lateral runout; determine needed action. **(P-1)**

V.D.12 Describe importance of operating vehicle to burnish/break-in replacement brake pads according to manufacturer's recommendations. **(P-1)**

We Support ASE Education Foundation

Directions: Before beginning this lab assignment, review the worksheet completely. Fill in the information in the spaces provided as you complete each task.

Tools and Equipment Required: Safety glasses, hands tools, shop towel and torque wrench

Procedure:

1. Check the level of the brake fluid in the master cylinder reservoir. ☐ Full ☐ Low
2. Raise the vehicle on a lift or place it on safety stands.
3. Remove the front wheel and tire assemblies.
4. Remove the caliper guide pins or disc pad retainers, and remove the disc pads or caliper.

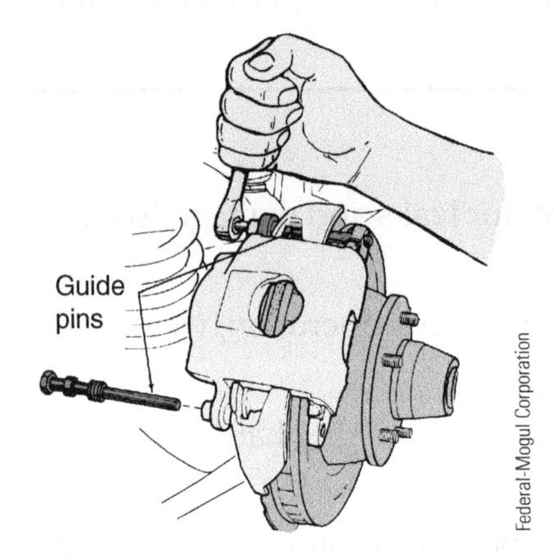

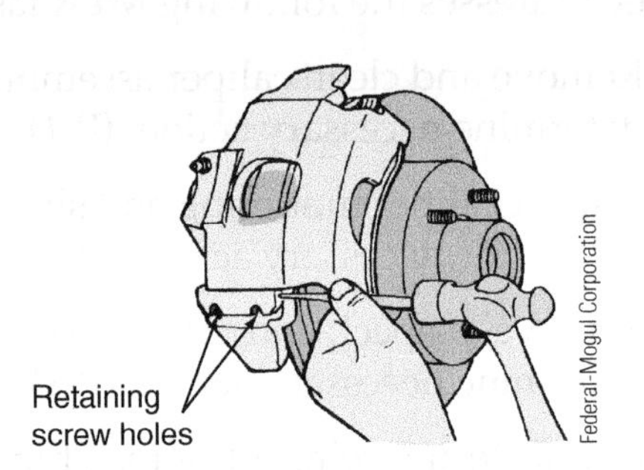

5. Do not allow the caliper to hang from its hose. Support it by hanging it from the suspension or the frame of the vehicle using a wire or bungee cord.
6. Inspect the condition of the rotors.

   ☐ Smooth ☐ Wavy ☐ Scored
7. Measure the rotor parallelism and runout as described in Worksheet #9-7, Inspect Front Disc Brakes.
8. Is the rotor reusable, or does it need to be machined or replaced?

   ☐ Reusable ☐ Service ☐ Replace

Wire

Caliper

Photo by Tim Gilles

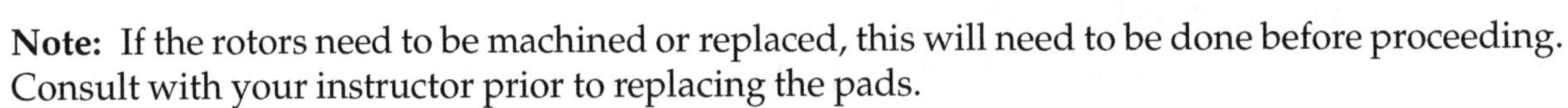

   Note: If the rotors need to be machined or replaced, this will need to be done before proceeding. Consult with your instructor prior to replacing the pads.
9. Inspect the caliper slides to be sure they are in good condition and lubricated.
10. Open the caliper bleeder screw and push the piston back into the caliper bore using a C-clamp or other suitable tool.

    Note: Tighten the bleeder screw when you are finished so that it is not accidentally left open.
11. Clean the caliper slides and install the caliper without the disc pads to check that it moves freely in its mount. Does it move freely? ☐ Yes ☐ No

12. Remove the caliper and install the disc pads. Then reinstall the caliper on its mount. Torque all bolts as required. Torque specification: ____________

13. Install the tires and wheel assemblies.

14. Lower the vehicle to the floor.

15. Torque the lug nuts and install the wheel covers. Torque specification: ____________

16. Before moving the vehicle depress the brake pedal several times. The pedal should be firm and at the correct height.

    The pedal height is ☐ Normal ☐ Low

    Note: Some manufacturers recommend burnishing/breaking-in brake pads after replacement/installation.

End of clamp against caliper

End of screw against outboard pad

17. Refer to the service information for the break in/burnishing procedure.

    Briefly describe the recommended procedure.

    ____ ☐ N/A

18. Record your recommendations for needed service or additional repairs and complete the repair order.

    ____

    ____

STOP

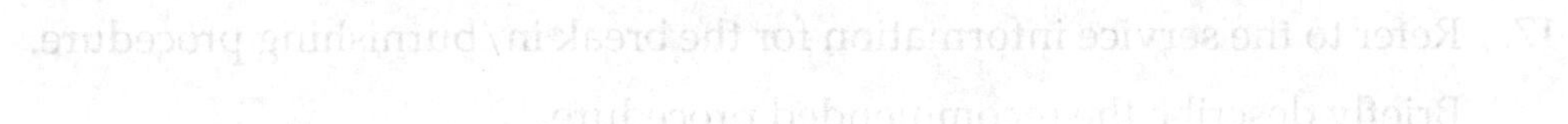

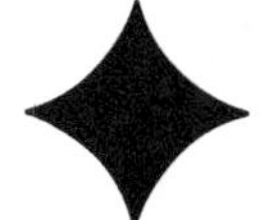

ASE Education Foundation Worksheet #9-9

REMOVE AND REPLACE A DISC BRAKE ROTOR

Name_______________ Class_______________

Score: ☐ Excellent ☐ Good ☐ Needs Improvement **Instructor OK** ☐

Vehicle year _______ **Make** _______________ **Model** _______________

Objective: Upon completion of this assignment, you will be able to replace a disc brake rotor.

ASE Education Foundation Correlation

This worksheet addresses the following **MLR** task:

V.D.6 Remove and reinstall/replace rotor. **(P-1)**

This worksheet addresses the following **AST/MAST** task:

V.D.7 Remove and reinstall/replace rotor. **(P-1)**

We Support ASE Education Foundation

Directions: Before beginning this lab assignment, review the worksheet completely. Fill in the information in the spaces provided as you complete each task.

Tools and Equipment Required: Safety glasses, fender covers, shop towels, hand tools, torque wrench

Procedure:

Note: Worksheet #9-7, Inspect Front Disc Brakes, should be completed before attempting to complete this worksheet. There are many different types of disc brake systems. Refer to service information for the correct method for removing and installing rotors for this vehicle.

Which service information are you using? _______________________________

CAUTION Before repairing the brakes of a vehicle with an ABS, consult the service information for precautions and procedures. Failure to follow procedures to protect ABS components during routine brake work can damage components and result in expensive repairs.

☐ Non-ABS

☐ ABS (refer to service information for precautions)

Removal

1. Raise the vehicle and remove the tire and wheel assembly. Inspect the rotor for damage.
2. Measure the rotor to determine serviceability.

   Minimum thickness specification: _______

   Measured size: _______

   Parallelism specification: _______

   Measured parallelism: _______

   Rotor runout specification: _______

   Measured runout: _______

   Is the rotor serviceable? ☐ Yes ☐ No

TURN

3. Remove the disc brake caliper. First loosen the bleeder screw. Then hold the rotor at the top and bottom and gently rock it to move the caliper pistons back into their bores to allow easier removal of the caliper. Retighten the bleed screw.
4. Remove the caliper fasteners and lift the caliper off the rotor. Do not allow it to hang from the brake hose. Wire it up or use a bungee cord to hold it.
5. Remove the rotor.

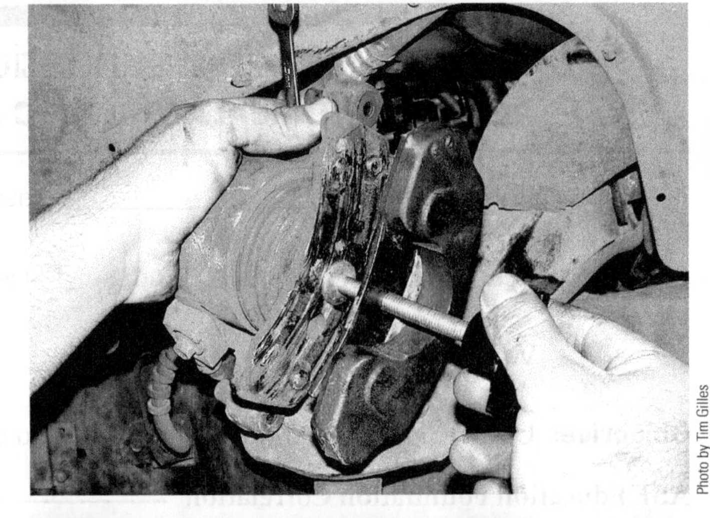

Installation

6. Replace the disc brake rotor.
7. Install the caliper and torque the fasteners to specification.

   Caliper bolt torque specification: ___________ ft.-lb
8. Install the tire and wheel assembly and torque the lug nuts.

   Lug nut torque specification: ___________ ft.-lb

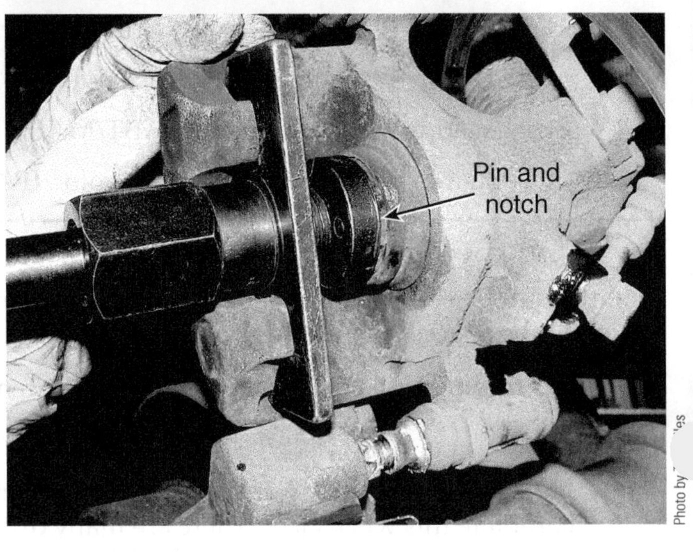

9. Install the wheel cover.
10. Check the brake fluid level and add fluid as necessary.

    Warning: Pump the pedal several times to test the brakes before moving the vehicle from the service bay following a brake repair. Failure to do so can result in a serious accident.

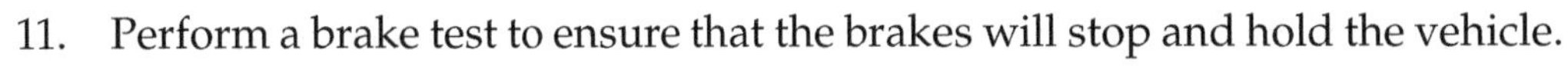

11. Perform a brake test to ensure that the brakes will stop and hold the vehicle.

    ☐ Brakes are good.

    ☐ Further repair is needed.

If you notice a problem, have your instructor check the brakes before driving the vehicle.

12. Before completing your paperwork, clean your work area, clean and return tools to their proper places, and wash your hands.
13. Record your recommendations for needed service or additional repairs and complete the repair order.

    ____

    ____

STOP

ASE Education Foundation Worksheet #9-10

REFINISH A DISC BRAKE ROTOR (OFF VEHICLE)

Name ______________________ Class ________________

Score: ☐ Excellent ☐ Good ☐ Needs Improvement **Instructor OK** ☐

Vehicle year ________ **Make** ______________ **Model** ________________

Objective: Upon completion of this assignment, you will be able to resurface a disc brake rotor using an off-car brake lathe.

ASE Education Foundation Correlation

This worksheet addresses the following **MLR** task:

V.D.8 Refinish rotor off vehicle; measure final rotor thickness and compare with specification. **(P-1)**

This worksheet addresses the following **AST/MAST** task:

V.D.9 Refinish rotor off vehicle; measure final rotor thickness and compare with specification. **(P-1)**

We Support ASE Education Foundation

Directions: Before beginning this lab assignment, review the worksheet completely. Fill in the information in the spaces provided as you complete each task.

Tools and Equipment Required: Safety glasses, hand tool, off-car brake lathe

Procedure:

1. Raise the vehicle on a lift or place it on jack stands.
2. Remove the tire and wheel assembly.
3. Remove the caliper fasteners and remove the caliper. Do not let the caliper hang from its hose. Support it by hanging it on a wire from the suspension or the frame of the vehicle.
4. Measure the rotor thickness. Thickness: _______ Specification: _______
5. Measure the rotor parallelism. Measurement: _______ Specification: _______
6. Measure the rotor for runout. Measurement: _______ Specification: _______
7. Is the rotor serviceable? ☐ Yes ☐ No
8. Remove the rotor from the vehicle.
9. Mount the rotor on the lathe.
10. Refinish the rotor.
11. Remove the rotor from the lathe. Do not touch the braking surface.
12. Measure the rotor thickness. Thickness: _____________ Specifications: _____________
13. Is the rotor still usable (within thickness limits)? ☐ Yes ☐ No

TURN

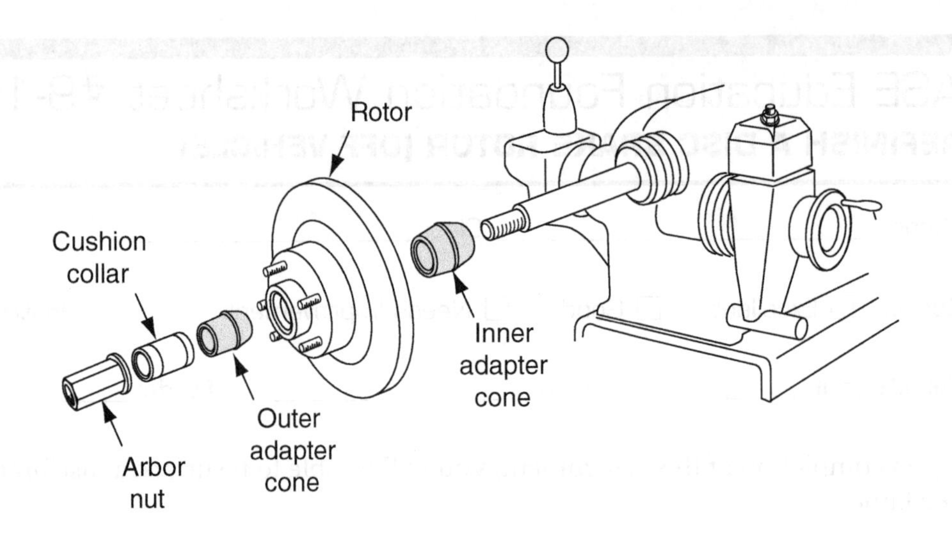

14. Install the rotor on the vehicle.
15. Install the caliper and disc pads.
16. Install the tire and wheel assembly.
17. Torque the lug nuts and install the wheel covers. Torque specification: ________
18. Before completing your paperwork, clean your work area, clean and return tools to their proper places, and wash your hands.
19. Record your recommendations for needed service or additional repairs and complete the repair order.

____

____

ASE Education Foundation Worksheet #9-11

REFINISH A DISC BRAKE ROTOR (ON VEHICLE)

Name_______________________________ Class______________________

Score: ☐ Excellent ☐ Good ☐ Needs Improvement **Instructor OK** ☐

Vehicle year __________ **Make** ____________________ **Model** ______________________

Objective: Upon completion of this assignment, you will be able to refinish a disc brake rotor using an on-car brake lathe.

ASE Education Foundation Correlation

This worksheet addresses the following **MLR** task:

V.D.7 Refinish rotor on vehicle; measure final rotor thickness and compare with specification. **(P-1)**

This worksheet addresses the following **AST/MAST** task:

V.D.8 Refinish rotor on vehicle; measure final rotor thickness and compare with specification. **(P-1)**

We Support
ASE | Education Foundation

Directions: Before beginning this lab assignment, review the worksheet completely. Fill in the information in the spaces provided as you complete each task.

Tools and Equipment Required: Safety glasses, hand tools, on-car brake lathe

Procedure:

1. Raise the vehicle and remove the tire and wheel assembly.
2. Remove the caliper fasteners and remove the caliper.

   Note: Do not allow the caliper to hang from its hose. Support it by hanging it on a wire or bungee cord from the suspension or the frame of the vehicle.

Photo by Tim Gilles

3. Measure rotor thickness. Thickness: __________ Specification: __________
4. Measure rotor parallelism. Measurement: __________ Specification: __________
5. Measure rotor runout. Measurement: __________ Specification: __________
6. Is the rotor serviceable? ☐ Yes ☐ No
7. Mount the brake lathe on the vehicle.

   Note: Before using an on-vehicle brake lathe, have your instructor demonstrate its use.

8. Adjust the lathe and refinish the rotor.
9. Remove the lathe from the vehicle.
10. Measure the rotor thickness. Thickness ________ Specifications ________
11. Is the rotor still within acceptable thickness specifications? ☐ Yes ☐ No
12. Reinstall the caliper and disc brake pads. Torque the caliper bolts to specification.

    Caliper mounting bolt torque specification: ________ ft/lb
13. Install the tire and wheel assembly and torque the lug nuts before installing the wheel covers.

    Lug nut torque specification ________ ft/lb
14. Before completing your paperwork, clean your work area, clean and return tools to their proper places, and wash your hands.
15. Record your recommendations for needed service or additional repairs and complete the repair order.

    ____

    ____

STOP

ASE Education Foundation Worksheet #9-12

PARKING BRAKE ADJUSTMENT

Name______________________________ Class______________________

Score: ☐ Excellent ☐ Good ☐ Needs Improvement **Instructor OK** ☐

Vehicle year __________ **Make** ____________________ **Model** ______________________

Objective: Upon completion of this assignment, you should be able to adjust a vehicle's parking brake.

ASE Education Foundation Correlation

This worksheet addresses the following **MLR** tasks:

V.F.2 Check parking brake system components for wear, binding, and corrosion; clean, lubricate, adjust and/or replace as needed. **(P-1)**

V.F.3 Check parking brake operation and parking brake indicator light system operation; determine necessary action. **(P-1)**

This worksheet addresses the following **AST/MAST** tasks:

V.F.3 Check parking brake system and components for wear, binding, and corrosion; clean, lubricate, adjust and/or replace as needed. **(P-1)**

V.F.4 Check parking brake operation and parking brake indicator light system operation; determine needed action. **(P-1)**

We Support ASE | Education Foundation

Directions: Before beginning this lab assignment, review the worksheet completely. Fill in the information in the spaces provided as you complete each task.

Tools and Equipment Required: Safety glasses, jack and jack stands or vehicle lift, hand tools, shop towel

Procedure:

1. Complete a rear drum brake adjustment before making any adjustment to the parking brake. Are the rear brakes correctly adjusted?

   ☐ Yes ☐ No

2. Apply the parking brake to ¼ of its full travel.
3. Put the transmission shift selector in Neutral.
4. Raise and support the vehicle.
5. Inspect the parking brake cables.

   ☐ Worn ☐ Damaged
   ☐ Rusted ☐ OK
   ☐ Binding

Parking brake cable
Intermediate lever
Pull-rod
Equalizer
Adjusting nut

TURN

6. Loosen the adjusting locknut on the parking brake equalizer bar.
7. Tighten the adjusting nut at the equalizer bar until the rear wheels will no longer turn when rotated by hand.
8. Release the parking brake lever.
9. Do both rear wheels turn freely?

   ☐ Yes ☐ No

10. Tighten the locknut on the adjuster.

Rear cables

Adjustment nut
(use ratcheting box wrench)

Hold front
cable

11. Lower the vehicle.
12. Check the operation of the parking brake.
13. Does the brake indicator light come on when the parking brake is applied with the key on?

    ☐ Yes ☐ No

14. Before completing your paperwork, clean your work area, clean and return tools to their proper places, and wash your hands.
15. Record your recommendations for needed service or additional repairs and complete the repair order.

    ____

    ____

ASE Education Foundation Worksheet #9-13

TEST A VACUUM-TYPE POWER BRAKE BOOSTER

Name_______________________ Class________________

Score: ☐ Excellent ☐ Good ☐ Needs Improvement **Instructor OK** ☐

Vehicle year ________ **Make** ______________ **Model** ________________

Objective: Upon completion of this assignment, you should be able to test a vacuum-type power brake booster.

ASE Education Foundation Correlation

This worksheet addresses the following MLR/AST/MAST tasks:

V.E.1 Check brake pedal travel with and without engine running to verify proper power booster operation.

V.E.2 Identify components of the brake power assist system (vacuum and hydraulic); check vacuum supply (manifold or auxiliary pump) to vacuum-type power booster. **(P-1)**

This worksheet addresses the following **AST/MAST** task:

V.E.3 Inspect vacuum-type power booster unit for leaks; inspect the check-valve for proper operation; determine needed action. **(P-1)**

We Support ASE Education Foundation

Directions: Before beginning this lab assignment, review the worksheet completely. Fill in the information in the spaces provided as you complete each task.

Tools and Equipment Required: Safety glasses, fender covers, shop towel, vacuum gauge, vacuum hose adapters

Procedure:

Brake Booster Operation Test

1. With the engine off, pump the brake pedal several times to deplete any vacuum from the booster reservoir. You will be able to hear the sound of air rushing into the booster from the passenger compartment.

   How many times did you depress the pedal before no more airflow was heard?

   ☐ None ☐ 1 ☐ 2 ☐ 3 ☐ 4 or more

2. Hold your foot on the pedal while starting the engine. If the power booster is operating correctly, the pedal will move about an inch closer to the floor after the engine starts.

   ☐ Pedal moves closer to the floor. ☐ Pedal height remains the same.

Problems with Braking Effort

When there is a problem with braking effort, there are several potential causes, including aftermarket tires and wheels with a larger circumference than the ones the brake system was designed for.

Testing for an Internal Booster Leak

3. To test for an internal booster leak, shut off the engine and apply steady pressure to the brakes. The pedal height should remain constant for at least 30 seconds. If the booster has an internal leak, the pedal will slowly rise during this test.

   ☐ Pedal height remains constant.

   ☐ Pedal slowly rises.

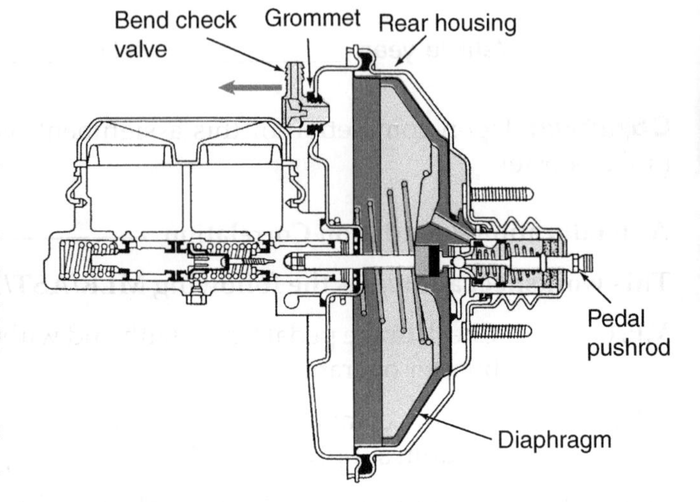

Testing the Check Valve:

4. If the check valve is bad, braking effort will vary according to the load on the engine. Test the check valve by carefully bending the valve against its rubber grommet. If the check valve is operating correctly, air will rush into the front part of the booster.
5. With the valve removed, you should be able to blow through it in one direction and it should seal from the other side.

Testing for Vacuum Supply:

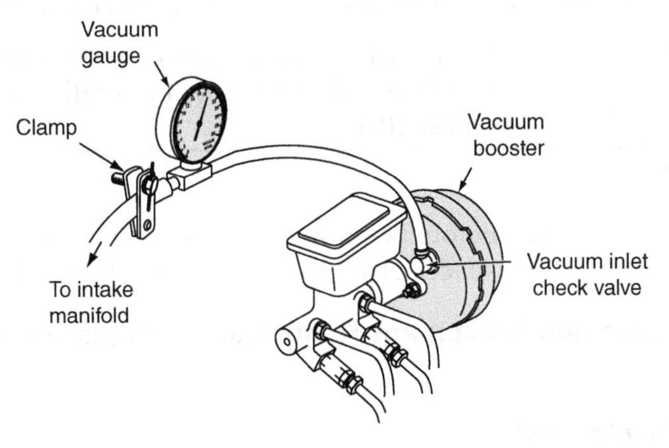

6. Disconnect the check valve hose from the brake booster and connect a vacuum gauge using a "T" connector.
7. Start the engine. Typical minimum acceptable engine vacuum with the engine idling is 15 inches of mercury (in. Hg).

   Vacuum reading at idle: _______ in. Hg
8. With the engine running, pinch off the vacuum hose on the engine side of the T connector. Vacuum should be maintained for at least 15 seconds without a loss of more than 1 in. Hg. If leakage is more than this, the check valve is defective or the booster has an internal leak.
9. Before condemning a booster, drive the vehicle with a vacuum gauge connected to the engine by a long hose. At cruising speed, apply the brakes several times and verify that the vacuum supply is sufficient.
10. Before completing your paperwork, clean your work area, clean and return tools to their proper places, and wash your hands.
11. Record your recommendations for needed service or additional repairs and complete the repair order.

____

____

STOP

ASE Education Foundation Worksheet #9-14

COMPLETE BRAKE INSPECTION WORKSHEET

Name ______________________ Class ________________

Score: ☐ Excellent ☐ Good ☐ Needs Improvement **Instructor OK** ☐

Vehicle year ________ **Make** ______________ **Model** ______________

Objective: Upon completion of this assignment, you should be able to inspect a vehicle's brake system.

ASE Education Foundation Correlation

This worksheet addresses the following **MLR** tasks:

V.B.3 Inspect brake lines, flexible hoses, and fittings for leaks, dents, kinks, rust, cracks, bulging, wear, and loose fittings/supports. **(P-1)**

V.C.4 Inspect wheel cylinders for leaks and proper operation; remove and replace as needed. **(P-1)**

V.D.3 Remove, inspect, and/or replace brake pads and retaining hardware; determine necessary action. **(P-1)**

V.D.5 Clean and inspect rotor, measure rotor thickness, thickness variation, and lateral runout; determine necessary action. **(P-1)**

This worksheet addresses the following **AST/MAST** tasks:

V.A.1 Identify and interpret brake system concerns; determine needed action. **(P-1)**

V.B.6 Inspect brake lines, flexible hoses, and fittings for leaks, dents, kinks, rust, cracks, bulging, wear, and loose fittings/supports; determine needed action. **(P-1)**

V.C.5 Inspect wheel cylinders for leaks and proper operation; remove and replace as needed. **(P-1)**

V.D.4 Remove, inspect, and/or replace brake pads and retaining hardware; determine needed action. **(P-1)**

V.D.6 Clean and inspect rotor, measure rotor thickness, thickness variation, and lateral runout; determine needed action. **(P-1)**

We Support ASE Education Foundation

Directions: Before beginning this lab assignment, review the worksheet completely. Fill in the information in the spaces provided as you complete each task.

Tools and Equipment Required: Safety glasses, shop towel

Procedure:

1. Master cylinder fluid level: ☐ OK ☐ Low ☐ Overfilled
2. Brake fluid appearance: ☐ Clean ☐ Dirty
3. Parking brake: ☐ OK ☐ Needs adjustment

TURN

4. Raise the vehicle, loosen the lug nuts, and remove the wheels.
5. Inspect the front brakes.
   a. Approximate brake pad thickness: ______ (in./mm)
   b. Rotor machined surface condition: ☐ Smooth ☐ Rough ☐ Grooved ☐ Rusty
   c. Rotor thickness: ______ (in./mm) Minimum specification: ______ (in./mm)
   d. Rotor parallelism: ______ OK ☐ Excessive
   e. Rotor runout: ______ (in./mm) Specification: ______ (in./mm)
   f. Caliper condition: ☐ OK ☐ Leaking ☐ Stuck
   g. Condition of the brake lines: ☐ OK ☐ Damaged
   h. Leaks? ☐ No ☐ Yes Location of leak: ______________________________
6. Inspect the rear brakes.

   What type of rear brakes are used on this vehicle? ☐ Drum ☐ Disc

   Rear Drum Brakes: If the vehicle has rear drum brakes, answer the following:

   Approximate brake shoe thickness: ____________ (in./mm)

   Drum machined surface condition: ☐ Smooth ☐ Rough ☐ Grooved ☐ Rusty

   Drum diameter: ____________ (in./mm) Specification: ____________ (in./mm)

   Drum out of round: ☐ Yes (List amount ____________ (in./mm) ☐ No

   Wheel cylinder condition: ☐ OK ☐ Leaking

   Condition of the brake lines: ☐ OK ☐ Damaged

   Leaks: ☐ No ☐ Yes Location of leak: ____________________

   Rear Disc Brakes: If the vehicle has rear disc brakes, answer the following:

   Brake pad thickness: ____________ (in./mm)

   Rotor machined surface condition: ☐ Smooth ☐ Rough ☐ Grooved ☐ Rusty

   Rotor thickness: ________ (in./mm) Minimum specification: ________ (in./mm)

   Rotor parallelism: ________ ☐ OK ☐ Excessive

   Rotor runout: ________ (in./mm)

   Caliper condition: ☐ OK ☐ Leaking ☐ Stuck

   Condition of the brake lines: ☐ OK ☐ Damaged

   Leaks: ☐ No ☐ Yes Location of leak ____________________

   Parking Brakes:

   What type of parking brake is used on this vehicle?

   ☐ Pad clamping ☐ Shoes in drum/drum-in-hat ☐ Other ____________

   What is the condition of the parking brakes? ☐ OK ☐ Service Required

   Inspect the parking brake cables and linkages. ☐ OK ☐ Service Required

7. Install the wheels and lower the vehicle.
8. Check brake light operation.

   ☐ OK ☐ Bad

   Key-on required? ☐ Yes ☐ No
9. Check the power booster. ☐ OK ☐ Bad ☐ N/A
10. Before completing your paperwork, clean your work area, clean and return tools to their proper places, and wash your hands.
11. Record your recommendations for needed service or additional repairs and complete the repair order.

    ____

    ____

STOP

ASE Education Foundation Worksheet #9-15

ADJUST A TAPERED ROLLER WHEEL BEARING

Name_______________________________ Class____________________

Score: ☐ Excellent ☐ Good ☐ Needs Improvement **Instructor OK** ☐

Vehicle year __________ **Make** __________________ **Model** ____________________

Objective: Upon completion of this assignment, you should be able to adjust a tapered roller wheel bearing.

ASE Education Foundation Correlation

This worksheet addresses the following **MLR** task:

V.F.1 Remove, clean, inspect, repack, and install wheel bearings; replace seals; install hub and adjust bearings. **(P-1)**

This worksheet addresses the following **AST/MAST** task:

V.F.2 Remove, clean, inspect, repack, and install wheel bearings; replace seals; install hub and adjust bearings. **(P-2)**

We Support
ASE | Education Foundation

Directions: Before beginning this lab assignment, review the worksheet completely. Fill in the information in the spaces provided as you complete each task.

Tools and Equipment Required: Safety glasses, jack and jack stands or vehicle lift, shop towel, adjustable pliers, diagonal cutters

Parts and Supplies: Cotter pins

Procedure:

1. Raise and support the vehicle. The vehicle is:

   ☐ Front-wheel drive ☐ Rear-wheel drive

   Note: Tapered roller bearings are found on the front of rear-wheel-drive vehicles and the rear of most front-wheel-drive vehicles. Consult the service information for specific adjusting information.

Photo by Tim Gilles

2. Remove the wheel cover or hubcap and remove the grease cap.

   Which tool was used?

   ☐ Grease cap pliers ☐ Adjustable joint pliers

3. Remove the cotter pin.

Photo by Tim Gilles

TURN

4. Use an adjustable wrench or adjustable joint pliers to tighten the spindle nut while turning the wheel. Tighten the nut until it is "snug" (25 ft.-lb).
5. Back off the spindle nut and retighten until *"zero-lash"* (1 ft.-lb) is reached. Rock the wheel as you tighten the nut until all looseness disappears.

   Note: It is difficult to use a torque wrench when adjusting wheel bearings. The following is a convenient way to check for proper adjustment. If the tabbed washer under the spindle nut can be moved easily with a screwdriver, the bearing is not too tight. End play can be 0.001" to 0.010". This is when the nut is backed off from $^1/_{16}$ to $^1/_8$ turn from snug (zero-lash).

6. Install the largest diameter cotter pin that will fit into the spindle hole. What size cotter pin is needed?

   ☐ $^5/_{32}$" ☐ $^1/_8$" ☐ $^3/_{16}$"

Photo by Tim Gilles

7. If the hole does not line up, see if there is another hole 90 degrees from it that does line up.

   If no hole lines up, *loosen* the nut until the cotter pin can be installed.

8. Bend and cut the cotter pin as demonstrated by your instructor.

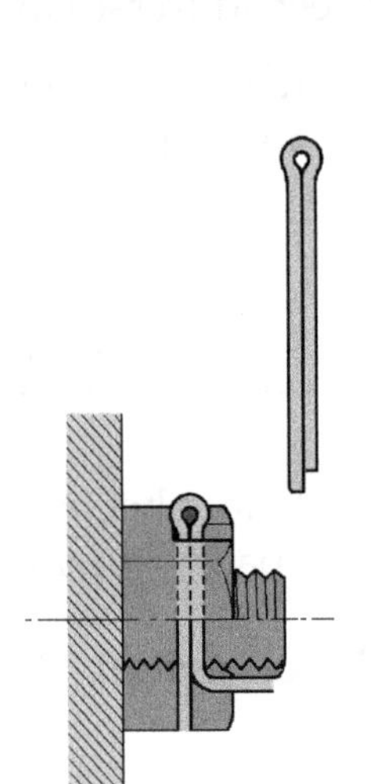

9. Reinstall the grease cap.
10. Install the wheel cover or hubcap.
11. Repeat steps 2 through 12 to adjust the other tapered wheel bearing.
12. After both wheel bearings are adjusted, lower the vehicle.
13. Before completing your paperwork, clean your work area, clean and return tools to their proper places, and wash your hands.
14. Record your recommendations for needed service or additional repairs and complete the repair order.

____

____

STOP

ASE Education Foundation Worksheet #9-16

REPACK WHEEL BEARINGS (DISC BRAKE)

Name____________________________ Class____________________

Score: ☐ Excellent ☐ Good ☐ Needs Improvement **Instructor OK** ☐

Vehicle year __________ **Make** ___________________ **Model** ____________________

Objective: Upon completion of this assignment, you should be able to repack tapered wheel bearings.

ASE Education Foundation Correlation

This worksheet addresses the following **MLR** tasks:

V.F.1 Remove, clean, inspect, repack, and install wheel bearings; replace seals; install hub and adjust bearings. **(P-1)**

V.F.5 Replace wheel bearing and race. **(P-2)**

This worksheet addresses the following **AST/MAST** tasks:

V.F.2 Remove, clean, inspect, repack, and install wheel bearings; replace seals; install hub and adjust bearings. **(P-2)**

V.F.6 Replace wheel bearing and race. **(P-3)**

We Support ASE | Education Foundation

Directions: Before beginning this lab assignment, review the worksheet completely. Fill in the information in the spaces provided as you complete each task.

Tools and Equipment Required: Safety glasses, jack and jack stands or vehicle lift, impact wrench, impact sockets, hand tools, hammer, punch, bearing packer, shop towel

Parts and Supplies: High-temperature grease, cotter pins

Procedure:

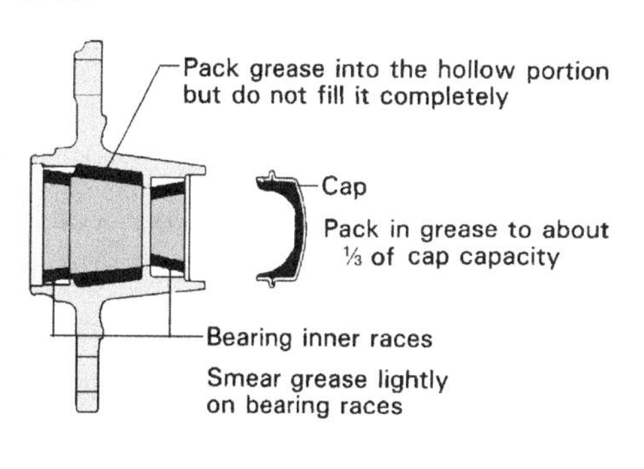

Pack grease into the hollow portion but do not fill it completely
Cap
Pack in grease to about ⅓ of cap capacity
Bearing inner races
Smear grease lightly on bearing races

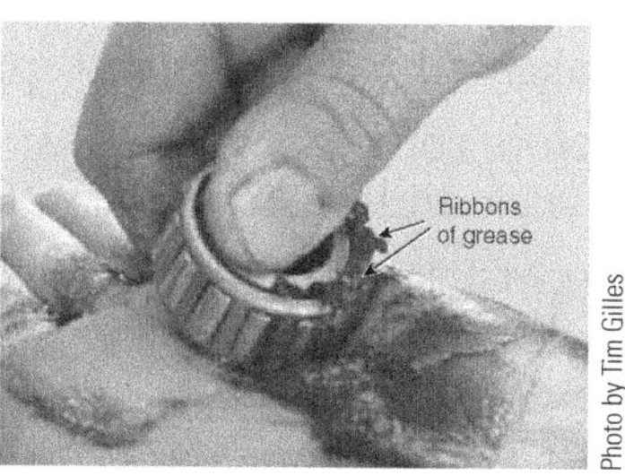

Ribbons of grease

Photo by Tim Gilles

Note: Complete Worksheets #9-8, Replace Front Disc Brake Pads, and #9-9, Remove and Replace a Disc Brake Rotor, before doing this worksheet.

Note: During a laboratory period, there may not be enough time to complete the repacking of both wheel bearings. Complete the bearing pack on one side of the vehicle at a time. If both sides are being packed at the same time, remember: *Do not interchange* the parts. Bearings "wear-mate" to the bearing cups. Interchanging the parts leads to premature bearing failure. When a bearing is replaced, its cup must be replaced also.

1. Check the service information for the correct procedure for disc brake caliper removal.
2. Raise and support the vehicle and remove the wheel cover or hubcap.
3. Remove the tire and wheel assembly, grease cap, and spindle nut.
4. Remove the disc brake caliper and rotor.

TURN

5. Reach through the rotor hub with a punch and remove the inner bearing and seal.
6. Wipe excess grease from the bearings and thoroughly clean them in solvent. Dry them with compressed air.

   Note: When drying a bearing with compressed air, hold the bearing cage so the bearing cannot spin. Spinning an unlubricated bearing will ruin the bearing, and it is dangerous.
7. Wipe old grease out of the hub.
8. Inspect the condition of the bearings and cups (races). Record your findings below:

   ☐ Reusable? ☐ Pitted? ☐ Cage damaged?

   ☐ Stained? ☐ Scored? ☐ Other (describe): ______
9. Thoroughly repack the bearings with high-temperature wheel bearing grease.

   What method was used to repack the bearings? ☐ By hand ☐ Bearing packer
10. Install the inner (large) bearing into its cup and carefully install the seal.
11. Clean the spindle and the machined surface where the seal rides before installing the rotor.
12. Carefully center the rotor as you rotate it over the spindle. Be careful not to damage the grease seal during installation.
13. Install the outer bearing and washer.

    Note: The washer is indexed to the spindle, preventing the rotating assembly from accidentally tightening or loosening the spindle nut.
14. Adjust the bearing as described in Worksheet #9-15, Adjust a Tapered Roller Wheel Bearing.
15. Reinstall the caliper. Torque the caliper bolts to specifications.

    Caliper bolt torque specification: ______ ft.-lb

Photo by Tim Gilles

Photo by Tim Gilles

Photo by Tim Gilles

16. Install the wheel, tighten the lug nuts, lower the vehicle to the ground, and retorque the lug nuts.

    Lug nut torque specification: ______ ft.-lb
17. Install the wheel covers or hubcaps.
18. Apply the foot brake halfway, repeatedly, until a firm pedal is felt.

    Pedal feel normal and firm? ☐ Yes ☐ No

CAUTION It is not unusual for the brake pedal to move all the way to the floor on the first application after the caliper has been removed. On vehicles with disc brakes, the piston is usually moved back in its bore when the caliper is removed. The pedal must be depressed to readjust the brakes. Always be sure to apply the brake pedal several times before moving the vehicle.

19. Before completing your paperwork, clean your work area, clean and return tools to their proper places, and wash your hands.
20. Record your recommendations for needed service or additional repairs and complete the repair order.

    ____

    ____

STOP

ASE Education Foundation Worksheet #9-17

REPLACE A TAPERED WHEEL BEARING

Name_______________________________ Class____________________

Score: ☐ Excellent ☐ Good ☐ Needs Improvement **Instructor OK** ☐

Vehicle year __________ **Make** __________________ **Model** ____________________

Objective: Upon completion of this assignment, you should be able to replace tapered wheel bearings.

ASE Education Foundation Correlation

This worksheet addresses the following **MLR** tasks:

V.F.1 Remove, clean, inspect, repack, and install wheel bearings; replace seals; install hub and adjust bearings. **(P-1)**

V.F.5 Replace wheel bearing and race. **(P-2)**

This worksheet addresses the following **AST/MAST** tasks:

V.F.2 Remove, clean, inspect, repack, and install wheel bearings; replace seals; install hub and adjust bearings. **(P-2)**

V.F.6 Replace wheel bearing and race. **(P-3)**

We Support
ASE | Education Foundation

Directions: Before beginning this lab assignment, review the worksheet completely. Fill in the information in the spaces provided as you complete each task.

Tools and Equipment Required: Safety glasses, jack and jack stands or vehicle lift, hand tools, hammer, wheel bearing punch, shop towel

Parts and Supplies: Wheel bearing, axle seal, grease

Note: When replacing the wheel bearings, it is always necessary to pack them with grease. Use the appropriate worksheets for help with the packing and adjustment procedures.

Photo by Tim Gilles

Procedure:

Complete Worksheet #9-16, Repack Wheel Bearings (Disc Brake), before attempting this worksheet.

1. Raise and support the vehicle, remove the tire and hub, and remove the wheel bearings from the hub.

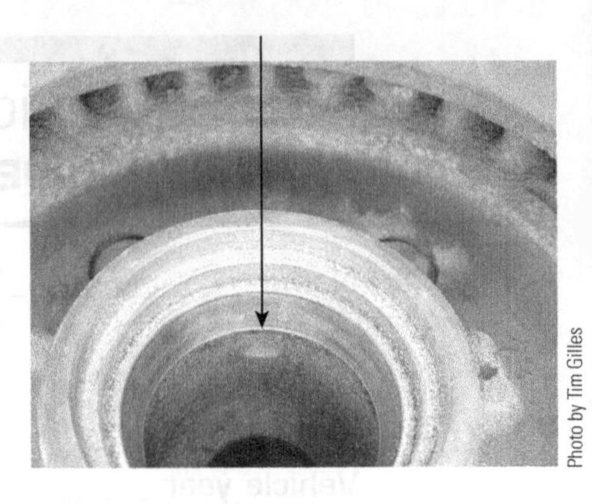
Photo by Tim Gilles

2. Locate the recesses in the hub on the inside of the bearing race.
3. Using a hammer and wheel bearing punch, drive the cup from the hub. Tap first on one side of the cup at the recess, then on the other side, until the race is removed.
4. Clean and inspect the hub.

   Is the hub damaged? ☐ Yes ☐ No

5. Install the new bearing cup.

   Note: If a tapered bearing cup driver is not available, grind a small amount off the outside of the old cup. Then use it as a driver to install the new cup in the hub. Tap the cup into the hub until it is bottomed in its bore.

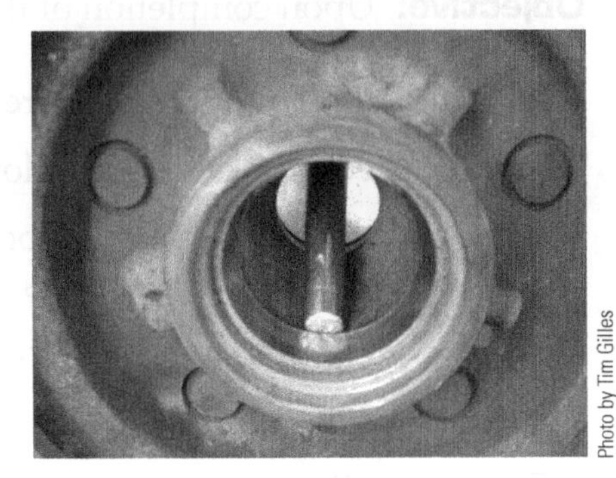
Photo by Tim Gilles

   What tool was used to install the new cup?

   ☐ Old cup ☐ Bearing cup driver

   Does the bearing cup fit tightly into the hub?

   ☐ Yes ☐ No

   Was the cup bottomed in its bore?

   ☐ Yes ☐ No

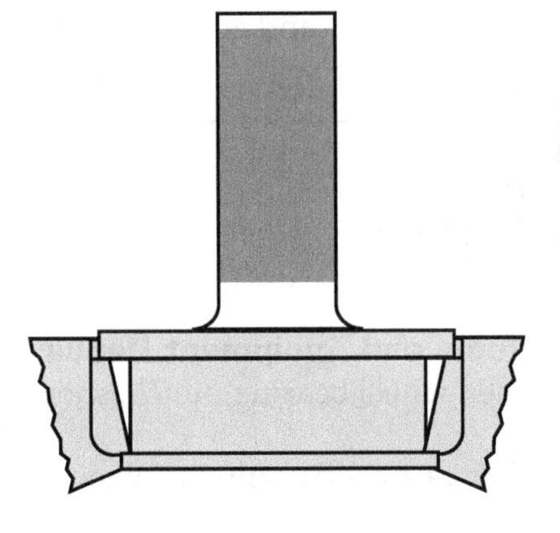

6. Pack the wheel bearing and reassemble it to the hub using Worksheet #9-16, Repack Wheel Bearings (Disc Brake), as a guide.
7. Before completing your paperwork, clean your work area, clean and return tools to their proper places, and wash your hands.
8. Record your recommendations for needed service or additional repairs and complete the repair order.

____

____

STOP

ASE Education Foundation Worksheet #9-18

BLEED A HYDRAULIC CLUTCH

Name ______________________ Class ________________

Score: ☐ Excellent ☐ Good ☐ Needs Improvement **Instructor OK** ☐

Vehicle year ________ **Make** ______________ **Model** ________________

Objective: Upon completion of this assignment, you should be able to bleed a hydraulic clutch.

ASE Education Foundation Correlation

This worksheet addresses the following **MLR** task:

III.B.1 Check and adjust clutch master cylinder fluid level; use proper fluid type per manufacturer specification **(P-1)**

This worksheet addresses the following **AST/MAST** tasks:

III.B.4 Bleed clutch hydraulic system. **(P-1)**

III.B.5 Check and adjust clutch master cylinder fluid level; check for leaks; use proper fluid type per manufacturer specification. **(P-1)**

We Support ASE Education Foundation

Directions: Before beginning this lab assignment, review the worksheet completely. Fill in the information in the spaces provided as you complete each task.

Tools and Equipment Required: Safety glasses, fender covers, shop towel, bleeder wrench, hose, container, brake fluid

Procedure:

1. Raise the hood and install fender covers on the vehicle.

CAUTION Brake fluid will damage vehicle paint. Always use fender covers and clean spills immediately.

Note: Use water to clean brake spills.

2. Remove the cap from the clutch master cylinder reservoir and fill with fluid as needed.

   What type of brake fluid is required? ☐ DOT 3 ☐ DOT 4 ☐ Other ________________

3. Raise and support the vehicle.
4. Inspect the clutch linkage, cables, adjuster, bracket bushings, and springs.

   ☐ OK

   List item(s) in need of attention: ________________________________

   ____

TURN

5. Locate and loosen the bleeder screw.

☐ Turns freely

☐ Frozen

6. Place a small hose over the bleed screw and direct it into a container to catch the fluid that drains while bleeding the clutch.
7. Have an assistant push down on the clutch pedal after you have loosened the bleeder screw. He or she should hold the pedal down until the bleeder is closed. Good communication is important to successfully completing this procedure.

   Does your helper understand the bleed procedure? ☐ Yes ☐ No

   Note: An alternate method for bleeding the clutch is to use gravity to move the fluid through the system. Since the clutch master cylinder is higher than the clutch slave cylinder and there usually are no valves in the system, bleeding can be done by simply opening the bleeder and waiting for the fluid to run through the system.

8. Regularly check the clutch master cylinder as you bleed the clutch.
9. After bleeding the clutch, check the clutch pedal free play. This is the amount of movement from the top of clutch pedal travel until resistance is felt.

   Measure and list the amount of clutch pedal free play. ______ inch

   Note: Some clutches are adjustable and others are not. Refer to the service information for the proper procedure and specifications. If the clutch is adjustable, a lack of free play can indicate a worn clutch.

10. Before completing your paperwork, clean your work area, clean and return tools to their proper places, and wash your hands.
11. Record your recommendations for needed service or additional repairs and complete the repair order.

    ____

    ____

STOP

ASE Education Foundation Worksheet #9-19

CONSTANT VELOCITY (CV) JOINT SERVICE AND REPLACEMENT

Name ______________________ Class ________________

Score: ☐ Excellent ☐ Good ☐ Needs Improvement **Instructor OK** ☐

Vehicle year ________ **Make** ______________ **Model** ________________

Objective: Upon completion of this assignment, you should be able to inspect constant velocity (CV) joints.

ASE Education Foundation Correlation

This worksheet addresses the following **MLR** task:

III.D.2 Inspect, service, and/or replace shafts, yokes, boots, and universal/CV joints. **(P-2)**

This worksheet addresses the following **AST/MAST** tasks:

III.D.1 Diagnose constant-velocity (CV) joint noise and vibration concerns; determine needed action. **(P-1)**

III.D.4 Inspect, service, and replace shafts, yokes, boots, and universal/CV joints. **(P-1)**

We Support ASE Education Foundation

Directions: Before beginning this lab assignment, review the worksheet completely. Fill in the information in the spaces provided as you complete each task.

Tools and Equipment Required: Safety glasses, shop towels, vehicle lift

Procedure:

1. Worn or damaged CV joints typically make noise during operation. If there is noise, how would you describe it?

   ☐ Click ☐ Clunk ☐ Scraping ☐ Other ☐ There is no noise.

2. Raise the vehicle on a lift or place it on jack stands.
3. Inspect the area around the CV joints.

   ☐ Area is clean. ☐ Grease is sprayed in the area.

4. Inspect the CV boots.

   ☐ Flexible

   ☐ Cracked

   ☐ Clamps tight

   ☐ Other damage? List: ______________________________

TURN

Note: Any signs of damage to the CV boots or noise from the CV joints will require the driveshaft to be removed and the CV joint to be disassembled for closer inspection.

5. Before completing your paperwork, clean your work area, clean and return tools to their proper places, and wash your hands.
6. Record your recommendations for needed service or additional repairs and complete the repair order.

____

____

ASE Education Foundation Worksheet #9-20

INSPECT AND REMOVE A REAR-WHEEL-DRIVE DRIVESHAFT

Name_______________________ Class_______________

Score: ☐ Excellent ☐ Good ☐ Needs Improvement **Instructor OK** ☐

Vehicle year ________ **Make** ______________ **Model** ______________

Objective: Upon completion of this assignment, you should be able to inspect and remove a rear-wheel-drive (RWD) driveshaft.

ASE Education Foundation Correlation

This worksheet addresses the following **MLR** task:

III.D.2 Inspect, service, and replace shafts, yokes, boots, and universal/CV joints. **(P-2)**

This worksheet addresses the following **AST/MAST** task:

III.D.4 Inspect, service, and replace shafts, yokes, boots, and universal/CV joints. **(P-1)**

We Support ASE | Education Foundation

Directions: Before beginning this lab assignment, review the worksheet completely. Fill in the information in the spaces provided as you complete each task.

Tools and Equipment Required: Safety glasses, shop towels, vehicle lift, hand tools, drain pan

Procedure: Before removing the driveshaft, give it a complete inspection. This will lead to the proper repair for the concern.

To Inspect a Driveshaft:

1. Raise the vehicle on a lift or position it firmly on jack stands.
2. Inspect the driveshaft for the following:
   a. Undercoating on the driveshaft? ☐ Yes ☐ No
   b. Missing balance weights? ☐ Yes ☐ No
   c. Obvious physical damage? ☐ Yes ☐ No
3. Grasp the driveshaft at the end and move it as you watch the U-joint for looseness.

   Is there any looseness? ☐ Yes ☐ No
4. Is there any rust around the U-joint cups? ☐ Yes ☐ No
5. At the front of the driveshaft, move the slip yoke up and down. Is there excessive movement at the transmission extension housing bushing?

   ☐ Yes ☐ No

TURN

6. Transmission mount condition:

   ☐ Oil-soaked ☐ Broken ☐ Damaged
   ☐ Good condition

To Remove a Driveshaft:

7. Mark the driveshaft so it can be replaced in the same position.
8. Unbolt the rear U-joint from its connection at the differential flange.
9. Pry the rear U-joint forward (away from the differential).
10. Install tape around the U-joint cups to prevent them from falling off.
11. If the driveshaft has a center support bearing, unbolt it.
12. Place a drain pan under the back of the transmission beneath the slip yoke.
13. Slide the driveshaft back, removing its slip yoke from the transmission. Some driveshafts are bolted to a flange on the transmission output shaft. In this case, it will be necessary to unbolt the universal joint from the transmission flange. It is a good practice to mark the driveshaft and flange before removal so it can be installed in its previously installed location.

To Install a Driveshaft:

14. Inspect all contact surfaces to see that they are clean.
15. Slide the slip yoke into the back of the transmission or bolt it to the transmission output flange, aligning the match marks first.
16. Align the rear U-joint with its flange and install the bolts.

    Are your match marks aligned? ☐ Yes ☐ No

    Is the U-joint fully seated against its flange? ☐ Yes ☐ No
17. Before completing your paperwork, clean your work area, clean and return tools to their proper places, and wash your hands.
18. Record your recommendations for needed service or additional repairs and complete the repair order.

    ____

    ____

STOP

ASE Education Foundation Worksheet #9-21

SERVICE AN AUTOMATIC TRANSMISSION

Name ______________________ Class ________________

Score: ☐ Excellent ☐ Good ☐ Needs Improvement **Instructor OK** ☐

Vehicle year _________ **Make** _______________ **Model** _________________

Objective: Upon completion of this assignment, you should be able to service an automatic transmission.

ASE Education Foundation Correlation

This worksheet addresses the following **MLR/AST/MAST** task:

II.B.4 Drain and replace fluid and filter(s); use proper fluid type per manufacturer specification. **(P-1)**

We Support ASE Education Foundation

Directions: Before beginning this lab assignment, review the worksheet completely. Fill in the information in the spaces provided as you complete each task.

Tools and Equipment Required: Safety glasses, shop towels, vehicle lift, hand tools, drain pan, transmission fluid, pan gasket, filter

Procedure:

Note: Worksheet #2-11, On-the-Ground Safety Checklist, should be completed before attempting this worksheet.

1. Obtain the fluid, filter, and gasket before starting any work on the vehicle.

   Do you have the required parts? ☐ Yes ☐ No

   Fluid type ________________ Number of quarts ________________

2. Open the hood and install fender covers.
3. Check the automatic transmission fluid level.

   ☐ Full ☐ Low

4. Check the fluid condition.

   ☐ Red/pink ☐ Brown ☐ Burnt ☐ Milky white

   ☐ Other: List ________________________

5. Raise the vehicle on a lift or place it on jack stands.
6. Position a drain pan under the transmission oil pan.
7. Loosen all of the transmission oil pan bolts, except two at one end of the pan. The two bolts will help control the fluid as it spills from the pan.

TURN

8. After most of the fluid has spilled from the pan, remove the last two bolts while carefully holding the oil pan. Pour the remainder of the fluid into the drain pan.
9. Inspect the contents of the pan and the magnet, if it has one.

   Was there any debris in the pan or on the magnet? ☐ Yes ☐ No ☐ N/A

   Type of debris: ☐ Black ☐ Aluminum ☐ Brass ☐ Plastic ☐ Iron/steel ☐ N/A

Courtesy of Ford Motor Company

10. Remove the filter and gasket.
11. Install the new filter and gasket. Tighten the bolts to the correct torque.

    Torque specification for filter: ______ in.-lb
12. Clean the oil pan and verify that its sealing surface is flat. Straighten as required.
13. Install the new pan gasket on the pan and insert at least two bolts to hold the gasket in place.
14. Install the pan on the transmission.

    Note: Start all of the bolts before tightening any of them!
15. Tighten all of the bolts to the recommended torque.

    Torque specification for pan bolts: ______ in.-lb
16. Lower the vehicle and add 2 quarts of fluid.
17. Start the engine and shift the transmission through its gear ranges.
18. Does the gear range indicator accurately indicate the gear range selected?

    ☐ Yes ☐ No

    Note: if the indicator is not accurate the transmission linkage will need to be adjusted or replaced. Refer to the service information for the correct adjustment/replacement procedure.
19. Check the fluid level and add fluid as needed until the transmission fluid level is correct.

    Note: Until the vehicle has been driven to warm the fluid, the fluid level should be approximately 2 quarts low.

    How much fluid was added in total? ______ quarts
20. Check the transmission for leaks. Leaks? ☐ Yes ☐ No
21. Before completing your paperwork, clean your work area, clean and return tools to their proper places, and wash your hands.
22. Record your recommendations for needed service or additional repairs and complete the repair order.

    ____

    ____

STOP

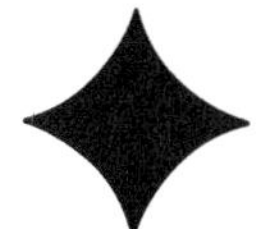

ASE Education Foundation Worksheet #9-22

REPLACE A REAR AXLE WITH A PRESSED-FIT BEARING

Name________________________________ Class________________________

Score: ☐ Excellent ☐ Good ☐ Needs Improvement **Instructor OK** ☐

Vehicle year __________ **Make** ___________________ **Model** ____________________

Objective: Upon completion of this assignment, you should be able to remove and replace a rear axle with a pressed-fit bearing.

ASE Education Foundation Correlation

This worksheet addresses the following **AST** tasks:

III.E.2.2 Remove and replace drive axle shafts. **(P-1)**

III.E.2.3 Inspect and replace drive axle shaft seals, bearings, and retainers. **(P-2)**

This worksheet addresses the following **MAST** tasks:

III.E.3.2 Remove and replace drive axle shafts. **(P-1)**

III.E.3.3 Inspect and replace drive axle shaft seals, bearings, and retainers. **(P-2)**

We Support ASE Education Foundation

Directions: Before beginning this lab assignment, review the worksheet completely. Fill in the information in the spaces provided as you complete each task.

Tools and Equipment Required: Safety glasses, jack and jack stands or vehicle lift, drain pan, hand tools, seal puller, slide hammer, shop towel

Parts and Supplies: Lubricant (possibly a bearing and/or seal)

This worksheet is to be used to remove an axle with a pressed-fit bearing. Refer to Worksheet #9-23, Replace a C-Lock-Type Rear Axle.

Procedure:

1. Raise and support the vehicle and remove the rear wheel.
2. Remove the brake drum or the caliper and rotor.

   ☐ Brake drum ☐ Caliper and rotor
3. Use a ratchet, extension, and socket to remove the retainer flange nuts and bolts.

   Number of retainer flange bolts

   ____

Slide hammer

TURN

Note: There is a hole provided in the axle flange for easy access to the bolts. Hold the nuts from the rear of the backing plate with a combination wrench.

Note: Sometimes it is necessary to remove a brake line and backing plate to remove the axle.

Retainer
Bearing
Retainer ring
Axle flange
Projected end of race

4. Remove the axle. Use a slide hammer if necessary.

   Slide hammer needed? ☐ Yes ☐ No

   Note: When using a slide hammer, if it feels really solid, either all of the bolts have not been removed or the axle is of the C-lock type. In either case, damage could result if the slide hammer is used.

5. Feel the axle bearing for roughness or damage. Does it need to be replaced?

   ☐ Yes ☐ No

   Note: Check with your instructor to see if the equipment is available to replace the bearing. If not, the axle will need to be sent to an automotive machine shop for bearing replacement.

   Was the bearing replaced? ☐ Yes ☐ No

6. If there is a separate seal, replace it now.

   Was the seal replaced? ☐ Yes ☐ No

   Note: When replacing a seal, always put a little grease on the sealing lip to protect it during initial startup.

   Lip of the seal lubricated? ☐ Yes ☐ No

7. Install the axle assembly. Slide the axle in until the splines align with the splines in the differential side gears. Be careful not to damage the axle seal.

8. Install the retainer flange nuts and bolts and torque them to specifications.

   Torque specification: ______ ft.-lb

9. Replace the brake drum and tire/wheel assembly.

10. Check the lubricant level in the differential and add as needed.

    How much lubricant was added? ______ pints ☐ None

11. Before completing your paperwork, clean your work area, clean and return tools to their proper places, and wash your hands.

12. Record your recommendations for needed service or additional repairs and complete the repair order.

    ____

    ____

STOP

ASE Education Foundation Worksheet #9-23

REPLACE A C-LOCK-TYPE REAR AXLE

Name______________________________ Class______________________

Score: ☐ Excellent ☐ Good ☐ Needs Improvement **Instructor OK** ☐

Vehicle year _________ **Make** ________________ **Model** __________________

Objective: Upon completion of this assignment, you should be able to remove and replace a C-lock-type rear axle.

ASE Education Foundation Correlation

This worksheet addresses the following **AST** tasks:

III.E.2.2 Remove and replace drive axle shafts. **(P-1)**

III.E.2.3 Inspect and replace drive axle shaft seals, bearings, and retainers. **(P-2)**

This worksheet addresses the following **MAST** tasks:

III.E.3.2 Remove and replace drive axle shafts. **(P-1)**

III.E.3.3 Inspect and replace drive axle shaft seals, bearings, and retainers. **(P-2)**

We Support ASE Education Foundation

Directions: Before beginning this lab assignment, review the worksheet completely. Fill in the information in the spaces provided as you complete each task.

Tools and Equipment Required: Safety glasses, jack and jack stands or vehicle lift, drain pan, impact wrench, impact sockets, hand tools, slide hammer, prybar, seal puller, shop towel, torque wrench

Parts and Supplies: Differential cover gasket, possibly an axle seal and/or axle bearing

This worksheet is to be used to remove the axles on a C-lock rear axle. Refer to Worksheet #9-22, Replace a Rear Axle with a Pressed-Fit Bearing.

Procedure:

1. Raise and support the vehicle, and remove the wheel, the brake drum, or rotor.
2. Drain the differential lubricant into a drain pan by removing the two lowest bolts on the cover.
3. Remove the differential cover bolts and remove the cover.
4. Remove the pinion shaft lock bolt and the pinion shaft. The shaft will slide out.

   Note: Once the pinion shaft has been removed, do not turn the axleshafts.

5. Push the axleshaft inward to the center of the vehicle until the C-lock is visible. Remove the C-lock.

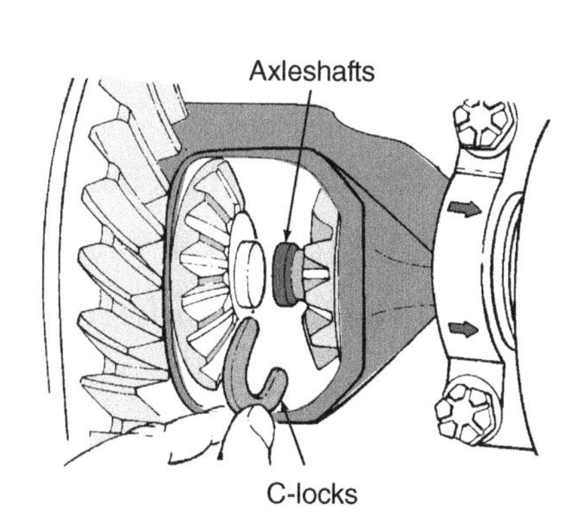

6. Slide the axle completely out of the housing, being careful not to damage the seal.

   Note: Allowing the axle to hang in the housing can damage the axle seal, resulting in a leak.

Hardened
Inner race area
Axle
bearing
Courtesy of Ford Motor Company

Seal and Bearing Replacement

7. If the seal requires replacement, remove it from the housing.

   Is the seal being replaced? ☐ Yes ☐ No

8. Inspect the axle bearing surface for wear or damage.

   Is the bearing surface worn? ☐ Yes ☐ No

   Is the axle damaged? ☐ Yes ☐ No

9. If the bearing needs to be replaced, pull it from the housing using a slide hammer.

   Is the bearing being replaced? ☐ Yes ☐ No

10. Install the new bearing until it is fully seated into the housing.

11. Install the new seal in the housing. Tap it into place until it is fully seated or flush with the housing.

12. Coat the sealing lip with grease or gear lubricant.

Puller adapter
Slide hammer

Reinstall Axles

13. Slide the axles carefully into the axle housing. Be careful not to damage the axle seal.
14. Slide the axles in until their splines engage the differential side gears.
15. Install the C-locks. Then pull the axles outward to seat the C-locks into the side gears.
16. Install the pinion shaft. Align the bolt hole and install the lock bolt. Torque it to specifications.

CAUTION Do not overtighten the lock bolt! These bolts break easily.

Pinion shaft lock bolt torque specification: ______ ☐ ft.-lb ☐ in.-lb

17. Clean *all* gasket material from the gasket surfaces.
18. Install a new differential cover gasket. Install the cover.
19. Fill the differential with the correct type of lubricant.

    What type of gear lubricant was used? ______________________________

    Does this differential require a friction modifier (limited slip) additive? ☐ Yes ☐ No

20. Replace the brake drum or rotor/caliper and reinstall the wheel.
21. Torque the lug nuts. Lug nut torque specification: ______ ft.-lb
22. Install the wheel cover or hubcap and lower the vehicle.
23. Before completing your paperwork, clean your work area, clean and return tools to their proper places, and wash your hands.
24. Record your recommendations for needed service or additional repairs and complete the repair order.

____

____

STOP

ASE Education Foundation Worksheet #9-24

FLUSH A POWER STEERING SYSTEM

Name______________________________ Class______________________

Score: ☐ Excellent ☐ Good ☐ Needs Improvement **Instructor OK** ☐

Vehicle year __________ **Make** ____________________ **Model** ______________________

Objective: Upon completion of this assignment, you should be able to flush a power steering system.

ASE Education Foundation Correlation

This worksheet addresses the following **MLR** task:

IV.B.3 Flush, fill, and bleed power steering system; use proper fluid type per manufacturer specification. **(P-2)**

This worksheet addresses the following **AST/MAST** task:

IV.B.10 Flush, fill, and bleed power steering system; use proper fluid type per manufacturer specification. **(P-2)**

We Support ASE Education Foundation

Directions: Before beginning this lab assignment, review the worksheet completely. Fill in the information in the spaces provided as you complete each task.

Tools and Equipment Required: Safety glasses, fender covers, shop towel, drain pan, power steering fluid

Procedure:

1. Open the hood and install fender covers.
2. Does the power steering system have a filter?

   ☐ Yes ☐ No

   Note: If the system has a filter it should be replaced when servicing/flushing the system.
3. Raise the vehicle on a lift or place it on jack stands, if necessary, to inspect the return and pressure hoses .
4. What is the condition of the power steering hoses?
   ☐ Ok ☐ Leaking ☐ Damaged
5. To flush the system, remove the return hose from the reservoir or pump. Plug the outlet.

Return hose

Courtesy of Federal-Mogul Corporation

6. Place the end of the return hose into a drain pan to catch the fluid as it is flushed from the system.
7. With the engine idling, turn the steering wheel from lock to lock (all the way in one direction, and then the other).
8. Shut the engine off and refill the reservoir with clean fluid.

9. List the type of fluid required for this power steering system: ______________________

10. Start the engine and wait until the fluid flows from the return hose, then shut the engine off. Repeat the cycle of refilling and flushing until the fluid coming from the return hose is clean and free from air bubbles.

11. Reinstall the return line to the power steering reservoir or pump and fill the reservoir with power steering fluid.

    Note: If the hoses are leaking or damaged it will be necessary to replace them.

12. Bleed the system of any remaining air by cycling the steering wheel lock to lock, holding it at each lock for 2–3 seconds.

13. Inspect the reservoir for bubbles. If bubbles are still present, repeat the bleeding procedure. If the fluid is foamy, allow it to sit for several minutes until the foam disappears before continuing the bleeding procedure.

14. Before completing your paperwork, clean your work area, clean and return tools to their proper places, and wash your hands.

15. Record your recommendations for needed service or additional repairs and complete the repair order.

    ____

    ____

STOP

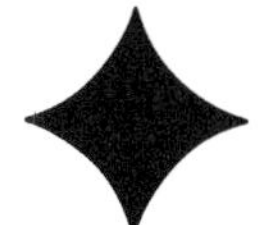

ASE Education Foundation Worksheet #9-25

INSPECT AND REPLACE STABILIZER BUSHINGS

Name ______________________ Class ________________

Score: ☐ Excellent ☐ Good ☐ Needs Improvement **Instructor OK** ☐

Vehicle year ________ **Make** ______________ **Model** ________________

Objective: Upon completion of this assignment, you should be able to inspect and replace stabilizer bushings.

ASE Education Foundation Correlation

This worksheet addresses the following **MLR** task:

IV.B.15 Inspect and replace front stabilizer bar (sway bar) bushings, brackets, and links. **(P-1)**

This worksheet addresses the following **AST/MAST** task:

IV.C.9 Inspect, remove, and/or replace front/rear stabilizer bar (sway bar) bushings, brackets, and links. **(P-3)**

We Support ASE Education Foundation

Directions: Before beginning this lab assignment, review the worksheet completely. Fill in the information in the spaces provided as you complete each task.

Tools and Equipment Required: Safety glasses, shop towels, vehicle lift, hand tools

Procedure:

1. Raise the vehicle on a lift or place it on jack stands so that *both* wheels are free to hang.

SAFETY NOTE: Both wheels must either be on the ground or in the air for this task. If one wheel is on the ground, there will be dangerous tension pushing upward against one of the stabilizer bushing retaining nuts.

2. Inspect the stabilizer bushings for the following:
   - a. Bushing walking out of the bracket ☐ Yes ☐ No
   - b. Large cracks or splits in bracket bushing ☐ Yes ☐ No
   - c. Bushing wallowed out ☐ Yes ☐ No
   - d. Large cracks or splits in link bushing ☐ Yes ☐ No
   - e. Loose or missing components ☐ Yes ☐ No
   - f. Obvious physical damage ☐ Yes ☐ No

To Replace the Stabilizer Bracket Bushing:

3. Remove the two bolts securing the bracket.
4. Inspect the bracket for cracks or damage. ☐ OK ☐ Needs replacement

TURN

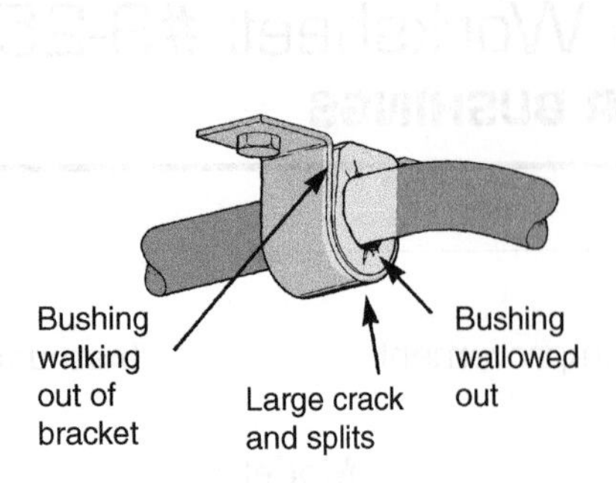

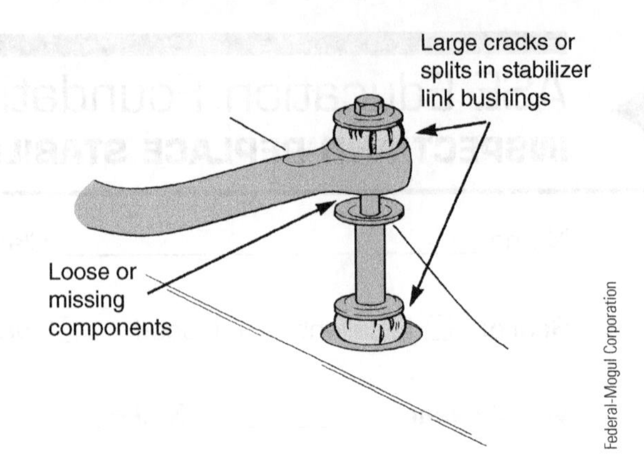

5. Remove the old bushing.
6. Install the new bushing.
7. Install and torque the bracket bolts. What is the torque Specification ____________________

To Replace the Stabilizer Link Pins and Bushings:

8. Note the order of the bushings, washers, and spacer on the old stabilizer link.
9. Remove the bolt, washers, bushings, and spacer.
10. Install the new parts in the correct order.
11. Tighten the bolt to the correct torque. What is the torque specification?____________________
12. Before completing your paperwork, clean your work area, clean and return tools to their proper places, and wash your hands.
13. Record your recommendations for needed service or additional repairs and complete the repair order.

____

____

ASE Education Foundation Worksheet #9-26

REPLACE SHOCK ABSORBERS

Name_______________________________ Class____________________

Score: ☐ Excellent ☐ Good ☐ Needs Improvement **Instructor OK** ☐

Vehicle year __________ **Make** ___________________ **Model** ____________________

Objective: Upon completion of this assignment, you should be able to replace a vehicle's shock absorbers.

ASE Education Foundation Correlation

This worksheet addresses the following **MLR** task:

IV.B.20 Inspect, remove, and replace shock absorbers; inspect mounts and bushings. **(P-1)**

This worksheet addresses the following **AST/MAST** task:

IV.D.1 Inspect, remove, and/or replace shock absorbers; inspect mounts and bushings. **(P-1)**

We Support ASE Education Foundation

Directions: Before beginning this lab assignment, review the worksheet completely. Fill in the information in the spaces provided as you complete each task.

Tools and Equipment Required: Safety glasses, fender cover, jack and jack stands or vehicle lift, hand tools, shock absorber wrench, shop towel

Parts and Supplies: Shock absorbers

Note: The following instructions are for standard shock absorbers. To replace Macpherson strut-type shock absorbers, refer to Worksheet #9-28, Remove and Replace a Macpherson Strut.

Procedure:

1. Raise the vehicle on a lift. Use a wheel-contact lift or safety stands so that the suspension does *not* hang free.

   Note: It is safer and easier to change the shocks when the wheels are not hanging free.

2. Which shocks are being replaced? ☐ Front ☐ Rear ☐ All four

   Note: Shocks should always be replaced in pairs, both fronts or both rears. It is not advisable to replace only one shock.

3. Locate the upper and lower shock mount bolts.

   Note: Sometimes the rear shock's upper mounting is located inside the trunk, or on hatchbacks, inside the passenger compartment. If working inside the vehicle, use fender covers to protect the paint.

Photo by Tim Gilles

TURN

Photo by Tim Gilles

4. Is a special shock tool needed to hold one end of the shock?

   ☐ Yes ☐ No

5. Spray the shock absorber fastener threads with penetrating oil.
6. Are the new shocks gas charged? ☐ Yes ☐ No

   Note: Gas shocks contain pressurized gas and do not require bleeding. They are shipped with a strap around them because they will expand to full travel when unrestrained. The strap should not be removed until after the shock is installed on the vehicle.

7. Remove the shock absorber and compare it to a new one. Are they the same length when fully compressed?

   ☐ Yes ☐ No

   If the new shock is not gas charged, compare the shocks when they are fully extended. Are they the same length when fully extended?

   ☐ Yes ☐ No

8. Does the old shock still resist movement when extended and compressed?

   ☐ Yes ☐ No

   If the shocks are gas charged, it is not necessary to bleed them; proceed to step #13.

   If they are not gas charged, then it will be necessary to bleed the shocks; proceed to step #9.

9. Extend the shock while it is in its normal vertical position.
10. Turn the shock over so that its top is down. Fully collapse the shock.
11. Repeat the process four or five times to work out any air trapped in the shock.
12. When does the new shock give more resistance?

    ☐ Compression ☐ Extension ☐ Equal resistance

13. Install the new shocks. Do not overtighten the nuts! Tighten them until the rubber bushings are just barely compressed.
14. Cut the retaining strap from gas-charged shocks. ☐ Yes ☐ No ☐ N/A
15. Lower the vehicle and bounce-test the shock absorbers.

    Do the shocks resist spring oscillation? ☐ Yes ☐ No

    Did you hear any unusual noise? ☐ Yes ☐ No

16. Before completing your paperwork, clean your work area, clean and return tools to their proper places, and wash your hands.
17. Record your recommendations for needed service or additional repairs and complete the repair order.

    ____

    ____

ASE Education Foundation Worksheet #9-27

MACPHERSON STRUT SERVICE

Name_______________________________ Class_____________________

Score: ☐ Excellent ☐ Good ☐ Needs Improvement **Instructor OK** ☐

Vehicle year __________ **Make** ___________________ **Model** _____________________

Objective: Upon completion of this assignment, you should be able to replace a Macpherson strut.

ASE Education Foundation Correlation

This worksheet addresses the following **MLR** tasks:

IV.B.7 Inspect front strut bearing and mount. **(P-1)**

IV.B.16 Inspect, remove, and/or replace strut cartridge or assembly; inspect mounts and bushings. **(P-1)**

This worksheet addresses the following **AST/MAST** task:

IV.C.10 Inspect, remove, and/or replace strut cartridge or assembly, strut coil spring, insulators (silencers), and upper strut bearing mount. **(P-3)**

We Support ASE | Education Foundation

Directions: Before beginning this lab assignment, review the worksheet completely. Fill in the information in the spaces provided as you complete each task.

Tools and Equipment Required: Safety glasses, fender covers, shop towel, hand tools

Procedure:

Inspect Strut Bearing

1. Before removing the strut, check the bearing at the top of the strut.
   a. While the vehicle is on the ground, steer the wheels to the right and then to the left. Did you hear any unusual bearing noises? ☐ Yes ☐ No
   b. If yes, the strut will need to be removed, disassembled, and the bearing carefully inspected.

Removing a Macpherson Strut

1. Open the hood and place fender covers on the vehicle.
2. Put one match mark on one of the bolts and another at its location on the top of the strut tower.

The center nut at the top of the strut must **NOT** be removed before the strut is removed from the vehicle and is safely installed in a spring compressor.

3. Remove two of the bolts holding the strut to the strut tower. Loosen the third but do not remove it.
4. Raise the vehicle on a lift or position it safely on jack stands.
5. Loosen the lug nuts and remove the tire and wheel assembly.
6. Disconnect the brake line from the strut assembly.
7. If the vehicle has antilock brakes, disconnect the ABS sensor.
8. How is the lower part of the strut held in place? Check one:

   ☐ Lower ball joint, castle nut

   ☐ Lower ball joint, pinch bolt

   ☐ Cam bolts

   ☐ Other (describe): ______________________

   Note: If cam bolts are used to hold the lower part of the strut to the suspension, put a match mark on the bolts before loosening them.
9. Disconnect the lower strut mounting and pry the suspension down to dislodge the strut from the control arm or steering knuckle.
10. Once the strut is loose, remove the third bolt from the top of the strut and lift the strut from the vehicle.

Installing a Macpherson Strut

11. Move the top of the strut under the fender and into the strut tower.
12. Align the match marks on the strut and the strut tower. Start one of the bolts.
13. Reconnect the lower part of the strut and torque the fasteners to specifications. If cam bolts are used, align the match marks before tightening the bolts.

    Torque specification: ______________________ ft.-lb
14. Install the remaining bolts to the top of the strut and torque them.

    Torque specification: ______________________ ft.-lb
15. Reconnect the brake line and bleed the brakes if the brake line was removed.

    Did you bleed the brakes? ☐ Yes ☐ No
16. Reconnect the ABS sensor, if necessary.
17. Install the wheel assembly and torque the lug nuts.

    Torque specification: ______________________ ft.-lb
18. Before completing your paperwork, clean your work area, clean and return tools to their proper places, and wash your hands.
19. Record your recommendations for needed service or additional repairs and complete the repair order.

    ____

    ____

STOP

ASE Education Foundation Worksheet #9-28

REMOVE AND REPLACE A MACPHERSON STRUT SPRING AND CARTRIDGE

Name_______________________ Class_______________

Score: ☐ Excellent ☐ Good ☐ Needs Improvement **Instructor OK** ☐

Vehicle year _________ **Make** _______________ **Model** _______________

Objective: Upon completion of this assignment, you should be able to replace a Macpherson strut cartridge.

ASE Education Foundation Correlation

This worksheet addresses the following **MLR** task:

IV.B.16 Inspect, remove, and/or replace strut cartridge or assembly; inspect mounts and bushings. **(P-2)**

This worksheet addresses the following **AST/MAST** task:

IV.C.10 Inspect, remove, and/or replace strut cartridge or assembly, strut coil spring, insulators (silencers), and upper strut bearing mount. **(P-3)**

We Support
ASE | Education Foundation

Directions: Before beginning this lab assignment, review the worksheet completely. Fill in the information in the spaces provided as you complete each task.

Tools and Equipment Required: Safety glasses, fender covers, shop towel, hand tools

Procedure: Before starting this worksheet refer to Worksheet # 9-28, Remove and Replace a Macpherson Strut, for the procedure for removing the strut from the vehicle.

Removing the Spring from a Strut

1. Mount the strut in a suitable holding fixture. Never use a vise to clamp the strut by its housing tube. This could distort the housing, causing the strut cartridge to bind.
2. Install the spring compressor on the spring and compress the spring.

CAUTION Always have proper supervision when using a spring compressor for the first time and follow the manufacturer's recommended procedures. Remember that the center nut at the top of the strut must **NOT** be removed before the strut is safely in a spring compressor.

3. After the spring is compressed, remove the nut from the strut piston rod.
4. Lift the upper strut mount, upper insulator, spring, and spring upper and lower insulator from the strut.
5. Inspect the strut bearing that is mounted in the upper strut mount. ☐ OK ☐ Needs replacement.

   Note: Most struts on older vehicles have replaceable cartridges. On newer vehicles the strut assembly is replaced, using the existing spring.

TURN

Replacing a Strut Cartridge (If Necessary)

6. Remove the strut cartridge retainer from the top of the strut.

   Which tool was used to remove the strut cartridge retaining nut?

   Check one: ☐ Socket ☐ Spanner ☐ N/A

7. Place a drain pan under the strut and remove the cartridge from the strut.
8. Clean the strut housing and dispose of the old strut oil.
9. Install the new strut cartridge and add the required strut oil.

   How much oil was added to the strut? ______ oz.

Reinstall the Spring

10. Install the lower insulator and check that it is properly seated.

    Insulator properly seated? ☐ Yes ☐ No

11. Install the spring bumper.
12. Pull the strut rod to full extension.

SHOP TIP A tie wrap tightened around the strut rod shaft will hold the shaft in full extension during assembly.

13. Install the compressed spring, carefully seating it in the lower insulator.
14. Install the upper insulator onto the spring.
15. Install the upper strut mount over the strut rod and onto the spring. Carefully position the end of the spring in its seat.
16. Slowly loosen the spring compressor while maintaining the alignment of the spring in the spring seats.
17. Install the strut rod nut and tighten to specifications. Torque specification: ______ ft.-lb

CAUTION Do not use an impact wrench, which can spin the strut rod and damage the seal.

18. Inspect the strut spring insulators and upper mount to confirm they are properly aligned and seated.

    All parts properly positioned? ☐ Yes ☐ No

19. Remove the spring compressor from the spring.
20. Refer to Worksheet #9-28, Remove and Replace a Macpherson Strut, for the procedure for reinstalling the strut assembly on the vehicle.
21. Before completing your paperwork, clean your work area, clean and return tools to their proper places, and wash your hands.

____

____

STOP

ASE Education Foundation Worksheet #9-29

CHECK BALL JOINT WEAR

Name______________________ Class______________

Score: ☐ Excellent ☐ Good ☐ Needs Improvement **Instructor OK** ☐

Vehicle year ________ **Make** ______________ **Model** ______________

Objective: Upon completion of this assignment, you should be able to check a vehicle's ball joints for wear.

ASE Education Foundation Correlation

This worksheet addresses the following **MLR** task:

IV.B.12 Inspect upper and lower ball joints (with or without wear indicators). **(P-1)**

This worksheet addresses the following **AST/MAST** task:

IV.C.5 Inspect, remove, and/or replace upper and/or lower ball joints (with or without wear indicators). **(P-2)**

We Support ASE Education Foundation

Directions: Before beginning this lab assignment, review the worksheet completely. Fill in the information in the spaces provided as you complete each task.

Tools and Equipment Required: Safety glasses, fender covers, jack and jack stands or vehicle lift, prybar, dial indicator, ball joint checker, shop towel

Procedure:

1. Locate the ball joint specifications in the service information.

   What is the radial play specification? ______

   What is the axial play specification? ______

2. Raise the car on a wheel ramp-type lift. If one is not available, it will be necessary to work on the ground.

3. Check the wheel bearing adjustment before testing ball joint looseness.

   ☐ Needs adjustment ☐ OK

4. Which is the load-carrying ball joint? Location: ☐ Upper ☐ Lower

   Note: When there are two control arms, the load-carrying joint is the one on the control arm that supports the spring.

5. Does the vehicle have wear indicator ball joints?

   ☐ Yes ☐ No

   Note: Some ball joints have a wear indicator on the grease fitting. Wear on these joints is inspected with the weight of the vehicle on the tires.

   Is any wear indicated? ☐ Yes ☐ No

TURN

6. To check a ball joint without a wear indicator, remove the load from the ball joint. Raise the vehicle with a jack placed on the *frame* or *lower control arm* as specified.

   The jack is on the: ☐ Frame ☐ Lower control arm

Check Ball Joint Axial Play

7. Raise the vehicle until the tire is off the ground just enough to be able to fit a prybar under it.
8. Pry the tire up and down while checking for vertical movement at the ball joint.

   Was any vertical movement noticed? ☐ Yes ☐ No
9. Measure and record the vertical movement:

   Measurement ______

Check Ball Joint Radial Play

10. Raise the vehicle until the tire is off the ground.
11. Grasp the tire at the top and bottom and try to move it alternately in and out at the top and bottom.

    Was any radial movement noticed?

    ☐ Yes ☐ No
12. Measure and record the radial movement.

    Measurement ______
13. Before completing your paperwork, clean your work area, clean and return tools to their proper places, and wash your hands.
14. Record your recommendations for needed service or additional repairs and complete the repair order.

____

____

STOP

ASE Education Foundation Worksheet #9-30

PREALIGNMENT AND RIDE HEIGHT INSPECTION

Name____________________________ Class____________________

Score: ☐ Excellent ☐ Good ☐ Needs Improvement **Instructor OK** ☐

Vehicle year __________ **Make** ____________________ **Model** ______________________

Objective: Upon completion of this assignment, you should be able to perform a prealignment inspection.

ASE Education Foundation Correlation

This worksheet addresses the following **MLR** task:

IV.C.1 Perform prealignment inspection and measure vehicle ride height; determine necessary action. **(P-1)**

This worksheet addresses the following **AST/MAST** task:

IV.E.2 Perform prealignment inspection and measure vehicle ride height; determine needed action. **(P-1)**

We Support ASE | Education Foundation

Directions: Before beginning this lab assignment, review the worksheet completely. Fill in the information in the spaces provided as you complete each task.

Tools and Equipment Required: Safety glasses, shop towels, vehicle lift, tire pressure gauge, dial indicator, ruler

Procedure:

1. Inspect the underside of the chassis for mud, snow, or other debris.

   Is there any foreign material stuck to the chassis? ☐ Yes ☐ No

2. Check the vehicle for excessive load.

   Does the trunk or passenger compartment contain heavy items that are not part of the normal vehicle load? ☐ Yes ☐ No

3. Fuel level: ☐ Full ☐ $\frac{3}{4}$ ☐ $\frac{1}{2}$ ☐ $\frac{1}{4}$ ☐ Empty

4. Perform a front and rear suspension bounce test to check the condition of the first stage of shock absorber operation.

   Note: This test does not confirm correct medium or high-speed operation of the shock absorber.

TURN

5. Measure the front and rear suspension ride height.

   Specified ride height: Front ______ Rear ______

   Measured front suspension ride height: Left ______ Right ______

   Measured rear suspension ride height: Left ______ Right ______

6. Raise the vehicle on a lift or position it safely on jack stands.
7. Check the condition of the tires. ☐ OK ☐ Worn ☐ Damaged
8. Inflate the tires to the recommended pressure shown on the tire placard.

   Tire pressure: Front ____________________ Rear ____________________

9. Rotate the front and rear tires while looking for radial and lateral runout.

   Check off any tire with excessive radial or lateral runout below:

   Excessive radial runout: Front: ☐ Right ☐ Left Rear: ☐ Right ☐ Left

   Excessive lateral runout: Front: ☐ Right ☐ Left Rear: ☐ Right ☐ Left

10. Inspect the control arms for damage and worn bushings.

    ☐ OK ☐ List any items in need of service: ____________________

11. Inspect the stabilized bar bushings and linkage.

    ☐ OK ☐ List any items in need of service: ____________________

12. Inspect all steering linkages and tie-rod ends for looseness.

    ☐ OK ☐ List any items in need of service: ____________________

13. Inspect the strut rod and track bar for damage or worn links or bushings.

    ☐ OK ☐ List any items in need of service: ____________________

14. Measure the steering wheel free play and list it here. ____________________
15. Inspect the coil springs, spring insulators, and suspension bumpers.

    ☐ OK ☐ List any items in need of service: ____________________

16. Inspect all shock absorbers for leaks and loose, worn mounting bushings and bolts.

    ☐ OK ☐ List any items in need of service: ____________________

17. Check front- and rear-wheel bearing end play.

    ☐ OK ☐ List any items in need of service: ____________________

18. Check service specifications for ball joint movement and list them below:

    Radial ______ Axial ______

19. Measure ball joint radial and axial movement and list below:

    Left lower ______ Left upper ______

    Right lower ______ Right upper ______

20. Before completing your paperwork, clean your work area, clean and return tools to their proper places, and wash your hands.
21. Record your recommendations for needed service or additional repairs and complete the repair order.

    ____

    ____

STOP

ASE Education Foundation Worksheet #9-31

CENTER THE STEERING WHEEL

Name________________________ Class__________________

Score: ☐ Excellent ☐ Good ☐ Needs Improvement **Instructor OK** ☐

Vehicle year _________ **Make** ________________ **Model** __________________

Objective: Upon completion of this assignment, you should be able to center a steering wheel.

ASE Education Foundation Correlation

This worksheet addresses the following **AST/MAST** task:

IV.E.3 Prepare vehicle for wheel alignment on alignment machine; perform four-wheel alignment by checking and adjusting front and rear wheel caster, camber and toe as required; center steering wheel. **(P-1)**

We Support Education Foundation

Directions: Before beginning this lab assignment, review the worksheet completely. Fill in the information in the spaces provided as you complete each task.

Tools and Equipment Required: Safety glasses, fender covers, jack and jack stands or vehicle lift, hand tools, steering wheel lock, shop towel

Procedure: The steering wheel is usually centered using the tie-rods, not by removing the steering wheel and putting it back on straight. When a toe-in adjustment is done, adjusting one of the tie-rods more than the other will cause the steering wheel to be off-center. If the steering wheel is not straight ahead when driving, follow this procedure to straighten it.

1. Count the number of turns the steering wheel makes as it is turned from lock to lock (include fractions).

   Number of turns lock to lock ______

2. Position the steering wheel halfway between the locks. It should be centered. If not, remove it and put it back on straight.

   Steering wheel centered? ☐ Yes ☐ No

3. Use a steering wheel lock to hold the steering wheel in the centered position.

4. From the front of the vehicle, sight down the tires on each side. The front tire should align with the rear one. This will give a rough estimate of the direction the tie-rods need to be turned.

Adjust each wheel to point left

Steering wheel position

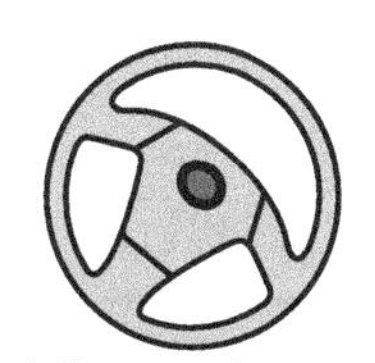

Adjust each wheel to point right

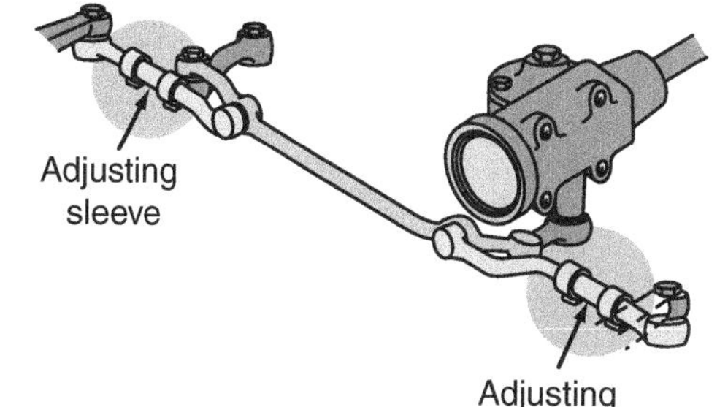

From which side can more of one rear tire than the other be seen? ☐ Right ☐ Left

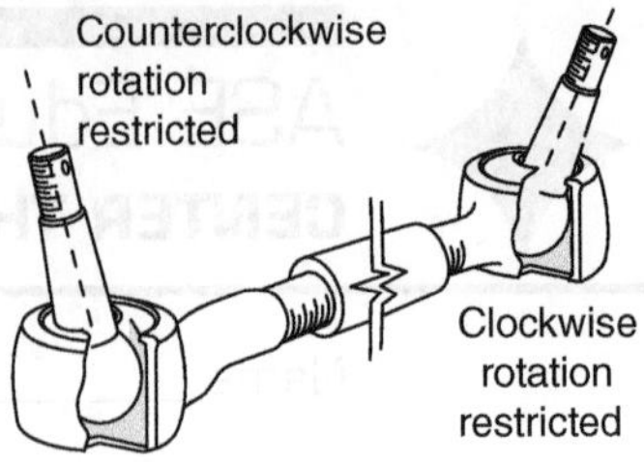

If the right rear tire is more visible when sighting down the tires, the steering needs to be adjusted to the left.

If the left rear tire is more visible when sighting down the tires, the steering needs to be adjusted to the right.

The steering linkage needs to be adjusted to the:

☐ Right ☐ Left

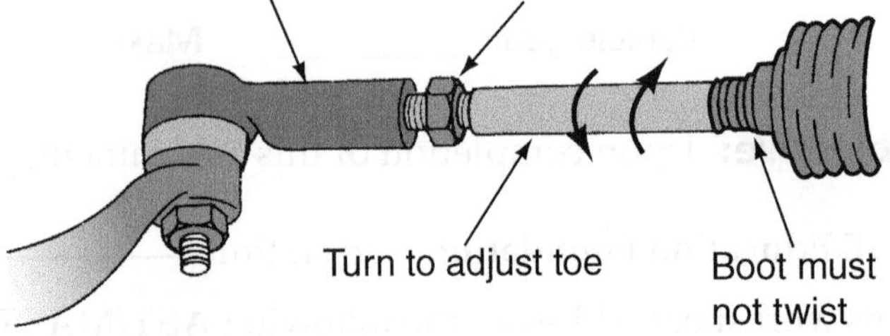

5. On rack and pinion steering, loosen the jam nut and turn the tie-rod. On parallelogram steering (with tie-rod ends on each side), loosen the clamps. Turn each tie-rod an equal amount in opposite directions. This maintains the current toe setting but moves the steering wheel position.

Note: The clamps on parallelogram tie-rods must be positioned so that they do not bind before retightening. See your instructor.

6. Test-drive the vehicle on a straight, level road.

   Is the steering wheel centered? ☐ Yes ☐ No

   If the steering wheel is off-center and points to the right, readjust the tie-rods so that the tires also are pointed to the right.

7. Before completing your paperwork, clean your work area, clean and return tools to their proper places, and wash your hands.

8. Record your recommendations for needed service or additional repairs and complete the repair order.

____

____

STOP

ASE Education Foundation Worksheet #9-32

INSPECT AND REPLACE FWD FRONT WHEEL BEARINGS

Name_______________________________ Class____________________

Score: ☐ Excellent ☐ Good ☐ Needs Improvement **Instructor OK** ☐

Vehicle year __________ **Make** ____________________ **Model** ______________________

Objective: Upon completion of this assignment, you should be able to inspect and replace FWD front wheel bearings.

ASE Education Foundation Correlation

This worksheet addresses the following **AST/MAST** tasks:

IV.C.6 Inspect, remove, and/or replace steering knuckle assemblies. **(P-3)**

IV.D.2 Remove, inspect, service and/or replace front and rear wheel bearings. **(P-1)**

We Support ASE | Education Foundation

Directions: Before beginning this lab assignment, review the worksheet completely. Fill in the information in the spaces provided as you complete each task.

Tools and Equipment Required: Safety glasses, fender covers, jack and jack stands or vehicle lift, hand tools, shop towels

Procedure:

1. Research the proper procedure for checking and replacing the front wheel bearing.

   Note: the following procedures are general. Follow the directions in the service information for best results.

2. Raise the vehicle on a lift or place it on jack stands.
3. Remove the tire and wheel assembly.
4. Remove the brake caliper.
5. Rotate the hub as you listen for noise and feel for roughness.

   Did you notice any roughness or hear any bearing noise? ☐ No ☐ Yes.

   If you answered yes, the bearing(s) will need to be replaced.

6. Remove the steering knuckle.
7. Remove wheel nut. It will be discarded and replaced, but keep it on hand for comparison with the new nut.
8. Remove the tie-rod end from the steering knuckle.
9. Disconnect the steering knuckle from the strut.
10. Disconnect the steering knuckle from the lower control arm
11. Slide the steering knuckle off the axleshaft and lift it from the vehicle.

Replace the Wheel Bearing(s)

12. Mount the steering knuckle in a vise.
13. If there is a seal, remove it.
14. Use the recommended press to remove the wheel bearing(s) from the hub.
15. Use the recommended press to install the new bearing(s) in the hub.

Install the Steering Knuckle

16. Position the steering knuckle onto the strut and torque the fasteners.

    Torque spec for steering knuckle to strut: ________ ft.lbs.
17. Slide the axleshaft into the steering knuckle and connect the lower ball joint to the steering knuckle. Torque the ball joint nut and install a new cotter key.

    Torque spec for lower ball joint: ________ ft.lbs.
18. Position and install the tie-rod end into the steering knuckle. Tighten the tie-rod nut to specifications and install a new cotter pin.

    Torque spec for tie-rod end nut: ________ ft.lbs.
19. Install a new axle nut and torque to specifications.

    Torque spec for axle nut: ________ ft.lbs.

    Note: Do not fully torque this nut until the vehicle is on the ground. Use the vehicle brakes to hold the axle while torquing the nut. Do not use an impact wrench. It can damage the bearing.
20. Install the disc rotor and brake caliper. Torque the brake caliper nut to specifications.

    Torque spec for brake caliper: ________ ft.lbs.
21. Install the tire and wheel assembly. What is the lug nut torque specification? ________.
22. Lower the vehicle
23. Before completing the paperwork, clean your work area, put the tools in their proper places, and wash your hands.
24. List your recommendations for future service and/or additional repairs and complete the repair order.

    ____

    ____

STOP

Part II

Lab Worksheets for ASE Maintenance and Light Repair

Service Area 10

Miscellaneous

ASE Education Foundation Worksheet #10-1

PREPARE FOR A TEST-DRIVE

Name_______________________________ Class_______________________

Score: ☐ Excellent ☐ Good ☐ Needs Improvement **Instructor OK** ☐

Vehicle year __________ **Make** ____________________ **Model** ______________________

Objective: Upon completion of this assignment, you should be able to prepare a vehicle for a test-drive.

ASE Education Foundation Correlation

This worksheet addresses the following **MLR** tasks:

I.C.1 Perform cooling system pressure and dye tests to identify leaks; check coolant condition and level; inspect and test radiator, pressure cap, coolant recovery tank, heater core, and galley plugs; determine necessary action. **(P-1)**

I.C.2 Inspect, replace, and/or adjust drive belts, tensioners, and pulleys; check pulley and belt alignment. **(P-1)**

IV.B.2 Inspect power steering fluid level and condition. **(P-1)**

IV.D.1 Inspect tire condition; identify tire wear patterns; check for correct tire size, application (load and speed ratings), and air pressure as listed on the tire information placard/label. **(P-1)**

This worksheet addresses the following **AST/MAST** tasks:

I.D.1 Perform cooling system pressure and dye tests to identify leaks; check coolant condition and level; inspect and test radiator, pressure cap, coolant recovery tank, heater core, and galley plugs; determine needed action. **(P-1)**

I.D.3 Inspect, replace, and/or adjust drive belts, tensioners, and pulleys; check pulley and belt alignment. **(P-1)**

IV.B.9 Inspect power steering fluid level and condition. **(P-1)**

IV.F.1 Inspect tire condition; identify tire wear patterns; check for correct tire size, application (load and speed ratings), and air pressure as listed on the tire information placard/label. **(P-1)**

We Support ASE Education Foundation

Directions: Before beginning this lab assignment, review the worksheet completely. Fill in the information in the spaces provided as you complete each task.

Procedure:

Unless a vehicle is unsafe to drive, a test-drive should always be done before performing repairs, and again when the repairs have been completed.

Before a test-drive, check the following items:

1. Open the hood and place fender covers on the fenders and front body parts.
   a. Check the oil level. ☐ Full ☐ Low
   b. Check the coolant level . ☐ Full ☐ Needs coolant
   c. Are there any coolant leaks? ☐ Yes ☐ No
   d. Check the power steering fluid level. ☐ OK ☐ Low
   e. Are there any power steering fluid leaks? ☐ Yes ☐ No
   f. Check the drive belt tension. ☐ OK ☐ Loose
2. Walk around the vehicle and inspect the condition of the tires.
   a. Damage to the sidewalls or tread area? ☐ Yes ☐ No
   b. Impact damage or bent rim? ☐ Yes ☐ No
   c. Condition of valve stems? ☐ Good ☐ Bad
   d. Adjust tire pressures to placard specification.
      Tire Placard Specification: Front ___ psi Rear _____psi
   e. Are the tires the correct size? ☐ Yes ☐ No
   f. What is the tire size specification for the vehicle? _______
   g. Are all of the tires the same brand and tread pattern? ☐ Yes ☐ No
3. Sit in the driver's seat and inspect the following:
   a. Depress the brake pedal. How does it feel? ☐ Spongy ☐ Firm
   b. Turn the key to the *ON* position. Does the malfunction indicator lamp illuminate?
      ☐ Yes ☐ No
   c. Check the fuel gauge. Is there enough fuel for the test-drive? ☐ Yes ☐ No
   d. Start the engine. Is there adequate oil pressure? ☐ Yes ☐ No
   e. Are any driver warning lights illuminated? ☐ Yes ☐ No
   f. With the engine on, turn the steering wheel from side to side. Does the power steering work equally in both directions? ☐ Yes ☐ No
4. Describe any problems you noticed during the pre-test-drive inspection.

____

____

Upon completion of the pre-test-drive inspection, get your instructor's approval before taking a vehicle on a test-drive.

SAFETY NOTE Before beginning a test-drive, adjust the seat and mirrors. Never drive a vehicle without using the seat belt.

STOP

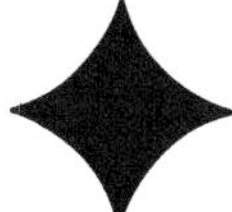

ASE Education Foundation Worksheet #10-2

PERFORM A TEST-DRIVE

Name________________________________ Class____________________

Score: ☐ Excellent ☐ Good ☐ Needs Improvement **Instructor OK** ☐

Vehicle year __________ **Make** ___________________ **Model** ____________________

Objective: Upon completion of this assignment, you should be able to perform a test-drive to check vehicle condition and identify problems.

ASE Education Foundation Correlation

This worksheet addresses the following **MLR** task:

V.B.1 Describe proper brake pedal height, travel, and feel. **(P-1)**

This worksheet addresses the following **AST/MAST** tasks:

II.A.1 Identify and interpret transmission/transaxle concerns, differentiate between engine performance and transmission/transaxle concerns; determine needed action. **(P-1)**

III.A.1 Identify and interpret drive train concerns; determine needed action. **(P-1)**

III.B.1 Diagnose clutch noise, binding, slippage, pulsation, and chatter; determine needed action. **(P-1)**

IV.B.3 Diagnose steering column noises, looseness, and binding concerns (including tilt/telescoping mechanisms); determine needed action. **(P-1)**

IV.B.4 Diagnose power steering gear (non-rack and pinion) binding, uneven turning effort, looseness, hard steering, and noise concerns; determine needed action. **(P-1)**

IV.B.5 Diagnose power steering gear (rack and pinion) binding, uneven turning effort, looseness, hard steering, and noise concerns; determine needed action. **(P-1)**

IV.C.1 Diagnose short and long arm suspension system noises, body sway, and uneven ride height concerns; determine needed action. **(P-1)**

IV.C.2 Diagnose strut suspension system noises, body sway, and uneven ride height concerns; determine needed action. **(P-1)**

IV.E.1 Diagnose vehicle wander, drift, pull, hard steering, bump steer, memory steer, torque steer, and steering return concerns; determine needed action. **(P-1)**

IV.F.2 Diagnose wheel/tire vibration, shimmy, and noise; determine needed action. **(P-1)**

IV.F.5 Diagnose tire pull problems; determine needed action. **(P-1)**

V.A.1 Identify and interpret brake system concerns; determine needed action. **(P-1)**

We Support ASE | Education Foundation

Directions: Before beginning this lab assignment, review the worksheet completely. Fill in the information in the spaces provided as you complete each task. Be sure to complete Worksheet #10-1, prior to doing this worksheet.

Procedure:

Do not attempt this worksheet without the knowledge and permission of your instructor.

TEST DRIVE

During the test-drive, a professional technician will check a vehicle for many driving conditions. The following are a few of the important checks that should be made. Read this material carefully and be sure you understand it before taking a test-drive.

Items to be checked during a test drive:

ENGINE PERFORMANCE. The vehicle should accelerate smoothly with enough power to climb hills.

1. TRANSMISSION OPERATION
   a. Automatic transmission—shifts smoothly and at the appropriate time during both acceleration and deceleration.
   b. Manual transmission—has smooth engagement of the clutch, shifts smoothly, and runs quietly.
2. BRAKES
   a. The brake pedal is firm, holds pressure.
   b. The vehicle stops straight, with no pulling to one side.
3. STEERING
   a. The vehicle should be easier to steer at higher speeds. Hard steering can be caused by binding parts, incorrect alignment, low tires, or a failure in the power steering system.
   b. The tires should not squeal on turns. Squealing tires could be due to a bent steering arm or low tire pressures.
4. NOISES
   a. Are there squeaks or clunks when going over bumps?
   b. These noises could be due to bad bushings, which can also cause changes in wheel alignment that can result in brake pull.
   c. Are there noises that change in pitch as the vehicle weaves to the left and then to the right?

      The outer wheel bearing turns faster during a turn. It will make more noise when turning to one direction than to the other.
5. SHIMMY OR TRAMP.

   Does the steering wheel shake from side to side?

   This could indicate a bent or out-of-balance wheel or excessive caster.
6. WANDER. Does the vehicle drift, requiring constant steering?

   This could indicate an incorrect caster alignment setting.

7. PULL. Does the vehicle pull to one side or the other at a cruise, under braking, or while accelerating?

   This could be due to problems with brakes, tires, or incorrect wheel alignment.

8. ROUGH RIDE. Does the vehicle have a rougher than normal ride?

   This could be due to tire pressures that are too high or a bent or frozen shock.

9. EXCESS BODY ROLL. Does the vehicle lean excessively to one side or the other during fast turns?

   This could be due to worn shock absorbers.

Upon completion of the pre-test-drive inspection, get your instructor's approval before taking a vehicle on a test-drive.

Before leaving on a test-drive, adjust the seat and mirrors. Never drive a vehicle without using the seat belt. Complete the Test Drive Checklist after returning from the test drive.

Test-Drive Checklist

☐ OK	☐ Problem		Does the vehicle go straight when you let go of the steering wheel?
☐ OK	☐ Problem		Accelerates smoothly?
☐ OK	☐ Problem		Good acceleration hot and cold, climbs hills easily?
☐ OK	☐ Problem	☐ N/A	Standard transmission shifts smoothly?
☐ OK	☐ Problem	☐ N/A	Automatic transmission shifts smoothly at the correct speed?
☐ OK	☐ Problem		Stops straight and fast?
☐ OK	☐ Problem		Rides smooth on the highway (tire balance, shocks, suspension)?
☐ OK	☐ Problem		No clunks when going over speed bumps?
☐ OK	☐ Problem		Quiet on highway (no wind leaks, no differential noise)?
☐ OK	☐ Problem		No smoke from exhaust on heavy acceleration or deceleration?
☐ OK	☐ Problem		Gauges appear accurate after warm-up?

STOP

ASE Education Foundation Worksheet #10-3

USED CAR CONDITION APPRAISAL CHECKLIST

Name ______________________ Class ______________

Score: ☐ Excellent ☐ Good ☐ Needs Improvement **Instructor OK** ☐

Vehicle year ________ **Make** ______________ **Model** ______________

Objective: Upon completion of this assignment, you should be able to inspect the condition of a used vehicle prior to purchase.

Directions: Fill in the information in the spaces provided as you inspect the vehicle.

Documents:

	OK	Inadequate
Registration documents		
Verification of seller's identity		
VIN matches registration		
Emission certification		

Kelly Blue Book or NADA value: Retail value ________ Wholesale value ________

Which did you use? ☐ Kelly Blue Book ☐ NADA ☐ Other ______________

Is this vehicle being sold by a dealer? ☐ Yes ☐ No

Is the seller the registered owner of the vehicle? ☐ Yes ☐ No ☐ N/A

Body/Paint Condition:

	OK	Bad
Rust: on roof, bottoms of doors, or fenders		
Paint:		
Consistent and bright?		
Faded evenly (no new repairs)		
Door jambs and under hood are same color?		
Paint on chrome strips or weatherstripping?		
Chrome in good condition?		
Dents: evidence of repair?		
Body fit		
Hidden welds on body joints		

Additional Comments: ____

____

____

Wear and Tear:

Odometer mileage ______________________

File of maintenance records is provided? ☐ Yes ☐ No

Wear and tear is consistent with odometer mileage? ☐ Yes ☐ No

Additional Comments: ______________________

Tires:

	OK	Bad
All are the same brand?		
All are the same size?		
Spare tire		

Additional Comments: ______________________

Electrical System:

	OK	Bad
Charging system operation		
Starting system operation		
Battery condition		
Gauges		
Radio		
Lights		
Heating		
Air conditioning		
Clock		
Horn		

Additional Comments: ______________________

Doors:

	OK	Bad
All doors shut tightly?		
Driver's door handle loose?		
Windows tight, roll easily?		
Electric windows working?		
Weather-stripping condition		

Additional Comments: ______________________

Interior Condition:

	OK	Bad
Upholstery		
Headliner		
Dashboard		
Carpets		
Accelerator and brake pedals worn?		

Additional Comments: ______________________________

Windshield:

	OK	Bad
Signs of leakage		
Frost around edges		
Rock chips		
Cracks		
Wipers:		
Condition		
Operation		
Washers		

Additional Comments: ______________________________

Mechanical Condition:

	OK	Bad
Condition of chassis		
Evidence of frame or collision damage		
Tires: alignment wear?		
Tires: amount of wear?		
Brake pedal height		
Brake pedal firmness		
Front brake lining condition		
Rear brake lining condition		
Steering wheel free play (less than 3")?		
Steering linkage		
Suspension parts		
Springs (ride height)		
Suspension bushings		
Shock absorbers		
Rubber snubbers		

Additional Comments: ______________________________

Leaks:

	OK	Bad
Engine oil		
Transmission fluid		
Differential oil		
Power steering fluid		
Brake fluid		
Coolant		

Additional Comments: ______________________________

Engine Operation:

	OK	Bad
Driver warning indicator (KOEO)		
Smooth Idle when the engine is cold?		
Smooth Idle when the engine is warm?		
Driver warning indicator off when engine is running		

Additional Comments: ______________________________

Transmission:

	OK	Bad
Stall test (5 sec.)		
Fluid level		
Fluid condition		

Additional Comments: ______________________________

Powertrain:

	OK	Bad
CV joint boots, Universal joints		
Clutch slippage		
Clutch free play		
Gear engagement		
Transmission noise		

Additional Comments: ______________________________

Engine:

	OK	Bad
Oil appearance		
Odometer mileage at last oil change ________		
Oil pressure at idle (warm)		
Engine noise when cold?		
Engine vacuum reading at idle		
Engine compression		
Radiator condition		
Hoses and belts		
Blowby (check PCV and air cleaner)		
Exhaust system condition		
Oil or soot in tailpipe?		
Exhaust smoke?		
Engine mounts		

Additional Comments: ____

Emission Controls:

	OK or N/A	Bad
Underhood emission label		
Air injection system		
Exhaust gas recirculation		
Catalytic converter		
Crankcase vent, PCV		
Evaporative controls		
Modifications?		
Have all the monitors run?		

Additional Comments: ____

Test-Drive: Use work sheet #10-2 Perform a Test Drive

Comments:____

____

____

STOP

ASE Education Foundation Worksheet #10-4

VEHICLE MAINTENANCE LOG

Enter information about your vehicle in the spaces below. Keep track of maintenance and repairs as they occur. Scan or photograph all receipts and repair orders or punch holes in them and keep them in your notebook.

Vehicle Information:

Make ________________________ Model ________________________

Year ________________________ Color ________________________

VIN ________________________

License plate number ________________________ State ________________________

Date purchased ________________________ Cost: $ ________________________

Purchased from ________________________

Odometer reading when purchased ________________________

Insurance Information:

Record your insurance information here and keep it in your files.

Insurance company ________________________

Agent ________________________

Address ________________________

Phone number ________________________

Policy number ________________________

Coverage ________________________

Maintenance and Repairs:

Record all services and repairs to your vehicle below.

Date	Description/Repair Shop	Cost

Total Cost ____________

Fuel Records:

Whenever you fill your fuel tank, make a record of the purchase and amount of fuel. To calculate the fuel mileage, divide the miles driven by the fuel used.

Date	Odometer Reading	Gallons/Liters Purchased	Price per Gallon/Liter	Total Cost	Miles per Gallon/Liter

Total Cost ____________

TURN

Fuel Records:

Record fuel use as it is purchased. To calculate the fuel mileage, divide the miles driven by the fuel used.

Date	Odometer Reading	Gallons/Liters Purchased	Price per Gallon/Liter	Total Cost	Miles per Gallon/Liter

Total Cost ________________

ASE Education Foundation Worksheet #10-5

PERSONAL TOOL INVENTORY

Name________________________ Class________________

Score: ☐ Excellent ☐ Good ☐ Needs Improvement **Instructor OK** ☐

Vehicle year ________ **Make** ______________ **Model** ________________

Objective: The following lists tools recommended for an entry-level technician. When you have completed this assignment, you should have a complete inventory of your personal tool set. This inventory will help you decide which additional tools to add to your set.

ASE Education Foundation Correlation

This worksheet addresses the following **Required Supplemental (RST)** task:

Tools and Equipment

Task 1: Identify tools and their usage in automotive applications.

We Support ASE Education Foundation

Directions: Place a check (✓) in the space next to the tools that you have in your collection.

Tools and Equipment Required: Personal tool set

Toolbox

- ☐ Top chest
- ☐ Rollaway
- ☐ Utility cart
- ☐ Portable box

Screwdrivers

Slot screwdrivers:

☐ 3" ☐ 4" ☐ 6" ☐ 8" ☐ 10" ☐ 12"

Phillips screwdrivers:

Standard length

☐ #1 ☐ #2 ☐ #3

☐ Short #2 ☐ Long #2

☐ Magnetic screwdriver

Torx® drivers:

☐ T8 ☐ T10 ☐ T15 ☐ T20

☐ T25 ☐ T30

☐ Other

Pliers

- ☐ 10" Multipurpose pliers
- ☐ Large multipurpose pliers
- ☐ 7" Locking pliers
- ☐ 10" Locking pliers
- ☐ Needle nose pliers
- ☐ Slip joint pliers
- ☐ Diagonal cutters
- ☐ Terminal pliers (wire/crimper)
- ☐ Other

Snap ring pliers

☐ Inside ☐ Outside ☐ Convertible

☐ Other

Hammers

- ☐ 16 oz. ball-peen hammer
- ☐ Plastic hammer
- ☐ Brass hammer
- ☐ Rubber mallet

TURN

☐ Dead blow hammer

☐ Other

Socket Drive Tools

☐ ⅜" Drive

☐ Ratchet

☐ Wobble extensions (15°) ☐ 3" ☐ 6"
☐ 12"

☐ Speed handle

☐ Impact driver

☐ ½" Drive

☐ Ratchet

☐ 6" extension

☐ 12" extension

☐ Flex handle (breaker bar)

Torque Wrenches

☐ ½" Drive ft.-lb

☐ ⅜" Drive ft.-lb

☐ ⅜" Drive in.-lb

Metric Sockets

⅜" Drive socket set

☐ 6 mm ☐ 7 mm ☐ 8 mm ☐ 9 mm

☐ 10 mm ☐ 11 mm ☐ 12 mm

☐ 13 mm ☐ 14 mm ☐ 15 mm

☐ 16 mm ☐ 17 mm ☐ 18 mm ☐ 19 mm

☐ Other

Deep Socket Set

☐ 6 mm ☐ 7 mm ☐ 8 mm ☐ 9 mm

☐ 10 mm ☐ 11 mm ☐ 12 mm

☐ 13 mm ☐ 14 mm ☐ 15 mm

☐ 16 mm ☐ 17 mm ☐ 18 mm ☐ 19 mm

☐ Other

Impact Socket Set

☐ 8 mm ☐ 9 mm ☐ 10 mm ☐ 11 mm

☐ 12 mm ☐ 13 mm ☐ 14 mm ☐ 15 mm

☐ 16 mm ☐ 17 mm ☐ 18 mm ☐ 19 mm

☐ Other

Universal Impact Socket Set

☐ 8 mm ☐ 9 mm ☐ 10 mm ☐ 11 mm

☐ 12 mm ☐ 13 mm ☐ 14 mm ☐ 15 mm

☐ 16 mm ☐ 17 mm ☐ 18 mm ☐ 19 mm

☐ Other

½" Drive Socket Set

☐ 10 mm ☐ 11 mm ☐ 12 mm ☐ 13 mm

☐ 14 mm ☐ 15 mm ☐ 16 mm ☐ 17 mm

☐ 18 mm ☐ 19 mm ☐ 20 mm ☐ 21 mm

☐ 22 mm ☐ 23 mm ☐ 24 mm ☐ 25 mm

☐ Other

Deep Socket Set

☐ 10 mm ☐ 11 mm ☐ 12 mm ☐ 13 mm

☐ 14 mm ☐ 15 mm ☐ 16 mm ☐ 17 mm

☐ 18 mm ☐ 19 mm ☐ 20 mm ☐ 21 mm

☐ 22 mm ☐ 23 mm ☐ 24 mm ☐ 25 mm

☐ Other

Impact Socket Set

☐ 10 mm ☐ 11 mm ☐ 12 mm ☐ 13 mm

☐ 14 mm ☐ 15 mm ☐ 16 mm ☐ 17 mm

☐ 18 mm ☐ 19 mm ☐ 20 mm ☐ 21 mm

☐ 22 mm ☐ 23 mm ☐ 24 mm ☐ 25 mm

☐ Other

Metric Wrenches

Combination wrenches (standard length)

☐ 7 mm ☐ 8 mm ☐ 9 mm ☐ 10 mm

☐ 11 mm ☐ 12 mm ☐ 13 mm ☐ 14 mm

☐ 15 mm ☐ 16 mm ☐ 17 mm ☐ 18 mm

☐ 19 mm ☐ Other

Flare-Nut Wrenches

☐ 8 mm ☐ 9 mm ☐ 10 mm ☐ 11 mm

☐ 12 mm ☐ 13 mm ☐ 14 mm ☐ 15 mm

☐ 16 mm ☐ 17 mm ☐ 18 mm ☐ 19 mm

☐ Other

TURN

Allen Wrench Set

☐ 0.7 mm ☐ 0.9 mm ☐ 2 mm ☐ 3 mm

☐ 4 mm ☐ 5 mm ☐ 6 mm ☐ 7 mm

☐ 8 mm ☐ 10 mm ☐ 12 mm ☐ 14 mm

☐ 17 mm ☐ 19 mm

☐ Other

Standard Sockets

Note: All vehicles currently manufactured use metric fasteners.

3/8" Drive socket set

☐ 1/4 ☐ 5/16 ☐ 3/8 ☐ 7/16 ☐ 1/2

☐ 9/16 ☐ 5/8 ☐ 11/16 ☐ 3/4 ☐ Other

Deep Socket Set

☐ 1/4 ☐ 5/16 ☐ 3/8 ☐ 7/16 ☐ 1/2

☐ 9/16 ☐ 5/8 ☐ 11/16 ☐ 3/4 ☐ Other

Impact Socket Set

☐ 5/16 ☐ 3/8 ☐ 7/16 ☐ 1/2 ☐ 9/16

☐ 5/8 ☐ 11/16 ☐ 3/4 ☐ Other

1/2" Drive Socket Set

☐ 3/8 ☐ 7/16 ☐ 1/2 ☐ 9/16 ☐ 5/8

☐ 11/16 ☐ 3/4 ☐ 13/16 ☐ 7/8

☐ 15/16

☐ 1 ☐ 1 1/16 ☐ 1 1/4 ☐ 1 5/16

☐ Other

Deep Socket Set

☐ 7/16 ☐ 1/2 ☐ 9/16 ☐ 5/8

☐ 11/16 ☐ 3/4 ☐ 13/16

☐ 7/8 ☐ 15/16

☐ 1 ☐ 1 1/16 ☐ 1 1/4 ☐ 1 5/16

☐ Other

Impact Socket Set

☐ 3/8 ☐ 7/16 ☐ 1/2 ☐ 9/16 ☐ 5/8

☐ 11/16 ☐ 3/4 ☐ 13/16 ☐ 7/8

☐ 15/16

☐ 1 ☐ 1 1/16 ☐ 1 1/4 ☐ 1 5/16

☐ Other

Standard Wrenches

Note: All vehicles currently manufactured use metric fasteners.

Combination Wrenches (standard length)

☐ 1/4 ☐ 5/16 ☐ 3/8 ☐ 7/16 ☐ 1/2

☐ 9/16 ☐ 5/8 ☐ 11/16 ☐ 3/4

☐ 13/16 ☐ 7/8 ☐ 15/16

☐ 1 ☐ 1 1/16 ☐ 1 1/4 ☐ 1 5/16

☐ Other

Flare-Nut Wrenches

☐ 5/16 ☐ 3/8 ☐ 7/16 ☐ 1/2 ☐ 9/16

☐ 5/8 ☐ 11/16 ☐ 3/4 ☐ 13/16 ☐ 7/8

☐ 15/16 ☐ 1 ☐ Other

Allen Wrench Set

☐ 1/16 ☐ 3/32 ☐ 7/64 ☐ 1/8 ☐ 9/64

☐ 5/32 ☐ 3/16 ☐ 7/32 ☐ 1/4 ☐ 5/16

☐ 3/8 ☐ 7/16 ☐ 1/2 ☐ 9/16 ☐ 5/8

☐ Other

Punch and Chisel Set

☐ Flat tip chisel ☐ Center punch

☐ Brass punch

☐ Pin punches: ☐ 1/16" ☐ 1/8" ☐ 1/4"

☐ Other

Files

☐ Mill ☐ Half round ☐ Round

☐ Other

Drill Index (Fractional sizes 1/16–1/2")

☐ Every 1/64? ☐ Every 1/32?

☐ Every 1/16?

Air Tools

☐ 1/2" drive impact wrench

☐ 3/8" drive impact wrench

- ☐ ⅜" drive ratchet
- ☐ Drill motor
- ☐ Blowgun
- ☐ Grinder

Spark Plug Tools

- ☐ Spark plug gauge
- ☐ ⅝" spark plug socket
- ☐ 13/16" spark plug socket
- ☐ Insulated pliers

Battery Tool Set

- ☐ Battery pliers
- ☐ Battery terminal puller
- ☐ Battery post cleaner
- ☐ Battery clamp spreader
- ☐ Side terminal cleaner
- ☐ 5/16 battery terminal wrench

Brake Tools

- ☐ Brake spoon
- ☐ Brake spring tool
- ☐ Brake retaining spring tool
- ☐ Wheel cylinder hone
- ☐ Disc brake piston removal tool

Miscellaneous

- ☐ Safety glasses
- ☐ Safety goggles
- ☐ Hacksaw
- ☐ Prybar
- ☐ Flexible magnetic pickup
- ☐ Scraper (putty knife)
- ☐ Wire brush
- ☐ Tire air pressure gauge
- ☐ Circuit tester (continuity tester)
- ☐ Feeler gauge set
- ☐ Adjustable long-handled mirror
- ☐ Flashlight
- ☐ Digital multimeter (DMM)
- ☐ Remote starter switch
- ☐ Compression gauge
- ☐ Vacuum fuel pressure gauge
- ☐ Oil filter wrench
- ☐ Creeper
- ☐ Tape measure

Additional Tools: List any additional tools that you have in your toolbox.

____

____

____

____

____

____

____

____

____

____

____

Part II

ASE Lab Preparation Worksheets

Appendix

Maintenance and Light Repair Technician Program Task List

MAINTENANCE AND LIGHT REPAIR (MLR)
Task List, effective January, 2017

ENGINE REPAIR

For every task in Engine Repair, the following safety requirement must be strictly enforced:

Comply with personal and environmental safety practices associated with clothing; eye protection; hand tools; power equipment; proper ventilation; and the handling, storage, and disposal of chemicals/materials in accordance with local, state, and federal safety and environmental regulations.

I. ENGINE REPAIR

A. General

P-1 1. Research vehicle service information, including fluid type, vehicle service history, service precautions, and technical service bulletins.

P-1 2. Verify operation of the instrument panel engine warning indicators.

P-1 3. Inspect engine assembly for fuel, oil, coolant, and other leaks; determine necessary action.

P-1 4. Install engine covers using gaskets, seals, and sealers as required.

P-2 5. Verify engine mechanical timing.

P-1 6. Perform common fastener and thread repair, to include: remove broken bolt, restore internal and external threads, and repair internal threads with thread insert.

P-2 7. Identify service precautions related to service of the internal combustion engine of a hybrid vehicle.

B. Cylinder Head and Valve Train

P-3 1. Adjust valves (mechanical or hydraulic lifters).

P-1 2. Identify components of the cylinder head and valve train.

C. Lubrication and Cooling Systems

P-1 1. Perform cooling system pressure and dye tests to identify leaks; check coolant condition and level; inspect and test radiator, pressure cap, coolant recovery tank, heater core, and galley plugs; determine necessary action.

P-1 2. Inspect, replace, and/or adjust drive belts, tensioners, and pulleys; check pulley and belt alignment.

P-1 3. Remove, inspect, and replace thermostat and gasket/seal.

P-1 4. Inspect and test coolant; drain and recover coolant; flush and refill cooling system; use proper fluid type per manufacturer specification; bleed air as required.

P-1 5. Perform engine oil and filter change; use proper fluid type per manufacturer specification; reset maintenance reminder as required.

P-1 6. Identify components of the lubrication and cooling systems.

ER Tasks	
P-1	12
P-2	2
P-3	1
	15

AUTOMATIC TRANSMISSION AND TRANSAXLE

For every task in Automatic Transmission and Transaxle, the following safety requirement must be strictly enforced:

Comply with personal and environmental safety practices associated with clothing; eye protection; hand tools; power equipment; proper ventilation; and the handling, storage, and disposal of chemicals/materials in accordance with local, state, and federal safety and environmental regulations.

II. AUTOMATIC TRANSMISSION AND TRANSAXLE

A. General

P-1 1. Research vehicle service information including fluid type, vehicle service history, service precautions, and technical service bulletins.

P-1 2. Check fluid level in a transmission or a transaxle equipped with a dip-stick.

P-1 3. Check fluid level in a transmission or a transaxle not equipped with a dip-stick.

P-2 4. Check transmission fluid condition; check for leaks.

P-1 5. Identify drive train components and configuration.

B. In-Vehicle Transmission/Transaxle

P-2 1. Inspect, adjust, and/or replace external manual valve shift linkage, transmission range sensor/switch, and/or park/neutral position switch.

P-1 2. Inspect for leakage at external seals, gaskets, and bushings.

P-2 3. Inspect, replace and/or align power train mounts.

P-1 4. Drain and replace fluid and filter(s); use proper fluid type per manufacturer specification.

C. Off-Vehicle Transmission and Transaxle

P-3 1. Describe the operational characteristics of a continuously variable transmission (CVT).

P-3 2. Describe the operational characteristics of a hybrid vehicle drive train.

AT Tasks	
P-1	6
P-2	3
P-3	2
	11

MANUAL DRIVE TRAIN AND AXLES

For every task in Manual Drive Train and Axles, the following safety requirement must be strictly enforced:

Comply with personal and environmental safety practices associated with clothing; eye protection; hand tools; power equipment; proper ventilation; and the handling, storage, and disposal of chemicals/materials in accordance with local, state, and federal safety and environmental regulations.

III. MANUAL DRIVE TRAIN AND AXLES

A. General

P-1 1. Research vehicle service information including fluid type, vehicle service history, service precautions, and technical service bulletins.

P-1 2. Drain and refill manual transmission/transaxle and final drive unit; use proper fluid type per manufacturer specification.

P-2 3. Check fluid condition; check for leaks.

P-1 4. Identify manual drive train and axle components and configuration.

B. Clutch

P-1 1. Check and adjust clutch master cylinder fluid level; use proper fluid type per manufacturer specification

P-1 2. Check for hydraulic system leaks.

C. Transmission/Transaxle

P-2 1. Describe the operational characteristics of an electronically-controlled manual transmission/transaxle.

D. Drive Shaft, Half Shafts, Universal Joints and Constant-Velocity (CV) Joints (Front, Rear, All, and Four-wheel drive)

P-2 1. Inspect, remove, and/or replace bearings, hubs, and seals.

P-2 2. Inspect, service, and/or replace shafts, yokes, boots, and universal/CV joints.

P-3 3. Inspect locking hubs.

P-2 4. Check for leaks at drive assembly and transfer case seals; check vents; check fluid level; use proper fluid type per manufacturer specification.

E. Differential Case Assembly

P-1 1. Clean and inspect differential case; check for leaks; inspect housing vent.

P-1 2. Check and adjust differential case fluid level; use proper fluid type per manufacturer specification.

P-1 3. Drain and refill differential housing.

P-1 4. Inspect and replace drive axle wheel studs.

MD Tasks	
P-1	9
P-2	5
P-3	1
	15

SUSPENSION AND STEERING

For every task in Suspension and Steering, the following safety requirement must be strictly enforced:

Comply with personal and environmental safety practices associated with clothing; eye protection; hand tools; power equipment; proper ventilation; and the handling, storage, and disposal of chemicals/materials in accordance with local, state, and federal safety and environmental regulations.

IV. SUSPENSION AND STEERING SYSTEMS

A. General

P-1 1. Research vehicle service information including fluid type, vehicle service history, service precautions, and technical service bulletins.

P-1 2. Disable and enable supplemental restraint system (SRS); verify indicator lamp operation.

P-1 3. Identify suspension and steering system components and configurations.

B. Related Suspension and Steering Service

P-1 1. Inspect rack and pinion steering gear inner tie rod ends (sockets) and bellows boots.

P-1 2. Inspect power steering fluid level and condition.

P-2 3. Flush, fill, and bleed power steering system; use proper fluid type per manufacturer specification.

P-1 4. Inspect for power steering fluid leakage.

P-1 5. Remove, inspect, replace, and/or adjust power steering pump drive belt.

P-2 6. Inspect and replace power steering hoses and fittings.

P-1 7. Inspect pitman arm, relay (centerlink/intermediate) rod, idler arm, mountings, and steering linkage damper.

P-1 8. Inspect tie rod ends (sockets), tie rod sleeves, and clamps.

P-1 9. Inspect upper and lower control arms, bushings, and shafts.

P-1 10. Inspect and replace rebound and/or jounce bumpers.

P-1 11. Inspect track bar, strut rods/radius arms, and related mounts and bushings.

P-1 12. Inspect upper and lower ball joints (with or without wear indicators).

P-1 13. Inspect suspension system coil springs and spring insulators (silencers).

P-1 14. Inspect suspension system torsion bars and mounts.

P-1 15. Inspect and/or replace front/rear stabilizer bar (sway bar) bushings, brackets, and links.

P-2 16. Inspect, remove, and/or replace strut cartridge or assembly; inspect mounts and bushings.

P-1 17. Inspect front strut bearing and mount.

P-1 18. Inspect rear suspension system lateral links/arms (track bars), control (trailing) arms.

P-1 19. Inspect rear suspension system leaf spring(s), spring insulators (silencers), shackles, brackets, bushings, center pins/bolts, and mounts.

P-1 20. Inspect, remove, and/or replace shock absorbers; inspect mounts and bushings.

P-2 21. Inspect electric power steering assist system.

P-2 22. Identify hybrid vehicle power steering system electrical circuits and safety precautions.

P-3 23. Describe the function of steering and suspension control systems and components, (i.e., active suspension, and stability control).

C. Wheel Alignment

P-1 1. Perform pre-alignment inspection; measure vehicle ride height.

P-1 2. Describe alignment angles (camber, caster and toe)

D. Wheels and Tires

P-1 1. Inspect tire condition; identify tire wear patterns; check for correct tire size, application (load and speed ratings), and air pressure as listed on the tire information placard/label.

P-1 2. Rotate tires according to manufacturers' recommendations including vehicles equipped with tire pressure monitoring systems (TPMS).

P-1 3. Dismount, inspect, and remount tire on wheel; balance wheel and tire assembly.

P-1 4. Dismount, inspect, and remount tire on wheel equipped with tire pressure monitoring system sensor.

P-1 5. Inspect tire and wheel assembly for air loss; determine necessary action.

P-1	6. Repair tire following vehicle manufacturer approved procedure.	
P-2	7. Identify indirect and direct tire pressure monitoring systems (TPMS); calibrate system; verify operation of instrument panel lamps.	
P-1	8. Demonstrate knowledge of steps required to remove and replace sensors in a tire pressure monitoring system (TPMS), including relearn procedure.	

SS Tasks	
P-1	29
P-2	6
P-3	1
	36

BRAKES

For every task in Brakes, the following safety requirement must be strictly enforced:

Comply with personal and environmental safety practices associated with clothing; eye protection; hand tools; power equipment; proper ventilation; and the handling, storage, and disposal of chemicals/materials in accordance with local, state, and federal safety and environmental regulations.

V. BRAKES

A. General

P-1 1. Research vehicle service information including fluid type, vehicle service history, service precautions, and technical service bulletins.

P-1 2. Describe procedure for performing a road test to check brake system operation, including an anti-lock brake system (ABS).

P-1 3. Install wheel and torque lug nuts.

P-1 4. Identify brake system components and configuration.

B. Hydraulic System

P-1 1. Describe proper brake pedal height, travel, and feel.

P-1 2. Check master cylinder for external leaks and proper operation.

P-1 3. Inspect brake lines, flexible hoses, and fittings for leaks, dents, kinks, rust, cracks, bulging, wear, and loose fittings/supports.

P-1 4. Select, handle, store, and fill brake fluids to proper level; use proper fluid type per manufacturer specification.

P-3 5. Identify components of hydraulic brake warning light system.

P-1 6. Bleed and/or flush brake system.

P-1 7. Test brake fluid for contamination.

C. Drum Brakes

P-1 1. Remove, clean, and inspect brake drum; measure brake drum diameter; determine serviceability.

P-1 2. Refinish brake drum and measure final drum diameter; compare with specification.

P-1 3. Remove, clean, inspect, and/or replace brake shoes, springs, pins, clips, levers, adjusters/self-adjusters, other related brake hardware, and backing support plates; lubricate and reassemble.

P-2 4. Inspect wheel cylinders for leaks and proper operation; remove and replace as needed.

P-1 5. Pre-adjust brake shoes and parking brake; install brake drums or drum/hub assemblies and wheel bearings; make final checks and adjustments.

D. Disc Brakes

P-1	1. Remove and clean caliper assembly; inspect for leaks and damage/wear; determine necessary action.
P-1	2. Inspect caliper mounting and slides/pins for proper operation, wear, and damage; determine necessary action.
P-1	3. Remove, inspect, and/or replace brake pads and retaining hardware; determine necessary action.
P-1	4. Lubricate and reinstall caliper, brake pads, and related hardware; seat brake pads and inspect for leaks.
P-1	5. Clean and inspect rotor and mounting surface, measure rotor thickness, thickness variation, and lateral runout; determine necessary action.
P-1	6. Remove and reinstall/replace rotor.
P-1	7. Refinish rotor on vehicle; measure final rotor thickness and compare with specification.
P-1	8. Refinish rotor off vehicle; measure final rotor thickness and compare with specification.
P-2	9. Retract and re-adjust caliper piston on an integral parking brake system.
P-1	10. Check brake pad wear indicator; determine necessary action.
P-1	11. Describe importance of operating vehicle to burnish/break-in replacement brake pads according to manufacturers' recommendation.

E. Power-Assist Units

P-2	1. Check brake pedal travel with, and without, engine running to verify proper power booster operation.
P-1	2. Identify components of the brake power assist system (vacuum and hydraulic); check vacuum supply (manifold or auxiliary pump) to vacuum-type power booster.

F. Related Systems (i.e., Wheel Bearings, Parking Brakes, Electrical)

P-1	1. Remove, clean, inspect, repack, and install wheel bearings; replace seals; install hub and adjust bearings.
P-2	2. Check parking brake system components for wear, binding, and corrosion; clean, lubricate, adjust and/or replace as needed.
P-1	3. Check parking brake operation and parking brake indicator light system operation; determine necessary action.
P-1	4. Check operation of brake stop light system.
P-2	5. Replace wheel bearing and race.
P-1	6. Inspect and replace wheel studs.

G. Electronic Brake, Traction Control, and Stability Control Systems

P-3	1. Identify traction control/vehicle stability control system components.
P-3	2. Describe the operation of a regenerative braking system.

BR Tasks	
P-1	29
P-2	5
P-3	3
	37

ELECTRICAL/ELECTRONIC SYSTEMS

For every task in Electrical/Electronic Systems, the following safety requirement must be strictly enforced:

Comply with personal and environmental safety practices associated with clothing; eye protection; hand tools; power equipment; proper ventilation; and the handling, storage, and disposal of chemicals/materials in accordance with local, state, and federal safety and environmental regulations.

VI. ELECTRICAL/ELECTRONIC SYSTEMS

A. General

P-1 1. Research vehicle service information including vehicle service history, service precautions, and technical service bulletins.

P-1 2. Demonstrate knowledge of electrical/electronic series, parallel, and series-parallel circuits using principles of electricity (Ohm's Law).

P-1 3. Use wiring diagrams to trace electrical/electronic circuits.

P-1 4. Demonstrate proper use of a digital multimeter (DMM) when measuring source voltage, voltage drop (including grounds), current flow, and resistance.

P-1 5. Demonstrate knowledge of the causes and effects from shorts, grounds, opens, and resistance problems in electrical/electronic circuits.

P-2 6. Use a test light to check operation of electrical circuits.

P-2 7. Use fused jumper wires to check operation of electrical circuits.

P-1 8. Measure key-off battery drain (parasitic draw).

P-1 9. Inspect and test fusible links, circuit breakers, and fuses; determine necessary action.

P-1 10. Repair and/or replace connectors, terminal ends, and wiring of electrical/electronic systems (including solder repair)

P-1 11. Identify electrical/electronic system components and configuration.

B. Battery Service

P-1 1. Perform battery state-of-charge test; determine necessary action.

P-1 2. Confirm proper battery capacity for vehicle application; perform battery capacity and load test; determine necessary action.

P-1 3. Maintain or restore electronic memory functions.

P-1 4. Inspect and clean battery; fill battery cells; check battery cables, connectors, clamps, and hold-downs.

P-1 5. Perform slow/fast battery charge according to manufacturers' recommendations.

P-1 6. Jump-start vehicle using jumper cables and a booster battery or an auxiliary power supply.

P-2 7. Identify safety precautions for high voltage systems on electric, hybrid-electric, and diesel vehicles.

P-1 8. Identify electrical/electronic modules, security systems, radios, and other accessories that require reinitialization or code entry after reconnecting vehicle battery.

P-2 9. Identify hybrid vehicle auxiliary (12v) battery service, repair, and test procedures.

C. Starting System

P-1 1. Perform starter current draw test; determine necessary action.

P-1 2. Perform starter circuit voltage drop tests; determine necessary action.

P-2 3. Inspect and test starter relays and solenoids; determine necessary action.

P-1 4. Remove and install starter in a vehicle.

P-2 5. Inspect and test switches, connectors, and wires of starter control circuits; determine necessary action.

P-3 6. Demonstrate knowledge of an automatic idle-stop/start-stop system.

D. Charging System

P-1 1. Perform charging system output test; determine necessary action.

P-1 2. Inspect, adjust, and/or replace generator (alternator) drive belts; check pulleys and tensioners for wear; check pulley and belt alignment.

P-2 3. Remove, inspect, and/or replace generator (alternator).

P-2 4. Perform charging circuit voltage drop tests; determine necessary action.

E. Lighting, Instrument Cluster, Driver Information, and Body Electrical Systems

P-1 1. Inspect interior and exterior lamps and sockets including headlights and auxiliary lights (fog lights/driving lights); replace as needed.

P-2 2. Aim headlights.

P-2 3. Identify system voltage and safety precautions associated with high-intensity discharge headlights.

P-1 4. Disable and enable supplemental restraint system (SRS); verify indicator lamp operation.

P-1 5. Remove and reinstall door panel.

P-3 6. Describe the operation of keyless entry/remote-start systems.

P-1 7. Verify operation of instrument panel gauges and warning/indicator lights; reset maintenance indicators.

P-1 8. Verify windshield wiper and washer operation; replace wiper blades.

EE Tasks	
P-1	26
	10
P-2	
P-3	2
	38

HEATING, VENTILATION, AND AIR CONDITIONING (HVAC)

For every task in Heating, Ventilation and Air Conditioning (HVAC), the following safety requirement must be strictly enforced:

Comply with personal and environmental safety practices associated with clothing; eye protection; hand tools; power equipment; proper ventilation; and the handling, storage, and disposal of chemicals/materials in accordance with local, state, and federal safety and environmental regulations.

VII. HEATING, VENTILATION, AND AIR CONDITIONING (HVAC)

A. General

P-1 1. Research vehicle service information, including refrigerant/oil type, vehicle service history, service precautions, and technical service bulletins.

P-1 2. Identify heating, ventilation and air conditioning (HVAC) components and configuration.

B. Refrigeration System Components

P-1 1. Inspect and replace A/C compressor drive belts, pulleys, and tensioners; visually inspect A/C components for signs of leaks; determine necessary action.

P-2 2. Identify hybrid vehicle A/C system electrical circuits and the service/safety precautions.

P-1 3. Inspect A/C condenser for airflow restrictions; determine necessary action.

C. Heating, Ventilation, and Engine Cooling Systems

P-1 1. Inspect engine cooling and heater systems hoses and pipes; determine necessary action.

D. Operating Systems and Related Controls

P-1 1. Inspect A/C-heater ducts, doors, hoses, cabin filters, and outlets; determine necessary action.

P-2 2. Identify the source of A/C system odors.

HA Tasks	
P-1	6
P-2	2
P-3	0
	8

ENGINE PERFORMANCE

For every task in Engine Performance the following safety requirement must be strictly enforced:

Comply with personal and environmental safety practices associated with clothing; eye protection; hand tools; power equipment; proper ventilation; and the handling, storage, and disposal of chemicals/materials in accordance with local, state, and federal safety and environmental regulations.

VIII. ENGINE PERFORMANCE

A. General

P-1 1. Research vehicle service information, including fluid type, vehicle service history, service precautions, and technical service bulletins.

P-2 2. Perform engine absolute manifold pressure tests (vacuum/boost); document results.

P-2 3. Perform cylinder power balance test; document results.

P-2 4. Perform cylinder cranking and running compression tests; document results.

P-2 5. Perform cylinder leakage test; document results.

P-1 6. Verify engine operating temperature.

P-1 7. Remove and replace spark plugs; inspect secondary ignition components for wear and damage.

B. Computerized Controls

P-1 1. Retrieve and record diagnostic trouble codes (DTC), OBD monitor status, and freeze frame data; clear codes when applicable.

P-1 2. Describe the use of the OBD monitors for repair verification.

C. Fuel, Air Induction, and Exhaust Systems

P-2 1. Replace fuel filter(s) where applicable.

P-1 2. Inspect, service, or replace air filters, filter housings, and intake duct work.

P-1 3. Inspect integrity of the exhaust manifold, exhaust pipes, muffler(s), catalytic converter(s), resonator(s), tail pipe(s), and heat shields; determine necessary action.

P-1 4. Inspect condition of exhaust system hangers, brackets, clamps, and heat shields; determine necessary action.

P-2 5. Check and refill diesel exhaust fluid (DEF).

EP Tasks	
P-1	8
P-2	7
P-3	0
	15

D. Emission Control Systems

P-2 1. Inspect, test, and service positive crankcase ventilation (PCV) filter/breather, valve, tubes, orifices, and hoses; perform necessary action.

P-1	125
P-2	40
P-3	10
Task Total	175
REQUIRED SUPPLEMENTAL TASKS	43
Grand Total	218

REQUIRED SUPPLEMENTAL TASKS

Shop and Personal Safety

1. Identify general shop safety rules and procedures.
2. Utilize safe procedures for handling of tools and equipment.
3. Identify and use proper placement of floor jacks and jack stands.
4. Identify and use proper procedures for safe lift operation.
5. Utilize proper ventilation procedures for working within the lab/shop area.
6. Identify marked safety areas.
7. Identify the location and the types of fire extinguishers and other fire safety equipment; demonstrate knowledge of the procedures for using fire extinguishers and other fire safety equipment.
8. Identify the location and use of eye wash stations.
9. Identify the location of the posted evacuation routes.
10. Comply with the required use of safety glasses, ear protection, gloves, and shoes during lab/shop activities.
11. Identify and wear appropriate clothing for lab/shop activities.
12. Secure hair and jewelry for lab/shop activities.
13. Demonstrate awareness of the safety aspects of supplemental restraint systems (SRS), electronic brake control systems, and hybrid vehicle high voltage circuits.
14. Demonstrate awareness of the safety aspects of high voltage circuits (such as high intensity discharge (HID) lamps, ignition systems, injection systems, etc.).
15. Locate and demonstrate knowledge of material safety data sheets (MSDS).

Tools and Equipment

1. Identify tools and their usage in automotive applications.
2. Identify standard and metric designation.
3. Demonstrate safe handling and use of appropriate tools.
4. Demonstrate proper cleaning, storage, and maintenance of tools and equipment.
5. Demonstrate proper use of precision measuring tools (i.e. micrometer, dial-indicator, dial-caliper).

Preparing Vehicle for Service

1. Identify information needed and the service requested on a repair order.
2. Identify purpose and demonstrate proper use of fender covers, mats.
3. Demonstrate use of the three C's (concern, cause, and correction).
4. Review vehicle service history.
5. Complete work order to include customer information, vehicle identifying information, customer concern, related service history, cause, and correction.

Preparing Vehicle for Customer

1. Ensure vehicle is prepared to return to customer per school/company policy (floor mats, steering wheel cover, etc.).

Workplace Employability Skills

Personal Standards (see Standard 7.9)

1. Reports to work daily on time; able to take directions and motivated to accomplish the task at hand.
2. Dresses appropriately and uses language and manners suitable for the workplace.
3. Maintains appropriate personal hygiene
4. Meets and maintains employment eligibility criteria, such as drug/alcohol-free status, clean driving record, etc.
5. Demonstrates honesty, integrity and reliability

Work Habits/Ethic (see Standard 7.10)

1. Complies with workplace policies/laws
2. Contributes to the success of the team, assists others and requests help when needed.
3. Works well with all customers and coworkers.
4. Negotiates solutions to interpersonal and workplace conflicts.
5. Contributes ideas and initiative
6. Follows directions
7. Communicates (written and verbal) effectively with customers and coworkers.
8. Reads and interprets workplace documents; writes clearly and concisely.
9. Analyzes and resolves problems that arise in completing assigned tasks.
10. Organizes and implements a productive plan of work.
11. Uses scientific, technical, engineering and mathematics principles and reasoning to accomplish assigned tasks
12. Identifies and addresses the needs of all customers, providing helpful, courteous and knowledgeable service and advice as needed.